W9-CGZ-492

Methods in Enzymology

Volume 115

DIFFRACTION METHODS FOR BIOLOGICAL MACROMOLECULES

Part B

METHODS IN ENZYMOLOGY

EDITORS-IN-CHIEF

Sidney P. Colowick   Nathan O. Kaplan

Methods in Enzymology

Volume 115

Diffraction Methods for Biological Macromolecules

Part B

EDITED BY

Harold W. Wyckoff

DEPARTMENT OF MOLECULAR
BIOPHYSICS AND BIOCHEMISTRY
YALE UNIVERSITY
NEW HAVEN, CONNECTICUT

C. H. W. Hirs

DEPARTMENT OF BIOCHEMISTRY,
BIOPHYSICS AND GENETICS
UNIVERSITY OF COLORADO HEALTH SCIENCES CENTER
DENVER, COLORADO

Serge N. Timasheff

GRADUATE DEPARTMENT OF BIOCHEMISTRY
BRANDEIS UNIVERSITY
WALTHAM, MASSACHUSETTS

1985

ACADEMIC PRESS, INC.

Harcourt Brace Jovanovich, Publishers

Orlando San Diego New York Austin
London Montreal Sydney Tokyo Toronto

ACADEMIC PRESS, INC.
Orlando, Florida 32887

United Kingdom Edition published by
ACADEMIC PRESS INC. (LONDON) LTD.
24–28 Oval Road, London NW1 7DX

LIBRARY OF CONGRESS CATALOG CARD NUMBER: 54-9110

ISBN 0–12–182015–7

PRINTED IN THE UNITED STATES OF AMERICA

85 86 87 88 9 8 7 6 5 4 3 2 1

Table of Contents

Section I. Primary Phasing

Section II. Modeling

Section III. Combined Model and Phase Refinement

Section IV. Presentation and Analysis of Structure

Contributors to Volume 115

Article numbers are in parentheses following the names of contributors.
Affiliations listed are current.

R. C. AGARWAL (8), *IBM Corporation, Thomas J. Watson Research Center, Yorktown Heights, New York 10598*

CYRUS CHOTHIA (28), *Medical Research Council, Laboratory of Molecular Biology, Cambridge CB2 2QH, England, and Christopher Ingold Laboratories, University College London, London WC1H OAJ, England*

J. DEISENHOFER (21), *Max-Planck-Institut für Biochemie, 8033 Martinsried, Federal Republic of Germany*

R. DIAMOND (18), *Medical Research Council, Laboratory of Molecular Biology, Cambridge CB2 2QH, England*

STEPHAN T. FREER (17), *M/A-COM Linkabit, Inc., San Diego, California 92121*

JONATHAN GREER (15), *Physical Biochemical Laboratory, Computer-Assisted Molecular Design, Abbott Laboratories, Abbott Park, Illinois 60064*

KARL D. HARDMAN (25), *Department of Medical Genetics, University of Toronto, Toronto, Ontario, M5S 1A8, Canada*

WAYNE A. HENDRICKSON (4, 19), *Department of Biochemistry and Molecular Biophysics, Columbia University, New York, New York 10032*

JAN HERMANS (13), *Department of Biochemistry, School of Medicine, University of North Carolina, Chapel Hill, North Carolina 27514*

N. W. ISAACS (8), *St. Vincent's Institute of Medical Research, Victoria Parade, Melbourne, Victoria, 3065, Australia*

JOËL JANIN (28), *Laboratoire de Biologie Physiochimique, Université de Paris-Sud, 91405 Orsay, France*

LYLE H. JENSEN (16), *Departments of Biological Structure and Biochemistry, University of Washington, Seattle, Washington 98195*

T. ALWYN JONES (12), *Department of Molecular Biology, Biomedical Center, S-751 24 Uppsala, Sweden*

EATON LATTMAN (5), *Department of Biophysics, School of Medicine, The Johns Hopkins University, Baltimore, Maryland 21205*

ARTHUR M. LESK (25), *Medical Research Council, Laboratory of Molecular Biology, Cambridge CB2 2QH, England*

BRIAN W. MATTHEWS (27), *Institute of Molecular Biology and Department of Physics, University of Oregon, Eugene, Oregon 97403*

D. C. PHILLIPS (9), *Laboratory of Molecular Biophysics, Department of Zoology, Oxford OX1 3PS, England*

S. J. REMINGTON (21), *Institute of Molecular Biology, University of Oregon, Eugene, Oregon 97403*

FREDERIC M. RICHARDS (10, 30), *Department of Molecular Biophysics and Biochemistry, Yale University, New Haven, Connecticut 06511*

DAVID C. RICHARDSON (14), *Department of Biochemistry, Duke University, Durham, North Carolina 27710*

JANE S. RICHARDSON (14, 23, 24), *Department of Biochemistry, Duke University, Durham, North Carolina 27710*

GEORGE D. ROSE (29), *Department of Biological Chemistry, Hershey Medical Center, Pennsylvania State University, Hershey, Pennsylvania 17033*

MICHAEL G. ROSSMANN (27), *Department of Biological Sciences, Purdue University, West Lafayette, Indiana 47907*

BYRON RUBIN (26), *Department of Chemistry, Emory University, Atlanta, Georgia 30322*

F. R. SALEMME (11), *E. I. Du Pont de Nemours & Co., Central Research and Development Department, Wilmington, Delaware 19898*

STEVEN SHERIFF (4), *Laboratory of Molecular Biology, National Institute of Arthritis, Diabetes, and Digestive and Kidney Diseases, National Institutes of Health, Bethesda, Maryland 20205*

JANET L. SMITH (4), *Department of Biochemistry and Molecular Biophysics, Columbia University, New York, New York 10032*

ROBERT A. SPARKS (3), *X-Ray Instruments Group, Nicolet Instrument Corporation, Madison, Wisconsin 53711*

W. STEIGEMANN (21), *Max-Planck-Institut für Biochemie, 8033 Martinsried, Federal Republic of Germany*

D. I. STUART (9), *Laboratory of Molecular Biophysics, Department of Zoology, Oxford, OX1 3PS, England*

JOEL L. SUSSMAN (20), *Department of Structural Chemistry, Weizmann Institute of Science, Rehovot, Israel, and Laboratory of Molecular Biology, National Institute of Arthritis, Diabetes, and Digestive and Kidney Diseases, National Institutes of Health, Bethesda, Maryland 20205*

J. SYGUSCH (2), *Département de Biochimie, Université de Sherbrooke, Sherbrooke, Quebec, Canada J1H 5N4*

LYNN F. TEN EYCK (22), *Interface Software, Cottage Grove, Oregon 97424*

A. TULINSKY (6), *Department of Chemistry, Michigan State University, East Lansing, Michigan 48824*

BI-CHENG WANG (7), *Biocrystallography Laboratory, VA Medical Center, Pittsburgh, Pennsylvania 15240, and Department of Crystallography, University of Pittsburgh, Pittsburgh, Pennsylvania 15260*

KEITH D. WATENPAUGH (1), *Physical and Analytical Chemistry Research, The Upjohn Company, Kalamazoo, Michigan 49001*

Preface

The aim of "Methods in Enzymology" volumes is to present as comprehensively as possible current techniques used in biochemistry, encompassing biological mechanisms, chemistry, and structure. In previous volumes, detailed coverage of solution physical–chemical techniques for the study of protein conformations, conformational changes, and interactions has been provided.

The two volumes on Diffraction Methods for Biological Macromolecules, Parts A and B, are devoted to a description of diffraction methods for biological macromolecules and assemblies. Different aspects of the methods involved in solving, presenting, and interpreting structure so that the reader can proceed knowledgeably and productively toward his goals are presented. We believe that an understanding of the fundamentals of each aspect of the overall method is both intellectually satisfying and practically important.

These two volumes have been divided according to the logical sequence of steps in structure determination. Part A is devoted to the experimental aspects of X-ray crystallography, starting from crystal growth and crystal handling, followed by methods of data collection. Part B includes analysis of the data, covering various aspects of phasing and refinement as well as the structures and methods for their analysis.

The goal which we hoped to attain was twofold: to give biochemists an introduction to the field of macromolecular structure determination, offering them guidance to pathways that are available to determine the structure of a protein, and to give practitioners of X-ray crystallography a comprehensive summary of techniques available to them, some of which are at the state-of-the-art level.

We wish to acknowledge with pleasure and gratitude the generous cooperation of the contributors. Their suggestions during the planning and preparation stages have been particularly valuable. Academic Press has provided inestimable help in the assembly of this material. We thank them for their many courtesies.

HAROLD W. WYCKOFF
C. H. W. HIRS
SERGE N. TIMASHEFF

METHODS IN ENZYMOLOGY

EDITED BY

Sidney P. Colowick and Nathan O. Kaplan

VANDERBILT UNIVERSITY
SCHOOL OF MEDICINE
NASHVILLE, TENNESSEE

DEPARTMENT OF CHEMISTRY
UNIVERSITY OF CALIFORNIA
AT SAN DIEGO
LA JOLLA, CALIFORNIA

METHODS IN ENZYMOLOGY

EDITORS-IN-CHIEF

Sidney P. Colowick and Nathan O. Kaplan

VOLUME XLVI. Affinity Labeling
Edited by WILLIAM B. JAKOBY AND MEIR WILCHEK

VOLUME XLVII. Enzyme Structure (Part E)
Edited by C. H. W. HIRS AND SERGE N. TIMASHEFF

VOLUME XLVIII. Enzyme Structure (Part F)
Edited by C. H. W. HIRS AND SERGE N. TIMASHEFF

VOLUME XLIX. Enzyme Structure (Part G)
Edited by C. H. W. HIRS AND SERGE N. TIMASHEFF

VOLUME L. Complex Carbohydrates (Part C)
Edited by VICTOR GINSBURG

VOLUME LI. Purine and Pyrimidine Nucleotide Metabolism
Edited by PATRICIA A. HOFFEE AND MARY ELLEN JONES

VOLUME LII. Biomembranes (Part C: Biological Oxidations)
Edited by SIDNEY FLEISCHER AND LESTER PACKER

VOLUME LIII. Biomembranes (Part D: Biological Oxidations)
Edited by SIDNEY FLEISCHER AND LESTER PACKER

VOLUME LIV. Biomembranes (Part E: Biological Oxidations)
Edited by SIDNEY FLEISCHER AND LESTER PACKER

VOLUME LV. Biomembranes (Part F: Bioenergetics)
Edited by SIDNEY FLEISCHER AND LESTER PACKER

VOLUME LVI. Biomembranes (Part G: Bioenergetics)
Edited by SIDNEY FLEISCHER AND LESTER PACKER

VOLUME LVII. Bioluminescence and Chemiluminescence
Edited by MARLENE A. DELUCA

VOLUME LVIII. Cell Culture
Edited by WILLIAM B. JAKOBY AND IRA PASTAN

VOLUME LIX. Nucleic Acids and Protein Synthesis (Part G)
Edited by KIVIE MOLDAVE AND LAWRENCE GROSSMAN

VOLUME LX. Nucleic Acids and Protein Synthesis (Part H)
Edited by KIVIE MOLDAVE AND LAWRENCE GROSSMAN

VOLUME 61. Enzyme Structure (Part H)
Edited by C. H. W. HIRS AND SERGE N. TIMASHEFF

VOLUME 62. Vitamins and Coenzymes (Part D)
Edited by DONALD B. MCCORMICK AND LEMUEL D. WRIGHT

VOLUME 63. Enzyme Kinetics and Mechanism (Part A: Initial Rate and Inhibitor Methods)
Edited by DANIEL L. PURICH

VOLUME 64. Enzyme Kinetics and Mechanism (Part B: Isotopic Probes and Complex Enzyme Systems)
Edited by DANIEL L. PURICH

VOLUME 65. Nucleic Acids (Part I)
Edited by LAWRENCE GROSSMAN AND KIVIE MOLDAVE

VOLUME 66. Vitamins and Coenzymes (Part E)
Edited by DONALD B. MCCORMICK AND LEMUEL D. WRIGHT

VOLUME 67. Vitamins and Coenzymes (Part F)
Edited by DONALD B. MCCORMICK AND LEMUEL D. WRIGHT

VOLUME 68. Recombinant DNA
Edited by RAY WU

VOLUME 69. Photosynthesis and Nitrogen Fixation (Part C)
Edited by ANTHONY SAN PIETRO

VOLUME 70. Immunochemical Techniques (Part A)
Edited by HELEN VAN VUNAKIS AND JOHN J. LANGONE

VOLUME 71. Lipids (Part C)
Edited by JOHN M. LOWENSTEIN

VOLUME 72. Lipids (Part D)
Edited by JOHN M. LOWENSTEIN

VOLUME 73. Immunochemical Techniques (Part B)
Edited by JOHN J. LANGONE AND HELEN VAN VUNAKIS

VOLUME 74. Immunochemical Techniques (Part C)
Edited by JOHN J. LANGONE AND HELEN VAN VUNAKIS

VOLUME 75. Cumulative Subject Index Volumes XXXI, XXXII, and XXXIV–LX
Edited by EDWARD A. DENNIS AND MARTHA G. DENNIS

VOLUME 76. Hemoglobins
Edited by ERALDO ANTONINI, LUIGI ROSSI-BERNARDI, AND EMILIA CHIANCONE

VOLUME 77. Detoxication and Drug Metabolism
Edited by WILLIAM B. JAKOBY

VOLUME 78. Interferons (Part A)
Edited by SIDNEY PESTKA

VOLUME 79. Interferons (Part B)
Edited by SIDNEY PESTKA

VOLUME 80. Proteolytic Enzymes (Part C)
Edited by LASZLO LORAND

VOLUME 81. Biomembranes (Part H: Visual Pigments and Purple Membranes, I)
Edited by LESTER PACKER

VOLUME 82. Structural and Contractile Proteins (Part A: Extracellular Matrix)
Edited by LEON W. CUNNINGHAM AND DIXIE W. FREDERIKSEN

VOLUME 83. Complex Carbohydrates (Part D)
Edited by VICTOR GINSBURG

VOLUME 84. Immunochemical Techniques (Part D: Selected Immunoassays)
Edited by JOHN J. LANGONE AND HELEN VAN VUNAKIS

VOLUME 85. Structural and Contractile Proteins (Part B: The Contractile Apparatus and the Cytoskeleton)
Edited by DIXIE W. FREDERIKSEN AND LEON W. CUNNINGHAM

VOLUME 86. Prostaglandins and Arachidonate Metabolites
Edited by WILLIAM E. M. LANDS AND WILLIAM L. SMITH

VOLUME 99. Hormone Action (Part F: Protein Kinases)
Edited by JACKIE D. CORBIN AND JOEL G. HARDMAN

VOLUME 100. Recombinant DNA (Part B)
Edited by RAY WU, LAWRENCE GROSSMAN, AND KIVIE MOLDAVE

VOLUME 101. Recombinant DNA (Part C)
Edited by RAY WU, LAWRENCE GROSSMAN, AND KIVIE MOLDAVE

VOLUME 102. Hormone Action (Part G: Calmodulin and Calcium-Binding Proteins)
Edited by ANTHONY R. MEANS AND BERT W. O'MALLEY

VOLUME 103. Hormone Action (Part H: Neuroendocrine Peptides)
Edited by P. MICHAEL CONN

VOLUME 104. Enzyme Purification and Related Techniques (Part C)
Edited by WILLIAM B. JAKOBY

VOLUME 105. Oxygen Radicals in Biological Systems
Edited by LESTER PACKER

VOLUME 106. Posttranslational Modifications (Part A)
Edited by FINN WOLD AND KIVIE MOLDAVE

VOLUME 107. Posttranslational Modifications (Part B)
Edited by FINN WOLD AND KIVIE MOLDAVE

VOLUME 108. Immunochemical Techniques (Part G: Separation and Characterization of Lymphoid Cells)
Edited by GIOVANNI DI SABATO, JOHN J. LANGONE, AND HELEN VAN VUNAKIS

VOLUME 109. Hormone Action (Part I: Peptide Hormones)
Edited by LUTZ BIRNBAUMER AND BERT W. O'MALLEY

VOLUME 110. Steroids and Isoprenoids (Part A)
Edited by JOHN H. LAW AND HANS C. RILLING

VOLUME 111. Steroids and Isoprenoids (Part B)
Edited by JOHN H. LAW AND HANS C. RILLING

VOLUME 112. Drug and Enzyme Targeting (Part A)
Edited by KENNETH J. WIDDER AND RALPH GREEN

VOLUME 113. Glutamate, Glutamine, Glutathione, and Related Compounds (in preparation)
Edited by ALTON MEISTER

VOLUME 114. Diffraction Methods for Biological Macromolecules (Part A) (in preparation)
Edited by HAROLD W. WYCKOFF, C. H. W. HIRS, AND SERGE N. TIMASHEFF

VOLUME 115. Diffraction Methods for Biological Macromolecules (Part B) (in preparation)
Edited by HAROLD W. WYCKOFF, C. H. W. HIRS, AND SERGE N. TIMASHEFF

VOLUME 116. Immunochemical Techniques (Part H: Effectors and Mediators of Lymphoid Cell Functions) (in preparation)
Edited by GIOVANNI DI SABATO, JOHN J. LANGONE, AND HELEN VAN VUNAKIS

VOLUME 117. Enzyme Structure (Part J) (in preparation)
Edited by C. H. W. HIRS AND SERGE N. TIMASHEFF

VOLUME 118. Plant Molecular Biology (in preparation)
Edited by ARTHUR WEISSBACH AND HERBERT WEISSBACH

VOLUME 119. Interferons (Part C) (in preparation)
Edited by SIDNEY PESTKA

VOLUME 120. Cumulative Subject Index Volumes 81–94, 96–101

VOLUME 121. Immunochemical Techniques (Part I: Hybridoma Technology and Monoclonal Antibodies) (in preparation)
Edited by JOHN J. LANGONE AND HELEN VAN VUNAKIS

VOLUME 122. Vitamins and Coenzymes (Part G) (in preparation)
Edited by FRANK CHYTIL AND DONALD B. MCCORMICK

VOLUME 123. Vitamins and Coenzymes (Part H) (in preparation)
Edited by FRANK CHYTIL AND DONALD B. MCCORMICK

Methods in Enzymology

Volume 115

DIFFRACTION METHODS FOR BIOLOGICAL MACROMOLECULES

Part B

Section I

Primary Phasing

[1] Overview of Phasing by Isomorphous Replacement

By KEITH D. WATENPAUGH

Introduction

The solution of the structure of macromolecules has been dependent on the success of determining the phases of diffraction data by the isomorphous replacement method. While some structures of closely related macromolecules have been determined by molecular replacement methods, the initial member of a macromolecular class has always been determined by using the differences observed in the diffraction intensities either through incorporating heavy atoms into the crystals or through anomalous scattering effects. The use of small variations of intensities in isomorphous structures to determine phases was initially applied in the determination of the alums structures by Cork in 1927.[1] Later Bokhoven *et al.*[2] solved the noncentrosymmetric structure of strychnine sulfate (or selenate), by using the centrosymmetric heavy-atom positions of the isomorphous crystals. In 1954, Green *et al.*[3] used heavy atoms to determine phases of centrosymmetric reflections in hemoglobin, but the real potential of isomorphous replacement methods in determining protein structures was recognized in 1956 with papers by Harker,[4] Perutz,[5] and Crick and Magdoff[6] using Argand diagrams or analytical expressions to solve for the phases with double isomorphous replacement. The possibility of using the anomalous scattering differences in noncentric structures was also being investigated as a means not only of determining absolute configuration, but also of determining phases.[7] In 1958, the use of heavy-atom isomorphous replacement led to the solution of the phases of myoglobin and the low-resolution determination of its structure.[8] Blow and Crick[9] introduced the idea of error theory into the treatment in 1959, and this

[1] J. M. Cork, *Philos. Mag.* **4,** 688 (1927).

[2] C. Bokhoven, J. C. Schoone, and J. M. Bijvoet, *Acta Crystallogr.* **4,** 275 (1951).

[3] D. W. Green, V. M. Ingram, and M. F. Perutz, *Proc. R. Soc. London, Ser. A* **225,** 287 (1954).

[4] D. Harker, *Acta Crystallogr.* **9,** 1 (1956).

[5] M. F. Perutz, *Acta Crystallogr.* **9,** 867 (1956).

[6] F. H. C. Crick and B. S. Magdoff, *Acta Crystallogr.* **9,** 901 (1956).

[7] J. M. Bijvoet, *Nature (London)* **173,** 888 (1954).

[8] J. C. Kendrew, G. Bodo, H. M. Dintzis, R. G. Parrish, and H. Wyckoff, *Nature (London)* **181,** 662 (1958).

[9] D. M. Blow and F. H. C. Crick, *Acta Crystallogr.* **12,** 794 (1959).

idea has formed the basis of nearly all subsequent macromolecular phase determinations.

Methods of Isomorphous Phasing

Although the more theoretical aspects of the isomorphous replacement method will not be covered in detail, nevertheless certain basic ideas must be understood concerning the relationship between structure factors and their phases. First, the diffraction data can be measured and reduced to structure factor amplitudes with unknown phases. The structure factor associated with a particular set of Miller indices is dependent on the relative positions of all of the atoms in the crystal and their scattering power. Second, if an additional atom or atoms of sufficient scattering power can be incorporated into the crystal lattice without significantly disturbing the original atomic positions, modified structure factor amplitudes may be experimentally obtained and are related as shown by the equation

$$\mathbf{F}_{PH} = \mathbf{F}_{P} + \mathbf{F}_{H} \quad (1)$$

or as shown by the geometric representation in Fig. 1. $\mathbf{F}_P$ refers to the structure factor of the parent or primary molecule, $\mathbf{F}_H$ to that of the heavy-atom contribution, and $\mathbf{F}_{PH}$ to that of the derivative. Each of these quantities is a complex number having both a magnitude F, and phase α. And third, if coordinates of the additional atom or atoms can be determined, it may be possible to determine the phases of the structure factors. This solution is shown geometrically in Fig. 2. It may also be found by solving two simultaneous equations of the form:

$$F_{PH}^2 = F_P^2 + F_H^2 + 2F_P F_H \cos(\alpha_P - \alpha_H) \quad (2)$$

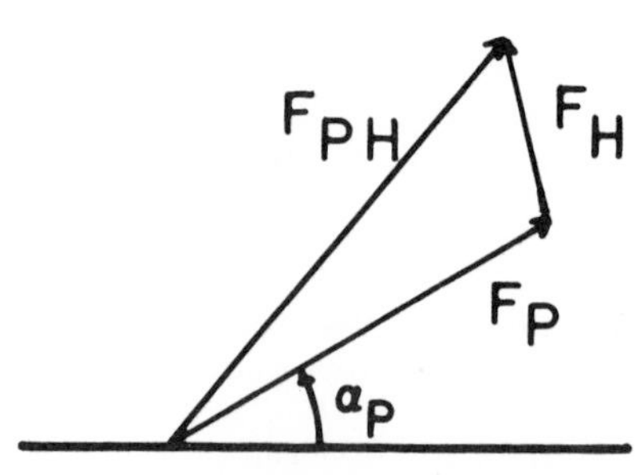

FIG. 1. Relationship among the vectors $\mathbf{F}_P$, $\mathbf{F}_H$, and $\mathbf{F}_{PH}$ for a reflection; α_P is the phases angle for the parent molecule.

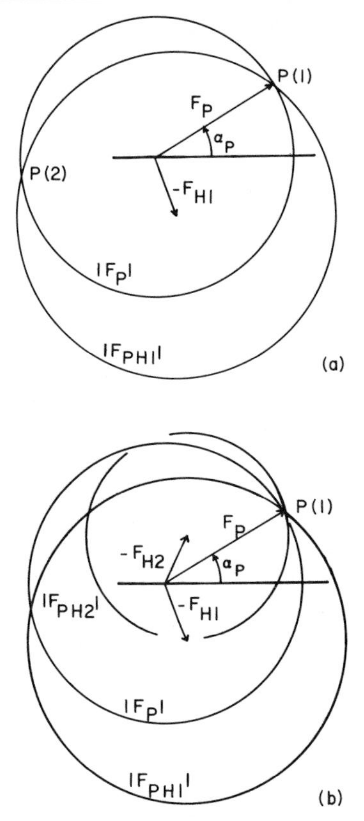

FIG. 2. Construction of circles to determine parent phase, α_P. (a) Heavy-atom scattering vector $\mathbf{F}_{H1}$ is subtracted, and circle denoting magnitude of first derivative F_{PH1} is drawn intersecting the parent circle F_P at two phase angles, *P(1)* and *P(2)*. (b) A second derivative with different heavy-atom vector, $\mathbf{F}_{H2}$, intersects parent at *P(1)* but not at *P(2)* to resolve the phase ambiguity.

The quantities F_P and F_{PH} are experimentally measured, while the quantities F_H and α_H are calculated from the determined heavy-atom positions. If the anomalous scattering component of the heavy atoms is large enough to produce significant differences between Friedel-related reflections in the derivative data, those differences may be used in a manner similar to two heavy-atom derivative data sets.

Such geometric or analytical solutions of the phases assume that the structure factors are perfectly measured and that there is perfect isomorphism. Since experimental and model errors are present, the circles (Fig. 2b) in the geometric representation will probably not intersect exactly and in some cases not intersect at all. Blow and Crick[9] suggested calculating the phase probabilities for a particular structure factor of a parent–derivative pair by the equation

$$P_{j(\mathrm{iso})}(\alpha) = \exp(-x_{j(\mathrm{iso})}^2/2E_{j(\mathrm{iso})}^2) \tag{3}$$

in which $E_{j(\mathrm{iso})}$ is an estimate of the cumulate error and $x_{j(\mathrm{iso})} = |F_{\mathrm{PH}} - F_{\mathrm{PH(calc)}}|$ is the calculated closure error (lack of closure) between the parent and the jth derivative. The calculated derivative structure factor is given by the equation

$$F_{\mathrm{PH(calc)}}^2 = F_{\mathrm{P}}^2 + F_{\mathrm{H}}^2 - 2F_{\mathrm{P}}F_{\mathrm{H}}\cos(\alpha - \alpha_{\mathrm{H}}) \tag{4}$$

The probability distribution (with an assumed E) for one derivative is shown in Fig. 3a and for a second in Fig. 3b. The joint distribution probability or product of the two probability distributions is shown in Fig. 3c. In the case of a single derivative and anomalous scattering, Blow and Rossmann[10] suggested using the equation

$$P_j(\alpha) = \exp\{-[(x_j^+)^2 + (x_j^-)^2]/2E^2\} \tag{5}$$

in which x^+ and x^- represent the closure error for the corresponding Friedel-related reflections, F_{PH}^+ and F_{PH}^-. North[11] pointed out that the nonanomalous scattering component and anomalous scattering component have different magnitudes of errors. Matthews[12] proposed treating the anomalous scattering by the separate probability:

$$P_{j(\mathrm{ano})}(\alpha) = \exp[-(x_j^+ - x_j^-)^2/2E_{\mathrm{ano}}^2] \tag{6}$$

This has the effect of greatly increasing the power that anomalous scattering has on determining the phases and improving the figures of merit. Combining the isomorphous (real component of the scattering) probability with the anomalous scattering probability to form the joint probability

$$P_j(\alpha) = P_{j(\mathrm{iso})}(\alpha)P_{j(\mathrm{ano})}(\alpha) \tag{7}$$

discriminates between the two equally probable phases for the parent phases calculated when only the real component of the heavy-atom scattering is used. Unfortunately, a single derivative with a single type (or dominate type) of anomalous scatterer gives an ambiguous determination of the heavy-atom positions between the correct model in the cell and a centrosymmetrically or mirror-related one. Both models give exactly the same statistical results when using a single derivative, but only one gives correct phases.

[10] D. M. Blow and M. G. Rossman, *Acta Crystallogr.* **14,** 901 (1961).
[11] A. C. T. North, *Acta Crystallogr.* **18,** 212 (1965).
[12] B. W. Matthews, *Acta Crystallogr.* **20,** 82 (1966).

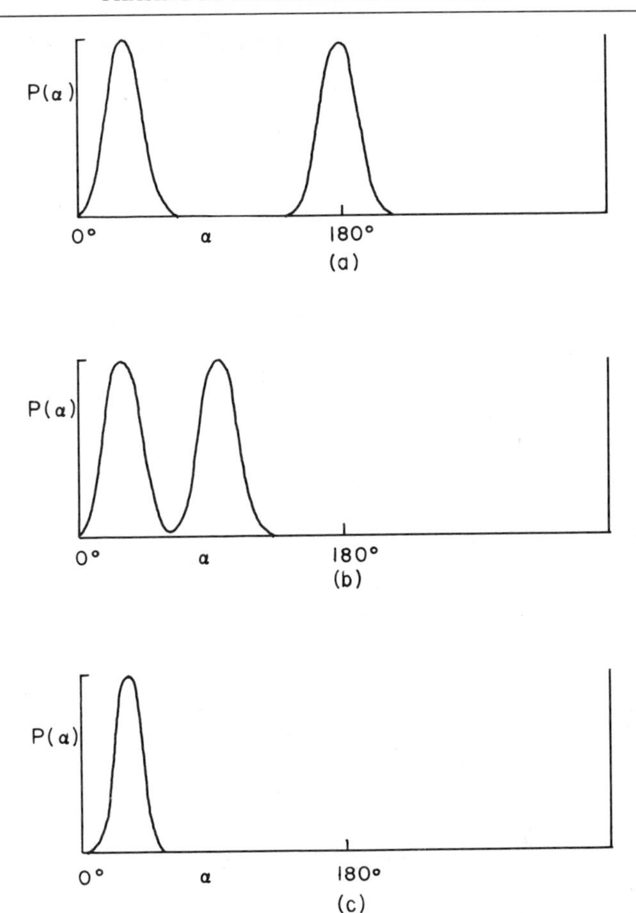

FIG. 3. Normalized phase probability distributions, $P(\alpha)$, with arbitrary closure error estimates. (a) $P(\alpha)$ for derivative PH1 of Fig. 2 with bimodal distribution. (b) $P(\alpha)$ for derivative PH2 of Fig. 2. (c) Joint probability density with unimodal distribution and maximum probability at α_P.

The use of the joint probability provides a means of combining any number of derivatives to yield a "most probable phase" and a "centroid phase" along with a "figure of merit." The "most probable phase" is that angle with the highest joint probability, while the "centroid (or best) phase" is the mean of the phase vectors weighted by their respective probabilities. The normalized figure of merit (m) can be approximated by

$$m = \left[\sum_i P(\Delta\alpha_i) \exp(i\ \Delta\alpha_i)\right] \Big/ \sum_i P(\Delta\alpha_i) \tag{8}$$

in which $\Delta\alpha_i = \alpha_{best} - \alpha_i$. In theory, at least, the arc cos m should be an estimate of the error in the phase angle.

A variation of the Blow–Crick method proposed by Hendrickson and Lattman[13] represents the phase probability from each derivative by the equation

$$P_j(\alpha) = \exp(K + A \cos \alpha + B \sin \alpha + C \cos 2\alpha + D \sin 2\alpha) \quad (9)$$

The coefficients K, A, B, C, and D for either isomorphous or anomalous information can readily be calculated not only from heavy-atom derivative data, but also from phase estimates for the parent data from other sources, such as structure factors calculated from a partial model of the macromolecule. Calculating the combined probability function from various derivatives and other sources is simply a matter of summing the corresponding coefficients. The use of these coefficients has been expanded on and error treatment discussed in additional references.[14,15]

While the Blow–Crick method or variations on it have been the primary isomorphous replacement phasing method for several decades now, new methods are being developed. Sygush[16] has developed what he calls the "minimum variance Fourier coefficient" (MVFC) method and this has been extended by Brigogne[17] to encompass more correlations between derivatives. These are discussed in a separate chapter. These methods use a least-squares approach in treating the phases in the complex plane. Initial phase estimates (either from the Blow–Crick method or a previous cycle) are refined to obtain a centroid phase and figure of merit.

Numerous computer programs are available for determining phases by isomorphous replacement methods and refining the heavy-atom sites. These programs are continually being modified in many different ways to meet the special requirements of whatever structure is being worked on. There are probably as many different weighting schemes, for example, as protein crystallography laboratories. Nevertheless, most of the programs can be traced back to three very early ones written in the laboratories of Dickerson, Rossman, and Perutz. A program written as a cooperative experiment of pooling a number of experienced protein crystallographers' knowledge has recently been completed.[18,19] This program uses the data

[13] W. A. Hendrickson and E. E. Lattman, *Acta Crystallogr., Sect. B* **B26,** 136 (1970).

[14] W. A. Hendrickson, *Acta Crystallogr., Sect. A* **A35,** 245 (1979).

[15] T. L. Blundell and L. N. Johnson, "Protein Crystallography," p. 375. Academic Press, London, 1976.

[16] J. Sygush, *Acta Crystallogr., Sect. A* **A33,** 512 (1977).

[17] G. Bricogne, *in* "Computational Crystallography" (Sayer, D. ed.), p 233. Oxford Univ. Press, New York, 1982.

[18] K. D. Watenpaugh, *Abstr., Am. Crystallogr. Assoc.* **8,** No. 1, 27 (1980).

[19] R. A. Alden, G. Bricogne, S. T. Freer, S. R. Hall, W. A. Hendrickson, P. Machin, R. J. Munn, A. J. Olson, G. N. Reeke, Jr., S. Sheriff, J. M. Stewart, J. Sygush, L.F. Ten Eyck, and K. D. Watenpaugh, *Computers Chem.* **7,** 137 (1983).

structure of the XTAL80 system of crystallographic programs,[20] is space group general, and is designed for intercomputer portability. It contains options to phase by the Hendrickson–Lattman modification of the Blow–Crick procedure, the MVFC method of Sygush, or the Bricogne modification. Preliminary tests have been encouraging on both known structures and structures that have had difficulties being solved by other programs.

Practical Aspects of Isomorphous Phasing

While the construction of Argand diagrams or the use of the Blow–Crick equations may seem simple, the actual process of deriving a set of useful phases can be frustrating due to the many decisions that must be made concerning correctness of the heavy-atom models, quality of the data, and questionable electron density maps. Unfortunately, the statistical parameters that can be calculated are insensitive indicators of the quality of the models or the set of phases, and the true test of the success of any phasing attempt is the production of an interpretable electron density map.

The first stage of the phasing process by the isomorphous replacement method is careful data collection, reduction, and scaling. All subsequent steps depend on the accurate measurement of small differences between relatively large numbers, and therefore small errors in deterioration scaling or absorption corrections can, for example, seriously affect the final results. Scaling between parent and derivative data sets depends on the amount of heavy-atom incorporation into the derivative crystal, something that is difficult to estimate. Applying scales and overall temperature factors to the derivatives that maintain constant values for the ratio $\Sigma F_{PH}/\Sigma F_P$ of slightly greater than unity for shells of increasing diffraction angle may be adequate for initially finding heavy-atom sites in Patterson maps. However, if the Friedel-related reflections have been collected, better scaling can be obtained by the method described by Singh and Ramaseshan.[21] Proper scaling is important throughout the phasing process, but can usually be determined more accurately during later stages. Errors in scaling of 0.01–0.02 can produce ghost peaks or holes at the heavy-atom sites larger than anything else in the final electron density maps.[22]

The second stage consists of the location and refinement of the heavy-atom model for each of the derivative data sets. Location of the heavy-

[20] S. R. Hall and J. M. Stewart, *Acta Crystallogr., Sect. A* **A36,** 979 (1980).

[21] A. K. Singh and S. Ramaseshan, *Acta Crystallogr.* **21,** 279 (1966).

[22] K. D. Watenpaugh, L. C. Sieker, L. H. Jensen, J. LeGall, and M. DuBourdieu, *Proc. Natl. Acad. Sci. U.S.A.* **69,** 3185 (1972).

atom positions either can be a trivial exercise in interpreting a "difference Patterson" to locate one or two dominate peaks, or it can be extremely frustrating. If numerous heavy atoms have been incorporated into the unit cell, the interpretation can become exceedingly difficult, and because of high R factors that are normally calculated for these models, along with the lack of chemical connectivity to assess them, one cannot initially be sure whether or not a model is correct. In many cases, the success of locating heavy-atom sites depends on the quality of the coefficients that are used in calculating the Patterson maps or used in other methods of locating their sites (e.g., Direct Methods discussed in another chapter). The coefficients for Patterson maps, $\Delta F^2_{\text{iso}} = |F_{\text{PH}} - F_{\text{P}}|^2$ and $\Delta F^2_{\text{ano}} = |F^+_{\text{PH}} - F^-_{\text{PH}}|^2$ are lower limit estimations of the F_{H} and the Bijvoet difference, in that they do not contain information about relative phases. Because they tend to be poor estimates, Patterson maps and electron density maps calculated using them can be very noisy and uninterpretable, and structure factor calculations using these estimates as observed quantities have large R factors (0.30–0.50). In fact, R factors in the same range can be obtained with totally incorrect models. Refinement of the heavy-atom models at this stage to obtain the best fit to the coefficients, with careful use of ΔF and $\Delta\Delta F$ maps to locate additional heavy-atom sites, avoids some of the problems resulting from heavy-atom refinement combined with phasing in the next stage. Modification of the coefficients to fit statistical distributions has recently been suggested by French and Wilson[23] and Lewis and Rees[24] and should not only give more interpretable Patterson maps but also aid in the least-squares refinement of the heavy-atom positions.

Better coefficients can be obtained if the anomalous and isomorphous components of the data are combined to calculate them.[21,25,26] In Matthews derivatization,[25] the term is

$$F^2_{\text{Mat}} = F^2_{\text{P}} + F^2_{\text{PH}} - 2F_{\text{P}}F_{\text{PH}}\{1 - [k(F^+_{\text{PH}} - F^-_{\text{PH}})/2F_{\text{H}}]^2\}^{1/2} \qquad (10)$$

in which k equals the ratio of the real to anomalous heavy-atom scattering and can be estimated from the difference between the parent and Friedel mates of the derivative. The value of k is not necessarily the theoretical value of the ratio and can also vary with diffraction angle. The R factors for correct heavy-atom models using the modified coefficients are usually in the range of 0.25–0.45.

[23] S. French and K. Wilson, *Acta Crystallogr., Sect. A* **A34,** 517 (1978).
[24] M. Lewis and D. C. Rees, *Acta Crystallogr., Sect. A* **A34,** 512 (1983).
[25] B. W. Matthews, *Acta Crystallogr.* **20,** 230 (1966).
[26] G. Kartha and R. Parthasarathy, *Acta Crystallogr.* **18,** 745 (1965).

In the third stage of the process, a second type of least-squares refinement cycles is combined with the phasing process. In this case the coefficients are derived not only from the magnitude of the structure factors, but also from the derived phases. Therefore, as the phases and the models improve, the estimates of the coefficients improve. However, incorrect models for the heavy atoms of a derivative will feed back incorrect bias into the phases. Hence, subsequent ΔF maps using these biased phases will make it difficult to improve the models. For this reason Dickerson *et al.*[27] suggested calculating ΔF maps using coefficients derived from phases not biased by the derivative in question. Also, Blow and Matthews[28] suggested refinement using phases which were not derived from the derivative being refined. However, it should be mentioned that in the majority of the applications of phasing and refinement, this practice has not been used. The MVFC procedure[22] seems to be less affected by the feedback than the Blow–Crick procedure. It converges in fewer cycles and the highly correlated parameters (occupancy and thermal) seem to be better behaved. It is the bias and feedback that limit the effectiveness of refinement at this stage if only a single derivative with its anomalous scattering is used to determine a set of phases.

Finally, in the fourth stage, electron density maps are calculated with the resulting phases. Generally, these electron density maps are calculated using the centroid or "best" phases and the structure factors weighted by their figures of merit. Inability to interprete the electron density map probably indicates that one should return to the second stage of the process. An interpretable but noisy map may be improved by redoing the third stage with careful checking of the heavy-atom models for the different derivatives.

Pitfalls

There are numerous ways in which the process of going from raw diffraction data to a set of phases and the resultant electron density map can be adversely affected. Some of these "pitfalls" affect the phases or their weighting in very subtle ways, while others cause totally incorrect results. The following list consists of pitfalls that are mentioned many times as having caused problems.

1. Occupancy of heavy-atom sites may vary from crystal to crystal or even with the exposure time to X rays. Merging of data from different

[27] R. E. Dickerson, M. L. Kapka, J. C. Varnum, and J. E. Weinzierl, *Acta Crystallogr.* **23,** 511 (1967).

[28] D. M. Blow and B. W. Matthews, *Acta Crystallogr., Sect. A* **A29,** 56 (1973).

crystals either should not be done or should be done with careful consideration of possible problems.

2. Crystal deterioration varies with respect to X-ray exposure time and diffraction angle, and maybe even with respect to reciprocal lattice directions.

3. When there is significant anomalous scattering, Friedel's law does *not* hold and transformations of indices should be done carefully.

4. When using anomalous scattering, all heavy-atom models must be in the correct absolute configuration or the anomalous contribution to the phasing will adversely affect the phases. If there is no anomalous scattering component to the isomorphous scattering, all derivatives must still be in the correct hand or all in the incorrect hand. If they are in the incorrect one, the structure determination will yield the wrong absolute configuration.

5. There are numerous scale factors between crystals and between parent and derivatives, for heavy-atom contributions, anomalous scattering, and diffraction angle (i.e., overall and individual temperature factors). These are highly correlated and should be watched and treated carefully. Heavy-atom occupancy scales are usually relative and not absolute. It must *not* be presumed that 1.00 is the maximum allowable occupancy, nor that it means fully occupied.

6. Sites not detected initially may appear in electron density or ΔF maps with improvements in the phases and need adding during later stages.

7. In the case of multiple derivatives, make sure that all sets have the same origin in the crystal unit cell.[29,30]

8. Closure errors vary from derivative to derivative, with respect to diffraction angle, and with respect to magnitude of structure factor intensities. The weighting of the probability distribution functions and their figures of merit are highly dependent on the estimates of the closure errors. It is important to consider experimental and model errors as separate quantities[31,32] and to consider the model errors as functions of diffraction angle, magnitudes, and data sets.

9. In the Blow–Crick method, derivatives with higher figures of merit tend to dominate over those with low ones. This can result in loss of phasing power by derivative data sets that may be good but due to initially poor heavy-atom models did not have high figures of merit.

[29] M. G. Rossmann, *Acta Crystallogr.* **13,** 211 (1960).

[30] L. K. Steinrauf, *Acta Crystallogr.* **16,** 317 (1963).

[31] L. F. Ten Eyck and A. Arnone, *J. Mol. Biol.* **100,** 3 (1976).

[32] M. G. Rossmann, *Acta Crystallogr., Sect. A* **A32,** 774 (1976).

Statistical Data and What They Mean

Most of the programs that are used to determine phases and refine heavy-atom parameters also put out a vast number of different statistics (R factors, closure errors, ratios between observed and calculated heavy-atom contributions, phase differences, etc.). The reason for so many different numbers to be listed is that there is no one sensitive measure of the quality of the goodness of the derived phases. Nevertheless, the different statistics may give some indication of where there are problems in the phasing procedure if resultant electron density maps are not interpretable or not as good as one would expect from the quality of the data.

The following are some of the statistics that commonly are listed by the phasing programs, usually as a function of $(\sin \theta)/\lambda$ and sometimes also as a function of the magnitude of the structure factors.

1. Figure of merit. This quantity has been previously defined [Eq. (8)], and in the case of Blow–Crick phasing it reflects the unimodality and sharpness of the phase probability distribution. There have been cases in which higher figures of merit were obtained from wrong models than from correct ones. In the cases of phasing with a single heavy-atom derivative using the anomalous scattering to help resolve the phase ambiguity, both possible hands give the same statistics and one is completely wrong.

2. R(Cullis). This R factor for the jth derivative is defined[33] as

$$R_j(\text{Cullis}) = \sum |(F_{\text{PH}} \pm F_{\text{P}}) - F_{\text{H}}| / \sum |F_{\text{PH}} - F_{\text{P}}| \tag{11}$$

It is usually defined for the centric reflections, but can be used for the acentric ones as well. Agreement in the sign of the difference should be included in the calculation of the statistic; otherwise it is overly optimistic.

3. R(Kraut). It is defined[34] as

$$R_j(\text{Kraut}) = \sum |F_{\text{PH}} - F_{\text{PH(calc)}}| / \sum |F_{\text{PH}}| \tag{12}$$

It is highly affected by the amount of substitution in the particular derivative. A lightly substituted derivative has a low R(Kraut) even when no heavy-atom contribution is included.

[33] A. F. Cullis, H. Miurhead, M. F. Perutz, M. G. Rossmann, and A. C. T. North, *Proc. R. Soc. London, Ser. A* **265,** 15 (1962).

[34] J. Kraut, L. C. Sieker, D. High, and S. T. Freer, *Proc. Natl. Acad. Sci. U.S.A.* **48,** 1417 (1962).

4. Isomorphous closure error. The mean square closure error for a particular data set is defined as

$$\langle x_{\text{iso}} \rangle_j^2 = \frac{\Sigma(F_{\text{PH}} - F_{\text{PH(calc)}})^2}{N} \quad (13)$$

in which N is the number of reflections summed over. Closure errors [usually calculated in regions of $(\sin\theta)/\lambda$] are generally used for weighting in the next cycle of phasing by the Blow–Crick procedure and in calculating figures of merit. Using closure errors for weighting and neglecting experimental errors may result in overly optimistic estimates of figures of merit and allow some derivatives to dominate the phasing process.

5. Anomalous closure error. The mean square anomalous closure error is the counterpart of the isomorphous (or real) closure error and is defined as

$$\langle x_{\text{ano}} \rangle_j^2 = \frac{\Sigma[(F_{\text{PH}}^+ - F_{\text{PH}}^-) - (F_{\text{PH(calc)}}^+ - F_{\text{PH(calc)}}^-)]^2}{N} \quad (14)$$

If this closure error becomes greater than half of the isomorphous closure error, the anomalous signal is contributing little to the phasing.

6. R(Modulus). It is defined[35] as

$$R(\text{Modulus}) = \sum |F_{\text{PH}} - F_{\text{PH(calc)}}| \Big/ \sum F_{\text{H}} \quad (15)$$

This is either the weighted or unweighted closure error divided by the weighted or unweighted heavy-atom contribution to the structure factors. As this value approaches or becomes greater than 1.0, it becomes impossible for the phases circles to intersect and phasing power is lost.

7. Phase difference between parent and heavy-atom model. It is defined[36] as

$$\Delta\alpha_j = \frac{\Sigma|\alpha_{\text{P}} - \alpha_{\text{H}}|}{N} \quad (16)$$

Since the phase of the heavy atoms and the parent molecule should be independent, this value should be 90°. It is sensitive to scaling errors between the parent and derivatives.

8. Agreement of sign between observed and calculated Friedel differences. This can give an indication that the Friedel-related reflections

[35] M. J. Adams, D. J. Haas, B. A. Jeffrey, A. McPherson, Jr., H. L. Mermall, M. G. Rossmann, R. W. Schevitz, and A. J. Wonacott, *J. Mol. Biol.* **41,** 159 (1969).

[36] E. J. Dodson, *in* "Crystallographic Computing Techniques" (F. R. Ahmed, K. Huml, and B. Sedláček, eds.), p. 259. Munksgaard, Copenhagen, 1976.

have been interchanged or that the heavy-atom model is incorrect for one of the derivatives in the multiple isomorphous derivative case.

Acknowledgments

The development and testing of the XTAL80 Multiple Isomorphous Phasing and Refinement Program were in part funded by GM 13366 from the National Institutes of Health. I would like to thank Dr. Ronald E. Stenkamp for useful criticism of the chapter.

[2] Least-Squares MIR Calculations: Minimum Variance Fourier Coefficient (MVFC) Refinement

By J. Sygusch

The interpretability of the calculated electron density map depends critically upon the protein phases used to calculate the map. The accuracy of the protein phases in turn depends upon the quality of the multiple isomorphous replacement (MIR) refinement. The minimum variance Fourier coefficient (MVFC) refinement[1] is an alternate approach to the conventional MIR refinement as originally proposed by Blow and Crick[2] and elaborated by others.[3–5] Protein phase refinement by the MVFC method is a least-squares method which instead of treating the protein phase as the refinable parameter refines the sine or cosine function of the phase. This simple transformation in parameter to be refined yields a considerable advantage in the subsequent electron density map calculations.

The trigonometric sine and cosine functions occur in the equations for both isomorphous replacement and electron density calculation. In the electron density calculations the Fourier coefficient is a linear function of the sine or cosine of the phase. The least-squares estimate from the MVFC refinement of the expected value of the trigonometric function thus inserts directly into the electron density calculations without transformation. Since estimates of protein phases are generally substantially in error, the Fourier coefficients are weighted by the standard deviation of

[1] J. Sygusch, *Acta Crystallogr.* **33,** 124 (1977).

[2] D. M. Blow and F. H. C. Crick, *Acta Crystallogr.* **17,** 794 (1959).

[3] R. E. Dickerson, J. C. Kendrew, and B. E. Standberg, *Acta Crystallogr.* **14,** 1188 (1961).

[4] A. C. T. North, *Acta Crystallogr.* **18,** 212 (1965).

[5] B. W. Matthews, *Acta Crystallogr.* **20,** 82 (1966).

the estimated sine or cosine of the phase obtained from the MVFC refinement. This approach avoids the necessity as in the Blow–Crick formalism to postulate an ad hoc probability distribution for the evaluation of the errors of the protein phase, the errors generally being required for the calculation of the best phase and figure of merit that is weighted in the electron density calculation. The estimation by least squares has the advantage that it enables the experimentally derived errors of the observed structure factors to contribute explicity not only in the determination of the expected value of the sine or cosine of the phase but also in the calculation of the associated standard deviation necessary for the weighting of the Fourier coefficients for electron density synthesis.

Theoretical Background

Refinement of the protein phases and of the heavy-atom parameters by the MVFC refinement minimizes differences between the observed and calculated structure factors and anomalous scattering differences.[6] The observational equations utilized for the heavy-atom derivative structure factor amplitudes are as follows:

$$|F_{Hi}^{c}|^2 = (X + a_{Hi})^2 + (Y + b_{Hi})^2 \tag{1}$$

where $|F_{Hi}^{c}|^2$ represents the calculated structure factor amplitude of the heavy-atom derivative; X the real component of the protein structure factor $= |F_p| \cos \alpha$; Y the imaginary component of the protein structure factor $= |F_p| \sin \alpha$; a_{Hi} the contribution of the real component of heavy-atom derivative to observed heavy-atom derivative structure factor amplitude; b_{Hi} similarly the imaginary component; H the subscript running over all reflection hkl; and i the subscript running over all derivatives.

In the case in which anomalous pairs have been measured, the refinement is carried out on the average $|F_{Hi}|^2$ and half-difference Δ_{Hi} of the Friedel pairs where

$$|F_{Hi}|^2 = \tfrac{1}{2}(|F_{Hi}^{+}|^2 + |F_{\bar{H}i}^{-}|^2) \tag{2}$$

$$\Delta_{Hi} = \tfrac{1}{2}(|F_{Hi}^{+}|^2 - |F_{\bar{H}i}^{-}|^2) \tag{3}$$

The advantage of the above formulation is that errors in the observations of anomalous differences Δ_{Hi} are frequently positively correlated or cancel such that the standard deviation of Δ_{Hi} can be smaller than the standard deviation of $|F_{Hi}|^2$.

[6] Familiarity with least-squares refinement procedures will be assumed in the following discussions.

The refinement of the protein phase and heavy-atom parameters is carried out by least-squares minimization of the following sum:

$$S = \sum_H \sum_i \omega_{Hi}(|F^{\mathrm{o}}_{Hi}|^2 - |F^{\mathrm{c}}_{Hi}|^2)^2 + \omega'_{Hi}(\Delta^{\mathrm{o}}_{Hi} - \Delta^{\mathrm{c}}_{Hi})^2 \tag{4}$$

where $|F^{\mathrm{o}}_{Hi}|^2$ and Δ^{o}_{Hi} represent the observed structure factor amplitude and anomalous difference, respectively, and ω_{Hi} and ω'_{Hi} are the corresponding weights.

In the most general case, the subscript i can be set to 0, which then reduces Eq. (4) to $(X^2 + Y^2 - |F_{\mathrm{p}}|^2)^2$. This simply restates that we can allow for variation not only in the phase angle but also in the modulus of the protein structure factor amplitude. In the MVFC refinement the modulus of the protein structure factor is not refined but is set equal to the observed protein structure factor amplitude.

To allow for experimental error in the observed protein structure factor the weights ω_{Hi} and ω'_{Hi} must be modified. Instead of defining

$$\omega_{Hi} = \frac{1}{\sigma^2(|F^{\mathrm{o}}_{Hi}|^2)} \tag{5}$$

and

$$\omega'_{Hi} = \frac{1}{\sigma^2(\Delta^{0}_{Hi})} \tag{6}$$

the weights become

$$\omega_{Hi} = \frac{1}{\sigma^2(|F_{Hi}|^2) + (\partial|F^{\mathrm{c}}_{Hi}|^2/\partial|F_{\mathrm{p}}|^2)^2\sigma^2(|F_{\mathrm{p}}|^2)} \tag{7}$$

and

$$\omega'_{Hi} = \frac{1}{\sigma^2(\Delta_{Hi}) + (\partial\Delta^{\mathrm{c}}_{Hi}/\partial|F_{\mathrm{p}}|^2)^2\sigma^2(|F_{\mathrm{p}}|^2)} \tag{8}$$

Compared to the weight obtained by Eq. (5) the contribution of experimental error in the protein structure factor amplitude will substantially alter the weight calculated by Eq. (7). The weight for anomalous scattering as given in Eq. (8), however, will not be significantly different numerically from the weight in Eq. (6), since the contribution of the protein modulus to the anomalous scattering differences is not substantial. The rate of convergence of the refinement process is considerably accelerated by using weights derived primarily from experimental error rather than weights which are recalculated after each cycle of refinement.[7]

[7] D. M. Blow and B. W. Matthews, *Acta Crystallogr., Sect. A* **A29,** 56 (1973).

The MVFC refinement treats the sine and cosine of the protein phase as a refinable parameter in a manner similar to the refinement of the heavy-atom parameters. The refinement procedure does not require any special strategy in refinement of the heavy-atom parameters and protein sine and cosine functions; all the parameters can be varied simultaneously. The sine and cosine of the phase of each reflection are refined subject to the constraint that the magnitude of $\sin^2$ phase + $\cos^2$ phase = 1. Otherwise the sine and cosine of the phase could potentially compensate for any systematic error such as nonisomorphism.

The starting values for the sine and cosine of each phase in the absence of initial value are chosen by evaluating each phase at regular points between 0 and 360° and then choosing the lowest value of the sum S [Eq. (4)] as the best approximation for the starting values. From then on the sine or cosine of the phase is treated as a refinable parameter and is adjusted only once per cycle. The structure of the normal equation matrix is diagonal in the sum with respect to the phases, and block diagonal in the sums for the refinement of the heavy-atom parameters, with each block of the matrix being constructed only from the sums referring to the parameters of a single heavy-atom derivative.

Figure of Merit

The equation for the calculation of the electron density is a linear sum in terms of Fourier coefficients:

$$\rho(\mathbf{r}) = \frac{1}{V} \sum_{\mathbf{H}} \mathcal{F}_{\mathrm{p}}(\mathbf{H}) \exp(2\pi_i \mathbf{H} \cdot \mathbf{r}) \tag{9}$$

where $\rho(\mathbf{r})$ represents the electron density evaluated at a point with real space coordinates $\mathbf{r}$; $\mathbf{H}$ the reciprocal space coordinates hkl; V the volume of the unit cell; $\alpha_{\mathbf{H}}$ the phase of the structure factor amplitude hkl; and

$$\mathcal{F}_{\mathrm{p}}(\mathbf{H}) = |F_{\mathrm{p}}(\mathbf{H})| \exp(i\alpha_{\mathbf{H}}) \tag{10}$$

In the presence of statistical fluctuations the calculated electron density with least error at each point $\mathbf{r}$ is a sum of statistically weighted Fourier coefficients. Mathematically and computationally, however, calculation of the weights associated with each Fourier coefficients at every point $\mathbf{r}$ is prohibitive. The best compromise is to calculate the electron density map with least overall error in the unit cell. The weights for the Fourier coefficients are then the familiar figures of merit. Provided the errors between the protein structure factor amplitudes and protein phase are not unduly correlated, the figure of merit m is defined as follows:

$$m = \frac{1}{[1 + \sigma^2(|F_p|)/|F_p|^2](1 + \sigma_T^2)} \tag{11}$$

where

$$\sigma_T^2 = \frac{S}{n - p} \frac{1 - \binom{s}{c}^2}{\Sigma_i\ \omega_{Hi}(\partial|F_{Hi}^c|^2/\partial\binom{c}{s})^2 + \omega'_{Hi}(\partial\Delta_{Hi}/\partial\binom{c}{s})^2} \tag{12}$$

where n is the total number of independent observations and p is the total number of refinable parameters including phases. $\partial|F_{Hi}^c|^2/\partial\binom{c}{s}$ represents the derivative of the heavy-atom structure amplitude with respect to sine (s) or cosine (c) function of the phase. The decision as to which trigonometric function is the independant variable is based upon which of the sums $(1 - s^2)$ or $(1 - c^2)$ is smaller. $\partial\Delta_{Hi}^c/\partial\binom{c}{s}$ is the derivative of heavy-atom anomalous difference with respect to the cosine or sine function of the phase. A high figure of merit depends upon three important factors. The experimental errors in the $|F_p|^2$ and $|F_{Hi}^o|^2$ should be as small as possible since then the weight ω_{Hi} is large and the ratio $\sigma^2(|F_p|)/|F_p|^2$ is small. The value of the overall least-squares fit, $S/(n - p)$, should be as close as possible to the ideal value of 1; that is, only statistical fluctuations remain in the data and not systematic error, such as nonisomorphism. Finally, each heavy-atom derivative or at least some of the heavy-atom derivatives contribute strongly to the phase determination, thereby assuring a large value for the derivatives defined above.

In the case of centric reflection in which only the sign of the phase is to be determined, the calculation of the derivatives $\partial|F_{Hi}^c|^2/\partial\binom{c}{s}$ required for the figure of merit calculation presents a problem. In the MVFC refinement the derivative is evaluated in the same manner as the acentric reflections, the centric phase distribution being considered to be a very narrow continuous distribution about each of the two possible phase values.

The figure of merit [Eq. (11)] decreases whenever the variance σ_T^2 or standard deviation σ_T of the phase function becomes large. Assuming that the overall least-squares fit is good, $S/(n - p)$ is small in Eq. (12), and the data are well measured, then a large variance is due to a shallow global minimum in the phase evaluation. Under this circumstance it is possible that the local minimum corresponding to the refined phase is not the global minimum. Consequently, for low figures of merit, the phase should be reevaluated at regular intervals about the phasing circle to determine whether the current local minimum is indeed the global minimum. The phase corresponding to the global minimum found should then be used for further phase refinement. It should be noted that the figure of merit can also be small if the first-order derivative $\partial|F_{Hi}^c|^2/\partial\binom{c}{s}$ vanishes at the point evaluated. In order to estimate the error, σ_T, it is essential that $\partial|F_{Hi}^c|^2/\partial\binom{c}{s}$

be nonvanishing. To circumvent this problem the series expansion of $|F_{Hi}|$ is carried out to second order only in the sine or cosine of the phase and the first derivative term is replaced by the second der˙ ative term $[\frac{1}{2}(\partial^2|F_{Hi}|^2/\partial(^c_s)^2)]^{1/2}$ in Eq. (12) whenever $\partial|F^c_{Hi}|^2/\partial(^c_s)$ becomes vanishing.

A meaningful figure of merit requires that a phase has been determined on the basis of at least several independent observations, at the very least a minimum of two observations to resolve the phase ambiguity. The MFVC refinement can refine a single isomorphous replacement (SIR) derivative without anomalous pairs. However the figure of merit is only meaningful for those reflections for which the difference between the two possible phases is small (normally a user-determined value) since in these cases the derivative used in the figure of merit calculation is not too different for the two possible solutions. It should be noted that this class of reflections, weak reflections included, having small phase ambiguity are also well suited for accurate refinement of heavy-atom positions.

Correlations

The most common correlations existing in MIR refinement are those cases in which heavy-atom derivatives share common sites of attachment to the protein. The rate of convergence of the refinement to the minimum in these cases is at best reduced somewhat and under extreme conditions the refinement may diverge. A slow convergence is brought about by a neglect of off-diagonal blocks in the normal equation matrix between heavy-atom derivatives as well as neglect of interactions between phases and heavy-atom parameters, the most serious of these being the neglect of the off-diagonal blocks. In general, a slower rate of convergence can be tolerated, provided the refinement does finally converge, and if necessary the appropriate off-diagonal blocks are computed in the normal equation matrix.

Neglect of correlations does, however, yield error estimates which are smaller than when correlations are taken into account. This is due to the fact that the presence of off-diagonal elements in the normal equation matrix tends to produce an increase in the diagonal elements of the inverted normal equation matrix which are used for parameters shift and error calculations. Consequently when correlations are neglected, parameters shifts and errors are reduced, resulting in slower convergence and artificially low parameter error estimates.

Artificially low error estimates in σ_T result in higher figures of merit which do not correspond to more interpretable maps. When heavy-atom derivatives share common sites the heavy-atom structure factor derivatives $\partial|F_{Hi}|^2/\partial(^c_s)$ in Eq. (12) resemble each other numerically, that is,

$$D_{ij} = \sum_H \sqrt{\omega_{Hi}} \left(\frac{\partial|F_{Hi}|^2}{\partial(^c_s)} - \left\langle \frac{\partial|F_{Hi}|^2}{\partial(^c_s)} \right\rangle \right) \sqrt{\omega_{Hj}} \left(\frac{\partial|F_{Hj}|^2}{\partial(^c_s)} - \left\langle \frac{\partial|F_{Hj}|^2}{\partial(^c_s)} \right\rangle \right) \neq 0 \tag{13}$$

where

$$\left\langle \frac{\partial|F_{Hi}|^2}{\partial(^c_s)} \right\rangle = \sum_H \frac{\partial|F_{Hi}|^2}{\partial(^c_s)} \Big/ \begin{matrix}\text{number of unique reflection for the} \\ i\text{th heavy-atom derivative}\end{matrix} \tag{14}$$

In this case the sum in Eq. (12) over all heavy-atom derivatives is no longer valid since information in one heavy-atom derivative is shared in part by another heavy-atom derivative. The calculation of σ_T^2 from Eq. (12) is based upon the assumption that the sum of Eq. (13) is zero or small. Thus, instead of summing over terms as in Eq. (12), the following sum should be taken in the denominator:[8]

$$\text{denominator} = \sum_i \left(\sqrt{\omega_{Hi}} \frac{\partial|F_{Hi}|^2}{\partial(^c_s)} - \sum_j^{i-1} r_{ij} \frac{\partial|F_{Hj}|^2}{\partial(^c_s)} \sqrt{\omega_{Hj}} \right)^2 \tag{15}$$

where

$$r_{ij} = D_{ij} \div \left\{ \left[\sum_H \omega_{Hi} \left(\frac{\partial|F_{Hi}|^2}{\partial(^c_s)} - \left\langle \frac{\partial|F_{Hi}|^2}{\partial(^c_s)} \right\rangle \right)^2 \right]^{1/2} \times \left[\sum_H \omega_{Hj} \left(\frac{\partial|F_{Hi}|^2}{\partial(^c_s)} - \left\langle \frac{\partial|F_{Hi}|^2}{\partial(^c_s)} \right\rangle \right)^2 \right]^{1/2} \right\} \tag{16}$$

The sums over H are taken over all unique reflections or locally such as a function of resolution. If anomalous scattering data is also utilized an identical treatment applies, and the derivative $\sqrt{\omega_{Hi}}(\partial\Delta_{Hi}/\partial(^c_s))$ in Eq. (13) is replaced by a like sum as in Eq. (15). Equation (15) removes the information duplicated between heavy-atom derivatives and thus reduces the magnitude of the sum in Eq. (13), thereby producing a better error estimate σ_T and consequently a lower but more realistic figure of merit.

Computer Programming

Implementation of the MVFC refinement can be easily carried out utilizing existing crystallographic computing software. An iterative least-squares structure factor refinement program is required. The accumula-

[8] J. Sygush, manuscript in preparation (1985).

tion of structure factors and structure factor derivatives is routine with the exception that additional derivatives of $(\partial|F_{Hi}|^2/\partial(^c_s))$, $(\partial\Delta_{Hi}/\partial(^c_s))$, and $\frac{1}{2}(\partial^2|F_{Hi}|^2/\partial(^c_s)^2)$ are accumulated in the normal equation matrix for the phase refinement. Calculation of heavy-atom parameter shifts and sine or cosine function shifts is straightforward. Anomalous scattering data, if present, are handled by use of the half-sum and half-difference definition in Eqs. (2) and (3). This avoids the usual cumbersome expression for anomalous differences. If heavy-atom derivatives do not share common sites the figure of merit calculation by Eqs. (12) and (13) is appropriate. If heavy-atom derivatives share common sites the formalism described by Eqs. (15) and (16) should be utilized. It is best that each of the factors r_{ij}, recalculated after every cycle, be broken down as a function of resolution since at high resolution common sites between heavy-atom derivatives are not necessarily identical.

Remarks

The MFVC refinement is not better than the quality of the data that are used for the refinement of the phases. The important advantage of the MVFC refinement is that experimental error is taken explicitly into account in the phase refinement and in the calculation of the figure of merit. It should be stated that more often than not systematic error or nonisomorphism is due in large part due to poorly measured or treated data rather than genuine nonisomorphous heavy-atom derivatives. The MFVC refinement is easy to use, in that all parameters including phases can be refined simultaneously. Heavy-atom derivatives sharing common sites can be allowed for both in the refinement of heavy-atom parameters, i.e., off-diagonal blocks can be added, and in figure of merit calculations. The MVFC refinement is capable of shifting heavy-atom positions to their correct position, which can be as much as 5–6 Å from the initial starting point. When all significant correlations have been taken into account, the convergence of the MVFC refinement is rapid. Experience has shown that the MVFC refinement is better than conventional refinement when few heavy-atom derivatives are available and when heavy-atom derivatives share common sites, with the proviso that all data have indeed been carefully measured and treated.

[3] Lease-Squares Refinement

By ROBERT A. SPARKS

Introduction

The least-squares problem can be defined in the following way. It is desired to fit a set of m observations, a_i, with a theoretical model expressed by the m equations:

$$g_i(x_1, x_2, \ldots, x_n) = a_i; \qquad i = 1, \ldots, m, \quad m > n \tag{1}$$

One wishes to find the values of a set of n ($n < m$) parameters x_i which give the "best" fit. In the least-squares formalism, Eq. (2) is minimized with respect to the n parameters:

$$d_i = g_i(x_1, x_2, \ldots, x_n) - a_i$$

$$\text{minimize} \quad \Phi = \sum_{i=1}^{m} w_i d_i^2, \qquad w_i = 1/\sigma_i^2 \tag{2}$$

To get the "best" fit, in a statistical sense, each of the equations is given a weight which is inversely proportional to the square of the estimated standard deviations of the observables. It is assumed that the random errors in the m observations are independent from one another. In the case of nonindependence, the following equation is minimized:

$$\text{minimize} \quad \Phi = \mathbf{d}^{\mathrm{T}} W \mathbf{d} \tag{3}$$

W is the inverse of an estimated covariance matrix of the observations and $\mathbf{d}$ is the vector of differences, $\mathbf{d}_i$, defined in Eq. (2). The formalism expressed in Eq. (3) is not often used because it is usually difficult to obtain reliable estimates of all of the elements of W. For the remainder of this chapter I will discuss methods for finding the minimum of Eq. (2).

If the functions g_i are linear (or can be made linear by a change of the variables) with respect to the parameters x_i, there is only one minimum. The method of solving the linear least-squares problem is discussed in many texts on numerical analysis. Most least-squares problems in crystallography are nonlinear and it will be the solution of these which I will discuss here. Because the equations are nonlinear it may be possible that there is more than one minimum. In most scientific problems it is expected (hoped?) that the "best" fit has a value for Φ which is much lower than all the other minima. It is this global minimum which is the goal of the least-squares calculations.

METHODS IN ENZYMOLOGY, VOL. 115

Simplex Method

One way of obtaining the minimum of Eq. (2) is the simplex method. Φ is evaluated at $n + 1$ points of a simplex in n-dimensional space. The coordinates of no one point of a simplex can be expressed as a linear combination of the coordinates of the other points (a two-dimensional simplex is a triangle; a three-dimensional simplex is a tetrahedron). The method shown here is due to Nelder and Mead[1] and consists of replacing one point at a time of the simplex by a process of reflection [Eq. (4)], expansion [Eq. (5)], and contraction [Eq. (6)].

Evaluate:

Evaluate Φ at $n + 1$ points in n-dimensional space:

$p_0, p_1, p_2, \ldots, p_n$

$\Phi_h = \max(\Phi_0, \ldots, \Phi_n)$

$\Phi_l = \min(\Phi_0, \ldots, \Phi_n)$

$\bar{p}$ = centroid of points p_i; $i = 0, n, i \neq h$

Test stop condition

Reflection:

$p^* = (1 + \alpha)\bar{p} - \alpha p_h$ $\quad$ α = reflection coefficient. Good value: $\alpha = 1$

if $\Phi^* < \Phi_l$ then try expansion.

if $\Phi_i < \Phi^* < \Phi_h$; $i = 0, n, i \neq h$ $\quad$ (4)

then p^* replaces p_h. Try contraction

if $\Phi^* > \Phi_h$ then try contraction

else p^* replaces p_h. Evaluate again

Expansion:

$p^{**} = \gamma p^* + (1 - \gamma)\bar{p}$ $\quad$ γ = expansion coefficient. Good value: $\gamma = 2$

if $\Phi^{**} < \Phi_l$ then replace p_h with p^{**}. Evaluate again $\quad$ (5)

else replace p_h with p^*. Evaluate again

Contraction:

$p^{**} = \beta p_h + (1 - \beta)\bar{p}$ $\quad$ β = contraction coefficient. Good value: $\beta = \frac{1}{2}$

if $\Phi^{**} < \Phi_h$ then replace p_h with p^{**}. Evaluate again $\quad$ (6)

else replace all p_i by $(p_i + p_l)/2$. Evaluate again

[1] J. A. Nelder and R. Mead, *Comput. J.* **7,** 308 (1965).

Stop condition:

Stop when $\sum_{i=0}^{n} (\Phi_i - \bar{\Phi})^2/(n + 1) < \varepsilon$

ε = arbitrary small constant

Else try reflection

The simplex adapts itself to the local landscape and contracts to the final minimum, although the method does not guarantee that the solution will not be a local minimum instead of the global minimum. The disadvantage is that the function Φ must be evaluated many times. Thus, for the examples of Nelder and Mead of two, three, and four variables, the function had to be evaluated more than 100 times. The method lends itself to those problems for which the evaluation of Φ is not time consuming and there are a small number of variables. Taupin[2] used the method to determine 2θ positions, integrated intensities, and half-widths for reflections which are overlapped in an X-ray powder diffraction pattern.

Newton–Raphson Method

At the minimum the first derivatives of Φ with respect to the variables must be equal to zero:

$$f_i(x_1, x_2, \ldots, x_n) = \partial\Phi/\partial x_i = 0, \qquad i = 1, \ldots, n \tag{7}$$

Note that this condition is true for all minima, maxima, and saddle points. The n functions f_i can be expanded about some approximate solution ($x = x_0$) in a Taylor's series. The second order and higher terms are neglected and the resulting n equations are linear with respect to the unknowns Δx_j:

$$(f_i)_{x=x_0} + \sum_{j=1}^{n} \left(\frac{\partial f_i}{\partial x_j}\right)_{x=x_0} \Delta x_j = 0 \tag{8}$$

$$-\mathbf{b} + A_{\mathrm{NR}}x = 0 \tag{9}$$

Equation (9) is the matrix representation of the n equations [Eq. (8)]. The system of linear equations is solved for the Δx_j values, which are added to the initial x_j and the process is repeated. If the initial guess is sufficiently close to the global minimum the process will converge to that minimum. Although the computation time for each iteration is long, the number of iterations required is much less than for the simplex method. The disad-

[2] D. Taupin, *J. Appl. Crystallogr.* **6,** 266 (1973).

vantage is that the initial approximation must be closer to the global minimum than is required for any of the points of the initial simplex in the simplex method.

Nonlinear Least Squares

The nonlinear least-squares method used by most crystallographers can be described as follows. The m observational equations [Eq. (1)] are expanded about an approximate solution in a Taylor's series:

$$d_i = g_i(x_1, x_2, \ldots, x_n) - a_i \approx (g_i)_{x=x_0} - a_i + \sum_{j=1}^{n} \left(\frac{\partial g_i}{\partial x_j}\right)_{x=x_0} \Delta x_j \quad (10)$$

The second and higher order terms are neglected and the resulting equations are then linear in the unknowns Δx_j. The function Φ then can be written as

$$\Phi = \sum_{i=1}^{m} w_i \left[(g_i)_{x=x_0} - a_i + \sum_{j=1}^{n} \left(\frac{\partial g_i}{\partial x_j}\right)_{x=x_0} \Delta x_j\right]^2 \quad (11)$$

At the minimum the first derivatives with respect to the Δx_j will be zero. The resulting n equations are called normal equations and are linear with respect to the new Δx_j:

$$\frac{\partial \Phi}{\partial \Delta x_j} = 2 \sum_{k=1}^{n} \sum_{i=1}^{m} w_i \left(\frac{\partial g_i}{\partial x_j}\right)\left(\frac{\partial g_i}{\partial x_k}\right) \Delta x_j + 2 \sum_{i=1}^{m} w_i(g_i - a_i) \left(\frac{\partial g_i}{\partial x_j}\right) = 0 \quad (12)$$

Equation (13) is the matrix representation of the n equations of Eq. (12).

$$A\mathbf{x} - \mathbf{b} = 0 \quad (13)$$

The system of linear equations is solved for the Δx_j values, which are then added to the initial x_j and the process is repeated. Note that the **b** term in Eq. (13) is identical to the **b** term in Eq. (9) of the Newton–Raphson procedure, but that the A matrix of Eq. (13) is not the same as the A_{NR} of Eq. (9). Since the functions of f_i in Eq. (7) are functions of g_i and the first partial derivatives of g_i, the A matrix in Eq. (9) will include partial second derivatives of g_i. The A matrix of Eq. (13) does not include these partial second derivatives. For this reason the computational time per iteration is longer for the Newton–Raphson method than for the nonlinear least-squares method. Because second-order terms are included in the Newton–Raphson procedure, it is expected that, in general, fewer iterations will be required for this method and, in general, the initial approximation can be further from the global minimum. The A matrix and $\mathbf{b}$ vector can be expressed as matrix products where the M matrix has the dimensions $m \times n$:

$$\begin{aligned} A &= M^{\mathrm{T}}M \quad \text{where} \quad m_{ij} = \sqrt{w_i}(\partial g_i/\partial x_j) \\ \mathbf{b} &= M^{\mathrm{T}}\mathbf{c} \quad \text{where} \quad c_i = \sqrt{w_i}(a_i - g_i) \end{aligned} \tag{14}$$

Atomic Parameter Least Squares

The crystallographic problem of determining the best set of atomic parameters which can be obtained from a set of experimental structure factors is usually solved with the nonlinear least-squares method described above. Φ is expressed as a function of observed and calculated structure factors:

$$\begin{aligned} \Phi &= \sum_{i=1}^{m} w_i(|F_{oi}| - |F_{ci}|)^2 \\ F_{ci} &= F_{\mathrm{calc}}(h) = G \sum_{r=1}^{p} f_T(h) \exp(2\pi i h x_r) \\ f_T(h) &= n_r f_r(h) q_r(h) \\ q_r(h) &= \exp\{-B_r[(\sin\theta)/\lambda]^2\} \end{aligned} \tag{15}$$

or

$$\begin{aligned} q_r(h) = \exp[&-2\pi^2(\mathsf{U}_r^{11}h^2a^{*2} + \mathsf{U}_r^{22}k^2b^{*2} + \mathsf{U}_r^{33}l^2c^{*2} \\ &+ 2\mathsf{U}_r^{23}klb^*c^* + 2\mathsf{U}_r^{31}lhc^*a^* + 2\mathsf{U}_r^{12}hka^*b^*)] \end{aligned}$$

where G is an overall scale factor, n_r is an occupancy factor for the rth atom, and $f_r(h)$ is a scattering factor. $q_r(h)$ is a temperature factor term where either B_r is an isotropic temperature factor or U_r^{ij} is a component of the symmetric tensor U_r, which describes anisotropic harmonic motion for the rth atom. Calculation of derivatives are described by Cruickshank[3] and will not be detailed here. Note that the sum over atoms in obtaining $F_{\mathrm{calc}}(h)$ is over all atoms in the unit cell. The parameters used in the nonlinear least-squares method must be independent and thus the calculated derivatives will be summations over the derivatives of the symmetry-related atoms. Consideration must be taken for atoms in special positions which have fewer independent parameters than those in general positions. Also, in polar space groups the origin is not fixed by symmetry elements in one or more dimensions, and in those cases one, two, or three

[3] D. W. J. Cruickshank, "Crystallographic Computing," p. 187. Munksgaard, Copenhagen, 1970.

position parameters must be held fixed (eliminated from the set of least-squares variables) for one atom. Clearly, the scale factor G and all of the occupancy parameters cannot be refined together because of the linear dependence of these parameters.

Constraints

If chemical information is known about parts of the crystal structure, it can be included in the least-squares equations as constraints. The kind of constraints that are often used are bond distances, bond angles, planarity of certain groups of atoms, rigid groups (i.e., the atoms within the group retain their positions with respect to each other), etc. Linear constraints on occupancy factors are also used. If analytical expressions as in Eq. (16) describing these constraints can be formulated, then derivatives can be evaluated:

$$x_i = s_i(v_1, v_2, \ldots, v_l), \qquad l < n \tag{16}$$

$$d\mathbf{x} = Q\, d\mathbf{v}, \qquad \text{where} \quad Q_{ij} = \partial x_i / \partial v_j \tag{17}$$

x_i is as shown in Eq. (1). v_j represents a new set of variables, some of which can be the same as some of the x_i. The number l of v_j values is less than the number n of x_i values. The matrix of new observational equations, Y, can now be formed:

$$Y = MQ \tag{18}$$

Y replaces the M of Eq. (14). After determining the best fit for the variables v_j, the atomic parameters x_i can be calculated using Eq. (16). Scheringer[4] has used this method to constrain atoms in rigid groups. The variables of a rigid group are three translational parameters and three rotational parameters. Because there are fewer variables, the amount of calculation time per iteration is decreased compared to a refinement of atomic parameters. A very important property observed by investigators who have used constraints is that the radius of convergence is larger than that for a refinement of atomic parameters. Thus, for benzene Scheringer showed that convergence was obtained for an initial model where the translational parameters were displaced by 0.37 Å and the rotational parameters were displaced by 60°.

[4] C. Scheringer, *Acta Crystallogr.* **16,** 546 (1963).

Restraints

Waser[5] introduced chemical information in a different way. He suggested reformulating the problem as follows:

$$\Phi = \sum_{i=1}^{m} w_i(|F_{oi}| - |F_{ci}|)^2 + \sum_{k=1}^{l} w_k[t_k(p_1, p_2, \ldots, p_n) - z_k]^2 \quad (19)$$

where p_j is an atomic parameter. In this formulation the analytical expressions of the constraints are added to the quantity to be minimized, Φ from Eq. (2). Each expression has a weight w_k assigned to it. Unlike the constraints described in the above section, these so-called restraints will not hold exactly at the minimum of Φ. How exactly will depend on the relative weights w_i assigned to the structure factors and the weights w_k assigned to the restraints. Unlike constraints, the number of variables is not decreased and hence computing time per iteration increases.

Investigators have found that restraints also improve the radius of convergence for a given problem. Restraints have an added advantage that certain chemical information (e.g., bond distances, bond angles) is known only with expected standard deviations and thus should not be made to hold exactly. Also, in some cases, for a given chemical property the expressions of t_k and their derivatives are easier to evaluate than the expressions of s_i and their derivatives.

Standard Deviations

An important property of a least-squares analysis is the ability to calculate variances and covariances (and hence standard deviations and correlation coefficients) of the variables. The variances and covariances are given by Eq. (20) where the a_{ij}^{-1} term is an element of the inverse of the matrix A of Eq. (13), used on the last iteration.

$$\mathrm{cov}(p_i, p_j) = a_{ij}^{-1} \sum_{k=1}^{m} w_k(|F_{ok}| - |F_{ck}|)^2/(m - n) \quad (20)$$

$$\mathrm{cov}(p_i, p_j) = \sigma_i^2$$

Goals for Refinement of Crystallographic Parameters

The primary goal for a crystallographic least-squares program is the adjustment of crystallographic parameters to produce the minimum for

[5] J. Waser, *Acta Crystallogr.* **16,** 1091 (1963).

the function Φ in Eq. (11) or Eq. (19). A secondary goal for a crystallographic least-squares program is the calculation of good estimates of the standard deviations for these parameters.

As with other good computer programs, a least-squares program should be easy to use. Since there are many input parameters, user mistakes are common. One of the easiest methods of minimizing such mistakes is the use of format-free input.

Since crystallographic least-squares problems can be poorly conditioned, the user should be given some measure of this conditioning. Subroutines which solve systems of linear equations or invert matrices can be made to indicate when a loss of precision has taken place. Such loss of precision is a good indication that one or more parameters are highly dependent on the values of the other parameters.

FORTRAN and ALGOL are the languages most often used for crystallographic least-squares programs because these programs are easiest for the user to modify. Modifications are usually made so the program can handle unusual parameters or unusual dependency between parameters. Parts of a least-squares program, such as formation of the normal equations, solution of linear equations, and matrix inversion, never need to change and so could be written in machine language to minimize execution time.

Efforts by Stewart[6] and colleagues to make crystallographic programs machine independent have been especially important with respect to least-squares programs since these programs are among the most complex used by crystallographers. A program written for machine independence has the advantage that a minimum amount of effort is needed to make the program run on a new computer and a minimum amount of overall program maintenance is needed.

Finally, all of these desirable features must be balanced with the need to minimize the cost of least-squares refinement. Hamilton[7] has shown that about 80% of the computer cost of structure determination is in least-squares refinement, and it is for this reason that so much effort has been put into approximations to the full-matrix technique.

Nonlinearity of Problem

The method of nonlinear least squares will not necessarily converge to the true minimum of the function Φ. It is possible for it to converge to any

[6] J. Stewart, private communication (1970).

[7] W. C. Hamilton, "Computational Needs and Resources in Crystallography," p. 9. Natl. Acad. Sci., Washington, D.C., 1972.

local critical point (minimum, maximum, or saddle point). Unfortunately, no easily computable criterion is available which could guarantee that the true minimum has been reached in a many-dimensional problem. Luckily, the crystallographic least-squares problem is usually well behaved. Usually, only one critical point will have a value of ψ [Eq. (21)] less than 3 or 4 when all of the normal parameters (atomic positions, anisotropic thermal, scale factor) have been included in the model:

$$\psi = \sum_{i=1}^{m} w_i(|F_{oi}| - |F_{ci}|)^2/(m - n) \tag{21}$$

There are cases, however, for which this is not true (e.g., a structure with a disordered atom close to a mirror must have a critical point for the Φ function with the atom on the mirror).

Assuming the least-squares procedure is leading to the true minimum, when should the refinement be stopped? Usually, crystallographers use a criterion that no more refinement cycles will be calculated when all of the last set of parameter shifts are less than some fraction (0.1–0.5) of the corresponding calculated standard deviations. Unfortunately, this criterion is not always reliable because it may happen that on successive iterations the shifts are small and always in the same direction, but yet the last set of parameters is still significantly different from those at the minimum. One should be very suspect of parameters which during refinement are constantly moving in small shifts in the same direction.

Finally, after the last cycle, a difference Fourier map should be calculated to ensure that all significant features of the electron density distribution are being fit with the model which is being refined.

Approximations to Full-Matrix Technique

The time required for a classic crystallographic full-matrix least-squares program is made up of the times for calculations of (1) structure factors and derivatives, (2) formation of normal equations, (3) solution of normal equations, and (4) completion of the inverse of the normal equations. These times are dependent on the number of reflections, the number of variables, the number of equivalent positions, and whether the structure has a center of symmetry. For an acentric 217-parameter problem with four equivalent positions and 4340 reflections, it was found that the percentages of time spent were, respectively, (1) 30%, (2) 61%, (3) 4%, and (4) 5%. Clearly, the formation of the normal equations is the most time-consuming part.

For these reasons approximations to the normal equations [Eq. (13)] are made for problems with a large number of parameters. It should be pointed out that when the vector **b** in Eqs. (13) or (9) is zero, convergence is complete. If this condition is met it does not matter whether all parts of A_{NR} [Eq. (9)] or A [Eq. (13)] have been calculated on each successive iteration. However, to obtain standard deviations using Eq. (20), the full matrix and its inverse are required. The "best" algorithm is the one that leads to the solution for the least amount of computer money.

In considering various algorithms, one must keep in mind not only the computing cost per iteration, but also how many iterations may be required compared to the classic full-matrix approach. Indeed, the convergence properties of a given algorithm may be so poor that a problem which would behave well with the classic approach may diverge with the given algorithm. What is not often realized is that the reverse may also be true. The convergence properties of the classic method can sometimes be improved upon. Using partial shifts may help a normally divergent problem to converge, will never cause a normally convergent problem to diverge, but may require more or less iterations to get to the minimum of Φ. So, if large shifts are calculated for some parameters, a good technique would be to use only a fraction of these shifts.

One of the best techniques is that of repeated use of the A matrix [Eq. (13)]. For example, if instead of calculating three cycles of full-matrix least squares, one cycle of full matrix followed by two cycles of calculation of only the **b** vector is performed. The shifts for the last two cycles are calculated with the use of the inverse matrix formed in the first cycle. In our example, a little over 30% (structure factor and derivative calculations) of the computer time for each of the last two cycles will be needed, or for the three cycles only 53% of the time necessary for the classical method. Clearly, this method works best when the shifts are small. Palenik[8] has used this method when the reliability factor R is less than 20% for a total of 20–50 structures and has found only one structure for which divergence took place yet convergence was obtained using the classic approach. In the few cases where the two methods were compared, Palenik found that the differences between the calculated parameter shifts were less than the corresponding calculated standard deviations for all parameters. Rollett[9] who also uses this method claims that it will work when the trial structure differs from the final structure by no more than 0.1 Å in position parameters and no more than 0.5 Å^2 in thermal parameters (B values).

In large problems, calculation of even one cycle of full-matrix least squares may be too costly. In these cases, various approximations to the

[8] G. J. Palenik, private communication (1974).

[9] J. S. Rollett, private communication (1974).

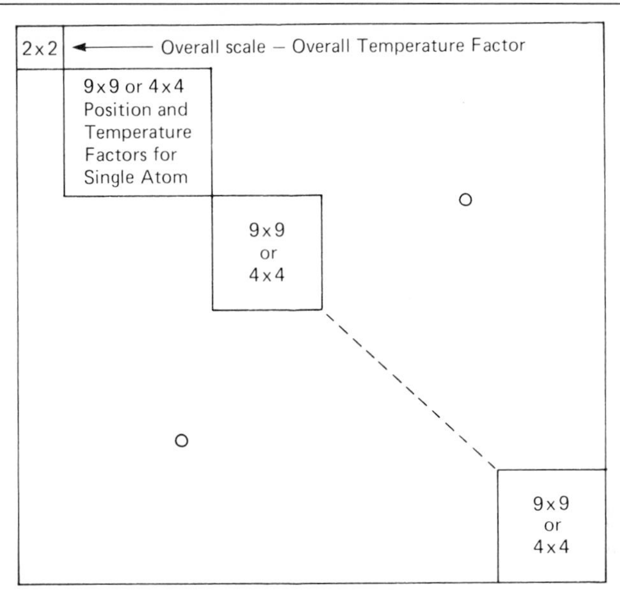

FIG. 1. Approximation to full matrix in which are calculated only off-diagonal elements between parameters within a given atom and the off-diagonal element between the overall scale factor and overall temperature factor.

A matrix are made. These approximations can be thought of as using the same linearizing assumption of the classic method followed by an additional approximation to solve the resulting linear problem. The block-diagonal approximation sets those off-diagonal elements of the full matrix equal to zero which one would expect to be small. Thus, a typical block-diagonal approximation would calculate only those off-diagonal elements between parameters within a given atom and the off-diagonal element between the overall scale factor and overall temperature factor (Fig. 1). Another method would calculate all off-diagonal elements between all position parameters, between all thermal parameters, and between thermal parameters and scale factor, but not those off-diagonal elements between thermal parameters and position parameters (Fig. 2).

An analysis of the effect of certain block-diagonal approximations on the linear problem was given by Sparks.[10] It was shown that if the normal equations are represented by the matrix in Eq. (22) and this is to be approximated by a block-diagonal set of normal equations [Eq. (23)], then

[10] R. A. Sparks, "Computational Methods and the Phase Problem in X-Ray Crystallography," p. 170. Pergamon, Oxford, 1961.

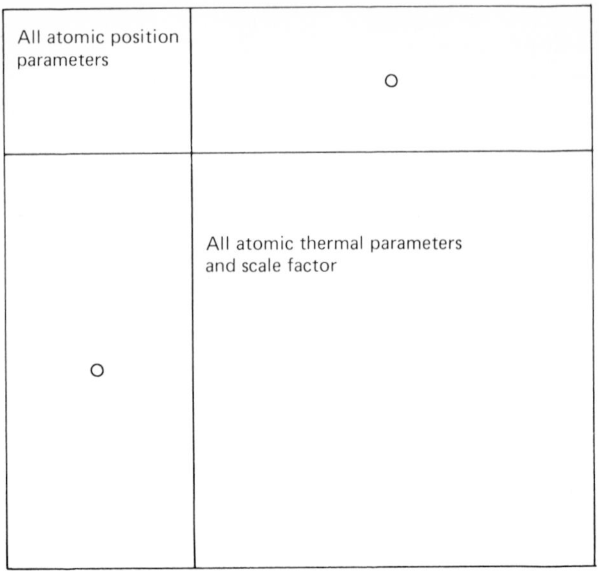

All atomic position parameters

0

All atomic thermal parameters and scale factor

0

FIG. 2. Approximation to full matrix in which are calculated all off-diagonal elements between all position parameters, between all thermal parameters, and between thermal parameters and scale factor.

the convergence properties are a function of the eigenvalues of the matrix in Eq. (24) where I is the identity matrix (ones on the diagonal; zeros off the diagonal).

$$A\mathbf{x} = \mathbf{b} \tag{22}$$

$$B\mathbf{y} = \mathbf{b} \tag{23}$$

$$L = I - B^{-1}A \tag{24}$$

For convergence all of the eigenvalues must be between -1 and $+1$. By using a partial shift factor η, convergence can always be assured because by replacing B^{-1} by ηB^{-1} the eigenvalues of matrix (24) can always be made to fall between -1 and $+1$ with the proper choice of η.

It is possible to solve the linear equations of Eq. (22) with as much accuracy as desired without ever forming the matrix A. It can be shown that Eq. (25) is valid if matrix in Eq. (24) has the convergence properties mentioned above:

$$\mathbf{x} = B^{-1}\mathbf{b} + [I - B^{-1}A]B^{-1}\mathbf{b} + [I - B^{-1}A]^2B^{-1}\mathbf{b} + \cdots \tag{25}$$

$$\mathbf{r}_0 = B^{-1}\mathbf{b} \tag{26}$$

The method would be to choose a block-diagonal approximation B for which a small amount of computer time is necessary to calculate B^{-1}.

Then calculate Eq. (26). Calculation of each successive term of Eq. (25) involves a multiplication of a vector by $I - B^{-1}A$. This can most easily be performed in the steps shown in Eq. (27):

$$\begin{aligned} &1. \quad s = Mr_i \\ &2. \quad t = M^{\mathrm{T}}s \\ &3. \quad v = B^{-1}t \\ &4. \quad r_{i+1} = r_i - v \end{aligned} \tag{27}$$

Steps 1 and 2 are by far the most time-consuming and involve a total of $2mn$ multiplications. Formation of the matrix A requires $mn(n + 1)/2$ multiplications. Thus, very roughly $1 + (n + 1)/4$ terms of Eq. (25) could be calculated in the time necessary to form matrix A. Using a B matrix with zeroes for off-diagonal terms between atoms, it was shown by Sparks[10] that this method coupled with an acceleration technique, to be described below, would give for a 64-parameter problem a solution whose shifts would be less than 1% different from those obtained by solving Eq. (22), but in only 0.65 of the time necessary for the classical method of solving Eq. (22). Even greater savings would be achieved with larger problems. A requirement for this method is that the $n \times m$ elements of the matrix M must be stored someplace. The most economical storage medium for this large number of elements would be magnetic tape or disk.

Another method closely related to the block-diagonal method is the Gauss–Seidel block approximation first proposed by Schomaker and Waser and described by Sparks.[10] This method is to obtain derivatives for one independent atom while calculating structure factors; solve the resulting nine equations in nine unknowns; correct the calculated structure factors forming a new set of nine equations for the next independent atom, and so forth. After the last atom parameters have been adjusted in this way, the scale factor is corrected by the same procedure, followed by a new adjustment of the first atom parameters, etc. This method is very much like including all of the off-diagonal elements below the diagonal of the full matrix without the need to actually calculate them. The method was programmed by Gilbert[11] for the Syntex XTL Structure Determination System. On one test structure, a solution which was obtained with one cycle of the full-matrix method took four cycles of the block-diagonal method and two cycles of the Gauss–Seidel block method. The calculation time per cycle for the Gauss–Seidel block method is comparable to twice that for a structure factor calculation. Hoard and Nordman[12] have

[11] M. Gilbert, "Operations Manual Syntex XTL Structure Determination System." Nicolet Instrument Corporation, Madison, Wisconsin, 1974

[12] L. G. Hoard and C. E. Nordman, *Acta Crystallogr., Sect. A* **A35,** 1010 (1979).

used this method with restraints and rigid group constraints for a 1104-atom protein structure.

Acceleration Techniques

As indicated above, it may be necessary to use a partial shift factor to make a given problem converge. For fast convergence (minimum number of cycles) an optimum shift factor should be used. Unfortunately, without knowledge of the eigenvalues of matrix of Eq. (24) it is not possible to know the optimum shift factor that should be used. After several cycles it is possible to calculate a good estimate of an optimum shift factor. Even better is a technique known as the conjugate gradient method[13] which determines on each cycle two shift parameters α and β used to calculate a linear combination of the two previous shift vectors and the current shift vector. This linear combination is the actual shift applied. The conjugate gradient method is popular for large linear problems and has been used on many nonlinear problems as well. A description of the various acceleration techniques and how they can be used for crystallographic least squares is given by Sparks[10] and Rollett.[14,15] Until recently, these methods were not used much because typically no more than two or three cycles of least squares would be used with all atoms having isotropic temperature parameters, followed by two or three more cycles with anisotropic temperature parameters, and finally two or three cycles with hydrogens added. Each change is really a change in the model being fitted and unfortunately two or three cycles are not enough to determine optimum shift parameters. Note, however, that if the algorithm described in Eqs. (26) and (27) is used to solve the linear problem, the conjugate gradient method becomes a very useful acceleration method. It is this method which was used to show that the 64-parameter linear problem could be solved in 0.65 of the time required for the classical method.

To use the conjugate gradient method instead of Eqs. (26) and (27) one forms the matrix N where

$$N^{\mathrm{T}}N = B^{-1} \tag{28}$$

The matrix N is upper triangular (elements below the diagonal are equal to zero) and is usually referred to as the square root of the approximation matrix B^{-1}. The following algorithm is used to determine the elements n_{ij} of N from the elements b_{ij}^{-1} of B^{-1}.

[13] M. R. Hestenes and E. Stiefel, *J. Res. Natl. Bur. Stand. (U.S.)* **49,** 409 (1952).

[14] J. S. Rollet, "Computing Methods in Crystallography," p. 73. Pergamon, Oxford, 1965.

[15] J. S. Rollett, "Crystallographic Computing," p. 167. Munksgaard, Copenhagen, 1970.

$$
\begin{aligned}
n_{ij} &= 0 \quad \text{for} \quad i > j \\
n_{11} &= \sqrt{b_{11}^{-1}} \\
n_{ij} &= b_{ij}^{-1}/n_{11} \quad \text{for} \quad j = 2, \ldots, n \\
n_{ii} &= \left(b_{ii}^{-1} - \sum_{l=1}^{i-1} n_{li}^2\right)^{1/2} \quad \text{for} \quad i = 2, \ldots, n \\
n_{ij} &= \left(b_{ij}^{-1} - \sum_{l=1}^{i-1} n_{li} n_{lj}\right) \Big/ n_{ii} \quad \text{for} \quad i = 2, \ldots, n, \quad j = i + 1, \ldots, n
\end{aligned}
\tag{29}
$$

where i is the row number and j is the column number. The 0th iteration for the conjugate gradient method is

$$
\begin{aligned}
r_0 &= N\mathbf{b} \\
y_0 &= 0
\end{aligned}
\tag{30}
$$

The first iteration is calculated according to the following algorithm:

$$
\begin{aligned}
s_0 &= N^{\mathrm{T}} r_0 \\
t_0 &= M s_0 \\
v_0 &= M^{\mathrm{T}} t_0 \\
w_0 &= N v_0 \\
\alpha_0 &= r_0^{\mathrm{T}} r_0 / r_0^{\mathrm{T}} w_0 \\
r_1 &= r_0 - \alpha_0 w_0 \\
y_1 &= y_0 + \alpha_0 r_0
\end{aligned}
\tag{31}
$$

Perform the following operations [Eqs. (32)–(34)] for $k = 1, \ldots, p$ until for the pth cycle $(y_{p+1} - y_p)^{\mathrm{T}}(y_{p+1} - y_p)$ is less than some previously chosen ε:

$$
\begin{aligned}
s_k &= N^{\mathrm{T}} r_k \\
t_k &= M s_k \\
v_k &= M^{\mathrm{T}} t_k \\
w_k &= N v_k
\end{aligned}
\tag{32}
$$

Solve the following two equations for the shift parameters α_k and β_k:

$$
\begin{aligned}
\alpha_k(r_k^{\mathrm{T}} w_k) - \beta_k(r_k^{\mathrm{T}} r_k) &= 0 \\
\alpha_k(r_{k-1}^{\mathrm{T}} w_k) + \beta_k(r_{k-1}^{\mathrm{T}} r_{k-1}) &= r_{k-1}^{\mathrm{T}} r_{k-1}
\end{aligned}
\tag{33}
$$

Form

$$
\begin{aligned}
y_{k+1} &= \alpha_k r_k + \beta_k y_k + (1 - \beta_k) y_{k-1} \\
r_{k+1} &= -\alpha_k w_k + \beta_k r_k + (1 - \beta_k) r_{k-1}
\end{aligned}
\tag{34}
$$

Finally, at the end of the pth cycle obtain the parameter shifts from

$$
x = N^{\mathrm{T}} y_{p+1} \tag{35}
$$

The time-consuming formation of A and A^{-1} was not performed in this procedure, yet by choosing ε sufficiently small, the parameter shifts x can be made to be as close to the solution of Eq. (22) as desirable. In Eqs. (30)–(35) **b**, r_k, y_k, s_k, t_k, v_k, w_k, and x are column vectors. N and M are matrices and α_k, β_k, and ε are scalar quantities.

Recently, Konnert[16] used the conjugate gradient method together with distance restraints. When forming the normal equation matrix only those off-diagonal terms which connect atoms positioned close together were calculated. All other off-diagonal terms were assumed to be zero. The resulting sparse normal equations can be solved very economically with the conjugate gradient method.

Agarwal[17] also used the conjugate gradient method. He calculated the **b** vector and the diagonal elements of the A matrix [Eq. (13)] using the fast Fourier transform method of Cooley and Tukey.[18] He assumed all off-diagonal elements to be equal to zero. This diagonal approximation method requires many more iterations than the bonded-block approximation of Konnert (see Sparks[10]), but because of the very efficient fast Fourier transform (FFT) calculations of the diagonal and **b** vectors, Agarwal's method is very economical.

Solution of Normal Equation and Matrix Inversion

It should be noted that matrix inversion is not required in order to solve a system of linear equations. In the example above, 5% of the time for a cycle of least squares was spent in completion of the matrix inversion. The inverse is required only when variances and covariances of the parameters are needed, that is, only on the last cycle of least-squares refinement.

Optimum Use of Hardware

Most of the crystallographic least-squares programs written for large computers store the normal equation matrix in core. These programs have a limit on the number of parameters. This limit is dependent on the amount of high speed random access memory available (e.g., Ibers[19] reports a limit of 240 parameters on his CDC 6400). Of course, it is possible

[16] J. H. Konnert, *Acta Crystallogr., Sect. A* **A32,** 614 (1976).
[17] R. C. Agarwal, *Acta Crystallogr., Sect. A* **A34,** 791 (1978).
[18] J. W. Cooley and J. W. Tukey, *Math. Comput.* **19,** 297 (1965).
[19] J. A. Ibers, "Computational Needs and Resources in Crystallography," p. 18, Natl. Acad. Sci., Washington, D.C., 1972.

to use a disk or drum for the storage of the normal equation matrix. The programming becomes more difficult because it is necessary to form the normal equation matrix in such a way that the time for disk transfers does not add greatly to the execution time.

It is possible to use small computers for the least-squares calculation. The algorithm for the Syntex E-XTL is the following. Derivative vectors for many reflections (several hundred) are written onto a disk file. When this file is filled, a block of the normal equation matrix (3072 elements) is transferred to core. This block is then updated with contributions from all of the vectors in the derivative file. Then that block of the normal equation matrix is written back on disk. The derivative matrix must be read for each 3072-element block of the normal equation matrix. Depending on the total number of reflections, the derivative file may have to be written more than once, and each time the normal equation matrix must be updated. Up to 500 parameters can be handled by the Syntex E-XTL Full Matrix Least Squares Program. The advantage of a small computer is that it can be designed for the specific application and thus can be more economical than a large general-purpose computer.[20]

Ill-Conditioning

A poorly conditioned least-squares problem is one which has one or more of the eigenvalues of A [Eq. (13)] close to or equal to zero. Usually, the crystallographic least-squares problem is well conditioned. However, in disordered structures, it can be that there is an apparent overlap of independent atoms and the problem then becomes poorly conditioned. The most common cause of ill-conditioning is due to the user trying to refine parameters which are linearly dependent. An example would be mistakes in deciding which thermal parameters are independent and what the linear dependencies are for other thermal parameters, for atoms on 3-fold, 4-fold, or 6-fold axes. Much computer time could be saved by having the least-squares program check for such errors before the calculations are started. As pointed out earlier, the program should give some indication of conditioning on each cycle of refinement.

Overall Strategy

The overall strategy for crystallographic least-squares refinement depends on the size and nature of the problem, but some techniques can be mentioned here.

[20] R. A. Sparks, "Minicomputers and Large Scale Computations," p. 94. Am. Chem. Soc., Washington, D.C., 1977.

At the start of refinement, the number of parameters should be minimized, by refining only position parameters and isotropic thermal parameters for the nonhydrogen atoms. In the early stages, the errors in the model are primarily due to poor phase angle information. It makes sense to eliminate those reflections whose calculated phases are unreliable, i.e., those with large $(\sin \theta)/\lambda$ or small $|F_c|/|F_o|$ or small $|F_o|/|F_c|$.

Early constraints or restraints will help the problem converge (the constraints or restraints could be later removed, if so desired). According to Rollett,[15] early block-diagonal or Gauss–Seidel block-diagonal methods may converge as well as the full-matrix method, because all methods are plagued by significant nonlinearities at this stage.

Later stages would introduce hydrogen atom parameters and anisotropic thermal parameters. As the phase angle information becomes more reliable, more reflections can be included. If the problem is small enough, a full-matrix calculation can be used but time and money will be saved by not recalculating the left-hand side for each cycle of refinement. If the problem is too large, some of the techniques mentioned in earlier sections must be used. If the block approximations must be used, a common technique is to make the blocks as large as possible. A final difference Fourier should always be calculated.

The methods of Hoard and Nordman,[12] Konnert,[16] and Agarwal[17] have been used for the refinement of protein structures.

Standard Deviation Calculations

As mentioned above, the standard deviations and correlation coefficients of the parameters are obtained from the inverse of the normal equations matrix. It is known that any diagonal or block approximation will cause the standard deviations to be underestimated. Schilling[21] found that the calculated standard deviations for a three-block approximation (nonhydrogen position parameters, nonhydrogen anisotropic thermal parameters, and hydrogen position and isotropic thermal parameters) were 25% less than those obtained from the inverse of the corresponding full matrix. I know of no method to estimate for a given case how big this effect will be.

Powder Diffraction

Using all the data points of a powder pattern as observations, many investigators have refined atomic parameters by the pattern-fitting struc-

[21] J. W. Schilling, Doctoral Thesis, p. 138. University of Michigan, Ann Arbor (1968).

ture refinement method of Rietveld.[22] Young *et al.*[23] described the power of this least-squares method together with the special problems involved.

Summary and Conclusions

Most of the methods mentioned in this chapter were known in the early 1960s. Because of the availability of large, fast computers, however, the most popular technique has been the full-matrix least-squares method. The most significant contributions to crystallographic least-squares calculations which have recently been developed have been for protein or polynucleotide structure refinement. It is clear that these new techniques and those of the early 1960s are more economical than the full-matrix least-squares method even for small-molecule structures.

[22] H. M. Rietveld, *J. Appl. Crystallogr.* **2,** 65 (1969).
[23] R. A. Young, P. E. Mackie, and R. B. von Dreele, *J. Appl. Crystallogr.* **10,** 262 (1977).

[4] Direct Phase Determination Based on Anomalous Scattering

By WAYNE A. HENDRICKSON, JANET L. SMITH, and STEVEN SHERIFF

Introduction

The scattering of X rays by matter arises from the acceleration of electrons in vibrations excited by the incident X-ray wave. It is to be expected, then, that resonance between these excitations and the natural frequencies of vibration of bound electrons in atoms will modulate the X-ray scattering and make it wavelength dependent. This indeed is the case. The resulting dispersive property is called anomalous scattering even though it is actually the norm for all but the lightest of atoms when one considers the wavelengths of interest in X-ray diffraction experiments. Perhaps more appropriately, this phenomenon is also sometimes referred to as "resonance scattering." In any case, the effects can be quite large—especially at the absorption edges of certain ionic species.

The potential usefulness of anomalous scattering in structure analysis was first recognized by Bijvoet.[1] He suggested that departures from

[1] J. M. Bijvoet, *Proc. Acad. Sci. Amsterdam* **B52,** 313 (1949).

Friedel's law, such as those observed in the classic work on zincblende by Coster, Knol, and Prins,[2] could be used to determine the absolute configuration of handed molecules and that anomalous scattering could also be used to resolve the phase ambiguity in the single isomorphous replacement experiment. The methodology for realizing the phasing potential of anomalous scattering in the isomorphous replacement method was worked out early in protein crystallography by Blow,[3] Rossmann,[4] Matthews,[5] North,[6] Kartha and Parthasarathy,[7] and others. The use of anomalous scattering as an adjunct to isomorphous replacement is now a standard of the field. In fact, rubredoxin[8] and several other protein structures have been solved from only a single isomorphous derivative. Anomalous scattering has also been used very effectively to locate native metal centers in protein crystals.[9–11]

Although these conventional uses of anomalous scattering—to determine the absolute configuration, to find the positions of native metal centers, and to supplement isomorphous replacement phasing—have considerable importance in macromolecular crystallography, they will not be dealt with directly in this chapter. Instead, the emphasis here is on more recent developments concerned with the use of anomalous scattering for the direct determination of phase angles. This use is direct in the sense that only the data from a single crystalline species are required.

There are two major analytic procedures for direct phasing from anomalous scattering. One of these uses the normal scattering contributions from the anomalous scatterers to resolve the ambiguity inherent in phase information from anomalous scattering at a single wavelength. The resolved anomalous phasing methods presented here derive from procedures developed in the structural analysis of crambin.[12] The second major category of direct anomalous phasing is based on measurements at multiple wavelengths. Such measurements can be made with the tunable radiation from synchrotron sources and can in principle lead to a definitive

[2] D. Coster, K. S. Knol, and J. A. Prins, *Z. Phys.* **63,** 345 (1930).
[3] D. M. Blow, *Proc. R. Soc., London, Ser. A* **247,** 303 (1958).
[4] M. G. Rossmann, *Acta Crystallogr.* **14,** 383 (1961).
[5] B. W. Matthews, *Acta Crystallogr.* **20,** 82 (1966).
[6] A. C. T. North, *Acta Crystallogr.* **18,** 212 (1965).
[7] G. Kartha and R. Parthasarathy, *Acta Crystallogr.* **18,** 745 (1965).
[8] J. R. Herriott, L. C. Sieker, L. H. Jensen, and W. Lovenberg, *J. Mol. Biol.* **50,** 391 (1970).
[9] J. Kraut, *J. Mol. Biol.* **35,** 511 (1968).
[10] P. Argos and F. S. Matthews, *Acta Crystallogr., Sect. B* **B29,** 1604 (1973).
[11] W. A. Hendrickson, G. L. Klippenstein, and K. B. Ward, *Proc. Natl. Acad. Sci. U.S.A.* **72,** 2160 (1975).
[12] W. A. Hendrickson and M. M. Teeter, *Nature (London)* **290,** 107 (1981).

solution of the phase problem. Calculations[13] and preliminary experiments[14,15] also indicate great practical promise. However, a successful application has yet to be performed for an unknown structure. Thus the description here is limited to some techniques that are expected to prove useful. A third category of procedures for the analysis of anomalous scattering data has also been advanced recently. The theory for integrating classic direct methods with anomalous scattering measurements has been developed by Hauptman[16] and by Karle,[17] but practical implementation is not yet in place.

Theoretical Foundation

Scattering Factors

X Rays are scattered by free electrons of atoms; nuclei are much too massive to be noticeably affected by electomagnetic radiation. Thus, it is conventional to normalize the scattering from an atom to be f times that from a single classic electron as found by J. J. Thompson. This atomic scattering factor decreases with scattering angle due to diffuseness of the electron cloud, but it is "normally" independent of the X-ray frequency (or wavelength). Here "normal" refers to the condition that the incident frequency be far from a natural frequency of an atomic electron. In fact, resonance effects are generally appreciable in diffraction experiments involving atoms heavier than oxygen, and thus anomalous scattering corrections are important.

A classic explanation of the anomalous scattering modifications is given by James,[18] but a quantitative treatment requires quantum mechanical calculations,[19] as is also true for the normal scattering factor, f°. The wavelength-dependent correction, or anomalous dispersion, is complex; that is, a phase shift δ accompanies the resonance scattering. Thus,

$$f = f^\circ + f^\Delta e^{i\delta} = f^\circ + \Delta f' + i\,\Delta f'' \tag{1}$$

where f^Δ is the magnitude of anomalous scattering and $\Delta f'$ and $\Delta f''$ are the real and imaginary components, respectively.

[13] J. C. Phillips and K. O. Hodgson, *Acta Crystallogr., Sect. A* **A36,** 856 (1980).
[14] W. Hoppe and V. Jakubowski, *in* "Anomalous Scattering" (S. Ramseshan and S. C. Abrahams, eds.), p. 437. Munksgaard, Copenhagen, 1975.
[15] L. K. Templeton, D. H. Templeton, R. P. Phizackerley, and K. O. Hodgson, *Acta Crystallogr., Sect. A* **A38,** 74 (1982).
[16] H. Hauptman, *Acta Crystallogr., Sect. A* **A38,** 632 (1982).
[17] J. Karle, *Acta Crystallogr., Sect. A* **A40,** 4 (1984).
[18] R. W. James, "The Optical Principles of the Diffraction of X-rays." Bell, London, 1948.
[19] D. T. Cromer and D. Liberman, *J. Chem. Phys.* **53,** 1891 (1970).

There is physical meaning to the mathematical distinction between real and imaginary parts. The imaginary component proves to be out of phase by π from the incident wave, and so the forward scattering depletes the primary beam in proportion to $\Delta f''$, i.e., absorption occurs. The absorption, and hence $\Delta f''$, rise sharply as the incident frequency approaches an atomic resonance. Experimentally determined scattering factors are generally in excellent agreement with calculated values except very near an absorption edge. As we will see in a later section, perturbations of molecular orbitals and electron scattering effects on absorptivity (EXAFS) are present as well as the strictly atomic X-ray scattering.

Structure Factors

The intensity measured in a diffraction experiment from a real (ideally imperfect) crystal is given by $I(\mathbf{h}) = K|F(\mathbf{h})|^2$ where K is a proportionality constant dependent on experimental conditions and $F(\mathbf{h})$ is the structure factor,

$$F(\mathbf{h}) = \sum_{j}^{\text{atoms}} f_j \exp(2\pi i \mathbf{h} \cdot \mathbf{x}_j - B_j s^2) \tag{2}$$

which is dependent on the positions $\mathbf{x}_j$, thermal displacements (represented by B_j for the isotropic case), and scattering factors of all atoms j in the unit cell. The diffraction maximum for the particular reflection with indices $\mathbf{h}$ occurs at a scattering angle θ for X rays of wavelength λ; thus, $s = (\sin\theta)/\lambda$.

Single-Wavelength Relationships. In the analysis of anomalous scattering data it is convenient to distinguish the contributions from different parts of the scattering factors. Different divisions are appropriate in different situations. In the case of experiments at a single wavelength we write

$$F(\mathbf{h}) = F'(\mathbf{h}) + F''(\mathbf{h}) \tag{3}$$

where F' relates to the real parts, $f' = f^\circ + \Delta f'$, and F'' relates to the imaginary parts, $f'' = \Delta f''$. In practical situations the anomalous scattering factors are negligible for the many light "normal" scatterers (H, C, N, O) of macromolecules, and only the few heavier atoms (e.g., S, Fe, Hg, Sm) in a structure contribute significantly to the anomalous scattering. The sum in Eq. (2) over just these significant anomalous scattering centers is designated $F''_A(\mathbf{h})$ for the imaginary part; the corresponding real contribution from these centers, $F'_A(\mathbf{h})$, is included in $F'(\mathbf{h})$. We designate the phase of $F'(\mathbf{h})$ to be ϕ, that of $F'_A(\mathbf{h})$ to be ψ, and that of $F''_A(\mathbf{h})$ to be $\psi + \omega$. Thus, $F'(\mathbf{h}) = |F'(\mathbf{h})| \exp(i\phi)$ and similarly for the others. It then follows from Eq. (2) that

$$F'(-\mathbf{h}) = |F'(\mathbf{h})| \exp(-i\phi)$$
$$F'_A(-\mathbf{h}) = |F'_A(\mathbf{h})| \exp(-i\psi)$$

and

$$F''_A(-\mathbf{h}) = |F''_A(\mathbf{h})| \exp[-i(\psi + \omega + \pi)]$$

It now suffices to adopt a shorthand notation whereby F^+, F^-, F', and F''_A stand for $F(\mathbf{h})$, $F(-\mathbf{h})$, $F'(\mathbf{h})$, and $F''_A(\mathbf{h})$, respectively.

The Friedel mates, **h** and $-\mathbf{h}$, are related by an inversion center in the lattice, but for acentric crystals including anomalous scatterers the diffracted intensities will differ. This difference can be found after substitution of the appropriate terms into Eq. (3) and performance of the indicated complex arithmetic to yield

$$|F^+|^2 = |F'|^2 + |F''_A|^2 + 2|F'||F''_A| \cos(\psi + \omega - \phi) \tag{4}$$
$$|F^-|^2 = |F'|^2 + |F''_A|^2 - 2|F'||F''_A| \cos(\psi + \omega - \phi) \tag{5}$$

Then,

$$|F^+|^2 - |F^-|^2 = 4|F'||F''_A| \cos(\psi + \omega - \phi) \tag{6}$$
$$|F'| = [\tfrac{1}{2}(|F^+|^2 + |F^-|^2) - |F''_A|]^{1/2} \tag{7}$$

In virtually all macromolecular cases, $|F''_A|$ will be small relative to $|F|$. In such circumstances $|F'| \simeq (|F^+| + |F^-|)/2$ and on substitution into Eq. (6), we have to very good approximation the Bijvoet difference

$$\Delta F \equiv |F^+| - |F^-| \simeq 2|F''_A| \cos(\psi + \omega - \phi) \tag{8}$$

In the event that all anomalous scatterers are of the same atomic species, $\omega = \pi/2$ and Eq. (8) simplifies to

$$\Delta F \simeq -2|F''_A| \sin(\psi - \phi) \tag{9}$$

which is the relationship used in the crambin analysis. Notice that in this case, $|F''_A| = (f''_A/f'_A)|F'_A|$.

Multiple-Wavelength Relationships. Inasmuch as scattering factors in general depend on the X-ray wavelength, so also do the structure factors. In a phase determination based on multiple-wavelength measurements it is essential to relate all data to a common reference point. We follow the lead of Karle[20] here and refer the analysis to the normal scattering terms. We also adopt Karle's notation by which wavelength is designated in a manner analogous with the nuclear isotope convention. Thus, ${}^{\lambda}F(\mathbf{h})$ is a structure factor for reflection **h** corresponding to the complete scattering

[20] J. Karle, *Int. J. Quantum Chem.* **7**, 357 (1980).

factors f at a particular wavelength λ, and ${}^{\circ}F(\mathbf{h})$ corresponds only to the normal scattering terms, f°. Then with reference to Eqs. (1) and (2),

$$ {}^{\lambda}F(\mathbf{h}) = {}^{\circ}F(\mathbf{h}) + \sum_{k} \left[\left(\frac{\Delta f'_k}{f^{\circ}_k} \right) + \left(\frac{i\,\Delta f''_k}{f^{\circ}_k} \right) \right] {}^{\circ}F_{A_k}(\mathbf{h}) \tag{10} $$

where ${}^{\lambda}F$ and ${}^{\circ}F$ include contributions from all atoms, whereas ${}^{\circ}F_A$ corresponds only to the kth kind of anomalous scatterer in the structure. It is then possible to express the measurable quantities, $|{}^{\lambda}F(\mathbf{h})|^2$, in terms of the separate normal scattering contribution and the atomic scattering factors for each different kind of anomalous scatterer.

It is instructive to examine the case of a single kind of anomalous scatterer. If we denote the phase of ${}^{\circ}F$ by ${}^{\circ}\phi$ and that of ${}^{\circ}F_A$ by ${}^{\circ}\phi_A$, then

$$ \begin{aligned} |{}^{\lambda}F(\mathbf{h})|^2 = |{}^{\circ}F(\mathbf{h})|^2 &+ a(\lambda)|{}^{\circ}F_A(\mathbf{h})|^2 \\ &+ b(\lambda)|{}^{\circ}F(\mathbf{h})|\,|{}^{\circ}F_A(\mathbf{h})| \cos[{}^{\circ}\phi(\mathbf{h}) - {}^{\circ}\phi_A(\mathbf{h})] \\ &+ c(\lambda)|{}^{\circ}F(\mathbf{h})|\,|{}^{\circ}F_A(\mathbf{h})| \sin[{}^{\circ}\phi(\mathbf{h}) - {}^{\circ}\phi_A(\mathbf{h})] \end{aligned} \tag{11} $$

where

$$ a(\lambda) = (f^{\Delta}/f^{\circ})^2 \tag{12a} $$
$$ b(\lambda) = 2(\Delta f'/f^{\circ}) \tag{12b} $$
$$ c(\lambda) = 2(\Delta f''/f^{\circ}) \tag{12c} $$

All wavelength dependence in Eq. (11) is embodied in coefficients, Eq. (12), which can be determined in advance of knowledge of the structure. Then from suitable measurements at $\pm\mathbf{h}$ and at multiple wavelengths, a system of linear equations [Eq. (11)] can be constructed to solve for $|{}^{\circ}F|$, $|{}^{\circ}F_A|$, and $({}^{\circ}\phi - {}^{\circ}\phi_A)$. In practice, for weak anomalous scatterers, this deterministic approach is apt not to be definitive and, if few wavelengths are used, estimates of uncertainty will be poor. Moreover, the scale of the algebraic problem increases rapidly with increase in kinds of anomalous scatterers: the number of terms goes as $(q + 1)^2$, where q is the number of kinds of anomalous scatterers. Thus a probabilistic analysis of the kind that has been used for single-wavelength problems may also be advantageous here.

Phase Equations

Single-Wavelength Case. From Eq. (8) it follows that there are two possible solutions for the desired phase angle ϕ, provided that ψ and ω are known from the structure of anomalous scatterers. Namely,

$$ \phi = \psi + \omega \pm \cos^{-1}[\Delta F/(2|F''_A|)] \tag{13} $$

In the absence of other information these alternatives are equally likely. Moreover, due to the substantial errors in ΔF, other phases may also be plausible, or, quite possibly, a formal solution will be precluded. This ambiguity and imprecision in phases can be taken into account by using a probability treatment of the kind introduced by Blow and Crick.[21] One can use an error model[22] whereby the probability distribution, $P_{\mathrm{ano}}(\phi)$, for phase information from anomalous scattering is

$$P_{\mathrm{ano}}(\phi) = N \exp\{-[\Delta F - 2|F''_{\mathrm{A}}| \cos(\psi + \omega - \phi)]^2/2E^2\} \qquad (14)$$

with a normalization factor N and standard error E. It is convenient to encode this information in the phasing coefficients of the simplified representation[23]

$$P(\phi) = N' \exp(A \cos \phi + B \sin \phi + C \cos 2\phi + D \sin 2\phi) \qquad (15)$$

The relevant coefficients from Eq. (14) are

$$A_{\mathrm{ano}} = 2\, \Delta F\, a''/E^2 \qquad (16a)$$
$$B_{\mathrm{ano}} = 2\, \Delta F\, b''/E^2 \qquad (16b)$$
$$C_{\mathrm{ano}} = -(a''^2 - b''^2)/E^2 \qquad (16c)$$
$$D_{\mathrm{ano}} = -2a''b''/E^2 \qquad (16d)$$

where $a'' = |F''_{\mathrm{A}}| \cos(\psi + \omega)$ and $b'' = |F''_{\mathrm{A}}| \sin(\psi + \omega)$.

The phase ambiguity presented by Eq. (13) must be resolved for the anomalous scattering to be of use. If the anomalous scatterers are sufficiently heavy, then phases for the total structure, ϕ, will tend to be close to those for the heavy-atom structure, ψ. One can then simply take the anomalous alternative that is nearer to the heavy-atom phase. However, this is not a sure process if the partial structure of anomalous scatterers is not a dominating influence. Uncertainties can better be taken into account with a treatment whereby phase information for the partial structure is based on the distribution given by Sim[24]:

$$P_{\mathrm{par}}(\phi) = N \exp[2Q|F'||F'_{\mathrm{A}}| \cos(\psi - \phi)/\langle F_{\mathrm{U}}^2 \rangle] \qquad (17)$$

Here $\langle F_{\mathrm{U}}^2 \rangle$ is the expected value of the scattering contribution from the unknown part of the structure; $|F'|$ and $|F'_{\mathrm{A}}|$ are the structure factor moduli as observed from the crystal and calculated from the anomalous scatterers, respectively; and Q is an arbitrary sharpening factor. The phasing coefficients, A_{par} and B_{par}, relevant to Eq. (17) are given elsewhere[23] and can also be derived readily.

[21] D. M. Blow and F. H. C. Crick, *Acta Crystallogr.* **12,** 794 (1959).

[22] W. A. Hendrickson, *Acta Crystallogr., Sect. A* **35,** 245 (1979).

[23] W. A. Hendrickson and E. E. Lattman, *Acta Crystallogr., Sect. B* **B26,** 136 (1970).

[24] G. A. Sim, *Acta Crystallogr.* **12,** 813 (1959).

[25] J. L. Smith and W. A. Hendrickson, *in* "Computational Crystallography" (D. Sayre, ed.), p. 209. Oxford Univ. Press (Clarendon), London and New York, 1982.

Multiple-Wavelength Case. In favorable circumstances an analysis of multiple-wavelength measurements in terms of a system of equations such as Eq. (11) should yield definitive values of ${}^{\circ}\phi - {}^{\circ}\phi_A$ and if the structure of anomalous scatterers can be interpreted from the set of derived $|{}^{\circ}F_A|$ values then ${}^{\circ}\phi_A$ can be calculated; thus, ${}^{\circ}\phi$ is determined. In principle it could suffice to use a set of measurements of different wavelength but all at $+\mathbf{h}$. However, such an approach would not take full advantage of the information content by which the phase information from Bijvoet differences is essentially orthogonal to that from dispersive differences due to wavelength variation. This can be seen by reference again to the single scatterer case. Using notation similar to that in Eq. (6) we find from Eq. (11) that from Bijvoet pairs

$$|{}^{\lambda}F^{+}|^2 - |{}^{\lambda}F^{-}|^2 = 2c(\lambda)|{}^{\circ}F||{}^{\circ}F_A| \sin({}^{\circ}\phi - {}^{\circ}\phi_A) \tag{18}$$

whereas differences resulting from two wavelengths λ_1 and λ_2 are given by

$$\begin{aligned} |{}^{\lambda_1}F|^2 - |{}^{\lambda_2}F|^2 = {} & [a(\lambda_1) - a(\lambda_2)]|{}^{\circ}F_A|^2 \\ & + [b(\lambda_1) - b(\lambda_2)]|{}^{\circ}F||{}^{\circ}F_A| \cos({}^{\circ}\phi - {}^{\circ}\phi_A) \end{aligned} \tag{19}$$

where $|{}^{\lambda}F|^2 = (|{}^{\lambda}F^{+}|^2 + |{}^{\lambda}F^{-}|^2)/2$. The respective dependence on sine and cosine functions of phase angles here is analogous to the orthogonality between anomalous scattering and isomorphous differences in the isomorphous replacement method. A noteworthy difference is the perfect isomorphism in this case.

Phasing Power

In contemplating diffraction experiments on a particular problem, it is useful to estimate the phasing power to be expected from the signals related to phase determination. The relevant expected values can be obtained from the intensity statistics relationship whereby

$$\langle |F| \rangle \propto \langle |F|^2 \rangle^{1/2} = \left(\sum f^2 \right)^{1/2} = M^{1/2} f \tag{20}$$

if all M atoms in the unit cell are of the same kind. From Eq. (8) it follows that the Bijvoet difference ratio at zero scattering angle is

$$\frac{\langle |\Delta F|_{\pm \mathbf{h}} \rangle}{\langle |F'| \rangle} = \sqrt{2} \left(\frac{N_A}{N} \right)^{1/2} \frac{\Delta f''_A}{Z_{\text{eff}}} \tag{21}$$

where N_A is the number of anomalous scatterers per molecule, N is the total number of nonhydrogen atoms in the molecule, and Z_{eff} is the effective average atomic number (~6.7 for proteins). The expected partial structure signal from the real components of the anomalous scatterers is, relative to the total scattering,

$$\frac{\langle|F'_A|\rangle}{\langle|F'|\rangle} = \left(\frac{N_A}{N}\right)^{1/2} \frac{f'_A}{Z_{eff}} \qquad (22)$$

Finally, from Eq. (19) after neglect of the first term it is possible to show that the dispersive difference ratio is approximately

$$\frac{\langle|\Delta F|_{\Delta\lambda}\rangle}{\langle|F'|\rangle} \simeq \frac{1}{\sqrt{2}}\left(\frac{N_A}{N}\right)^{1/2} \frac{|{}^{\lambda 1}\Delta f' - {}^{\lambda 2}\Delta f'|}{Z_{eff}} \qquad (23)$$

The expected diffraction ratios for some macromolecules of interest are shown in the table in comparison with those for crambin which was solved directly by resolved anomalous scattering. These have all been done at zero scattering angle with Cu K_α radiation. The diffraction ratios are generally enhanced at higher angles since there is little angular dependence of anomalous scattering. It is also worth noting that the scattering strength of a cluster of n anomalous scatterers is enhanced by a factor of $n^{1/2}$ if data are limited to spacings insufficient to resolve the individual atoms.

Resolved Anomalous Phasing Applications

Procedures developed initially for the crambin analysis have been further advanced in the course of other applications in our laboratory. This has led us to a system of computer programs and a rather standard

EXPECTED DIFFRACTION RATIOS

Molecule	Normal scatterers	Anomalous scatterers	$\frac{\langle\|\Delta F\|_{\pm h}\rangle}{\langle\|F'\|\rangle}$ (%)	$\frac{\langle\|F'_A\|\rangle}{\langle\|F'\|\rangle}$ (%)
Sulfur-rich proteins				
Crambin	400	6S	1.4	29
Snake neurotoxin	500	8S	1.4	29
Metalloproteins				
Hi PIP	700	4Fe/8S	4.9	36
Hemerythrin	1000n	2nFe	3.0	16
Oligonucleotides				
d(CpGpCpGpCpG)	300	10P	1.5	38
Heavy-metal complexes				
Myoglobin	1300	1Hg	4.3	30
Lysozyme	1100	1U	8.0	36
100,000 mol. wt. protein with Hg cluster				
Low resolution	~7000	Hg_4	6.8	47
High resolution	~7000	4Hg	3.7	26

set of procedures for resolved anomalous phasing based on formulae presented above. An outline of these methods is given below. Additional details are given in Refs. 12 and 25.

Diffraction Measurements

An optimal use of anomalous scattering effects depends upon accurate measurements of small differences between relatively large numbers. For example, one seeks to measure Bijvoet differences [Eq. (8)] where an average ΔF is typically only 1–8% of $|F|$ (see table). However, if appropriate data collection strategies are employed, it is often possible to measure these differences more precisely than the $|F|$ values themselves can be measured. In particular, (1) if the Bijvoet differences can be measured from the same crystal, then major scaling errors can be eliminated; (2) if the Bijvoet pairs are measured close together in time, then radiation damage errors can be minimized; and (3) if the Bijvoet differences are taken directly from Friedel pairs (rather than symmetry equivalents) and in such a way that diffracting rays trace essentially the same paths through the crystal for both mates, then absorption effects for the two measurements will be nearly equivalent for favorable crystal morphologies.

A commonly used strategy for collecting Bijvoet pairs on a diffractometer involves measuring a row of $F(\mathbf{h})$ reflections or a block of about 20 $F(\mathbf{h})$ data and then repeating these same measurements for the $F(-\mathbf{h})$ set. Two methods have been used to assure that diffraction paths will be essentially equivalent for the front and back side of the reflecting planes. The $F(\mathbf{h})$ data are measured in the usual bisecting configuration at goniometer angles of $(2\theta, \phi, \chi)$ and then the $F(-\mathbf{h})$ data are measured either at $(-2\theta, \phi, \chi)$ or at $(2\theta, \phi + \pi, -\chi)$. Photographic film measurements tend to be less accurate than those from a diffractometer. However, if mirror-related Bijvoet pairs are obtained on the same films, systematic errors can be reduced.

It is, of course, important to accumulate sufficient counts that counting precision will not be limiting. As a rule of thumb, if F values must be known with 1% accuracy then at least 2500 counts should be collected, $\sigma_F/F = \frac{1}{2}\sigma_I/I = \frac{1}{2}(N^{1/2}/N)$.

Many of the residual experimental errors in anomalous differences can be eliminated numerically. Local scaling procedures can be very effective in minimizing systematic errors due to uncorrected absorption differences. We have found it to be useful to parameterize the local scaling factors to be anisotropic and smoothly varying. In the case of crambin the local scaling of Bijvoet pairs was crucially important. Factors ranging from 0.97 to 1.01 eliminated errors that otherwise swamped the signal.

Comparison of the rms anomalous differences for centric and acentric data reveals immediately the magnitude of the observed anomalous scattering signal, which can in turn be compared with the expected value [Eq. (21)].

Location of Anomalous Scatterers

The philosophical underpinning of anomalous scattering methods for structure analysis lies in the possibility to make a small problem out of a large one. There generally are only a few anomalous scatterers and these can be located from their unique diffraction signal. Once the substructures of anomalous scatterers is known, one then seeks to develop the much more complicated structure of normal scatterers by taking advantage of the diffraction interference between normal and anomalous scatterers.

The usual method for locating anomalous scattering centers is by interpretation of $(\Delta F)^2$ Patterson functions.[4] Direct methods provide an alternative possibility. Although $(\Delta F)^2$ differs by a trigonometric factor [Eq. (8)] from the desired coefficient, $(F''_A)^2$, these difference Pattersons are quite similar to ones from the proper coefficients. However, they are very sensitive to erroneous measurements. It is important to screen for unreasonable values. To do so one can calculate the distribution of $|\Delta F|$ values and reject as outliers those few anomalous differences greater than about five times the rms value. We also find that the error-prone weak data tend to produce especially unreliable anomalous differences and thus also reject reflections with $|F^+|$ or $|F^-|$ less than about $5\sigma_F$.

Once a tentative anomalous scatterer model has been found, it is important to refine this structure. As shown in work on myohemerythrin[11] and crambin,[12] this can be done accurately and effectively if only the largest differences are included in the refinement. Since the distribution of structure factors is governed by intensity statistics, as $|\Delta F|$ becomes large the trigonometric factor must approach ± 1 so that $|\Delta F| \simeq 2|F''_A|$. We have obtained quite reliable results using the strongest third of the anomalous differences (excluding outliers) in least-squares refinement of the anomalous scatterer structure. Examination of difference Fourier maps after preliminary refinement can be helpful in finding additional sites and correcting mistakes.

Phase Determination

When a refined model for the structure of anomalous scatterers has been achieved, the components needed for phase determination are also in place. However, before probability distributions can be calculated with Eqs. (14) and (17), certain scaling and error analyses must be performed.

Since the probability formula for partial structure information [Eq. (17)] mixes observed and calculated magnitudes, it is important that the absolute scale factor be known. In the case of native anomalous scatterers of precisely known stoichiometry, this scale factor can in principle be obtained from the refinement against $|\Delta F|$ values. In such calculations it is important to recognize the bias from the low signal-to-noise ratio and effects of the trigonometric factor. Perhaps a procedure to follow in general, and certainly so in instances of variable stoichiometry, is to base the absolute scaling on Wilson intensity statistics. When the absolute scale and overall thermal parameters have been determined, then F'_A and $\langle F_U^2 \rangle$ can be calculated and used in Eq. (17) with properly scaled $|F'|$ values.

All that is needed for the anomalous phasing after the calculation of F'_A, and hence ψ, is an estimation of E, the standard error. Since measurement error is a particularly important component in this situation, we compose E from the errors in ΔF and a residual lack-of-closure error. Thus, $E = (\sigma_{\Delta F}^2 + E_0^2)$ with E_0 estimated from averages in $(\sin \theta)/\lambda$ ranges.

There remains the problem of combining the phase probability distributions from Eqs. (14) and (17). One approach is the straightforward multiplication of $P_{ano}(\phi)$ and $P_{par}(\phi)$. This works well, but concern about undue bias toward the "heavy-atom" phases led us to develop a scheme for probabilistic choices[26] that proves to be very effective. For bimodal distributions, choice is made between the alternatives of maximal probability in $P_{ano}(\phi)$, provided that $P_{par}(\phi)$ discriminates well. If a choice is made the intrinsic figure of merit is assigned. When $P_{ano}(\phi)$ is sharply unimodal this distribution is used directly. For other acentric reflections the multiplicative combination is used. Of course, for centric reflections or those where the anomalous scattering data are wholly unreliable, only the $P_{par}(\phi)$ information pertains.

Structure Interpretation

The process of interpretation of electron density distributions calculated by Fourier transformation is essentially the same here as with other means of phase determination. However, there is one special difference. In general the structure of anomalous scatterers deduced from the $|F''_A|$ signal could be in either hand. Nothing in the anomalous scattering and partial structure information from those scattering centers alone decides between enantiomers, but the resulting phases will not be sensible unless

[26] W. A. Hendrickson, *Acta Crystallogr., Sect. B* **B27**, 1474 (1971).

based on the correct choice. Although independent information, e.g., from direct methods, can be brought to bear on this choice, it is also often possible to decide on the basis of chemical reasonableness of the features in maps calculated with the alternative phase sets.

It should be noted that the primary limitation on accuracy of the anomalous scattering information comes from measurement errors and hence this is within experimental control. However, limitations of the accuracy of partial structure information are inherently statistical. If the partial structure is rather weak, the phase ambiguity will not be reliably resolved. In such cases supplemental information derived from a preliminary interpretation of the electron density map can be combined with $P_{ano}(\phi)$ to yield an improved map. For example, if some features can be interpreted as an atomic model, then the larger partial structure can be used to produce a model resolved map. This proved to be crucial in the work on myohemerythrin.[25] Noncrystallographic symmetry and solvent constraints provide another source of supplemental information. Molecular averaging based on the 3-fold symmetry of trimeric hemerythrin[27] was important in that structure determination.

Multiple-Wavelength Analysis

The definitive character of the mathematical analysis for multiple-wavelength measurements and the excellent tunability of radiation from synchrotron sources make this a very promising method. However, the instrumentation to exploit this promise is not yet fully ready and successful applications have not yet been reported. Thus, this section will be limited to a few of the unique aspects in the analysis of planned experiments.

Since synchrotron beam time is scarce, most diffraction experiments are designed to make simultaneous measurements, either on photographic film or on electronic area detectors. Diffractometer techniques for measuring Bijvoet pairs are not suitable. If a mirror plane is present in the diffraction pattern and in the physical crystal, then Bijvoet pairs of reflections that experience essentially the same absorption paths can be measured simultaneously. In general, it is also possible to measure Friedel mates by alternating frames through rotation of π about a normal to the X-ray beam. The monochromator must be rather rapidly tunable so that different wavelength sets can be measured close together in time. Other special features of synchrotron data measurements include normalization

[27] J. L. Smith, W. A. Hendrickson, and A. W. Addison, *Nature (London)* **303,** 86 (1983).

for variation in incident beam intensity due to decay of current in the storage ring and appropriate polarization corrections.

The problem of optimal choice of wavelengths of measurement is also unique to these experiments. This has been addressed by numerical simulation[13] and theoretically.[28] One possible good solution is to use three wavelengths: (1) at the minimum $\Delta f'$ value, (2) at the maximum $\Delta f''$ value, and (3) at a remote point with relatively larger $\Delta f'$ (i.e., less negative). For iron these might be 1.740, 1.738, and 1.5 Å, respectively.

In fact the exact choice of wavelengths must depend upon experiment, since the precise positions of absorption edges depend critically on the chemical state of the metal. Fortunately the necessary information is readily accessible by way of X-ray absorption spectroscopy since the atomic absorption coefficient, μ, is directly related to $\Delta f''$. From James,

$$\Delta f''(\omega) = (mc/4\pi e^2)\omega\mu(\omega) \tag{24}$$

where m and e are the electron mass and charge, respectively, c is the speed of light, and $\omega = 2\pi c/\lambda$ is the circular frequency of the X-radiation. The real part of the anomalous scattering is related to the imaginary part by the Kramers–Kronig transformation. Thus,

$$\Delta f'(\omega) = \frac{2}{\pi}\int_0^\infty \frac{\omega'\,\Delta f''(\omega')}{(\omega^2 - \omega'^2)}\,d\omega' \tag{25}$$

X-Ray absorption measurements of $\mu(\omega)$ for the spectral region of interest can be made directly from the macromolecular crystal by fluorescence techniques. These data are needed to determine the actual values of $\Delta f'$ and $\Delta f''$ at the measurement wavelengths as well as for the selection of wavelengths. Sharp features at the absorption edge and EXAFS ripples in the extended region beyond the edge are not taken into account in the theoretical calculations of atomic scattering factors. The calculated and experimental scattering factors should agree at points remote from edges and this can be used for normalization in the data processing. Similarly, this is true for adjusting results of limited-range numerical integrations of Eq. (25).

A final point of special importance in multiple-wavelength anomalous analyses concerns the actual absorption of diffracted X rays. The dispersive effects are expected to involve small differences between intensities at different wavelengths. Since absorption coefficients, both for the anomalous scatterers and also for the light atoms, vary with wavelength, inaccuracies in absorption corrections will contribute noise to these differences. Clearly careful corrections are indicated, but local scaling

[28] R. Narayan and S. Ramaseshan, *Acta Crystallogr., Sect. A* **A37,** 636 (1981).

should also help. Relationships such as can be derived from Eq. (18) that produce measurable quantities which must be wavelength invariant should provide a solid basis for such scaling.

Acknowledgments

This work was supported in part by NIH Grants GM-29548 and GM-34102. S.S. was supported by an NRC Research Associateship.

[5] Use of the Rotation and Translation Functions

By EATON LATTMAN

Introduction

Frequently in protein crystallography one wishes to determine the crystal structure of a molecule when the structure of a similar or homologous molecule is already known. The structures of inhibitor or effector complexes with an enzyme may be of interest, for example, or the variation with species of the details of conformation. In the most favorable cases the crystal of experimental interest is isomorphous with the known one, and the analysis can proceed directly by difference Fourier methods. More commonly, however, isomorphism does not pertain, and one must either do the new structure from scratch or seek other ways of using the known structure to facilitate the target structure's determination. The only successful technique for this purpose has been termed by Rossmann[1] the "molecular replacement method." In essence it involves generating a preliminary model of the target crystal structure by orienting and positioning the search molecule within the unit cell of the target crystal so as best to account for the observed diffraction pattern. Phases can then be calculated from this model and combined with the observed amplitudes to give an approximate Fourier synthesis of the structure of interest. This in turn can be subject to any of the several forms of refinement to provide a final, accurate structure of the target crystal.

If there is one molecule in the crystallographic asymmetric unit, then six parameters—three translations and three angles—fully describe how

[1] M. G. Rossmann, ed., "The Molecular Replacement Method," Int. Sci. Rev. Ser., No. 13. Gordon & Breach, New York, 1972. This work contains reprints of many important papers plus comments by the editor.

the search molecule is placed in the unit cell. In principle one could simply do a search on these six variables to determine the model that gives best agreement between F_o and F_c. In general such a search is beyond the capacity of even the fastest computers, although in special cases in which symmetry reduces the number of parameters such searches have been successful.[2] Rossmann and Blow[3] realized that this six-dimensional search could be reduced to a sequence of two three-dimensional searches in which first the orientation and then the position of the search molecule are determined.

The Rotation Function Representation

The rotational search is carried out by looking for agreement between the Patterson functions of the search and target structures as a function of their relative orientation. A function to evaluate this agreement was originally defined by Rossmann and Blow as

$$R(\mathbf{C}) = \int_U P_1(\mathbf{x}) P_2(\mathbf{Cx})\, dV \tag{1}$$

Here P_1 and P_2 are Patterson functions, $\mathbf{C}$ is a rotation operator that rotates the coordinate system of P_2 with respect to P_1, and U is a volume of integration, usually spherical, centered at the origin. A maximum in the rotation function $R(\mathbf{C})$ indicates a potential orientation for the search molecule in the target crystal.

For mathematical convenience it is useful to have the integration extend over all space. If P_2 is the Patterson function of a crystal, one can define $U(\mathbf{x})$ to be one inside the sphere U and zero elsewhere. Equation (1) then becomes

$$R(\mathbf{C}) = \int_{\substack{\text{all}\\ \text{space}}} P_1(\mathbf{x}) U(\mathbf{x}) P_2(\mathbf{Cx})\, dV \tag{2}$$

If P_2 is the Patterson function P_M of an isolated molecule, as is generally the case for the search methods we are discussing, then it falls to zero automatically past a radius D, where D is the maximum dimension of the molecule. Equation (1) then appears as

$$R(\mathbf{C}) = \int_{\substack{\text{all}\\ \text{space}}} P_1(\mathbf{x}) P_M(\mathbf{Cx})\, dV \tag{3}$$

[2] J. M. Baldwin, *J. Mol. Biol.* **136,** 103 (1980).

[3] M. G. Rossmann and D. M. Blow, *Acta Crystallogr.* **15,** 26 (1962).

Parseval's theorem is a statement of the conservation of power[4] and relates the integral of the product of two functions to the integral of their Fourier transforms. Application of this theorem to Eqs. (2) or (3) gives an equivalent reciprocal space representation of R. Apart from constants of proportionality

$$R(\mathbf{C}) = \sum I_{\text{obs}}(\mathbf{h}) F_{\text{M}}^2(\mathbf{Ch}) \tag{4}$$

Here $I_{\text{obs}}(\mathbf{h})$ are the intensities of the Bragg reflections from crystal 1, the target crystal in the search procedures. $F_{\text{M}}^2(\mathbf{Ch})$ is the Fourier transform of $U(\mathbf{x})P_2(\mathbf{Cx})$ in Eq. (2) or $P_{\text{M}}(\mathbf{Cx})$ in Eq. (3), and is a continuous function defined over all reciprocal space. These two representations provide independent insights about the proper use of the rotation function. The source of the signal in R is most readily seen through Eqs. (2) or (3), by mentally dividing up the Patterson vectors into two groups: vectors within a single molecule (self-vectors) and vectors between molecules (cross-vectors). The self-vector set is independent of the crystal into which the molecule is packed. It is determined only by the molecular structure. $R(\mathbf{C})$ is therefore large when a self-vector set in P_2 or P_{M} is rotated so as to superimpose upon a similar self-vector set in P_1. In Eq. (3) P_{M} has by definition only one self-vector set, and R will have a maximum at the orientation of each molecule in the target crystal. If there is one molecule in the asymmetric unit, then these maxima arise at space-group-related orientations.

In reciprocal space the situation is very similar. If we consider for simplicity a crystal with two molecules in the unit cell, then

$$I(\mathbf{h}) = [\mathbf{F}_1(\mathbf{h}) + \mathbf{F}_2(\mathbf{h})][\mathbf{F}_1^*(\mathbf{h}) + \mathbf{F}_2^*(\mathbf{h})] = F_1^2 + F_2^2 + \mathbf{F}_1\mathbf{F}_2^* + \mathbf{F}_1^*\mathbf{F}_2$$

where $\mathbf{F}_1$ and $\mathbf{F}_2$ are the Fourier transforms of the two molecules (i.e., the same molecule in two orientations) and F_1 and F_2 the corresponding moduli. Thus R will have a maximum when F_{M}^2 is rotated so as to be superimposed upon F_1^2 or F_2^2.

Sampling and Resolution

In either Patterson or reciprocal space $R(\mathbf{C})$ is evaluated on a grid in the angular variables used to represent $\mathbf{C}$. In addition the integral in Eq. (3) is normally reduced to a summation over a grid in P_1. In Patterson space the cost of calculating the rotation function is governed by the resolution of the data used to synthesize P_1 and P_{M}, the number of grid points in P_1 needed for the calculation, and the size of the steps taken in orientation space. These factors are interrelated, and some consideration

[4] J. W. Goodman, "Introduction to Fourier Optics," p. 10. McGraw-Hill, New York, 1968.

is needed in making proper choices. Selection of the resolution is discussed later, but we choose 6 Å here for definiteness. This limit implies that, according to the Shannon sampling theorem, P_1 must be sampled on a grid 3 Å on edge if the integral in Eq. (3) is to be accurately represented as a sum over grid points. P_2 or P_M enters into the calculation with nonintegral grid point coordinates, and must therefore be sampled frequently enough so that linear interpolation between grid points is accurate. A frequently quoted criterion for this is a spacing of one-third the resolution, or 2 Å. The radius D is determined by the size of $\mathbf{P}_M$, but this value can be decreased slightly without ill effect, since P_M is very small near its periphery. The value of D determines the step in angle space. This step should be large enough to superimpose adjacent grid points at the periphery of P_1, but not larger. This point is discussed below in the section on angular variables.

Similar criteria apply in reciprocal space. The $I_{obs}(\mathbf{h})$ are defined and stored only at reciprocal lattice points. The number of I values is therefore determined by the resolution. F_M^2 is a continuous function which must be finely enough sampled for linear interpolation. It is synthesized by Fourier components from all points in the Patterson function P_M. Those points in P_M farthest from the origin contribute the highest frequency terms to F_M^2, i.e., determine its "resolution," so that the Shannon theorem suggests a minimum interval of $\frac{1}{2}D$, where D is the maximum radius of P_M in angstroms. The interpolation criterion suggests a finer spacing of $\frac{1}{3}D$ Å^{-1}. Evidence has shown that in practice sampling need not be this fine. The Shannon criterion arises as follows: If we crystallize P_M in a unit cell of dimension $2D$, the copies of P_M at each origin will not mix with each other. There will be no aliasing. The diffraction pattern of this crystal is F_M^2 sampled at the spacing $\frac{1}{2}D$. However, the function P_M is very weak near its outer edges, since there are very few long vectors in the Patterson functions. Hence, only a very small inaccuracy is introduced if some mixing at the edges is allowed.

It is worth noting that core storage can be saved by a factor of 4 (at least according to the Shannon criterion) if one is willing to store $\mathbf{F}$, the Fourier transform of an individual molecule, instead of F^2. An individual molecule is half the dimension of P_M in each direction, so that the Shannon criterion gives a sampling of $1/D$ in reciprocal space. This gives 2^3 or 8 times fewer points, but each $\mathbf{F}$ is complex so that the net saving in storage is a factor of 4. One also saves, of course, in the generation of the transform. The rotation function program will take longer to execute, however, since F^2 will have to be made from $\mathbf{F}$ many times. Since the electron density of the molecule does not fall off near the edge as for P_M, it is more difficult to violate the Shannon condition successfully in this case.

It is worth noting that if one has a long slender molecule there is no real reason to put it in a cubic unit cell for the calculation of F_M^2 or P_M. So long as the Shannon criterion is satisfied in each dimension separately the sampling will be adequate. It is clear that each individual case must be examined to decide which is the cheaper model of calculation. Most packages of computer programs are designed for reciprocal space use, so that, without reprogramming, many constraints already exist.

Origin Removal

The Patterson function P_1 has large peaks at the origin. If D is larger than one of the lattice constants of P_1, then for at least some orientations P_M will overlap these origin peaks and make large but meaningless contributions to R. In Patterson space one can simply delete the origin(s) of P_1. In reciprocal space the problem is more intricate. The coefficients $I(\mathbf{h})$ must be modified so as to remove the origin of P_1. The normal technique for doing this is to subtract the average value of I, $\bar{I}(\mathbf{h})$, from $I(\mathbf{h})$ to give the modified coefficient $I'(\mathbf{h})$. However, it is customary to leave out of the summation in Eq. (3) terms for which $I(\mathbf{h})$ is very small, since they cannot contribute significantly. The $I'(\mathbf{h})$ values for the same reflections, however, become large negative, and must be included if origin removal is to be preserved. One must therefore delete those values of I' closest to zero, taking care that the sum of all I' actually used is equal to zero. Since $P(0)$ is the sum of the I', this will ensure origin removal.

For very elongated molecules, origin overlap can affect a significant fraction of orientation space. It is possible to do a crude packing analysis that will map out these forbidden orientation regions, thus improving the signal-to-noise ratio of R. If one performs the rotation function comparing P_M with a modified P_1 in which only the origin peaks are included and all else is left out, then a nonzero value of R implies origin overlap and therefore a forbidden value of the orientation. Such points in orientation space can be omitted from the real rotation function calculation.

Angular Variables

It has not yet been specified how the rotations performed by the matrix $\mathbf{C}$ are described. The most common system makes use of the Eulerian angles θ_1, θ_2, θ_3 ($\boldsymbol{\theta}$). In the usual variant of these, θ_1 is a rotation about the z axis of an orthogonal coordinate system with positive values defined as counterclockwise rotations looking down the positive z axis toward the

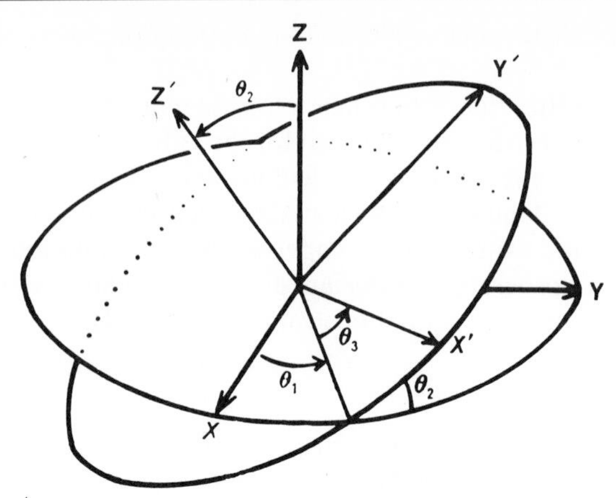

FIG. 1. Definition of Eulerian angles: the unprimed system represents the fixed or laboratory coordinate system. The rotation θ_1 about the z axis is made first, moving the x and y axes. The rotation θ_2 about the moved x axis is made second, rotating the y and z axes. The rotation θ_3 about the moved z axis is made last. (Reproduced from Ref. 3 with permission.)

origin. θ_2 is a rotation about the moved x axis, and θ_3 is a rotation about the moved z axis. These rotations are illustrated in Fig. 1, and the matrix **C** in terms of these angles is given in Table I. These rotations are difficult to visualize in space, but have useful analytical and symmetry properties. Another difficulty with these angles is that when θ_2 is small, θ_1 and θ_3 are rotations about nearly parallel axes, and are strongly coupled. Rotation functions plotted on a Cartesian grid in θ_1 and θ_3 display a distorted, streaky character in this region. Use of the pseudo-orthogonal Eulerian angles[5]

$$\theta_+ = \theta_1 + \theta_3, \qquad \theta_- = \theta_1 - \theta_3, \qquad \theta_2 = \theta_2$$

can remedy these difficulties (Figs. 2 and 3). R is plotted in sections of constant θ_2. The interval between sections, $\Delta\theta_2$, is a constant and equal to Δ, the actual angular step desired in the map. $\Delta\theta_+$ is given by $\Delta/\cos(\theta_2/2)$ and $\Delta\theta_-$ by $\Delta/\sin(\theta_2/2)$. As mentioned earlier Δ is the rotation necessary to superimpose adjacent lattice points in $I(\mathbf{h})$ at the outer edge of the data sphere used. For Patterson space calculations this becomes instead the rotation needed to superimpose adjacent grid points at the edge of P_1.

When P_1 and/or P_M have rotational symmetry, the rotation function also has symmetry, and the full range of angles need not be explored. The symmetry of the rotation function appears in a particularly simple way

[5] E. E. Lattman, *Acta Crystallogr., Sect. B* **B28**, 1065 (1972).

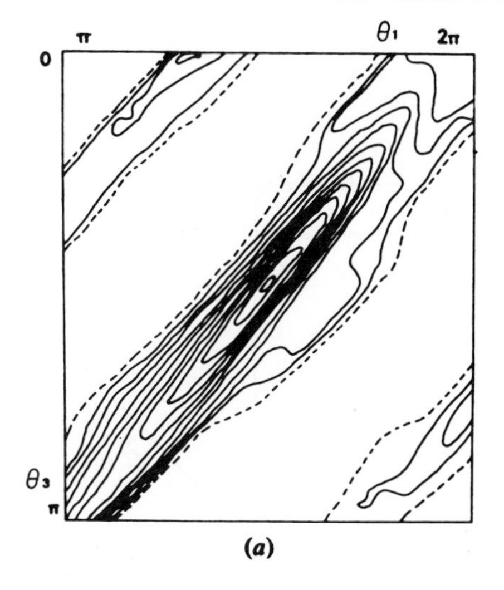

(*a*)

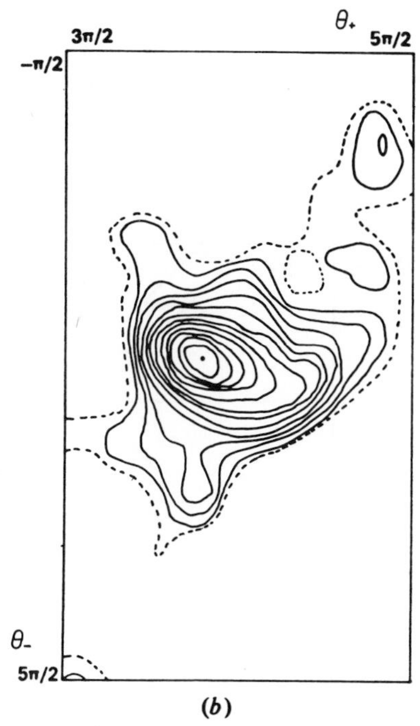

(*b*)

FIG. 2. Use of the θ_{+}–θ_{-} angular system: (a) shows a section of $\theta_2 = 15°$ from a rotation function comparing a molecule of sperm whale myoglobin with a crystal of the same preparation. Note the strongly streaked character of the peak. (b) shows the same section recalculated using the θ_{+}–θ_{-} system, and demonstrates improvement.

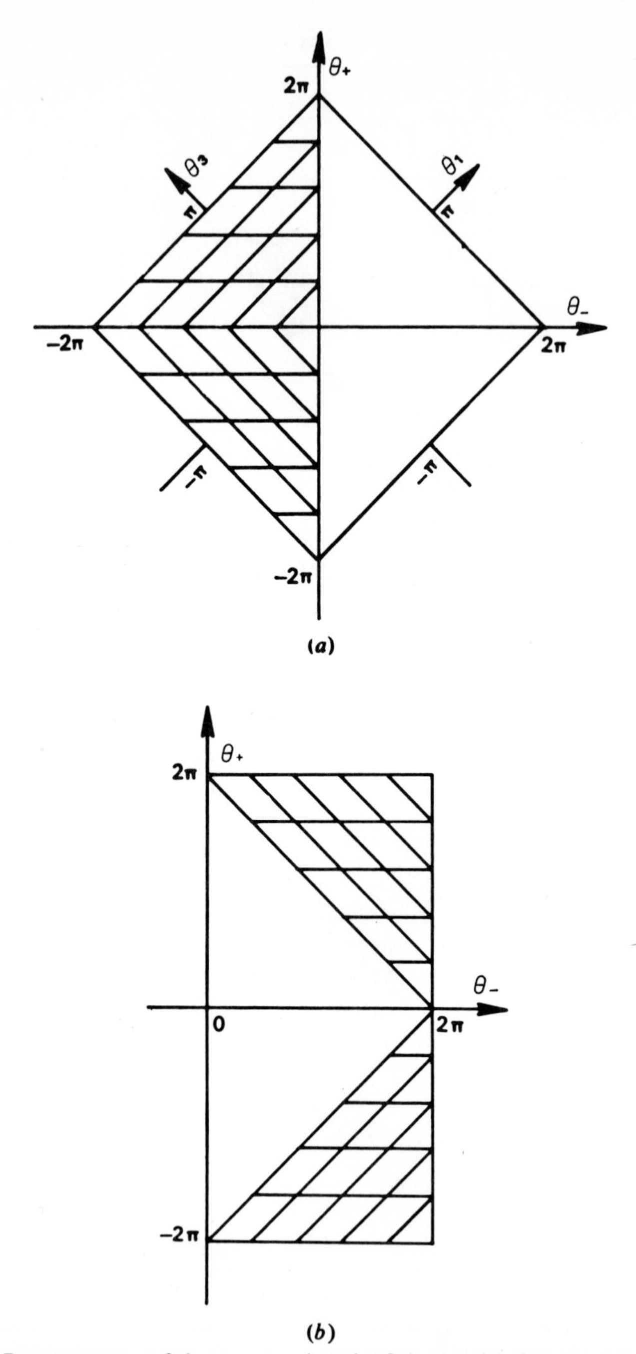

FIG. 3. Rearrangement of the asymmetric unit of the rotation function for θ_+–θ_-: (a) shows a square asymmetric unit in the θ_1–θ_3 plane. With respect to the θ_+–θ_- axes, also shown, it is inconveniently oriented for plotting: (b) the hatched triangles in (a) have been moved by identity translations of $\pm 2\pi$ in θ_1 and θ_3 to provide a rectangular area in θ_+–θ_-.

TABLE I
MATRIX **C** IN TERMS OF EULERIAN ANGLES θ_1, θ_2, AND θ_3

$-\sin\theta_1\cos\theta_2\sin\theta_3 + \cos\theta_1\cos\theta_3$	$\cos\theta_1\cos\theta_2\sin\theta_3 + \sin\theta_1\cos\theta_3$	$\sin\theta_2\sin\theta_3$
$-\sin\theta_1\cos\theta_2\cos\theta_3 - \cos\theta_1\sin\theta_3$	$\cos\theta_1\cos\theta_2\cos\theta_3 - \sin\theta_1\sin\theta_3$	$\sin\theta_2\cos\theta_3$
$\sin\theta_1\sin\theta_2$	$-\cos\theta_1\sin\theta_2$	$\cos\theta_2$

when Eulerian angles are used, and indeed corresponds in most cases to a space group symmetry in the periodic angular variables. These matters have been discussed by Main *et al.*[5a] Tables which simplify the use of these results have been compiled by Rao and Hartsuck.[6]

The other common system is the spherical polar angles. They are based on the theorem that any rotation can be accomplished by an appropriate spin about a properly chosen axis. The angles ϕ and ψ specify the longitude and colatitude of this axis and χ the spin about it. This system is useful whenever the order or direction of rotation axis can be anticipated. These angles are illustrated in Fig. 4, and the matrix elements of **C** for this system are given in Table II.

Computer Programs

In conventional programs the rotation function is calculated point by point on a grid spanning the asymmetric unit of angle space. For reasons of history, and of compatibility with the translation function, most programs operate in reciprocal space. They evaluate the summation in Eq. (3) directly using linear interpolation to find the required values of $F_M^2(\mathbf{Ch})$. One convenient package of programs, available from the author, comprises routines for generating F_M^2, for sorting and removing the origin from the $I(\mathbf{h})$, and for calculating the plotting R. It is called RATFINC.

The Fast Rotation Function

Some years ago Crowther[7] described how, by expanding the Patterson function in spherical harmonics, the rotation function could be calculated by fast Fourier transform for the variables θ_1 and θ_3; the calculation is reinitialized for each θ_2. This program has some limitations in terms of the

[5a] Tollin, P., Main, P., and Rossmann, M. G. *Acta Crystallogr.*, **20,** 404 (1966).
[6] S. N. Rao and J. H. Hartsuck, *Acta Crystallogr., Sect. A* **A36,** 878 (1980).
[7] R. A. Crowther, *in* "The Molecular Replacement Method" (M. G. Rossmann, ed.), Int. Sci. Rev. Ser., No. 13, p. 10. Gordon & Breach, New York, 1972.

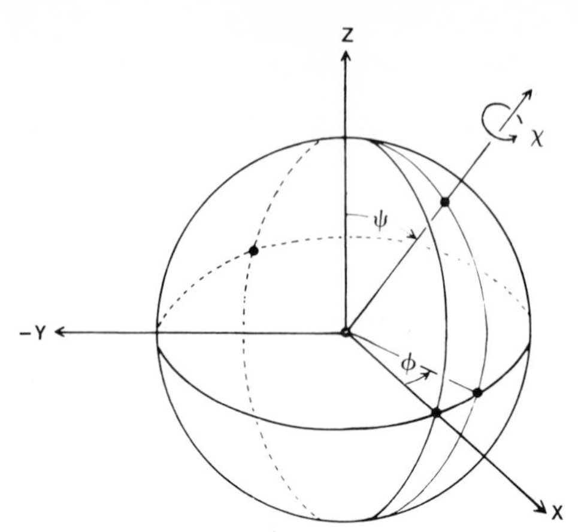

FIG. 4. Definition of spherical polar angles. The direction of the rotation axis is defined by the angles ψ, measured from the z axis, and by the angle ϕ measured in the x–y plane from the x axis. This amounts to giving the longitude and colatitude of the tip of the rotation axis. The rotation itself is given by the spin χ about that axis. The labeling of these axis is often permuted for convenience in dealing with certain space groups. (Reproduced from Ref. 3 with permission.)

radius of the Patterson function used, and in the resolution of the $I(\mathbf{h})$, but it is enormously faster than conventional programs when it is applicable. The program is available for distribution.[8] It is beyond the scope of this chapter to review the highly mathematical derivation of the fast rotation function; however, a related but much simpler calculation, the fast rotation function comparing two function $f(x,y)$ and $g(x,y)$ within a circle in the plane, gives the flavor of the method. Only one angle ϕ is needed; it is a rotation about a normal to the plane passing through the origin. Thus,

$$R(\phi) = \int f(x,y)g(\mathbf{C},x,y)\, dV \tag{6}$$

Using the polar coordinates $x = r \cos \theta$, $y = r \sin \theta$ in Eq. (6), we have

$$R(\phi) = \int f(r,\theta)g(r,\theta - \phi)r\, dr\, d\theta \tag{7}$$

Note that the rotation ϕ becomes a simple shift of ϕ in the angular variable. Both f and g are periodic in the angular variable ϕ and can be written

[8] Contact Dr. R. A. Crowther, M.R.C. Laboratory of Molecular Biology, Hills Road, Cambridge, England.

TABLE II

MATRIX **C** IN TERMS OF SPHERICAL POLAR ANGLES θ AND ψ AND ROTATION ANGLE χ[a]

$\cos\chi + \sin^2\psi\cos^2\phi(1-\cos\chi)$	$\sin\psi\cos\psi\cos\phi(1-\cos\chi) - \sin\psi\sin\phi\sin\chi$	$-\sin^2\psi\cos\phi\sin\phi(1-\cos\chi) - \cos\psi\sin\chi$
$\sin\psi\cos\psi\cos\phi(1-\cos\chi) + \sin\psi\sin\phi\sin\chi$	$\cos\chi + \cos^2\psi(1-\cos\chi)$	$-\sin\psi\cos\psi\sin\phi(1-\cos\chi) + \sin\psi\cos\phi\sin\chi$
$-\sin^2\psi\sin\phi\cos\phi(1-\cos\chi) + \cos\psi\sin\chi$	$-\sin\psi\cos\psi\sin\phi(1-\cos\chi) - \sin\psi\cos\phi\sin\chi$	$\cos\chi + \sin^2\psi\sin^2\phi(1-\cos\chi)$

[a] The order of the rows corresponds to a setting with the b axis the north pole.

as the Fourier series

$$f(r,\theta) = \sum_n A_n(r)\exp(in\theta), \qquad g(r,\theta-\phi) = \sum_m B_m(r)\exp[im(\theta-\phi)] \tag{8}$$

Inserting these into Eq. (7) we have

$$R(\phi) = \int r\,dr \left\{\sum\sum A_n B_m \int \exp[i(\mathrm{m}+\mathrm{n})\theta]\exp(-im\phi)\,d\theta\right\} \tag{9}$$

The integral over θ vanishes unless $m = -n$, when

$$R(\phi) = \sum \exp(in\phi) \int A_n B_{-n} r\,dr$$

Letting

$$C_n = \int A_n B_{-n}\, r\,dr$$

and substituting in Eq. (8), we find that

$$R(\phi) = \sum_n C_n \exp(in\phi) \tag{9}$$

Thus in the plane the rotation function R can be expressed as a Fourier series in the angular variable ϕ, in which the coefficients are calculated only once from the Fourier transforms of the original functions f and g. In the three-dimension fast rotation function, expansions in spherical harmonics and the associated Lengendre polynomials are utilized, in lieu of the Fourier transform, and manipulation of the expansion coefficients as a function of $\boldsymbol{\theta}$ is required. This results in complex mathematical development.

Translation Function

Once the orientation of a test molecule (and therefore of its symmetry mates) is known, the position of the molecule must be found. Many versions of a translational search have been proposed[9,10] but the most commonly used is the T_1 function of Crowther and Blow.[11] This function can also be interpreted in terms of the Patterson function or in reciprocal space. Consider a crystal having the symmetry of the space group P2, with two molecules in the unit cell. The calculated intensity $F_c^2(\mathbf{h})$ is given by

$$\mathbf{F}_c(\mathbf{h})\mathbf{F}_c^*(\mathbf{h}) = (\mathbf{F}_{M1} + \mathbf{F}_{M2})(\mathbf{F}_{M1}^* + \mathbf{F}_{M2}^*) \tag{10}$$

where $\mathbf{F}_{M1}$ and $\mathbf{F}_{M2}$ are the values of the molecular transforms of the two molecules at point $\mathbf{h}$. They are related by the symmetry of the crystal so that in this case $\mathbf{F}_{M2}(h,k,l) = \mathbf{F}_{M1}(-h,k,-l)$.

Translation functions build models of the target crystal by moving properly oriented molecules about in the unit cell. For each position of the test molecule, say $\mathbf{x}$, one can calculate the intensities for the resultant model and see how they agree with the $I_{obs}(\mathbf{h})$. Crowther and Blow initially defined

$$\mathrm{T}(\mathbf{x}) = \sum_{\mathbf{h}} F_c^2(\mathbf{h},\mathbf{x}) I_{obs}(\mathbf{h}) \tag{11}$$

To simplify this computation one can express F_c^2 in terms of the molecular transform $\mathbf{F}_M$, as in Eq. (10) above. The phase of $\mathbf{F}_M$ depends upon the position of the molecular center. If the stored values represent the transform when the center of mass of the molecule is at the origin, then the value of the transform when the molecular center is shifted to point $\mathbf{x}$ is $F_M \exp(2\pi i\mathbf{h}\cdot\mathbf{x})$. Thus for our chosen crystal

$$|F_c^2(\mathbf{h})|^2 = [\mathbf{F}_{M1}(\mathbf{h}) \exp(2\pi i\mathbf{h}\cdot\mathbf{x}_1) + \mathbf{F}_{M2}(\mathbf{h}) \exp(2\pi i\mathbf{h}\cdot\mathbf{x}_2)] \\ \times [\mathbf{F}_{M1}^*(\mathbf{h}) \exp(-2\pi i\mathbf{h}\cdot\mathbf{x}_1) + \mathbf{F}_{M2}^*(\mathbf{h}) \exp(-2\pi i\mathbf{h}\cdot\mathbf{x}_2)]$$

Here $\mathbf{x}_1$ and $\mathbf{x}_2$, the putative positions of the molecular centers, are related by crystal symmetry, and $\mathbf{F}_{M1}$ and $\mathbf{F}_{M2}$ are related by reciprocal space symmetry as above. Substituting these expressions into Eq. (11) one obtains

$$T(\mathbf{x}_1, \mathbf{x}_2) = \sum I_{obs}(\mathbf{h})\{F_{M1}^2 + F_{M2}^2 + \mathbf{F}_{M1}\mathbf{F}_{M2}^* \exp[2\pi i\mathbf{h}(\mathbf{x}_1 - \mathbf{x}_2)] \\ \times \mathbf{F}_{M1}^*\mathbf{F}_{M2} \exp[-2\pi i\mathbf{h}(\mathbf{x}_1 - \mathbf{x}_2)]\} \tag{12}$$

[10] P. Tollin, *Acta Crystallogr.* **21,** 613 (1966); D. Langs, *Acta Crystallogr., Sect. A* **A31,** 543 (1975); D. B. Litvin, *ibid.* **33,** 62 (1977).

[11] R. A. Crowther and D. M. Blow, *Acta Crystallogr.* **23,** 544 (1967).

Interpreting Eq. (12) in light of the Patterson function we see that the individual terms represent different classes of vectors in the search Patterson function. F_{M1}^2 and F_{M2}^2 are the coefficients determining vectors within individual molecules 1 and 2; they are of no interest since the position of the molecule does not affect them. The $\mathbf{F}_{M1}\mathbf{F}_{M2}^*$ are the coefficients for the set of vectors running from molecule 2 to molecule 1, and similarly for the last term. To simplify the calculation and to allow unique identification of peaks, only the term in $\mathbf{F}_{M1}\mathbf{F}_{M2}^*$ is retained in the translation function T_1. It is also possible to improve the signal-to-noise ratio by removing the coefficients for the self-vectors from I_{obs}. These vectors are not functions of molecular positions, but add much background. Removal of these can be effected approximately by writing

$$I_{cross} = I_{obs} - k(F_{M1}^2 + F_{M2}^2)$$

and

$$T_1(\mathbf{x}_2 - \mathbf{x}_1) = \sum_{\mathbf{h}} I_{cross}(\mathbf{h})\mathbf{F}_{M1}\mathbf{F}_{M2}^* \exp[-2\pi i\mathbf{h}(\mathbf{x}_2 - \mathbf{x}_1)] \qquad (13)$$

Here k scales the observed and calculated intensities together and the self-vectors from the search molecules have been subtracted from the observed Patterson function to leave, approximately, an observed cross-vector set. Equation (13) is now an ordinary Fourier summation that can be evaluated by any standard program. If all goes well, T_1 will have a unique peak at a point $\mathbf{x}_2 - \mathbf{x}_1$ that corresponds to the vector from 2 to 1. The problem of determining $\mathbf{x}_1$ from this peak is strictly identical to that of interpreting the Patterson function of a crystal with one atom in the asymmetric unit.

A number of simplifications often pertain. Frequently the crystal symmetry places restrictions on $\mathbf{x}_2 - \mathbf{x}_1$. In our special case the two vectors must have the same y component, because of the 2-fold axis along that direction. Thus only the section $y = 0$ has to be calculated in T_1. In general, of course, there are more than two molecules in the unit cell. In this case one must calculate a T_1 for every independent pair of molecules in the unit cell. In the orthorhombic example discussed below there are three such, one for each 2-fold axis. These three maps taken together provide some redundancy, since there are two independent chances to find each intermolecular vector. For high-symmetry space groups there is a lot of noise in the translation function, and it is important to calculate all the independent pairs.

Computer Programs

As noted above the function T_1 can be evaluated by a simple Fourier summation. Determining the coefficients, however, involves a structure factor calculation based on the test molecule properly oriented in the unit cell of the target crystal. This calculation can be done by any conventional program. However, if the molecular transform $\mathbf{F}_M$ has been previously calculated, the values of $\mathbf{F}^*_{M1}\mathbf{F}_{M2}$ and so on can be made very rapidly by properly sampling and phase shifting this function. This procedure is followed by RATFINC.

Difficulties and Recent Progress

Superficially it might seem that translation functions should be more successful than rotation functions, since the former correlate complete diffraction patterns or Patterson functions. In practice it is quite common for an apparently clear rotation function solution to provide no clear or unique result in the translation function. T_1 seems to be very sensitive to misorientation of the search molecule by a few degrees. Thus rotation function peaks that are perturbed from their true positions by noise may be effectively useless in T_1. One useful trick is to recalculate the rotation function in the vicinity of major peaks, by using a full set of higher resolution data. Frequently peaks are sharpened and shifted slightly by this process. The choice of resolution is a rather difficult problem. Almost universally low-order reflections, with spacings larger than 10–25 Å, are omitted from the calculation, since they contain more information about solvent vectors than about protein vectors. The cutoff at high resolution depends upon the degree to which the search molecule is similar to the target molecule; the greater the similarity the higher the resolution data that can be used. The cost, however, also goes up rapidly with the resolution, so that most searches have used limits of 4–6 Å as the high-resolution cutoff. As mentioned earlier, it is also important to calculate all the independent T_1 values for the space group one is dealing with.

In a recent paper Harada and co-workers[12] have provided what appear to be a number of improvements in the translation function. They have introduced a function

$$\mathrm{T_H}(\mathbf{x}) = \mathrm{TO}(\mathbf{x})/\mathrm{O}(\mathbf{x})$$

where

$$\mathrm{TO}(\mathbf{x}) = \sum_{\mathbf{h}} I_{\mathrm{obs}}(\mathbf{h}) I_{\mathrm{c}}(\mathbf{x},\mathbf{h}) \Big/ \sum_{\mathbf{h}} I^2_{\mathrm{obs}}(\mathbf{h})$$

$$\mathrm{O}(\mathbf{x}) = \sum_{\mathbf{h}} I_{\mathrm{c}}(\mathbf{h},\mathbf{x}) \Big/ N \sum_{\mathbf{h}} F^2_{\mathrm{M}}(\mathbf{h})$$

[12] Y. Harada, A. Lifchitz, J. Berthou, and P. Jollès, *Acta Crystallogr., Sect. A* **A37,** 273 (1981).

TO(**x**) is basically identical to the function T of Crowther and Blow, except for the normalization factor ΣI_{obs}^2. O(**x**) is a packing function that measures interpenetration of molecules. The integral over the unit cell of the squared electron density function is equal, by Parseval's theorem, to the sum over all reflections of F_c^2. If molecules do not interpenetrate, this number is also given by $N \Sigma F_M^2$, where F_M^2 is the squared transform of the molecule. As molecules interpenetrate, however, the integral of squared electron density rises, and O(**x**) becomes larger than one since the denominator is constant. The quotient of TO and O therefore deweights the translation function when interpenetration is occurring. The authors show an impressive figure in which large spurious peaks in TO are wiped out in T_H. The authors also note that the use of normalized structure factors improves the utility of T_H.

Finally, the authors present an elegant algorithm for the Fourier transform-based calculation of TO using the position variable **x** explicitly, instead of the intermolecular vector, such as $\mathbf{x}_2 - \mathbf{x}_1$, that usually appears. They make no mention of the availability of computer programs.

Examples

Self-Rotation Function of Glutathione Peroxidase

Glutathione peroxidase is a tetramer containing four identical subunits, and has a molecular mass of 84,000 Da. Crystals of this molecule have the symmetry of the space group C2, with a = 90.4 Å, b = 109.5 Å, c = 58.2 Å. The angle β is 99°. There is one molecule in the crystallographic asymmetric unit. The possibility that the molecule possessed the molecular (point) symmetry 222 was clear, and the investigators performed a rotation function using spherical polar angles with $\chi = 180$, corresponding to all possible 180° rotations.[13] The result is shown in Fig. 5. This is a projected view of the sphere shown in Fig. 4 which defines the polar angles. We are looking down on the north pole, which is the b axis in this setting. The large peak at the origin marks the crystallographic 2-fold axis along b. Two additional peaks are seen near ψ = 90°. They are separated by 90° in ϕ, and therefore demonstrate the 222 pseudo-symmetry of the Patterson function. The rotation function was in this case calculated using a program that directly compares Patterson functions. A hollow sphere of inner radius 5 Å and outer radius 30 Å was compared with itself. One of the 2-fold molecular axes is coincident with the crystallographic 2-fold

[13] R. Ladenstein, O. Epp, K. Bartels, A. Jones, R. Huber, and A. Wendel, *J. Mol. Biol.* **134,** 199 (1979).

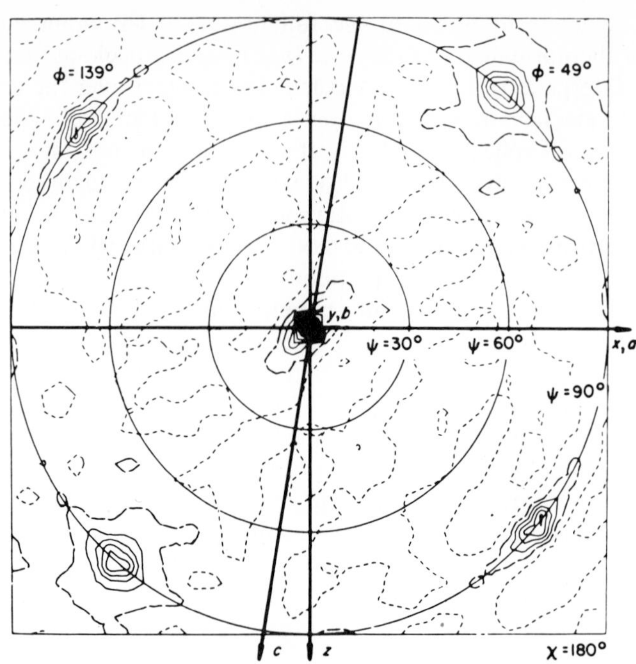

FIG. 5. Stereogram of the $\chi = 180°$ section of the self-rotation function of glutathione peroxidase, illustrating potential 2-fold rotations. The radius is proportional to ψ, while the polar angle is identical with ϕ. The coordinate system has been orthogonalized by placing x parallel to the unit cell axis **a**, y in the **a–b** plane, and z parallel to **c***. The origin peak has been normalized to 100, and the map is contoured in steps of 10 with the mean at 25.5 being marked by the dashed line contour. Solid contours are above the mean and dotted contours are below it. (Reproduced from Ref. 13 with permission.)

axis, meaning that there is half a tetramer in the asymmetric unit. The possibility that there was a whole tetramer in the asymmetric unit, with a 2-fold axis nearly parallel to b, existed, since this rotation function peak would have been masked by the origin peak. It was excluded by inspection of the $(u, 0, w)$ section of the Patterson function, which must have a large pseudo-origin peak if this is the case.

Molecular Replacement—Straightforward

Carboxypeptidase B and carboxypeptidase A are pancreatic proteases with considerable homology in amino acid sequence. The structure of carboxypeptidase A was determined by isomorphous replacement in the laboratory of Lipscomb.[14] The atomic model of carboxypeptidase A using

[14] D. C. Rees, M. Lewis, R. B. Honzatko, W. N. Lipscomb, and K. D. Hartman, *Proc. Natl. Acad. Sci. U.S.A.* **78,** 3408 (1981).

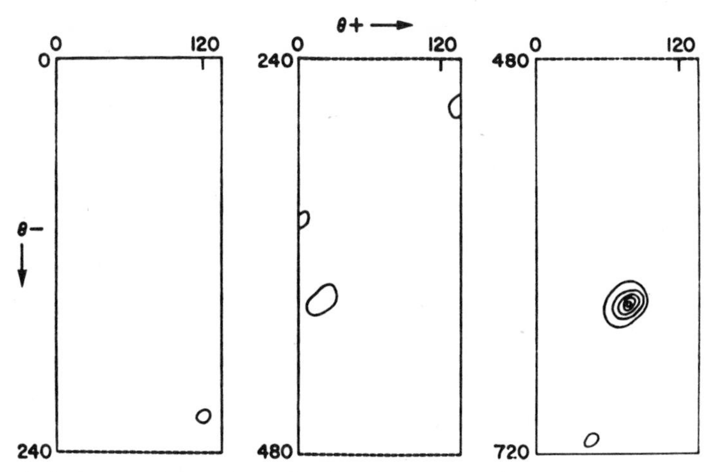

FIG. 6. Carboxypeptidase B rotation function: section shown contains the maximum value, $\theta_2 = 104°$. Contours start at an arbitrary level of 60 and go in steps of 10 to the normalized maximum of 100. The range of θ_- is 0–720°, so the section is divided into thirds for ease in display.

all main-chain atoms and C_β served as the search molecule in a molecular replacement study of the crystal structure of carboxypeptidase B.[15] Carboxypeptidase B crystallized in the space group $P3_1$ (or $P3_2$) with $a = b =$ 59.4 Å and $c =$ 104.8 Å, and buoyant density evidence suggested that there was one molecule in the asymmetric unit. The Fourier transform of the carboxypeptidase A molecule was calculated by placing it in a triclinic cell with $a = b = c =$ 120 Å and all angles equal 90°. A highly optimized conventional structure factor calculation routine was used. The rotation function [Eq. (4)] was calculated using reflections with spacings between 12 and 5.5 Å. The quasi-orthogonal Eulerian angles θ_+ and θ_- were used. Figure 6 shows the section of R having the main peak. The map is remarkably clean in this section and elsewhere. The second largest peak is three standard deviations below the one shown. The true sampling interval Δ in the map was about 8°, but a subsequent scan of the major peak was done in 3° steps to define it better.

The translation function T_1 of Crowther and Blow was then calculated. An ambiguity exists in this function because of the ambiguity in space group. The authors define molecule 2 to be related to molecule 1 by a clockwise rotation of 120°, looking down the z axis. The corresponding intermolecular vector must have a component along z of 1/3 or 2/3, depending on whether $P3_1$ or $P3_2$ is the correct space group. Figure 7 shows

[15] M. F. Schmid, E. E. Lattman, and J. R. Herriott, *J. Mol. Biol.* **84,** 97 (1974).

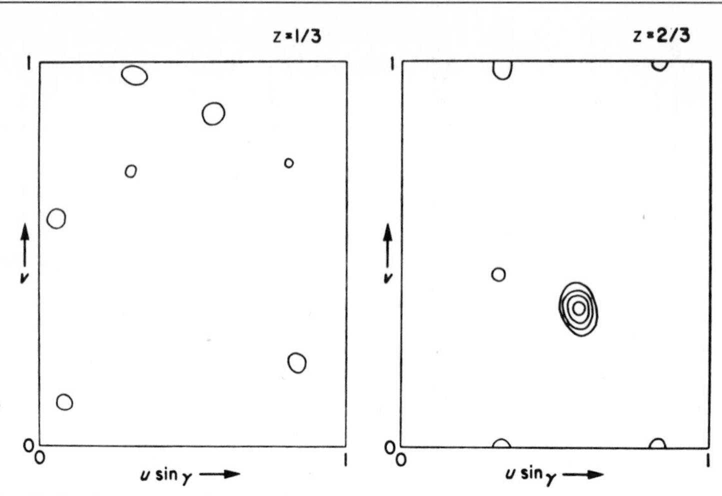

FIG. 7. Section at $z = 1/3$ and $2/3$ through the carboxypeptidase B translation function. The significant maximum clearly occurs at $z = \frac{2}{3}$.

sections at $z = 1/3$ and $z = 2/3$ from T_1. The section at $z = 2/3$ clearly has the major peak, and the space group is therefore determined to be $P3_2$. (The data are also consistent with an enantiomorphic carboxypeptidase B molecule made of D-amino acids in the space group $P3_1$, but this possibility was deemed unlikely.) The position of the peak in this section was readily interpreted to give the actual molecular position, and the structure was solved. The structure was eventually refined at 2.8 Å resolution, thus eliminating any bias that may have resulted from the choice of model.[16]

The results of the searches in this study are extraordinarily clear, but it is not obvious why this should be the case. Structures with closer homologies than carboxypeptidase A and carboxypeptidase B have produced molecular replacement results that are much poorer.

Molecular Replacement—Challenging

The crystal structure of a new form of human deoxyhemoglobin was determined by Ward and co-workers[17] using the atomic model of horse deoxyhemoglobin as a test molecule. The crystals belong to the space group $P2_12_12_1$ with $a = 97$ Å, $b = 99$ Å, and $c = 65$ Å. There is a whole hemoglobin tetramer in the asymmetric unit. The transform of the horse hemoglobin tetramer was evaluated using the above-mentioned structure

[16] M. F. Schmid and J. R. Herriott, *J. Mol. Biol.* **103,** 175 (1976).
[17] K. B. Ward, B. C. Wishner, E. E. Lattman, and W. E. Love, *J. Mol. Biol.* **98,** 161 (1975).

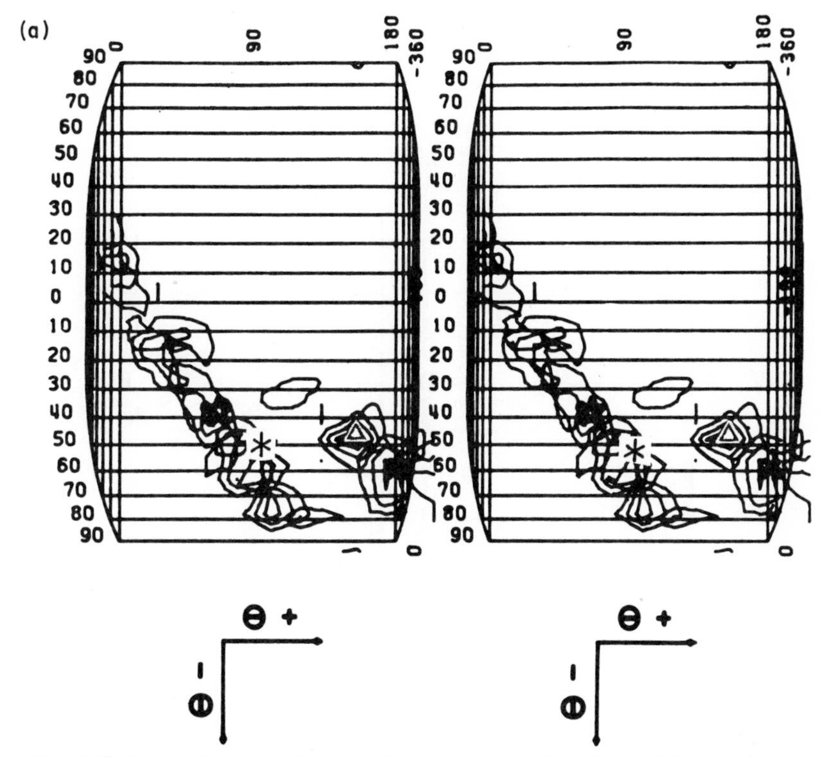

FIG. 8. Stereoscopic views of rotation function maps of deoxyhemoglobin A using θ_+–θ_- angles. The linear distance between grid points is proportional to the angular step Δ, and is thus the same on all levels. Since the angular ranges explored along θ_+ and θ_- are also the same on all levels, while the increments in these angles vary with θ_2 (see text), each level of constant θ_2 contains a different number of sample points. The level $\theta_2 = 0$ contains only one row of constant (and indeterminate θ_-). Levels of increasing θ_2 contain an increasing number of rows of constant θ_-, with the maximum number occurring at $\theta_2 = 180°$. Similarly, the number of points in each row decreases with increasing θ_2 until at $\theta_2 = 180°$ the rows contain but one point. Values of θ_2 appear at the upper and lower left-hand corner of each level. The ranges of θ_+ and θ_- are given across the top and along the right side, respectively, of the uppermost level. (a) Calculation using the horse hemoglobin tetramer as a search molecule and 30% of the crystal data in the range 0.1–0.16 Å^{-1}. Contours begin at 2σ above the mean (σ is the root mean square fluctuation of the map) and are in intervals of σ. The triangle marks the largest peak, and the asterisk indicates the final correct orientation. (b) Rotation function using the $\alpha_1\beta_1$ dimer as a test molecule, and 30% of the crystal data. Orientations of the two dimers in the asymmetric unit are revealed by density near A and B. (c) The region B of (b) sampled on a fine grid using the tetramer as a test molecule and all of the crystal data. The asterisk in this map marks the correct position.

(continued)

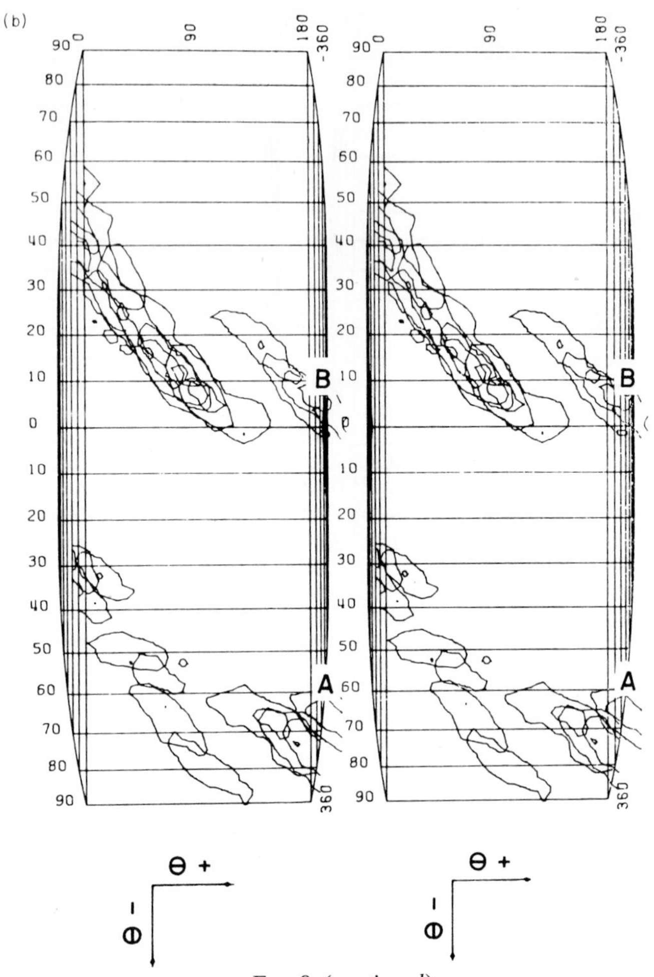

FIG. 8. (*continued*)

factor program with the tetramer placed in a triclinic cell having all edges 200 Å and all angles 90°. The resolution limit was 6 Å and an overall temperature factor of 20 Å^2 was applied. The transform of the dimer was similarly calculated at 5.5 Å resolution at sampling intervals of 1/128 Å. The rotation function was calculated using the $\theta_+ - \theta_-$ system with a true step size of 10°. About 30% of the reflections within the range 1/10 to 1/6 Å^{-1} were used in the calculation, and the origin was removed by subtraction of $\bar{I}$ from all reflections. Figure 8a shows a stereo view of this rotation function map. Details are given in the legend. The largest peak, marked

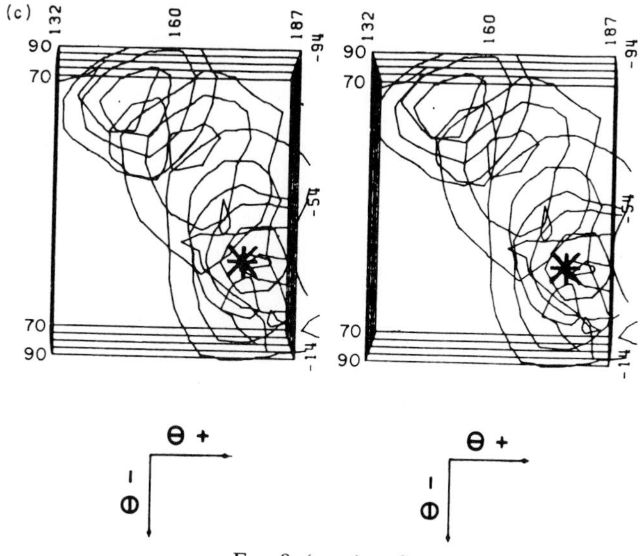

FIG. 8. (*continued*)

by a triangle, was used in a translation function calculation, but no family of peaks giving consistent values for the molecular position was found. To examine which of the other peaks in the rotation function might be the correct one, a rotation function was performed using the $\alpha_1\beta_1$ dimer as the test molecule. This map has an intrinsically lower signal-to-noise ratio, but proved useful because of the redundancy of the hemoglobin tetramer structure. The tetrameric dyad is collinear with the c axis of the transform. Therefore peaks in the dimer rotation function corresponding to the $\alpha_1\beta_1$ dimer and $\alpha_2\beta_2$ dimer orientations in the target crystal should differ by 2π in θ_3. This is equivalent to a translation of 2π along θ_- because of additional symmetry present in the rotation function. The correct orientations were seen to correspond to points A and B (Fig. 8b) since only these regions contained large values of R related by the predicted translation of 2π. Region B was reexamined in a tetramer rotation function using 100% of the reflections and steps of 5° and then, in a smaller volume, of 2°. Under these conditions the peak marked by the asterisk in Fig. 8c was the absolute maximum on the map, and a translation function calculation based on this orientation revealed a consistent set of peaks. These are seen in Fig. 9. The authors note that an orientation only 8° away from the correct one gave an uninterpretable translation function.

In order to improve their solution the authors undertook a rigid-body refinement of the structure generated by the orientation and position from

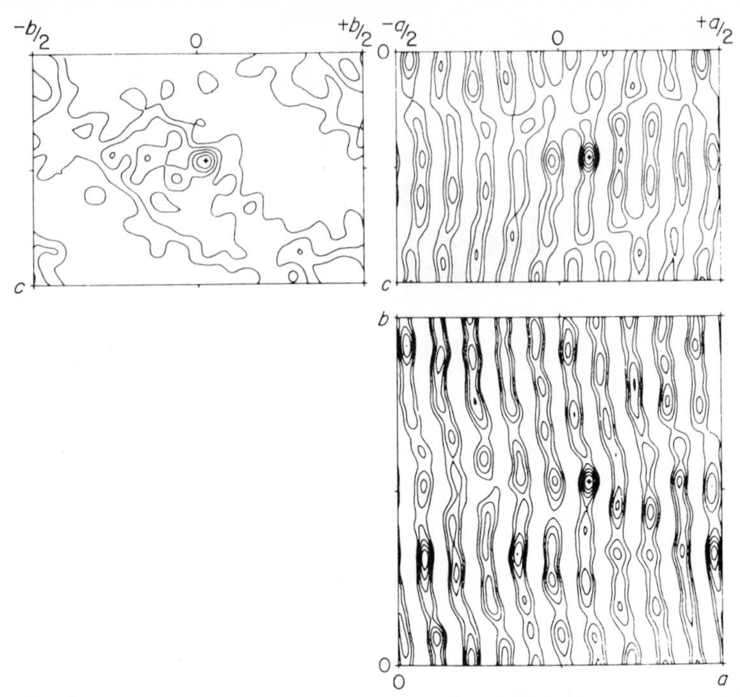

FIG. 9. Three Harker sections of the translation function for human deoxyhemoglobin. Each section is scaled to a maximum value of 100, and the average value is zero. Contours begin at zero and are spaced 20 units apart. The largest peaks on each section are marked with a +. These peaks form a consistent set since the two estimates of x, the two of y, and the two z all agree.

the searches. The difference between observed and calculated structure factors was minimized as a function of the three components of translation and the three Eulerian angles. The R factor dropped from 0.441 to 0.411 during this procedure. Because of the noisy character of the searches the correctness of the solution was of concern. To provide verification structure factors were calculated from the final model using the horse hemoglobin coordinates with the heme group deleted. A difference Fourier synthesis using calculated phases and the difference between observed and calculated amplitudes revealed peaks where the heme groups were known to be. Since the heme was not included in the model, it can only have reappeared in the difference map because the calculated phases and amplitudes were roughly correct.

This investigation illustrates the need to use all pieces of information to overcome noise in the search functions, and to view the results of such searches with due skepticism.

Final Comments

This survey of the most commonly used search procedures has attempted to clarify many practical matters in the application of the methods. A certain aura of mystery seems to surround search methods, so that they are less frequently used than is perhaps appropriate. If this guide helps in any way to dispell this aura, the author will be amply rewarded.

[6] Phase Refinement/Extension by Density Modification

By A. TULINSKY

Introduction

It is possible to improve an otherwise approximate or mediocre set of phase angles, and hence the corresponding electron density, by appropriately modifying the latter and carrying out its Fourier inversion to obtain a better set of phases. The method can be applied cyclicly, and is particularly suitable for improving multiple isomorphous replacement (MIR) phases at fixed resolution in protein structure determination, and for generating phase angle information beyond that resolution and refining it without recourse to an assumed atomic model. The method was originally introduced by Hoppe and Gassmann[1] to "correct" phases of partially known organic structures, and later it was even tested in a difference electron density application involving chemical bonding effects in a hypothetical structure.[2] The method has also been extended to obtain the phases of all the unique reflections of *Staphylococcus aureus* nuclease at 2.0 Å resolution (~9900) from the most reliable MIR phases (~4500)[3] and to the extension and refinement of phase angles from 2.8 Å resolution (~9700 reflections) to 1.8 Å resolution (~19,000 reflections) with α-chymotrypsin (α-CHT).[4] Finally, the method has been spectacularly successful in virus crystallographic applications in which density modification is essentially achieved by molecular averaging over noncrystallographic

[1] W. Hoppe and J. Gassmann, *Acta Crystallogr., Sect. B* **B24,** 97 (1968).

[2] T. Ito and I. Shibuya, *Acta Crystallogr., Sect. A* **A33,** 71 (1977).

[3] D. M. Collins, M. D. Brice, T. F. M. La Cour, and M. J. Legg, *in* "Crystallographic Computing Techniques" (F. R. Ahmed, K. Huml, and B. Sedlāček, eds.), p. 330. Munksgaard, Copenhagen, 1976.

[4] N. V. Raghavan and A. Tulinsky, *Acta Crystallogr., Sect. B* **B35,** 1776 (1979).

symmetry elements. The modification methods to be considered in this chapter will be those other than molecular averaging.

An inverse Fourier transform of an electron density distribution produces the Fourier coefficients (amplitudes and phase angles) that give the electron density distribution. A crystallographically determined density will always correspond to some resolution limit which formally depends on the wavelength of the shortest Fourier wave contributing to the distribution (d_{min} spacing). The inverse transform of the density beyond this resolution is identically zero. However, if the electron density distribution is modified prior to Fourier inversion, the amplitudes and phases obtained are different from those used to generate the unaltered distribution and, furthermore, they are no longer zero beyond the resolution limit. The density modification process can be expressed as

$$\rho_M(\mathbf{r}) = M(\rho)\rho(\mathbf{r})$$

where $\rho_M(\mathbf{r})$ is the modified electron density distribution, $\rho(\mathbf{r})$ is the electron density, and $M(\rho)$ is the modification function. Thus, by suitably choosing $M(\rho)$, it is possible to improve approximate phase angles and to determine new phases beyond the resolution limit of the original electron density.

Since the original phase angles are approximate, $M(\rho)$ is chosen in a way such that $\rho_M(\mathbf{r})$ results in a better electron density than $\rho(\mathbf{r})$. The phases of this new density can then be applied to observed structure amplitudes, which are different from the amplitudes of the inverse transform of $\rho_M(\mathbf{r})$, to produce another electron density distribution. In fact, phase angles beyond the original resolution can also be calculated. This new electron density can then be modified, initiating a cyclic process. If the resolution limit is kept constant, the cyclic process corresponds to phase refinement; if the resolution limit is increased to include additional observed amplitudes at higher resolution and calculated phase angles of $\rho_M(\mathbf{r})$, the process corresponds to phase extension and refinement. In either case, an atomic model is not presupposed, rendering these refinement/extension procedures objective and free of assumptions about structure.

Choosing a Modification Function

There are a number of different ways in which $M(\rho)$ can be chosen so that $\rho_M(\mathbf{r})$ is in principle a better electron density than $\rho(\mathbf{r})$. The simplest of these is to make use of the criterion of positivity,[6] so that

$$\begin{aligned} M(\rho) &= 1 \quad \text{for} \quad \rho(\mathbf{r}) > 0 \\ M(\rho) &= 0 \quad \text{for} \quad \rho(\mathbf{r}) \leq 0 \end{aligned}$$

[6] G. Kartha, *Acta Crystallogr., Sect. A* **A25,** S87 (1969).

TABLE I
SOME USEFUL MODIFICATION FUNCTIONS

$M(\rho)$	Application[a]	Comments	Ref.
$\frac{(1-2T)}{(1-T)} + \frac{(1+T)\rho}{(1-T)} - \frac{\rho^2}{(1-T)}$	D,R	T = threshold; $\rho > T$ enhanced, $\rho < T$ suppressed; first terms of Taylor expansion	1
Linear form of above	D	Faster convergence	7
"Molecular" envelope	D,R	Suppressing density beyond molecular region	8
ρ	D,R	"Atomic" envelope; Sayre's equation	9
ρ^n, $n > 1$	D,R	Related forms of "atomic" envelope	8
Zero for $\rho \leq 0$; ρ for $\rho > 0$	D	Positivity followed by squaring	10
$3\rho - 2\rho^2$ for $\rho > 0$; zero for $\rho \leq 0$	D	Taylor expansion above with $T = 1/2$	11
1 for $\rho \geq$ q; ρ = q for $\rho <$ q	D	Similar to positivity	12
1, inside molecular boundry; $\langle\rho\rangle/\rho$, outside molecular boundry; 0.1, for $\rho < 0$	D	Positivity and constancy of solvent	12a

[a] D, Direct space application; R, reciprocal space relationships among F values are also possible.

In general, all modification functions will strive to enhance the correct features of an electron density (larger positive regions) and to suppress the remainder at the expense of the former. A list of other modification functions which have been applied in various contexts is given in Table I.[7–12a]

Some density modifications lead directly to relationships among structure factors in reciprocal space (Table I). For instance, $M(\rho) = \rho(\mathbf{r})$ [i.e., $\rho_M(\mathbf{r}) = \rho^2(\mathbf{r})$] results in Sayre's equation.[9] In cases in which explicit

[7] W. Hoppe, J. Gassmann, and K. Zechmeister, *in* "Crystallographic Computing" (F. R. Ahmed, S. R. Hall, and C. P. Huber, eds.), p. 26. Munksgaard, Copenhagen, 1970.

[8] J. Gassmann, *in* "Crystallographic Computing Techniques" (F. R. Ahmed, K. Huml, and B. Sedlāček, eds.), p. 144. Munksgaard, Copenhagen, 1976.

[9] D. Sayre, *Acta Crystallogr.* **5,** 60 (1952).

[10] A. N. Barrett and M. Zwick, *Acta Crystallogr., Sect. A* **A27,** 6 (1971).

[11] D. M. Collins, *Acta Crystallogr., Sect. A* **A31,** 388 (1975).

[12] V. I. Simonov, *in* "Crystallographic Computing Techniques" (F. R. Ahmed, K. Huml, and B. Sedlāček, eds.), p. 138. Munksgaard, Copenhagen, 1976.

[12a] R. W. Schevitz, A. D. Podjarny, M. Zwick, J. J. Hughes, and P. B. Sigler, *Acta Crystallogr., Sect. A* **A37,** 669 (1981).

relationships are not possible, the modification function must be applied directly to the electron density (Table I); the inverse Fourier transform of the modified density then gives new values of phase angles. With the advent of fast, large-scale computers and computer programs based on fast Fourier transform (FFT)[13] algorithms,[14,15] it has become practical to apply modification functions to protein electron density maps ($\sim 4 \times 10^5$ points) and to calculate the inverse transform (~20,000 coefficients) in a matter of minutes.[4]

Modification functions which sharpen or effectively increase the formal resolution of an electron density distribution require that $\rho_M(\mathbf{r})$ be calculated on a grid of smaller intervals than that required for calculation of the initial electron density, to accurately represent the set of structure factors corresponding to $\rho_M(\mathbf{r})$. Shannon's sampling theorem[16] indicates that an electron density must be calculated at intervals smaller than or equal to $d_{min}/2$ to faithfully produce structure factors to that resolution. Thus, for $M(\rho) = \rho(\mathbf{r})$ or $\rho_M(\mathbf{r}) = \rho^2(\mathbf{r})$, which is "sharper" than $\rho(\mathbf{r})$, twice as many points are required in each direction to give an accurate calculation of structure factors for phase refinement or extension of resolution.[17] Although such modifying functions can be avoided since they increase computing times, Shannon's criteria are of paramount importance and must be observed for accurately and efficiently extending phase angles to higher resolution.

In order to keep the largest features of an electron density from growing without bound, it is convenient to modify the density where the largest feature has been normalized to unity. For structures of like atoms, this is equivalent to factoring the scattering factor along with an overall thermal parameter from the structure factor equation. Few real structures comply to such a case, especially partially known structures or protein structures for which the maximum peak heights can differ significantly from the average peak height. In such cases, cyclic application of a modification function will enhance the largest regions and lead to the spurious growth of these at the expense of the remainder of the map. Heavy-atom-containing structures must certainly be treated as a special case which in a mild form can consist of a protein with sulfur atoms.

[13] J. W. Cooley and J. W. Tukey, *Math. Comput.* **19,** 297 (1965).

[14] C. R. Hubbard, C. O. Quicksall, and R. A. Jacobson, "The Fast Fourier Algorithm and Programs ALFF, ALFFDP, ALFFPROJ, ALFFT and FRIEDEL." Ames Laboratory, Ames, Iowa, 1971.

[15] L. F. TenEyck, *Acta Crystallogr., Sect. A* **A29,** 183 (1973).

[16] C. E. Shannon, *Proc. IRE* **37,** 10 (1949).

[17] B. Gold and C. M. Radar, "Digital Processing of Signals," p. 166. McGraw-Hill, New York, 1969.

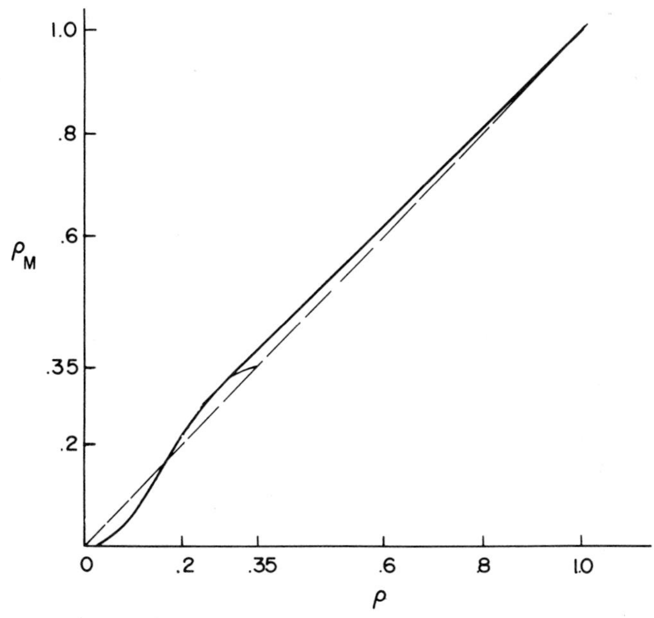

FIG. 1. Modification $3\rho^2 - 2\rho^3$ for $\rho_0 > 0$ with $\chi = 0.35$. Broken line is no modification or $\rho_M(\mathbf{r}) = \rho(\mathbf{r})$. (Reprinted from Ref. 4.)

One way of dealing with the problem has been suggested by Collins *et al.*,[3] using $M(\rho) = 3\rho - 2\rho^2$, and it has been successfully applied to rubredoxin,[3] staphylococcal nuclease,[3] α-CHT,[4] and KDPG aldolase.[18] This was accomplished by multiplying $M(\rho)$ by a value χ, where $\chi/2$ is the value of barely significant normalized density[3,4] and smoothly joining the modification function near $\rho \simeq \chi$ with the tangent from the point $\rho = \rho_M = 1$, as shown in Fig. 1. The value of $\chi = 0.35$ was used in the phase extension/refinement of α-CHT[4] and it suppressed density below 0.65 e Å^{-3} [$\sim 3\sigma(\rho)$] of the 1.8 Å resolution map, where the largest point in the density was 3.7 e Å^{-3}. The maximum density of the 2.8 Å resolution map was 2.5 e Å^{-3} whereas the remainder averaged about 1.5 e Å^{-3} with $\sigma(\rho) = 0.18$ e Å^{-3}. In the latter case of phase refinement only, χ was taken as 0.5, which suppressed regions below 0.6 e Å^{-3}. In both cases, maximum

[18] I. M. Mavridis, M. H. Hatada, A. Tulinsky, and L. Lebioda, *J. Mol. Biol.*, **162,** 419–444 (1982).

enhancement was only about 10–15% near the 1.0 e $Å^{-3}$ level with the highest regions of electron density remaining essentially unaltered. Since the peak–noise ratio increased with higher resolution (from 4.2 to 5.7), a corresponding adjustment was made in the χ value. This will generally be true of phase extension.

Phase Refinement/Phase Extension

The cyclic nature of phase refinement by density modification or phase refinement/phase extension is illustrated schematically with a flow chart in Fig. 2. Although in principle, the most probable MIR phases should be used to initiate the process with a protein structure determined by the MIR method, in practice "best" MIR phases can be substituted, especially in well-determined MIR solutions. After calculating the inverse transform of $\rho_M(\mathbf{r})$, the calculated amplitudes $|F_c'|$ are scaled to the observed ones $|F_o|$ to obtain new values, $|F_c|$, and a new electron density is computed using $(2|F_o| - |F_c|)$ as Fourier amplitudes and α_c as phase angles. Some measure of control can be imposed on the process by only

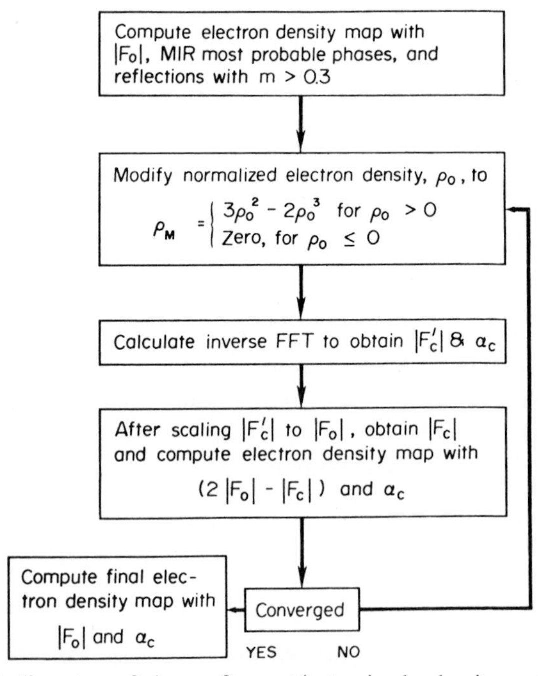

FIG. 2. Cyclic nature of phase refinement/extension by density modification.

including reflections satisfying $|F_c|/|F_o| > C$ in the electron density calculation, where C is an appropriately chosen acceptance criterion reflecting the reliability of phase angles. As will be seen later in this section, this is an important consideration in the initial few cycles of phase extension, because the $|F'_c|$ beyond the d_{min} spacing of the electron density map are necessarily small in the early stages of extension. Most of the poorly determined MIR phases (e.g., $m < 0.3$, where m in the figure of merit determined by the MIR method) improve vastly in one cycle of fixed resolution phase refinement and the whole process converges within two to three cycles. Thereafter, higher resolution phase angles can be calculated and applied to corresponding observed amplitudes to give a higher resolution map and to initiate phase extension refinement.

Progress of Refinement

A number of informative quantities and relationships may be used to follow the progress of the refinement. Some of these are listed in Table II along with certain constants pertinent to the refinement (χ and C). It should be noted here that it is the change of the magnitude of these quantities which is important and not the values of the magnitudes themselves. The latter are intimately dependent upon $M(\rho)$ and will be smaller as the amount of density modification decreases. As Collins *et al.*[3] pointed out so correctly, "improvements in the conventional sense" of R, $\Delta\alpha$, and $\Delta\alpha_\omega$ could be attained (shift to smaller values) quite irrespective of any improvement in the corresponding phase set (e.g., simply by reducing the amount of modification). Finally, it can also be seen from Table II that the extension of phase angles is essentially complete in four to five cycles (94% assigned) using the relatively mild modification function of $M(\rho) = 3\rho - 2\rho^2$ with $\chi = 0.35$ and $M(\rho) = 0$ for $\rho \leq 0$.

It is also useful and informative to monitor some of the foregoing quantities as a function of scattering angle or resolution (Fig. 3). For instance, it can be seen from Fig. 3 that the R values at 2.8 Å resolution remain essentially constant after extension of phases to 1.8 Å resolution. Moreover, the initially high value of overall R (~0.40) is primarily due to the high-order reflections (approximately half the total number). This results in part from the abrupt change in scale factor between 2.8 Å and higher resolution data. The magnitudes of the phase-extended reflections of the early cycles are generally quite small and are zero for the Fourier inversion of the 2.8 Å resolution map. Since the latter is only slightly modified, it gives rise to nonzero, but generally only small, structure factors beyond 2.8 Å resolution. Because of this, the use of an acceptance criterion (e.g., C) becomes practically a necessity. The manner in which

TABLE II

QUANTITIES MONITORED DURING PROGRESS OF EXTENSION/REFINEMENT OF α-CHT[a]

Quantity	Cycle number										
	0	1	2	3	4	5	6	7	8	9	10
R[b]	—	0.206	0.131	0.399	0.364	0.332	0.276	0.218	0.182	0.158	0.147
$\langle \Delta\alpha \rangle$/cycle	—	12.3	7.7	4.6	5.9	7.5	6.5	6.9	5.8	4.7	3.8
$\langle \Delta\alpha_\omega \rangle$/cycle[c]	—	9.3	5.5	3.4	4.4	5.2	5.0	5.3	4.4	3.6	3.0
k[d]	—	1.13	1.10	1.20	1.10	1.04	1.01	1.00	1.00	1.00	1.01
χ	—	0.50	0.50	0.50	0.35	0.35	0.35	0.35	0.35	0.35	0.35
$\rho_{max}\chi/2$ e Å^{-3}	—	0.62	0.64	0.68	0.48	0.52	0.59	0.65	0.65	0.65	0.65
$C = \|F_c\|/\|F_o\|$	—	0.15	0.15	0.30	0.25	0.20	0.15	0.15	0.15	0.15	0.15
Phases assigned	8400	9677	9745	11,129	13,720	15,800	17,720	18,589	18,896	19,001	19,066
Fraction assigned	0.425	0.490	0.493	0.563	0.694	0.800	0.897	0.941	0.956	0.962	0.965
$\langle \|\alpha_{MIR} - \alpha_c\| \rangle$[e]	—	9.8	13.9	16.6	19.6	21.6	23.3	24.8	26.0	27.0	28.0

[a] Reprinted from Ref. 4.

[b] $R = \Sigma||F_o| - |F_c||/\Sigma|F_o|$. This is computed for all data.

[c] $\langle \Delta\alpha_\omega \rangle = \Sigma|F_o|\,\Delta\alpha/\Sigma|F_o|$.

[d] $k = \Sigma|F_o||F_c'|/\Sigma|F_c'|^2$. This is computed for all data.

[e] Tabulated values are the modulus-weighted average and are based on the 8400 reflections in the starting set.

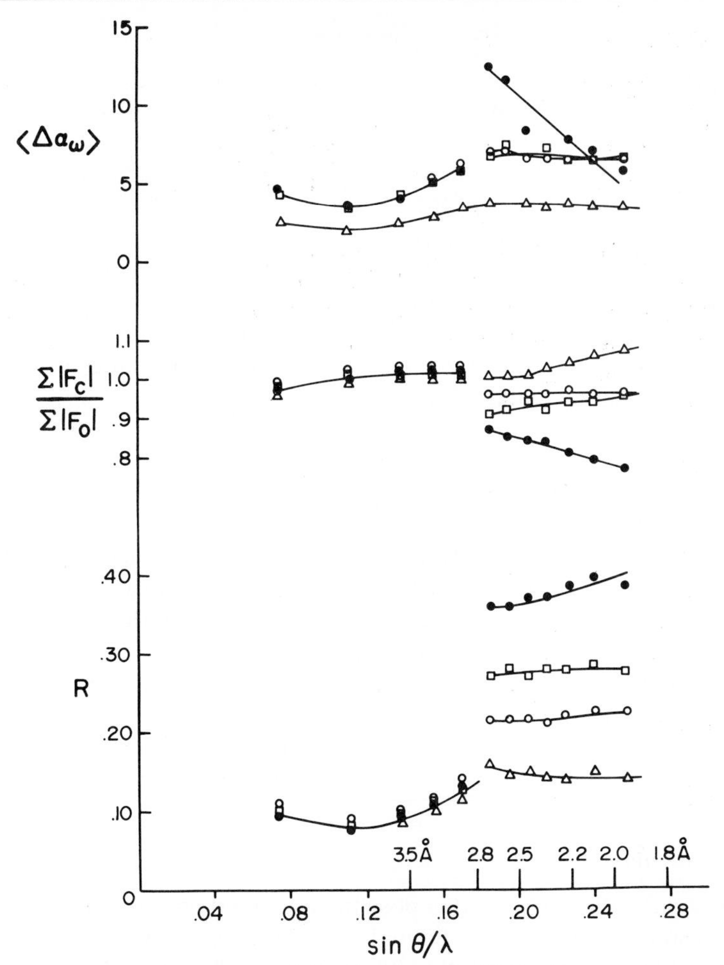

FIG. 3. Summary of phase extension refinement of α-CHT with respect to resolution. Quantities as in Table II; (●) cycle 5, (□) cycle 6, (○) cycle 7, (△) cycle 10; only phased reflections included. (Reprinted from Ref. 4.)

the apparent discontinuities disappear with further phase extension and density modification can also be seen clearly from Fig. 3. Finally, the phase changes (from zero) of the first cycle of phase extension average about 18–20°. At convergence of phase extension, the phase changes between cycles of both the high- and low-order reflections tend to become similar at about 3°.

MIR–Extension–Refinement Comparisons

The average change in phase angle with respect to MIR phases should converge to a value somewhat less than that expected of the average figure of merit, $\langle m \rangle$. In the case of α-CHT,[4] $\langle |\alpha_{MIR} - \alpha_c| \rangle$ was about 28° at convergence while $\langle m \rangle = 0.76$ implies an average error of about 40° in MIR phases. In fact, 67% of the MIR phases changed by less than 40°. A bivariate analysis of m and $|\alpha_{MIR} - \alpha_c|$ of the final cycle of the α-CHT extension/refinement showed that the phase change was small when m was large; moreover only 11 of the 320 centrosymmetric 2.8 Å resolution MIR signs changed during the entire phase extension.[4] Another bivariate analysis of $|F_o|$ and $|\alpha_{MIR} - \alpha_c|$ showed that the phase change was generally small for large values of $|F_o|$.

Solvent Considerations with Proteins

Any density modification function for proteins can be improved by taking interstitial solvent into account.[12a] This can be accomplished by replacing solvent density beyond the first or second solvation sphere of a protein with that of the average value of the density in these regions in a manner similar to that used in virus applications.[19,20] This has the effect of attenuating the solvent density fluctuation and can be made as a function of the distance from the protein surface. Otherwise, solvent distant to the protein, which is at best questionable, will be weighted equally to that of more reliable indications near the surface of the protein.

Requirements

Although it is clear that it is possible to extend phase angles beyond a given resolution based on a limited set of phases[3,6,21] it is equally clear that, in order to accomplish this, the initial set of known phases must be sufficiently large.[3,21] Phase angle refinement by density modification at fixed resolution does not have any such restrictions and significant improvement in phase angles can be accomplished at any resolution through a few cycles of refinement.[18] Thus, MIR "best" electron density maps can always be expeditiously and objectively improved by density modification prior to interpretation.

The relationship between the size of the starting phase set and the phase extension possible from it is much less obvious. On the one hand,

[19] P. Argos, G. C. Ford, and M. G. Rossmann, *Acta Crystallogr., Sect. A* **A31,** 499 (1975).
[20] G. Bricogne, *Acta Crystallogr., Sect. A* **A32,** 832 (1976).
[21] D. Sayre, *Acta Crystallogr., Sect. A* **A30,** 180 (1974).

the starting set appears to have to be at least as large as the set of phases (for observed amplitudes) to be determined, while on the other, it appears that sufficient resolution must also be present initially for density modification to lead to logical consequences and correct results at higher resolution. Thus, if the electron density of a phenyl ring has an indication of a depression at its center at lower resolution, then proper modification combined with higher order data will lead to a hole in the density corresponding to the benzene ring. On the other hand, if the depression is not present initially, phase angle extension by density modification might be incapable of producing the fine detail. Similar but less obvious cases will pervade a lower resolution map so that the higher the initial resolution, the better the case for meaningful phase extension and reliability of the higher resolution map, especially on approaching atomic resolution.

In the case of rubredoxin, Collins *et al.*[3] have shown that phase extension by density modification from 2.0 to 1.5 Å resolution leads to results that are remarkably similar to the least-squares model refined phase set.[22] Starting at 2.5 Å resolution did not substantially affect the final results; however, the same was not true with an initial 3.0 Å resolution starting set. Interestingly, the application of Sayre's equation to the 3.0 Å resolution data of rubredoxin also did not lead to meaningful phase extension.[21] These considerations led Collins *et al.*[3] to suggest that, for high-resolution data sets, a useful rule might be a starting phase set of at least 3 unique phases per atom in the asymmetric unit. This is supported by our work on α-CHT dimer, in which phases were successfully extended from 2.8 to 1.8 Å resolution (9745 initial phases per 3528 atoms extended to 19,066 phases).[4]

Phasing Noncentrosymmetrical Difference Maps

Use of the difference Fourier technique[23] is widespread in protein crystallography to obtain information such as (1) the positions of heavy atoms in isomorphous derivatives, (2) the location of active sites through binding studies of inhibitors or substrate-like molecules, and (3) the nature of the conformational changes accompanying protein interactions with other molecules in processes such as change of pH or onset of denaturation.[24] The difference electron density map, $\Delta\rho$, is customarily approxi-

[22] K. D. Watenpaugh, L. C. Sieker, J. R. Heriott, and L. H. Jensen, *Acta Crystallogr., Sect. B* **B29,** 943 (1973).

[23] T. L. Blundell and L. N. Johnson, "Protein Crystallography," Chapter 14. Academic Press, New York, 1976.

[24] A. Tulinsky, *in* "Biomolecular Structure, Conformation, Function and Evolution" (R. Srinivasan, ed.), Vol. 1, p. 183. Pergamon, Oxford, 1980.

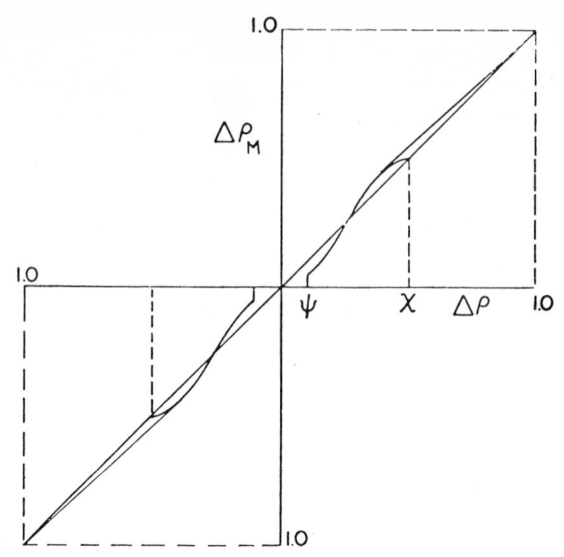

FIG. 4. Modification for difference electron density based on $3\rho^2 - 2\rho^3$ for protein derivative phase formation and refinement. χ and ψ as defined in text; straight line is no modification or $\Delta\rho_M = \Delta\rho$.

mated using isomorphous derivative amplitudes, $|F_{PD}|$, and native protein phases, α_P, according to

$$\Delta\rho = \frac{1}{V}\sum_{\mathbf{S}} (|F_{PD}| - |F_P|) \exp(i\alpha_P) \exp(-2\pi i \mathbf{S}\cdot\mathbf{r})$$

where $|F_P|$ is the amplitude of the protein.

As previously mentioned, with the development of fast Fourier algorithms, it has become possible to perform a Fourier inversion of such maps in a matter of minutes and to use the phase angles thereof to approximate the derivative phase angle, α_{PD}, by

$$\alpha_{PD} = \tan^{-1}[(B_P + B_D)/(A_P + A_D)]$$

where the subscript D denotes the contribution of the difference electron density.[25] The phase angles α_{PD} can then be used to form new difference density coefficients of the form

$$[|F_{PD}| \exp(i\alpha_{PD}) - |F_P| \exp(i\alpha_P)]$$

to calculate a more nearly correct difference map. In fact, the process can be recycled as in phase refinement/extension until the difference density contribution becomes stationary.[25]

[25] L. D. Weber, Ph.D. Dissertation, Michigan State University, Ann Arbor (1978).

Although a modification function is not a necessity in the difference Fourier application, nonetheless, a suitably chosen modification can accelerate convergence and enhance the final results. A modification function for difference electron density maps based on that of Collins[11] was used by Weber[25] in studying the difference maps of the fluorescent probe 1-anilino-naphthalene 8-sulfonate (ANS) bound to α-CHT and it is shown in Fig. 4. Since both positive and negative density is of interest in this case, the modification function is antisymmetrical about the origin. In addition, a background cutoff of ψ was introduced such that $|\Delta\rho| \leq \psi$ is modified to be zero, thus removing low-level features almost certain to be random noise. In practice, it was found that a suitable value of ψ was about 2–3 rms $\Delta\rho$, where rms $\Delta\rho = [\Sigma(\Delta\rho)^2/N]^{1/2}$ and N is the number of points in the summation of the difference map. It was also noted that convergence was generally rapid and complete after three to four cycles of phase calculation following such procedures.

As expected, "derivative-phased" difference Fourier maps result in higher peak heights[23,26] which proved to be about double those of conventional maps.[25] When the original map is modified, the largest features tend to enhance somewhat at the expense of lesser ones.[25] In some cases there is even a relatively rapid enhancement of some originally barely significant density (e.g., slightly greater than ψ).[25] At present, it is difficult to assess the full significance of such enhancement without more detailed study.

In conclusion, it should be noted that phase refinement by Fourier transformation of difference electron density peaks was also described by Ito and Shibuya[2] in studying chemical bonding effects of a hypothetical structure. They used the following modification function:

$$\Delta\rho(\text{peak}) = \Delta\rho(\mathbf{r} \leq \mathbf{r}_{\text{mx}})$$
$$|\Delta\rho(\text{peak})| > |\Delta\rho|_{\text{noise}}$$

where $|\Delta\rho|_{\text{noise}}$ is the background noise and only the density greater than $|\Delta\rho|_{\text{noise}}$ was considered within a radius $\mathbf{r}_{\text{mx}}$ of the peak position. Starting with phases of a spherical free-atom model, the hypothetical aspherical structure of $NaNO_2$ was reproduced well after five cycles of iteration.

Acknowledgments

This review was written during the course of a sabbatical stay as an Alberta Heritage Foundation for Medical Research Visiting Scientist at the Department of Chemistry, University of Calgary, Calgary, Alberta. I would like to thank the Department of Chemistry and Dr. K. Ann Kerr in particular for their cordial hospitality. I would also like to thank Dr. P. W. Codding for critically reading this chapter.

[26] V. Luzzati, *Acta Crystallogr.* **6**, 142 (1953).

[7] Resolution of Phase Ambiguity in Macromolecular Crystallography

By BI-CHENG WANG

Introduction

The problem of phase ambiguity in the isomorphous replacement method was first described in 1949 by Bijvoet.[1] Despite many attempts to resolve this problem, a general method to overcome it without additional experimental data or special information concerning the structure has not yet been found. Because of this problem the structure determination of a macromolecule using only two sets of diffraction data is very difficult, if not impossible. Accordingly, the discovery of a general solution is of considerable theoretical and practical importance.

In this chapter the philosophy and the practical aspects of such a solution and the results from applications to known and unknown structures are presented. The present procedure is applicable to single isomorphous replacement (SIR) data without anomalous scattering information as well as to pure single-wavelength anomalous scattering (SAS) data alone, and it does not require the presence of noncrystallographic symmetry. Test results show that it is very effective; not only can it remove the phase ambiguity but it will also refine the phases at the same time. In actual application to SIR data the method has even produced results which appear to be superior to those obtained from the multiple isomorphous replacement method. Part of the results described here have been reported at the Twelfth International Congress of Crystallography[2] and at the 1982 Winter and Summer Meeting of the American Crystallographic Association (Gaithersburg, Maryland and San Diego, California).

Summary of Previous Approaches and Background

One of the situations in which the phase ambiguity problem occurs is when the isomorphous replacement method is applied to a noncentrosymmetric structure using only one isomorphous data set. This is the single isomorphous replacement (SIR) method. In such a case the available data will give two possible solutions, one true and one false for each phase angle. In the absence of adequate means for resolving the ambiguity the

[1] C. Bokhoven, J. C. Schoone, and J. M. Bijvoet, *Acta Crystallogr.* **4,** 275 (1951).
[2] B. C. Wang, *Acta Crystallogr., Sect. A* **A37,** Suppl. C-11 (1981).

best one could do for protein (hereafter means a macromolecule) structure determination was to use both of the solutions. Blow and Rossmann[3] proposed to use the average vectors of the structure factors of both solutions. Their method, which is equivalent to that of a double-phased synthesis[1,4] and the β-isomorphous synthesis,[5] was shown to give an interpretable Fourier map of myoglobin at 2 Å resolution.[3] However, the false solutions for the phase angles can introduce a high level of noise,[3,4,6] which can often prevent correct interpretation of the Fourier map. Therefore, without some way to remove the phase ambiguity the SIR method has seldom been used in protein structure determination.

The phase ambiguity, of course, can be resolved by using more data. Bijvoet and colleagues pointed out that the ambiguity can be removed if a second isomorphous derivative[1] becomes available or if the anomalous scattering signal[7] of the heavy atom can be measured. These suggestions have become the basis of multiple isomorphous replacement (MIR), which was used first by Perutz and co-workers[8] for protein structure determination, and now has become the most powerful method for determining phases in protein crystallography. However, the MIR method is not without shortcomings. The need for two or more derivatives has often created difficulty in structure determination since the preparation of isomorphous derivatives is one of the most time-consuming steps in protein crystallography. Although in theory the anomalous scattering signal from the heavy atom can be used to resolve the phase ambiguity, in practice the method, SIRAS (single isomorphous replacement with anomalous scattering), has been used only occasionally because it requires the diffraction data to be measured to a degree of accuracy that is often unobtainable.

Resolution of phase ambiguity by the direct method using the tangent formula has also been attempted, but the results were not encouraging[9-11] since only a modest improvement in phases was achieved. The most interesting attempt for resolving SIR phase ambiguity without additional data has been the use of noncrystallographic symmetry suggested by

[3] D. M. Blow and M. G. Rossmann, *Acta Crystallogr.* **14,** 1195 (1961).

[4] G. Kartha, *Acta Crystallogr.* **14,** 680 (1961).

[5] G. N. Ramachandran and S. Raman, *Acta Crystallogr.* **12,** 957 (1959).

[6] G. N. Ramachandran and R. Srinivasan, "Fourier Methods in Crystallography." Wiley (Interscience), New York, 1970.

[7] J. M. Bijvoet, *Nature (London)* **173,** 888 (1954).

[8] D. W. Green, V. M. Ingram, and M. F. Perutz, *Proc. R. Soc. London, Ser. A* **225,** 287 (1954).

[9] C. L. Coulter, *Acta Crystallogr., Sect. B* **B27,** 1730 (1971).

[10] J. E. Weinzierl, D. Eisenberg, and R. E. Dickerson, *Acta Crystallogr., Sect. B* **B25,** 380 (1969).

[11] W. A. Hendrickson, *Trans. Am. Crystallogr. Assoc.* **9,** 61 (1973).

Bricogne.[12] His approach has been successfully used in the structure determination of at least one protein, hemagglutinin,[13] which has 3-fold noncrystallographic symmetry.

Phase ambiguity also exists if anomalous scattering data from a single wavelength are used for phase determination.[7,14] However, the phase ambiguity problem here is not as serious as that from using SIR data. This is because the heavy-atom phase is no longer located midway between the two possible solutions. Therefore, it may be used to predict which of the two solutions is most likely to be correct. How accurately can a heavy-atom phase be used to make such a prediction has been studied theoretically by Parthasarathy[15] in terms of probabilities. The greater the percentage of the heavy-atom contribution is to the total scattering intensity, the greater is the probability. The Fourier synthesis using the phases selected by the heavy-atom phases has been named the quasi-anomalous dispersion synthesis[6] and it has been successfully used to solve a number of small-molecule structures (see, for example, Ref. 6). Recently Hendrickson and Teeter[16] demonstrated elegantly that the same principle used in the quasi-anomalous dispersion method can also be used in a protein structure determination. However, it should be pointed out that their method, the resolved-anomalous phasing[16,17] like the quasi-anomalous dispersion method, in theory is expected to resolve only a certain portion of the total phase ambiguity, even when the data are completely error free. This is because the method depends on how accurately a protein phase can be approximated by the phase of the partial structure (heavy atom). The accuracy, of course, depends critically[15] on the relative contribution of the heavy atom to the total diffraction intensity of the crystal. This contribution, which varies from protein to protein, is relatively small in an average macromolecule. Therefore their method can offer some correct phases but it will have to rely on other methods for improvement before an unambiguous electron density map can be synthesized.

Philosophy of the Present Approach

For convenience of explaining the present approach we first take a set of SIR data as an example and assume that the data are reasonably accurate and that the problems we are facing are related to the situation of

[12] G. Bricogne, *Acta Crystallogr., Sect. A* **A32,** 832 (1976).

[13] I. A. Wilson, J. J. Skehel, and D. C. Wiley, *Nature (London)* **289,** 366 (1981).

[14] S. Raman, *Acta Crystallogr.* **12,** 964 (1959).

[15] S. Parthasarathy, *Acta Crystallogr.* **18,** 1028 (1968).

[16] W. A. Hendrickson and M. M. Teeter, *Nature (London)* **290,** 107 (1981).

[17] J. Smith and W. A. Hendrickson, *in* "Computational Crystallography" (D. Sayre, ed.), p. 209. Oxford Univ. Press (Clarendon), London and New York, 1982.

phase ambiguity. As mentioned earlier, a SIR map calculated based on the method of Blow and Rossmann[3] is equivalent to a double-phased Fourier map.[14] Because of the linearity property of Fourier transforms, the SIR map can be regarded as a superposition of two Fourier maps (Fig. 1a), one being a correctly phased protein electron density map, and the other being a map produced with false phases. It is known that when the heavy atoms in the isomorphous derivative are not in special positions or related by a center of symmetry, the Fourier map produced by the false phases will contain no structural information and will be in a form of general background.[3,4,6] These background noises are generally lower than the correct protein density in the map. This observation is the basis for the present approach.

Contrary to most previous phasing methods which start by attempting to eliminate false solutions in reciprocal space, the present approach begins with a noise filtering (or image enhancement) process in direct space. As will be discussed in a later section a procedure has been developed to define objectively the molecular boundary from a noisy electron density map. Then, the densities within the protein region are raised by a constant value and the very weak and negative densities are removed. Outside the molecular envelope the density is smoothed to a constant level. This procedure is to enhance the gross image of the protein and to approximate the reality that electron density is positively defined and protein crystals consist of the ordered (protein) and disordered (solvent) regions. This enhanced image then becomes the partial structure which in turn is used for resolving the phase ambiguity. Since this partial structure will *always* be a large percentage of the total scattering density, the phase ambiguity associated with SIR data can therefore be *overcome with internal information inherent in the data,* and a set of SIR data is in principle sufficient for a unique structure determination of the macromolecule.

The principle outlined has been implemented via a series of computer programs and has been applied to a known and several unknown structures. In the actual calculations the protein gross image obtained from direct space filtering is Fourier inverted into reciprocal space. The Fourier inversion results are then used to construct a reciprocal space filter for the removal of reciprocal space errors (false solutions). Since it is not always possible to remove all the errors in any single step and there exists also the possibility of systematic and random errors in the original diffraction data as well as in the heavy-atom parameters, the filtering procedures have been carried out between direct and reciprocal space in an iterative manner (Fig. 2) until the results converge. The present approach may therefore be called the iterative single isomorphous replacement (ISIR) method.

It is also known[7,14] that for a set of single anomalous scattering (SAS)

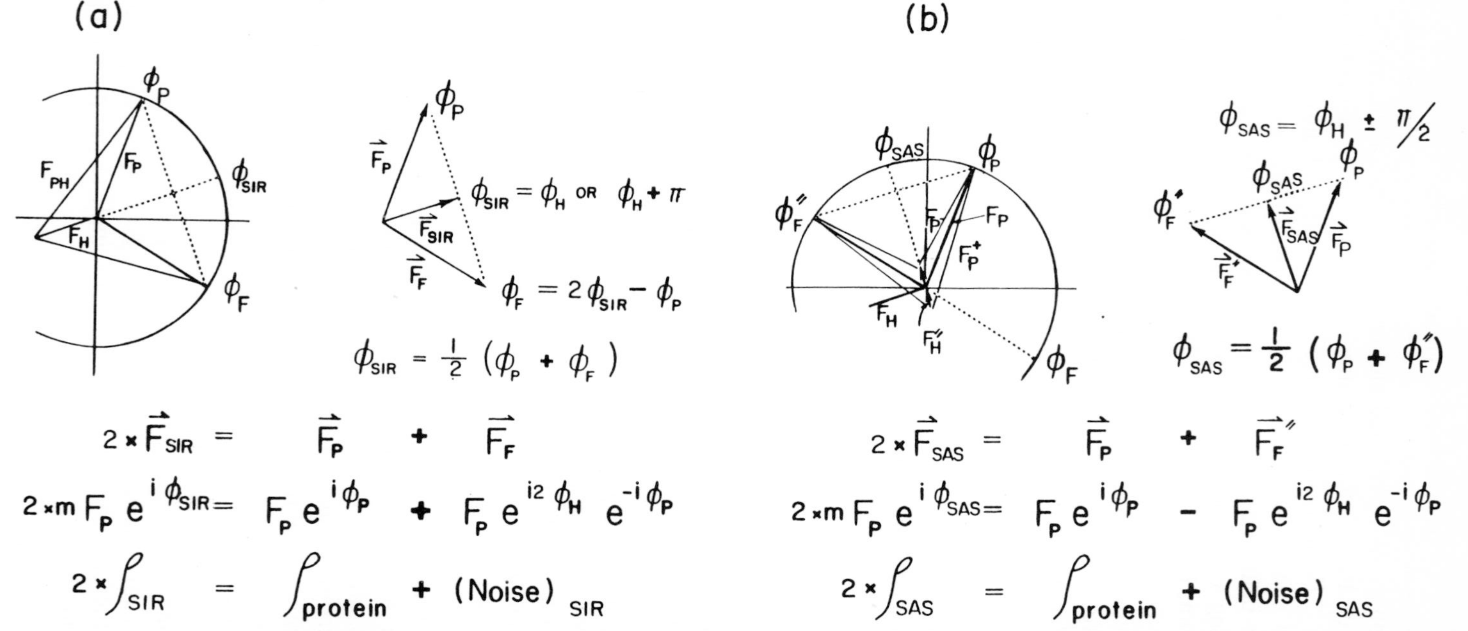

FIG. 1. (a) The Harker construction for phase calculation by the SIR method and how a SIR map can be considered as a superposition of two Fourier maps, one being a correctly phased protein density and the other being noise. (b) The Harker construction for phase calculation by the single anomalous scattering (SAS) method and how an SAS map can be considered as a superposition of two maps.

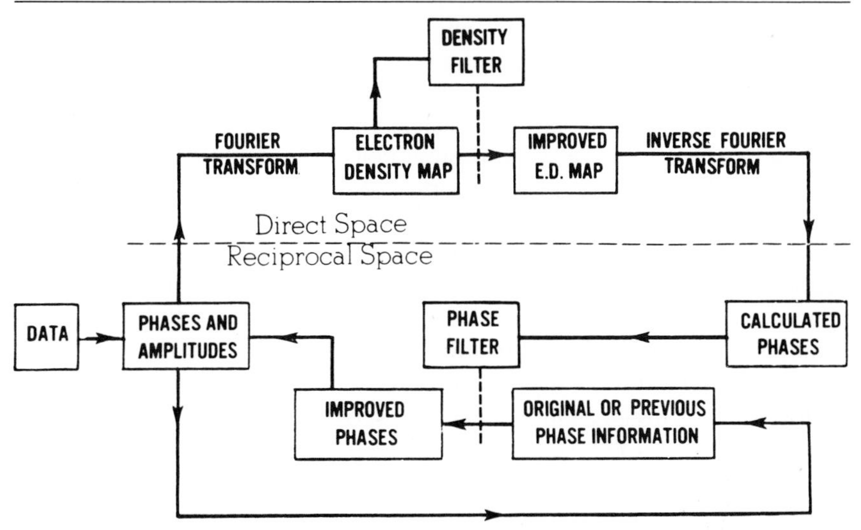

FIG. 2. A schematic diagram representing the various procedures in the present approach.

pairs phase ambiguity can occur in a similar manner as in SIR data. An unresolved SAS map will also contain two maps, the protein and the noise (Fig. 1b). Therefore it is evident that the principle used for solving the problem in SIR data should also be applicable to the SAS data. The use of such a principle for phasing a macromolecule structure with SAS data alone is referred to as the iterative single anomalous scattering (ISAS) method in this chapter and will also be discussed.

The philosophy of the present approach is summarized in Fig. 2, which represents a complete cycle of filtering of errors in both direct and reciprocal space. It is interesting to note that Fig. 2 can also be used to explain most of the iterative Fourier processes commonly used in crystallography. For example, acceptance of only chemically reasonable structural fragments from preliminary maps is a form of direct space filtering, and use of only strong reflections or weighting structure factors by their associated figures of merit are forms of reciprocal space filtering. Other examples are the various applications for the improvement or extension of MIR phases,[18–24] and on the use of noncrystallographic symmetry for

[18] D. M. Collins, *Acta Crystallogr., Sect. A* **A31,** 388 (1975).

[19] R. W. Schevitz, A. D. Podjarny, M. Zwick, J. J. Hughes, and P. B. Sigler, *Acta Crystallogr., Sect. A* **A37,** 669 (1981).

[20] T. N. Bhat and D. M. Blow, *Acta Crystallogr., Sect. A* **A38,** 21 (1982).

[21] D. M. E. Szebenyi, S. K. Obendorf, and K. Moffat, *Nature (London)* **294,** 327 (1981).

[22] D. A. Agard and R. M. Stroud, *Acta Crystallogr., Sect. A* **A38,** 186 (1982).

phase determinations.[12–13,25–27] It should be pointed out that the philosophy here should not be described as a density modification process for two reasons. First, the process involves both direct and reciprocal space. Second, the word *modification* is misleading since what has been emphasized here is not to modify the protein density map but to suppress the noise, since by doing so the undistorted protein density will be revealed automatically. Otherwise the true image of a protein will never be recovered, as it will always be in a modified form, even if the noise is removed. Furthermore, the word *modification* implies that the density is incomplete or incorrect and does not acknowledge the important concept that the true image of a protein is already in the map, and it is only the noise which obscures its detection. Since the awareness and utilization of this concept are the basis of the present approach, the term *filtering of errors* is preferred for a precise description of the process.

The practical aspects of carrying out the concepts described here will be introduced in the next section. These practical aspects actually bear some resemblances to that of Bricogn's direct space molecular replacement method[12] for the resolution of SIR phase ambiguity using noncrystallographic symmetry and the density modification procedures for the improvement of MIR phases.[18–24] However, in the present approach the noncrystallographic symmetry, which is the major source of geometric constraint in Bricogne's approach, is no longer required. Instead, a great deal of emphasis is placed on the molecular boundary and the partial structural information within the molecular envelope. The structure itself provides information for making selections between the true and the false solutions. The use of a Fourier process in an iterative manner is a means for reaching a complete structure from a partial structure. Therefore the philosophy present here is consistent with that of the Fourier method for structure determination, which has been used since the beginning of crystallography.

The Seven Steps in the Process

Computationally the present process has been divided into seven steps. The functions of these steps and how they were handled in our studies are presented below.

---

[23] E. Cannillo, R. Oberti, and L. Ungaretti, *Acta Crystallogr., Sect. A* **A39,** 68 (1983).
[24] N. V. Raghavan and A. Tulinsky, *Acta Crystallogr., Sect. B* **B35,** 1776 (1979).
[25] P. Argos, G. C. Ford, and M. G. Rossmann, *Acta Crystallogr., Sect. A* **A31,** 499 (1975).
[26] M. G. Rossmann and R. Henderson, *Acta Crystallogr., Sect. A* **A38,** 13 (1982).
[27] I. Rayment, *Acta Crystallogr., Sect. A* **A39,** 102 (1983).

Step 1. Prepare a data file which contains the phase probability coefficients from the SIR data[28] or from the SAS data (see the ISAS method described below). Here we assume that the heavy-atom (or anomalous scatterer) parameters have been accurately determined by conventional methods[29] prior to preparation of the file. However, if the heavy-atom parameters are not well defined, one can still start the process with the available information and use the resulting phases to help locate any additional heavy-atom sites. After all possible sites are found and refined one should repeat this step and proceed to the final run for the best results.

Step 2. Calculate a Fourier map with the available structure factor information. Here the F_{000} term is excluded from the summation as it will be incorporated later (step 4). A grid spacing of approximately one-third of the data resolution was found to be adequate.

Step 3. Locate the molecular boundary from the SIR or SAS map obtained in step 2 and use it to produce a mask which identifies both protein and solvent regions (a binary 0,1 map will do). To accomplish this we use an automated procedure which is described later.

Step 4. Filter the errors in direct space. Before removing the errors from the map, a constant density, ρ^c, is added to the entire electron density map such that the average solvent density after adding such a constant will follow the following constraint:

$$(\rho^c + \langle\rho_{sol}\rangle)/(\rho^c + \rho_{max}) = S \quad (1)$$

where $\langle\rho_{sol}\rangle$ and ρ_{max} are, respectively, the average solvent density and the maximum protein density in the Fourier map, and S is a constant.

Theoretically, if the phases are known the ratio, S can be estimated from a knowledge of the protein and solvent composition, the mean thermal factor, and the resolution of the data. However, this estimated value of S or ρ^c based on the perfect phases is not necessary the best value for providing optimal convergence when the phases are not perfect. Therefore, we chose to define S as an empirical parameter which we adjust to optimize convergence for a particular structure at a given resolution. Ideally S should be evaluated and compiled from a number of solved structures so that optimal values can be used for the determination of other unknown structures with similar conditions. At the present time only the structural data for Bence-Jones protein Rhe[30] have been used. The optimal values of S for Rhe whose crystals were grown from a solu-

[28] W. A. Hendrickson and E. E. Lattman, *Acta Crystallogr., Sect. B* **B26,** 136 (1970).

[29] T. L. Blundell and L. N. Johnson, "Protein Crystallography." Academic Press, New York, 1976.

[30] B. C. Wang, C. S. Yoo, and M. Sax, *J. Mol. Biol.* **129,** 657 (1979).

tion containing 2 *M* $(NH4)_2SO_4$ were found to be 0.06 and 0.25, respectively, for 3 and 6 Å data. These values have been used as guidelines when the process was applied to other unknown structures with similar protein and solvent composition. Interestingly these optimal *S* values are considerably smaller than those obtained when the correct value of ρ^c (F_{000}/V) is inserted into Eq. (1). The corresponding calculated values are 0.17 and 0.66, respectively, for Rhe at 3 and 6 Å resolution.

After ρ^c is evaluated from Eq. (1), the removal of the noise in the map is carried out in the following manner: (1) ρ^c is added to the protein region of the map and any negative density remaining in this region is considered an error and set to zero; and (2) the density in the solvent region is substituted with the new average density, $\rho^c + \langle\rho_{sol}\rangle$. With the removal of almost the entire noise in the solvent region and part of the noise in the protein region, the map becomes the partial structure of the protein under investigation. Note that this partial structure will *always* be a large percentage of the total scattering density. This constitutes direct space filtering and we must emphasize the fact that no modification of the protein density has occurred other than truncation of very weak or negative density.

Step 5. Calculate new structure factor amplitudes and phases by Fourier inversion.

Step 6. Filter the errors in reciprocal space. Here the treatment is equivalent to the conventional phase combination procedure[31] except that we view the physical meaning of this step differently. Bricogne's adaptation[12] of Sim's weighting scheme[32] has been used to produce the image phase probability distribution $P_c(\phi)$:

$$P_c(\phi)\alpha \exp[2|F_o||F_c|/(\langle|F_o^2 - F_c^2|\rangle) \cos(\phi - \phi_c)]$$

which we view as a phase filter. Here ϕ_c is an image phase, ϕ is a general phase angle, and F_o and F_c are the observed and calculated structure factor amplitudes, respectively. $P_c(\phi)$ is then multiplied with an incoming signal $P_o(\phi)$ (the original bimodel SIR phase probability distribution) to produce a filtered phase probability distribution, $P_{new}(\phi)$:

$$P_{new}(\phi) = P_o(\phi)P_c(\phi)$$

During our studies the filter $P_c(\phi)$ was updated in each iteration but the signal $P_\sigma(\phi)$ was always the original SIR or SAS phase probability distribution.

[31] D. M. Blow and F. H. C. Crick, *Acta Crystallogr.* **12,** 794 (1959).

[32] G. A. Sim, *Acta Crystallogr.* **12,** 813 (1959).

Although viewing the phase combination procedure as a filtering process is not essential for the successful application of the process, it emphasizes the novel approach to the problem. The emphasis here is that the correct solution in both direct and reciprocal space may be obtained simply through the removal of noise or errors. In other words, our approach may be regarded as a signal (image) enhancement process through error reduction (filtering of errors in direct and reciprocal space).

Step 7. Repeat the process. New phase angles and their figures of merit are calculated based on the corrected (filtered) phase probability distribution. Then the process is repeated from steps 2 to 7, excepting step 3, which does not have to be repeated in each iteration (see later section). Convergence is generally achieved in four to eight iterations.

Molecular Boundary from a Noisy Electron Density Map

The general practice for defining the molecular boundary for a macromolecule has traditionally involved human interpretation of an electron density map. But the high level of noise associated with a SIR or SAS map has made any reliable interpretation very difficult if not impossible. In addition, transfer of the interpreted boundary information, even if it is correct, into a computer is often tedious and time-consuming. These difficulties, however can now be overcome by an automated method[33] which will delineate the boundary between the molecule and solvent even in the presence of substantial phase error.

Since the method's technical aspects and test results are being published elsewhere[33] we describe here only the basic principles involved. The method is based on the assumption that except in certain unfavorable cases the population and strengths of signals detected for the protein region are higher than that of the solvent. Therefore it is possible in most cases to evaluate each grid point in the map in terms of its "probability" of being in the protein envelope by a transformation of density involving surrounding grid points. In this method a new map is constructed such that the density at each grid point is proportional to the weighted sum of the diffraction signals (densities above a certain background level) within a sphere of radius R from that grid point in the original map,

$$\rho_j' = K \sum_i^R w_i \rho_i \qquad \begin{cases} w_i = 1 - r_{ij}/R, & r_{ij} < R \ \text{and} \ \rho_i > 0 \\ w_i = 0, & r_{ij} > R \ \text{or} \ \rho_i < 0 \end{cases} \tag{2}$$

[33] B. C. Wang, *Acta Crystallogr.* (in press).

where r_{ij} is the distance between grid points i and j, K is an arbitrary constant, and ρ contains no ρ^0 term. Note that the "negative" densities (below-averaged signals) are omitted by giving them zero weights. The best range of summation, R, has been found to vary with resolution of the data, being 9 and 12 Å for 3 and 6 Å data, respectively. The result of this summation is a map in which the errors produced by the false phases, which are in a form of statistical variations, will effectively sum to a constant and therefore have little effect on the new map. The resulting new map has large homogeneous connected volumes of relatively high and low density. The molecular boundary is then revealed by setting a threshold density level chosen so that the volume of density (number of grid points) falling below the threshold corresponds to the volume fraction known to contain solvent.

The method has proved[33] to be both accurate and convenient. Since it can tolerate large random errors in the map the method is well suited for analyzing SIR and SAS maps in which the phase ambiguity is the major source of error.

Application to Single Isomorphous Replacement Data

Application to the Rhe Structure

Data used in this study were the same native and Au-derivative data of Bence-Jones protein Rhe collected several years ago for the determination of its structure[30] at 3 Å resolution. The space group of Rhe is $P2_12_12$ and the cell parameters are a = 54.6, b = 52.2, and c = 42.6 Å. The solvent content in the crystal is 51%.

The 3 Å SIR pairs which consisted of 2040 reflections were processed according to the procedure described in step 1. These were the same reflections used in the structure determination[30] of Rhe by the MIR method, and the heavy-atom parameters used in preparing the master data file were those determined earlier. A SIR map (64 × 60 × 48 grid points) was calculated using a fast Fourier algorithm. Then, the boundary between the molecule and solvent regions was determined by the automated procedure described in step 3 and a solvent mask (filter) was produced. The noise in the map was then filtered. The filtered map was Fourier inverted into reciprocal space and the original bimodal phase probability distributions were filtered to suppress the false solutions. This ended one iteration cycle. The progress of the process was then monitored by comparison with a set of least-squares refined phases obtained from a refinement of the Rhe structure at 1.6 Å resolution.[34] The mean phase difference, $\langle\Delta\phi\rangle = \langle|\phi_c - \phi_{ls}|\rangle$, where ϕ_c and ϕ_{ls} are the calculated and least-squares refined phases, respectively, was 50.3° for the original phases. At the end of the first iteration, $\langle\Delta\phi\rangle$ dropped to 39.9°. At this time another Fourier map using the improved phases was calculated and a new solvent mask was made. A visual comparison of the new and the original solvent mask showed that they were very similar. The process then proceeded with the original solvent mask for three more cycles. At the end of fourth cycle the mean phase difference dropped to 33.4°. Then a new solvent mask was produced using the improved phases. With this new mask (second) the process was repeated starting from the *original* SIR phases for four iterations. At the end of the fourth cycle the mean phase difference dropped to 32.8°. Then a third solvent mask was made using the further improved phases and again the process was repeated starting from the original SIR phases. With this third mask the process was iterated for eight cycles. Each mask calculation took 12 min CPU time on our DEC-10 computer. The total CPU time was less than 60 min for the entire phasing process for 3 Å Rhe data.

While the mean, $\langle\Delta\phi\rangle$, and the weighted mean, $\langle mF\ \Delta\phi\rangle$, phase differences (where m and F are the figure of merit and the structure factor amplitude, respectively) produced by the MIR method were 32.4° and 25.6°, the corresponding values obtained by the ISIR method at the end of filter 3, cycle 8 were 31.8° and 23.8°. The MIR results were produced by using two sets of isomorphous derivative and one set of anomalous scattering data. A comparison of the Fourier maps using the SIR, ISIR, MIR, and the least-squares refined phases is shown in Fig. 3.

Application to Unknown Structures

To date, over a dozen unknown structures have been processed in this laboratory by the method, most of which were sent to us from other laboratories. These structures include exfoliation ($B2_1$; our laboratory), Bence-Jones protein Pav ($P2_12_12_1$; our laboratory), Bence-Jones protein Loc ($P2_12_12_1$; Schiffer, Argonne National Laboratory), proteinase K ($P4_33_12$; Pahler/Saenger, Freie Universitat Berlin), transferrin ($P2_1$; Abola, University of Pittsburgh), troponin C ($P3_121$; Rao/Sundaralingam, University of Wisconsin), hemoprotein subunit, sulfide reductase ($P2_12_12_1$; McRee/Richardson, Duck University), glutaminase-asparaginase (I222; Wlodawer/Sjolin/Ammon/Murphy, NBS/University of Maryland), and DNA *Eco*RI endonuclease (P321; Frederick/Grable/Rosen-

[34] W. Furey, B. C. Wang, S. C. Yoo, and M. Sax, *J. Mol. Biol.* **167,** 661 (1983).

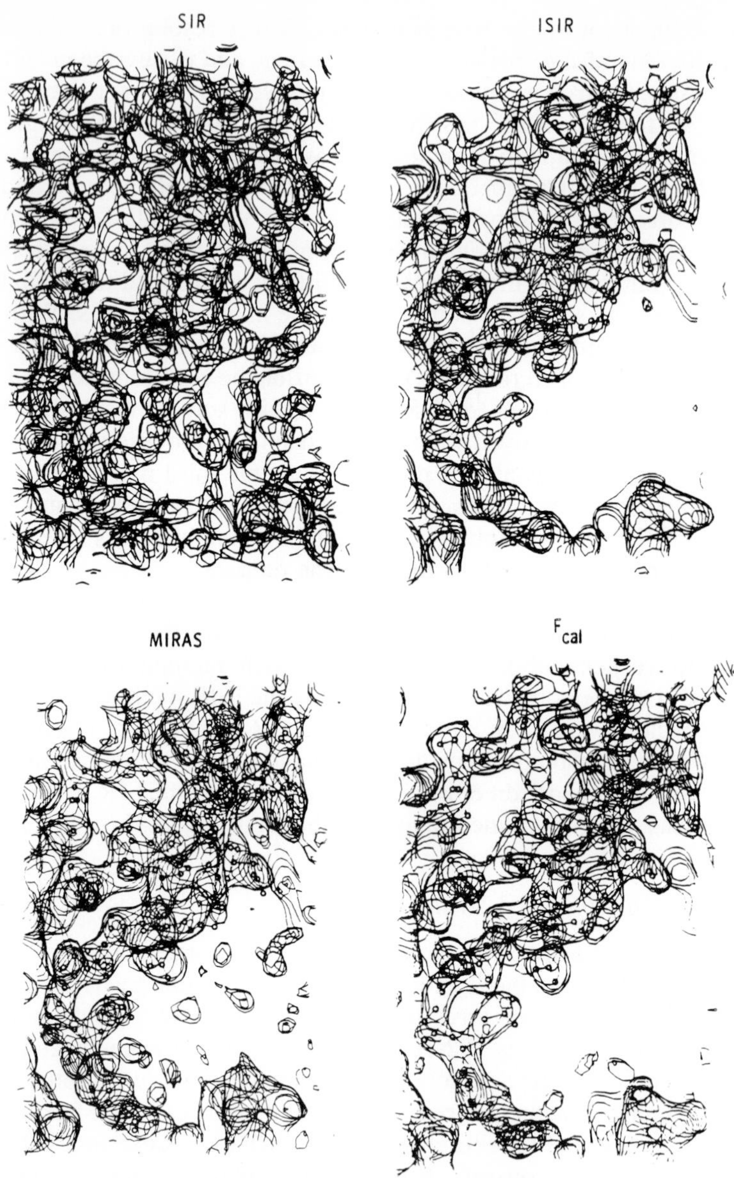

FIG. 3. A superposition of 1.6 Å refined atomic positions of Rhe with the 3 Å Fourier maps obtained from the various phasing processes. The F_{cal} map was synthesized based on the calculated F and ϕ values obtained from the 1.6 Å refined structure. The maps shown here are slices of density 12 Å thick, 21 Å wide, and 32 Å long. Each of these maps contains more than 25 residues (about one-quarter of the molecule).

berg). A report on the structure of DNA *Eco*RI endonuclease based on the application of the ISIR method has recently been published.[35]

While the detailed results of the above applications will be published in the future, a brief evaluation of the results for the two Bence-Jones proteins, Pav and Loc, is presented here. Both Pav and Loc are λ-type immunoglobulin light chains with molecular weights of approximately 55,000. Both proteins were crystallized in space group $P2_12_12_1$. While Pav has unit cell constants a = 94.6, b = 92.6, and c = 72.7 Å, Loc has a = 149.3, b = 72.4, and c = 46.5 Å. The solvent content in Pav is 64% and in Loc it is 55%.

Pav. The SIR pairs were the 4.5 Å native and Pt-derivative data. Only the major Pt site was correctly identified from the difference Patterson map at the beginning and the ISIR process started with this site. After four iterations (no change of solvent mask) the improved phases were used to produce a difference Fourier map where six possible minor sites were found. Five of the six possible sites were explainable from a reexamination of the original difference Patterson map and they converged to reasonable parameters by a least-squares refinement process. With the new information about the heavy atoms, the data file described in step 1 was recalculated and the ISIR process was repeated. During the first iteration a new solvent mask (the second) was generated and was used for the next three iterations. At the end of the fourth iteration the solvent mask (the third) was then regenerated. At this point it was believed that both the heavy-atom parameters and the solvent mask had converged and were more reliable than those used initially. With the use of the third solvent mask the process was repeated starting from the original SIR phases. The process converged in eight iterations.

With the new phases, an ISIR map of Pav was calculated and clearly showed four distinct domains of concentrated density. Each domain was then separately interpreted by a procedure similar to that of Navia *et al.*[36] using the α-carbon coordinates of another Bence-Jones protein Mcg[37] as a model. Figure 4 shows a fit of the α-carbon backbone of the two constant regions of Mcg to the ISIR density of Pav. Detailed results of the Pav structure derived from this ISIR map will be reported elsewhere.[38]

Loc. The 5 Å data of native Loc and its Pt derivative were supplied by Dr. M. Schiffer of Argonne National Laboratory. When the SIR pairs

[35] C. A. Frederick, J. Grable, M. Melia, C. Samudzi, L. Jen-Jacobson, B. C. Wang, P. Greene, H. Boyer, and J. Rosenberg, *Nature* (*London*) **309,** 327 (1984).

[36] M. A. Navia, D. M. Segal, E. A. Padlan, D. R. Davies, N. Rao, S. Rudikoff, and M. Potter, *Proc. Natl. Acad. Sci. U.S.A.* **76,** 4071 (1979).

[37] M. Schiffer, R. L. Girling, K. R. Ely, and A. B. Edmunson, *Biochemistry* **12,** 4620 (1973).

[38] J. Rose, S. Swaminathan, D. Yang, W. Furey, Jr., C. S. Yoo, B. C. Wang and M. Sax (in preparation).

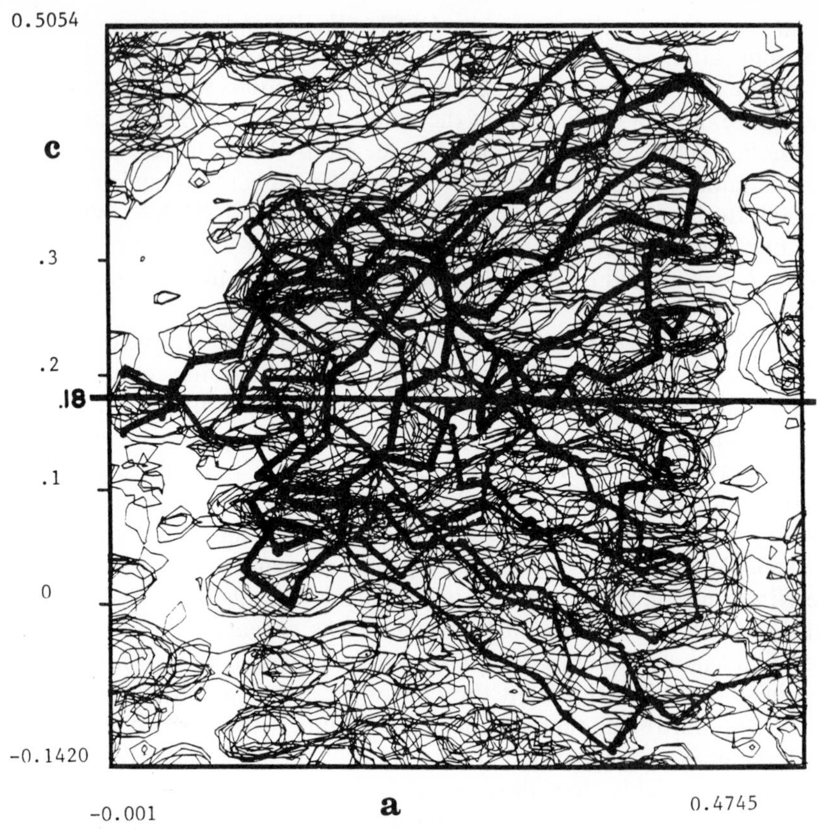

FIG. 4. α-Carbon structure of C_L dimer superimposed on the electron density projection of Pav.

were sent to us, a MIR map based on three derivatives (also 5 Å) was already calculated and a model of the molecule was built. Along with the SIR data heavy-atom parameters for one major and two minor sites were given and the ISIR process proceeded with that information. With the first solvent mask the process was iterated for four cycles. At the end of the fourth cycle a difference Fourier map was calculated and it revealed two more minor sites which were not known previously. The centric R(Cullis) dropped from 0.55 to 0.51 when the two new Pt sites were incorporated. Then the process was repeated, including recalculation of solvent masks in a similar manner to that described above for Pav. The process also converged in eight iterations for the final run.

To compare the MIR and ISIR results, electron density maps were calculated at Argonne National Laboratory using both phase sets; sections of these maps are shown in Fig. 5. The ISIR map appears less noisy

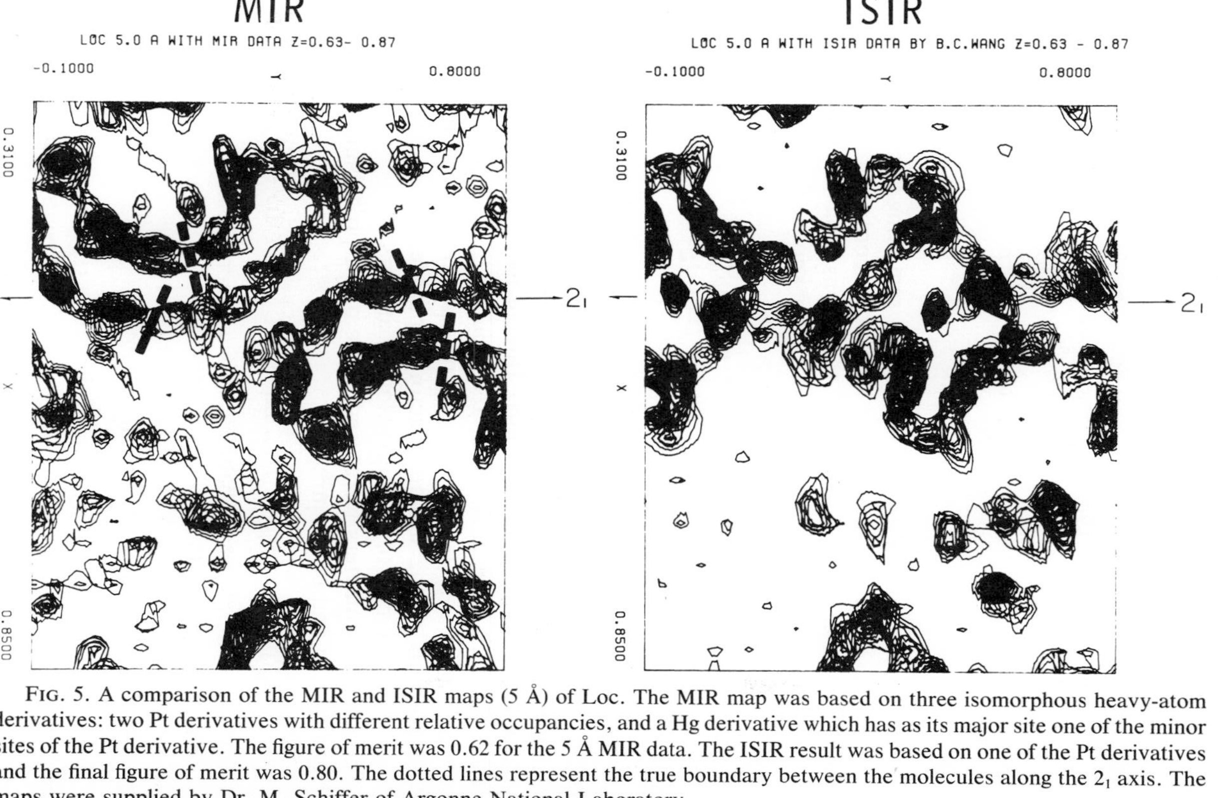

FIG. 5. A comparison of the MIR and ISIR maps (5 Å) of Loc. The MIR map was based on three isomorphous heavy-atom derivatives: two Pt derivatives with different relative occupancies, and a Hg derivative which has as its major site one of the minor sites of the Pt derivative. The figure of merit was 0.62 for the 5 Å MIR data. The ISIR result was based on one of the Pt derivatives and the final figure of merit was 0.80. The dotted lines represent the true boundary between the molecules along the 2_1 axis. The maps were supplied by Dr. M. Schiffer of Argonne National Laboratory.

in the solvent region, and in some parts of the ISIR map the boundary of the molecule is better defined than in the MIR map. For example, in Fig. 5 the true boundaries of the molecule (indicated by the dotted lines) are less clear in the MIR map than in the ISIR map.

The structure of Loc has now been determined from a 3.5 Å ISIR map and refined to a crystallographic R factor of 0.27.[38a]

Application to Single Anomalous Scattering Data

The Phase Probability Distribution

Equations for calculating the anomalous scattering phase probability distribution used in the study were derived according to the principle given by Hendrickson.[39] The explicit equations which can be conveniently used for computation to the best of our knowledge have not been previously reported and are presented below:

$$A_{\text{ano}} = 2[F_{\text{p}}(+) - F_{\text{p}}(-)]a/E^2 \tag{3}$$

$$B_{\text{ano}} = 2[F_{\text{p}}(+) - F_{\text{p}}(-)]b/E^2 \tag{4}$$

$$C_{\text{ano}} = -(a^2 - b^2)/E^2 \tag{5}$$

$$D_{\text{ano}} = -2ab/E^2 \tag{6}$$

where $a = \delta \cos \alpha$, $b = \delta \sin \alpha$, and E is obtained from a polynomial fit by least squares to approximate the root mean squares differences of the structure factor amplitudes of centric Bijvoet pairs as a function of the average F_{p} in F_{p} ranges. The variables on the right-hand sides of the equations are $F_{\text{p}}(+), F_{\text{p}}(-)$, the observed structure factor amplitudes of a Bijvoet pair; and δ, the structure factor amplitude, and its phase angle α for the imaginary part of the anomalous scatterers.

The above parameters were used in the phase probability expression of Hendrickson and Lattman[28]:

$$P(\phi) = k \exp(A_{\text{ano}} \cos \phi + B_{\text{ano}} \sin \phi + C_{\text{ano}} \cos 2\phi + D_{\text{ano}} \sin 2\phi)$$

where $P(\phi)$ is the phase probability, ϕ is a general phase angle, and k is the normalization constant.

Application to a Derivative Crystal of Rhe

The anomalous scattering data of the Au derivative of Rhe had been used in this study. The data were those collected several years ago and no

[38a] C. H. Chang, M. T. Short, F. A. Westholm, F. J. Stevens, B. C. Wang, W. Furey, Jr., A. Soloman and M. Schiffer. *Biochemistry* (1985) (in press).

[39] W. A. Hendrickson, *Acta Crystallogr., Sect. A* **A35**, 245 (1979).

special treatment such as anisotropic local scaling (or even measurement at $\pm 2\theta$) had been used. The master data file was prepared using the anomalous scatterer (Au) parameters determined earlier[30] and the equations described above. The average figure of merit for the acentric reflections was 0.38. The centric reflections were assigned an arbitrary phase value (zero) and figure of merit of zero. This was step 1 of the ISAS method and it is the only step which differs from that of the ISIR method. Then the ISAS process proceeded in a similar manner as for the ISIR process except that map-inverted phases were assigned to the centric reflections. The solvent mask was also updated every four cycles and the third mask was used for the final eight iterations (again restarting from the original SAS phases). The final average figure of merit was 0.77. A comparison of the SAS and ISAS maps for the same region shown in Fig. 3 and an illustration of two smaller regions of the ISAS map are given in Fig. 6. Judging from the ISAS maps it is certain that the Rhe structure could have been determined using only the Au-derivative data alone. It is noteworthy that the absolute occupancy of the major site of Au is only 48% and the molecular weight of Rhe in an asymmetric unit is 12,500.

Since the diffraction data of the native crystal were not included in the analysis, obviously it did not matter whether or not the Au derivative was isomorphous to the Rhe native crystal. Therefore the method should be applicable to any good diffracting crystal containing a heavy atom with a measurable anomalous scattering signal.

Phasing Power of the ISAS Method

A set of error-free data based on the anomalous scattering of the two sulfur atoms (the only disulfide) in the Rhe native crystal was calculated for Cu radiation as the test data. The diffraction ratios based on calculated F values (not from Wilson's statistics) are

$$R_1 = \frac{\langle F_p(+) - F_p(-) \rangle}{\langle F_p \rangle} = 0.6\% \qquad \text{and} \qquad R_2 = \frac{\langle F_{\text{sulfur}} \rangle}{\langle F_p \rangle} = 6.7\%$$

Note that these ratios are considerably smaller then the corresponding values for the structure determination of crambin[16] ($R_1 = 1.4\%$ and $R_2 = 29\%$). We estimate based on the statistics given by Parthasarathy[15] that at this low ratio of R_2 the reliability for correctly resolving the phase ambiguity would be less than 60% if the quasi-anomalous dispersion[7] or the resolved-anomalous scattering[16,17] method was used. In other words more than 40% of the reflections would be incorrectly resolved by those methods even when the data are error free. This is because those methods depend on phases of the sulfur atoms to approximate phases of the pro-

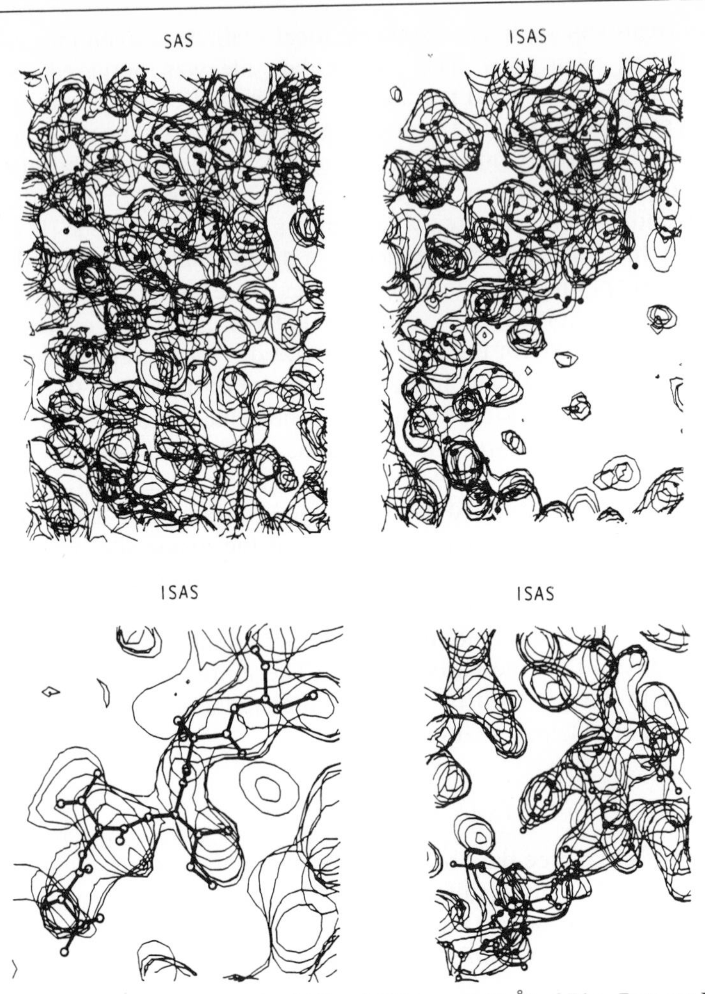

FIG. 6. Top: A comparison of the SAS and ISAS maps (3 Å) of Rhe. Bottom: Further illustrations showing the quality of the ISAS maps (3 Å) by comparing it with the refined atomic positions for residues 18 to 22 (left) and 81 to 89 (right) of Rhe. The mean percentage differences, $\langle \Delta F/F \rangle$, among the Bijvoet pairs of the SAS data were 3.2% (noise) and 4.8% (noise + signal) for the 266 available centric and 1721 acentric reflections, respectively.

tein. Therefore it would be interesting to see how well the ISAS method can do, since its resolving power depends on a partial structure consisting of the entire protein region, and hence, the applicability of this method should not depend on the relative size of the anomalous scatterers, provided that the anomalous signal can be measured. Another reason for carrying out the test is that although the expected signal ($R_1 = 0.6\%$) is small at the moment, this signal can change if a different X-ray source is used. For example, if the data are collected with Cr radiation the ratio will be 1.2% which should be measurable in view of the experience in crambin (R_1 = 1.4%). So the test case here is not all that unrealistic for a protein structure determination if the method indeed works.

The master data file was prepared using a procedure similar to that for the Au-derivative data of Rhe, except that the E values [Eqs. (3)–(6)] were calculated as one-fourteenth (1/14) of those used in the Au-derivative run. This choice of E values was to make the average figure of merit (0.38) of the error-free SAS data fall into a range expected for an experimental SAS set. The 3 Å SAS data consisted of 2044 acentric and 635 centric reflections and all were used in the ISAS process. During the process the map-inverted phases were assigned to each of the centric reflections, and the solvent mask was updated every four iterations. Whenever the solvent mask was updated, the ISAS process was restarted from the beginning except for the third (final) run, which ended after eight cycles. In Fig. 7 the sequence of recovery of the original structure from the SAS data during the process is shown. It is interesting to note that the map improved more when the filter was changed than after more iterations. While a nearly complete recovery of the original structure (compare with the F_{cal} map) was achieved at the end of the final run (filter 3, cycle 8), the majority of the recovery occurred after filter 1, cycle 1, i.e., the first cycle.

Concluding Remarks

The test results shown here indicate that the applicability of the ISAS method indeed is not hindered by the relatively small size of the anomalous scatterers, provided the data are accurate. The ability to recover the majority of the total structure in just one cycle justifies calling the process a solution to the phase ambiguity problem and not merely a phase refinement or improvement process. The fact that the map improved more when the filter was changed than after more iterations indicates that the automated process[33] which objectively defined the molecular boundary was the most crucial step in the procedure. The phasing power illustrated above for the ISAS method should exist in the ISIR process also, since both share the same principle of recovering the phase information from the diffraction data.

It should be emphasized however, that the method's strength and its main purpose are to make phase selections in order to resolve phase ambiguity. If neither of the phase choices are valid then obviously the method cannot provide reasonable phase. Hence, the method cannot overcome systematic errors in the data due to nonisomorphism or serious errors from other sources. Therefore, prior to the application of this pro-

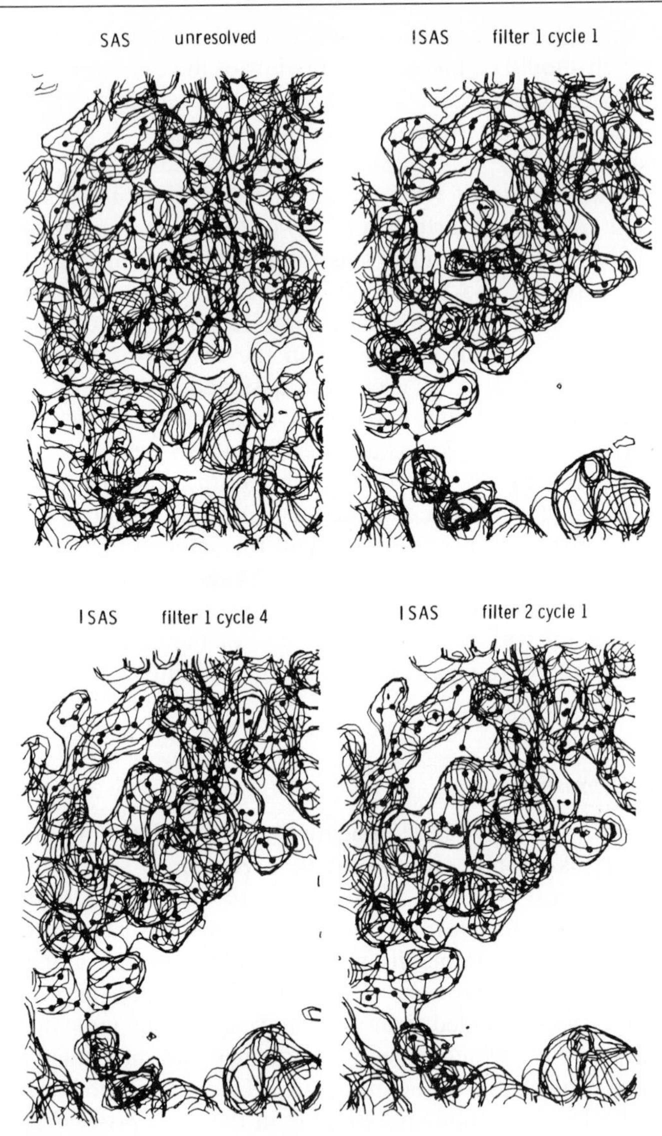

FIG. 7. The sequence of recovery of the original structure for the test data (see text) during the ISAS process. Note that the map improved more when the filter was changed than after more iterations and the majority of the recovery occurred at the end of the first cycle (filter 1, cycle 1).

cess one should make every effort to obtain a set of SIR or SAS data as reliable as possible. Also, the position of the heavy atoms or anomalous scatterers should be determined to a reasonable accuracy and they should not be related to each other by an exact center of symmetry.

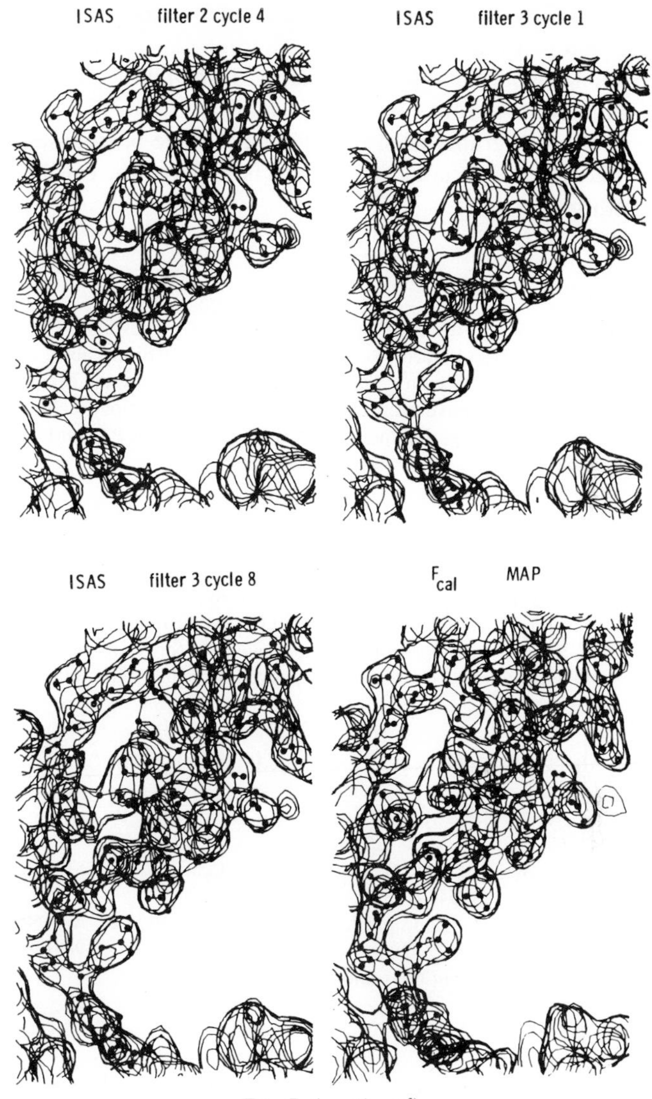

FIG. 7. (*continued*)

The method presented here has demonstrated a remarkable effectiveness for overcoming the phase ambiguity problem which has been the major obstacle in the use of SIR or SAS data alone for macromolecular studies. In view of the above finding and the test results on known and unknown structures, we see that the best strategy for a macromolecular structure determination in the future is not to aim for more isomorphous derivatives but to concentrate on one "good" derivative and to improve the quality of its SIR or SAS data, because, as long as one can accurately measure the data, a macromolecular structure can be determined from SIR or SAS data even when the occupancy of the heavy atom or the size of the anomalous scatterer is relatively small.

The computer programs used in our studies have been assembled into a program package. This package, which is in Fortran, can be obtained from the author.

Acknowledgments

I thank William Furey, Jr. for his Fourier programs and suggestions, Chung Soo Yoo (deceased, August 31, 1983) and Martin Sax for their interest and discussions, and Marianne Schiffer for her electron density maps. This work was supported by the Medical Research Service of the Veterans Administration and by National Institute of Health Grants AM-18827 and AM-25447. The computing facilities at the University of Pittsburgh's computing center are also acknowledged.

NOTE ADDED IN PROOF: The ISAS method has been used recently to solve the crystal structure of Cd,Zn metallothionein,[40] using only the anomalous scattering signal of the Cd atoms in the native data. Recent publications resulted from the use of ISIR–ISAS programs in the author's laboratory, including that of troponin C[41] and histone octamer[42] by combined SIR and SAS data. Recently an approach in reciprocal space for resolving phase ambiguity has been favorably demonstrated by using an improved probability formula in direct methods.[43–45]

[40] W. F. Furey, Jr., A. H. Robbins, L. L. Clancy, D. R. Winge, B. C. Wang, and C. D. Stout (in preparation).

[41] M. Sundaralingam, R. Bergstrom, S. Strasburg, S. T. Rao, P. Roychowdhury, M. Greaser, and B. C. Wang, *Science* **227,** 945 (1985).

[42] R. W. Burlingame, W. E. Love, B. C. Wang, R. Hamlin, N.-H. Xuong, and E. N. Moudrianakis, *Science* **228,** 546 (1985).

[43] H.-F. Fan, F.-S. Han, J.-Z. Qian, and J.-X. Yao, *Acta Crystallogr.* **A40,** 489 (1984).

[44] H.-F. Fan, F.-S. Han, and J.-Z. Qian, *Acta Crystallogr.* **A40,** 495 (1984).

[45] J.-X. Yao and H.-F. Fan, *Acta Crystallogr.* **A41,** 284 (1985).

[8] Free Atom Insertion and Refinement as a Means of Extending and Refining Phases

By N. W. ISAACS and R. C. AGARWAL

The fast Fourier transform (FFT) method of structure factor least-squares refinement developed by Agarwal[1] provides a fast, accurate, and economical method of protein structure refinement. Calculations on test structures show that with a suitable weighting scheme the method has a radius of convergence of about 0.75 Å.[1] In some recent refinements of proteins using this method the average shift of main-chain atoms during refinement was 0.5 Å and that of side chain atoms was 0.6 Å.[2,3]

Although a low-resolution protein electron density map does not provide sufficient detail to enable the positioning of atoms, it does restrict the volume within which an atom may be placed. Given the polymeric nature of a protein structure, it is expected that most points within the more concentrated density of a low-resolution map will be less than 0.77 Å (half of a C–C bond distance) from a true atomic position. Least-squares structure refinement should then move an atom placed at any point in the stronger density toward a true position, resulting in an improvement in the calculated phases.

The free atom refinement method[4] consists of three steps: (1) free atom insertion into a low-resolution map phased by multiple isomorphous replacement (or other) methods; (2) least-squares refinement of these atomic positions and temperature factors; and (3) calculation of new phases from the refined positions to obtain an improved electron density map at higher resolution. This cycle of operations may be repeated, with the new map as a starting point, until the best electron density map is obtained.

Free Atom Insertion

The free atoms may be placed automatically in the density by a computer program taking care that (1) the atoms are placed in more concentrated regions of density, (2) no atom has more than three close neighbors, and (3) the atoms placed in each region should have a density which approximates to the density of the map. At low resolution it is incorrect to place atoms on the peaks of highest density since such peaks may occur between atomic sites. A preferable procedure to follow in placing the free atoms may be to define upper and lower limits of density and to place atoms in densities within those limits rather than at positions of maximum density. By varying the upper and lower bounds, atoms may be placed in the higher density regions initially and then placed in lower density in subsequent passes.

No atom should have more than three neighbors closer than a prede-

[1] R. C. Agarwal, *Acta Crystallogr., Sect. A* **A34,** 791 (1978).
[2] N. W. Isaacs and R. C. Agarwal, *Acta Crystallogr., Sect. A* **A34,** 782 (1978).
[3] E. N. Baker and E. J. Dodson, *Acta Crystallogr., Sect. A* **A36,** 559 (1980).
[4] R. C. Agarwal and N. W. Isaacs, *Proc. Natl. Acad. Sci. U.S.A.* **74,** 2835 (1977).

termined distance, of ~1.2 Å for larger peaks and 1.4 Å for smaller peaks. For the purposes of the refinement all the atoms should be treated as nitrogen and be assigned isotropic temperature factors which reflect the level of the electron density at that position. A table of the value of the electron density of a nitrogen atom with a range of B values may be precomputed for use in the assignment of temperature factors.

Refinement of the Free Atoms

The positions and temperature factors of the free atoms are refined using a least-squares procedure, initially with low-resolution structure factor data (e.g., 2.25 Å) and then increasing the data to 2 Å or higher resolution. At the beginning of the refinement, higher weights should be given to low-resolution data. This is best achieved with a weighting scheme of the type[1] $w_{hkl} = [(2 \sin \theta)/\lambda]^{-n}$, where $n = 1.5$ initially and is gradually relaxed in later cycles of refinement to give unit weights. The coordinates should be refined before the temperature factors. Although the normal R factor[5] will decrease rapidly, there is a danger that overrefinement could lead to a concentration of the electron density at a few atom sites. If the newly phased electron density map appears to be oversharpened, a repeat of the least-squares refinement with fewer cycles may give a better result. It is preferable to cycle a few times through the procedure of atom insertion, refinement, and map calculation, rather than to refine a single set of free atoms to obtain the lowest R factor.

Calculation of the Phase-Refined Electron Density Map

The results of tests with insulin data[4] shown in the table indicate that the phases calculated after free atom refinement are more accurate for the larger $|F_{obs}|$ data. This should be taken into account when the new electron density map is calculated with the free-atom-refined phases alone or with these phases recombined with isomorphous phases, using procedures similar to those discussed by Bricogne.[6] Even without such phase recombination or weighting, the quality of the phase-extended maps may be sufficient to allow for an interpretation. The 2 Å phase-extended map of insulin shown in Fig. 1 was computed with phases calculated after free atom refinement and without any weighting of the observed structure factors.[4]

[5] The crystallographic R factor is defined as $R = (\Sigma||F_{obs}| - |F_{calc}||)/\Sigma|F_{obs}|$.

[6] G. Bricogne, *Acta Crystallogr., Sect. A* **A32,** 832 (1976).

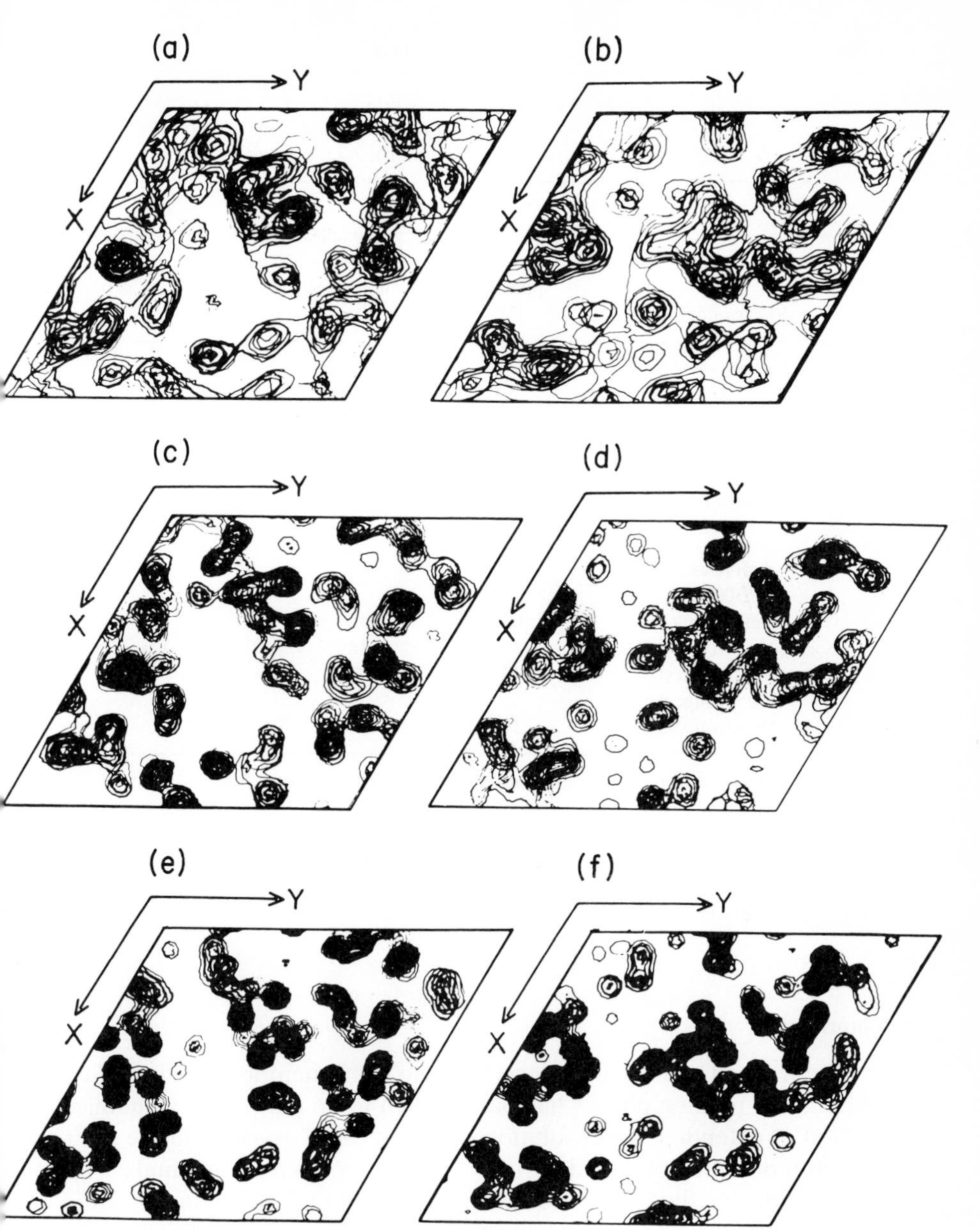

FIG. 1. Composite electron density maps, computed with different phase sets, of 2-zinc insulin. (a) and (b) are sections of a 3 Å resolution map phased by isomorphous replacement methods. (c) and (d) are sections of a 2 Å resolution map phased from the 3 Å map by a free atom insertion and refinement. For comparison, (e) and (f) show sections of a 1.5 Å map with phases calculated from the refined structure. The maps in the left column show sections $Z = -1.4$ to -0.5 Å and those in the right column show sections $Z = 0.9$ to 2.8 Å. (Reproduced by permission of the publishers from Agarwal and Isaacs.[4])

AVERAGE PHASE ERROR ($|\overline{\Delta\phi}|$) BETWEEN PHASES CALCULATED FROM THE LEAST-SQUARES-REFINED INSULIN STRUCTURE AND VARIOUS OTHER PHASE SETS CALCULATED FOR 2-ZINC INSULIN[a]

Phases compared	1.9 Å MIR phases (°)	1.5 Å phases, Sayre's method[b,c] (°)	1.5 Å phases, free atom insertion and refinement (°)
1.5 Å phases			
All 13,400 reflections			70
All 10,000 reflections phased by Sayre method[c]		55	68
1.9 Å phases			
All 6300 reflections	60	52	65
2000 strongest $\|F_{obs}\|$	48	38	47
1000 strongest	43	33	39
500 strongest	37	32	32
250 strongest	34	30	27

[a] Reproduced by permission of the publishers from Agarwal and Isaacs.[4]

[b] D. Sayre, *Acta Crystallogr., Sect. A* **A28,** 210 (1972).

[c] J. F. Cutfield, E. J. Dodson, G. G. Dodson, D. C. Hodgkin, N. W. Isaacs, K. Sakabe, and N. Sakabe, *Acta Crystallogr., Sect. A* **A31,** S21 (1975).

Comments

The method of free atom insertion and refinement as a means of extending and refining phases has been shown to be a fast and economical method of extending the resolution of a map[4] or of assisting in the interpretation of a map obtained by isomorphous replacement methods.[7] The limits of the method have not been fully explored either in the initial placement of the free atoms or in the combination of the refined phases with isomorphous phases. Modifications to the method could include (1) the use of a molecular envelope to exclude solvent regions from the positioning of free atoms, and (2) the use of multiple starting sets of free

[7] P. R. Evans and P. J. Hudson, *Nature (London)* **279,** 500 (1979).

atoms from which a common set of sites is obtained after refinement. Such sites should provide a closer approximation to real atom positions.

Acknowledgments

N. W. I. is a Senior Research Fellow of the National Health and Medical Research Council of Australia.

[9] On the Derivation of Dynamic Information from Diffraction Data

By D. I. STUART and D. C. PHILLIPS

Recently it has become clear that single-crystal diffraction methods, once widely thought to be incapable of supplying dynamic information about proteins, are in fact a most powerful tool in such studies. That crystallography can supply quantitative results of some precision concerning intramolecular motion has been convincingly demonstrated for small molecules.[1] While the analysis of such results for macromolecules is largely at a more qualitative stage we look forward to rapid developments in the field.

For reasons that are discussed below X-ray diffraction can detect motion over the complete time scale of atomic, molecular, and lattice vibrations as well as disorder. Information on the actual time scale cannot, however, be extracted from the Bragg diffraction and so the technique is complementary to the temporally sensitive but spatially almost blind spectroscopic techniques. As a consequence of this limitation it will be impossible to detect certain rare but perhaps interesting phenomena (such as the rotational flipping of aromatic side groups) in our electron density maps. Although at present the refinement of protein molecules with an anisotropic thermal model is unusual, such refinement of heavy-atom parameters is common and therefore we discuss the significance of the anisotropic thermal parameters in some detail.

Statistical Theory of Thermal Motion

Ideally the thermal motion of atoms in crystals should be described in terms of the normal modes of vibration of the entire crystal, but this

[1] K. N. Trueblood and J. D. Dunitz, *Acta Crystallogr., Sect. B.* **B39,** 120 (1983).

formally correct and comprehensive approach cannot usually be followed in crystal structure analysis because it is too cumbersome and too difficult to apply to any but the simplest systems.[2] Instead, the enormously simplified statistical model is used, as described elsewhere in this volume, in which the atoms in the structure are assumed to move independently, that is, with no coupling between them. It is interesting to note that this approximation does not impair our model for the intensities of the Bragg reflections although it prevents any sensible analysis of the non-Bragg, thermal diffuse scattering.[3] Furthermore the motion of the individual atoms is nearly always assumed to be harmonic and, very often, isotropic, largely because the description of more complex motion requires the use of more parameters than can be derived reliably from the available diffraction data.

Let the instantaneous displacement of an atom from its "rest" position be given by the vector $\mathbf{u}$. We may write the projection of this displacement onto a particular reciprocal lattice vector $\mathbf{d}^*$ as $\mathbf{u} \cdot \mathbf{d}^*$. The thermal vibrations of the atoms, which have a frequency of about 10^{13} $\sec^{-1}$, are slow compared to the X-ray frequency, which is about 10^{18} $\sec^{-1}$. Consequently, to the X rays the atoms appear to be stationary and so we will build up during our diffraction experiment a time average of many instantaneous "frozen" states. Furthermore we will also take a spatial average over the different unit cells of the crystal. Clearly these two effects are indistinguishable. In what follows we often use the term *motion* to refer to this atomic smearing.

If we assume that the atoms are independent then we may perform the averaging over the atoms separately, which will give us the probability density distribution for the atom.

The Anisotropic Case

If we assume harmonic motion, that is, the restoring force on an atom is proportional to its distance from its rest position, the probability density distribution for each atom will be ellipsoidal. This is referred to as the "thermal ellipsoid" for the atom. This ellipse is described by the mean square displacements, $\langle u_1^2 \rangle$, $\langle u_2^2 \rangle$, $\langle u_3^2 \rangle$, along the orthogonal principal axes $\mathbf{X}_1$, $\mathbf{X}_2$, $\mathbf{X}_3$. These axes are arbitrarily positioned with respect to the crystal axes $\mathbf{x}_1^c$, $\mathbf{x}_2^c$, $\mathbf{x}_3^c$.

The effect in reciprocal space is to attenuate the scattering from the atom by a factor corresponding to the Fourier transform of the probability

[2] B. T. M. Willis and A. W. Pryor, "Thermal Vibrations in X-ray Crystallography." Cambridge Univ. Press, London and New York, 1975.

[3] C. Scheringer, *Acta Crystallogr., Sec. A* **A28,** 512 (1972).

distribution, namely the temperature factor, which is given by[4]

$$T^{hkl} = \exp[-2\pi^2\langle(\mathbf{u}^c \cdot \mathbf{d}^*)^2\rangle] \tag{1}$$

This may be written as

$$T^{hkl} = \exp[-2\pi^2\mathbf{d}^{*T}\langle\mathbf{u}^c\mathbf{u}^{cT}\rangle\mathbf{d}^*] \tag{2}$$

or

$$T^{hkl} = \exp[-2\pi^2\mathbf{d}^{*T}\mathbf{U}^c\mathbf{d}^*] \tag{3}$$

where $\mathbf{U}^c$ is the mean square displacement tensor, and is symmetric:

$$\mathbf{U}^c = \begin{pmatrix} \langle u_1^c u_1^c\rangle & \langle u_1^c u_2^c\rangle & \langle u_1^c u_3^c\rangle \\ \langle u_2^c u_1^c\rangle & \langle u_2^c u_2^c\rangle & \langle u_2^c u_3^c\rangle \\ \langle u_3^c u_1^c\rangle & \langle u_3^c u_2^c\rangle & \langle u_3^c u_3^c\rangle \end{pmatrix} \tag{4}$$

Equation (3) may be expanded to give

$$\begin{aligned} T^{hkl} = \exp\{-2\pi^2[&U_{11}^c(ha^*)^2 + U_{12}^c(ha^*)(kb^*) \\ &+ U_{13}^c(ha^*)(lc^*) + \cdots]\} \end{aligned} \tag{5}$$

If we make the substitutions

$$\begin{aligned} U_{11}^c &= \beta_{11}/2\pi^2\mathbf{a}^{*2} \\ U_{12}^c &= U_{21} = \beta_{12}/2\pi^2\mathbf{a}^*\mathbf{b}^* \\ &\vdots \end{aligned}$$

Eq. (5) becomes the standard anisotropic temperature factor expression:

$$T^{hkl} = \exp[-(\beta_{11}h^2 + \beta_{22}k^2 + \beta_{33}l^2 + 2\beta_{12}hk + 2\beta_{23}kl + 2\beta_{31}lh)] \tag{6}$$

This formulation is to be preferred to an alternative expression used in some analyses:

$$T^{hkl} = \exp[-(\beta_{11}h^2 + \beta_{22}k^2 + \beta_{33}l^2 + \beta_{12}hk + \beta_{23}kl + \beta_{31}lh)] \tag{7}$$

The latter formulation obscures the relationship with the convenient matrix notation.

The Isotropic Case

If the mean square displacements in all directions are the same, $\langle u^2\rangle$, then

$$\langle u^2\rangle = \langle u_1^2\rangle = \langle u_2^2\rangle = \langle u_3^2\rangle \tag{8}$$

[4] R. W. James, "The Optical Principles of the Diffraction of X-rays." Bell, London, 1962.

In this case $\langle(\mathbf{u} \cdot \mathbf{d}^*)^2\rangle$ reduces to $\langle u^2\rangle d^{*2}$ and we write the temperature factor as

$$T^{hkl} = \exp(-2\pi^2\langle u^2\rangle d^{*2}) \tag{9}$$

The usual form of the isotropic temperature factor is

$$T^{hkl} = \exp(-\tfrac{1}{4}Bd^{*2}) \tag{10}$$

i.e., $B = 8\pi^2\langle u^2\rangle$.

There has been some confusion over the meaning of $\langle u^2\rangle$. Note that it is *not* the mean square displacement of the atom from its rest position which is given instead by

$$\langle u_{\text{Tot}}^2\rangle = \langle u_1^2\rangle + \langle u_2^2\rangle + \langle u_3^2\rangle$$
$$\langle u_{\text{Tot}}^2\rangle = 3\langle u^2\rangle$$

Thus the total mean square displacement of an atom is given by

$$\langle u_{\text{Tot}}^2\rangle = 3B/8\pi^2 \tag{11}$$

It is sometimes useful to define an "equivalent isotropic" temperature factor:

$$\langle u^2\rangle = \tfrac{1}{3}[(U_{11}^{\circ})^2 + (U_{22}^{\circ})^2 + (U_{33}^{\circ})^2] \tag{12}$$

where the relationship between $\mathbf{U}^{\circ}$ and $\mathbf{U}^{c}$ is as defined below.

Determination of the Directions and Magnitudes of the Principal Axes of Vibration

If the matrix $\mathbf{U}^c$ is transformed to the orthogonal axial system $\mathbf{X}_1$, $\mathbf{X}_2$, $\mathbf{X}_3$, described above, then it will reduce to the diagonal form with $\langle u_1^2\rangle$, $\langle u_2^2\rangle$, $\langle u_3^2\rangle$, as the diagonal elements. This may be conveniently achieved in two stages. The first stage is needed only when the crystallographic axes are nonorthogonal.[5,6]

We define the matrix $\mathbf{A}$:

$$\begin{pmatrix} \sin\gamma & 0 & (\cos\beta - \cos\alpha\cos\gamma)/\sin\gamma \\ \cos\gamma & 1 & \cos\alpha \\ 0 & 0 & \{\sin^2\alpha - [(\cos\beta - \cos\alpha\cos\gamma)/\sin\gamma]^2\}^{1/2} \end{pmatrix}$$

[5] H. Lipson and W. Cochran, "The Determination of Crystal Structures." Bell, London, 1968.

[6] D. W. J. Cruickshank, D. E. Pilling, A. Bujosa, F. M. Lovell, and M. R. Truter, *in* "Computing Methods and the Phase Problem in X-Ray Crystal Analysis" (R. Pepinsky, J. M. Robertson, and J. C. Speakman, eds.), p. 32. Pergammon, Oxford, 1961.

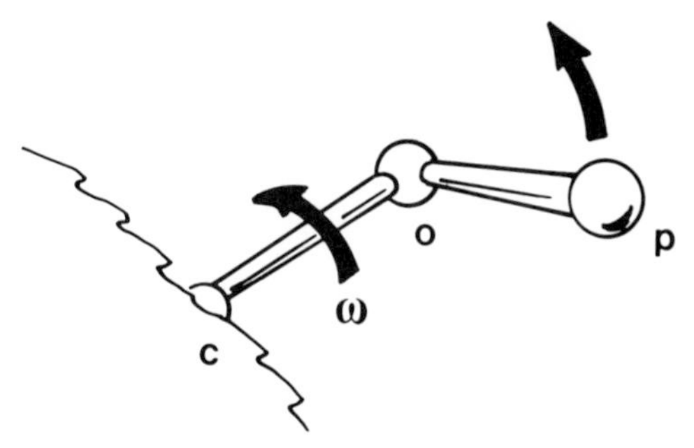

FIG. 1. Librational motion.

which may be used to convert crystallographic coordinates to a standard orthogonal axial system, defined with $\mathbf{X}_2^{\circ}$ coincident with $\mathbf{X}_2^{c}$, $\mathbf{X}_1^{\circ}$ the projection of $\mathbf{X}_1^{c}$ onto the plane perpendicular to $\mathbf{X}_2^{c}$, and $\mathbf{X}_3^{\circ}$ perpendicular to both $\mathbf{X}_1^{\circ}$ and $\mathbf{X}_2^{\circ}$.

Further we define the diagonal matrix $\mathbf{D}$:

$$\begin{pmatrix} \sin\alpha/N & 0 & 0 \\ 0 & \sin\beta/N & 0 \\ 0 & 0 & \sin\gamma/N \end{pmatrix}$$

where $N^2 = 1 + 2\cos\alpha\cos\beta\cos\gamma - \cos^2\alpha - \cos^2\beta - \cos^2\gamma$. $\mathbf{U}^{c}$ may now be converted to the orthogonal axial system:

$$\mathbf{U}^{\circ} = (\mathbf{AD})\mathbf{U}^{c}(\mathbf{AD})^{\mathrm{T}}$$

$\langle u_1^2\rangle$, $\langle u_2^2\rangle$, $\langle u_3^2\rangle$ are now the eigenvalues of $\mathbf{U}^{\circ}$ and the principal axes of the ellipse, $\mathbf{X}_1$, $\mathbf{X}_2$, $\mathbf{X}_3$, are the eigenvectors.

Anharmonic Motion

The anisotropic harmonic model of motion has in general been found surprisingly adequate. In practice, however, the atomic motion is known to be more complex than this even in small molecules and it seems certain that anharmonic motions must occur in protein molecules, where the atoms and groups of atoms have markedly asymmetric environments. Intramolecular libration (rotational vibration) of an atom or group of atoms about a covalent bond is a simple illustration of this (Fig. 1).

Clearly in this case the probability density function (and hence its Fourier transform) becomes much more complex. Johnson[7] has shown that the addition of 10 third-order terms to the 6 second-order terms

[7] C. K. Johnson, *Acta Crystallogr., Sect. A* **A25,** 187 (1969).

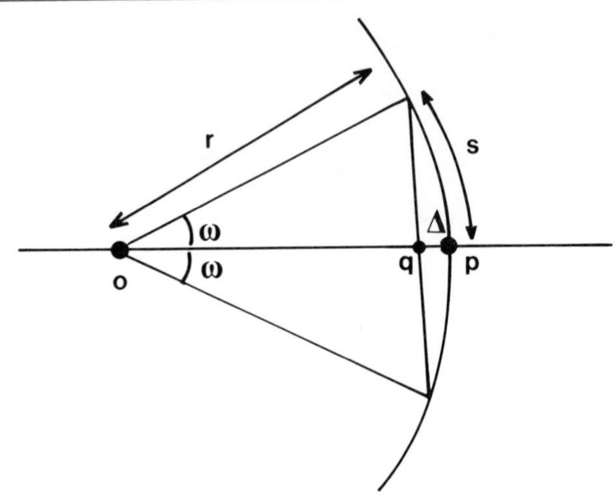

FIG. 2. Construction illustrating the shortening of apparent bond lengths (Δ) caused by librational motion.

described above allows a good description of such curved distributions. Unfortunately there is little hope of refining 10 further parameters per atom for most proteins!

Atomic motion about an arc also has the effect of reducing the apparent bond length when it is modeled by an ellipse. Figure 2 illustrates this. If the mean square displacement of the atom p from its rest position is s^2, the center of gravity will be approximately at q, and then

$$\Delta = r(1 - \cos \omega)$$

If $r = 1.5$ Å and $\omega = 20°$ (such a value would not be uncommon even for a small-molecule study) the apparent bond shortening using the harmonic model will be some 0.09 Å. Cruickshank[8,9] has expressed this effect in terms of a generalized matrix formulation.

Apparent Thermal Motion in Crystals

Thermal Vibrations

In general two types of thermal vibrations in molecular crystals can be distinguished, the external modes which involve the concerted move-

[8] D. W. J. Cruickshank, *Acta Crystallogr.* **9,** 747 (1956).
[9] D. W. J. Cruickshank, *Acta Crystallogr.* **14,** 896 (1961).

ments of neighboring molecules and the internal modes which involve intramolecular vibrations.[10] In relatively small molecules the atoms are connected together mainly by covalent bonds, whereas the molecules are linked in the crystal lattice by relatively weak interactions, e.g., van der Waals forces and hydrogen bonds. Consequently the thermal vibrations in such crystals correspond mainly to motion of the molecules as rigid bodies and intramolecular vibrations are of relatively little importance. As Scheringer[10] has shown, however, the intramolecular modes make a larger contribution as the number of atoms in the molecules increases. His analysis (Eq. 19 in Ref. 10) gives, for example, an estimate of the mean square amplitude of intramolecular vibration of carbon atoms in the lysozyme crystal as 0.2 $Å^2$, which is similar to the observed values. Further work is needed to establish that Scheringer's analysis, applicable to small molecules such as anthracene, can be extended to proteins whose conformations largely depend on weak interactions of the kind that also hold together the molecules in the crystal. Furthermore the special nature of protein crystals means that the protein–protein contacts in the crystal are usually very limited, with the molecules being largely suspended in their liquor of crystallization. There is now considerable other evidence, both experimental and theoretical, of intramolecular motion in proteins (see Petsko and Ringe[11] and references therein).

Static Disorder and Conformational Variability

A fundamental difficulty is that apparent thermal parameters obtained by conventional X-ray analysis also include the effects of static disorder within the crystals. As we have discussed above, indistinguishable displacements may be produced by thermal motion and by static disorder (sometimes known as frozen heat motion) if corresponding atoms in the different unit cells throughout the crystal take up slightly different mean positions. In a protein crystal such disordering may arise because corresponding molecules in different unit cells adopt different positions or orientations (due, e.g., to the inclusion of impurities), but it may also arise because individual molecules take up different conformations. Provided that the resultant atomic displacements are less than the nominal resolution achieved in the electron density map, their effect in a study at one temperature will be distinguishable from the effects of true thermal motion.[12] In practice, however, matters are more complicated. It has been

[10] C. Scheringer, *Acta Crystallogr., Sect. A* **A28,** 516 (1972).
[11] G. A. Petsko and D. Ringe, *Annu. Rev. Biophys.* **13** (in press).
[12] J. D. Dunitz, *Ciba Found. Symp.* **60,** new ser., 181 (1978).

well known since the earliest studies on myoglobin[13] that individual side chains may take up radically different conformations in different molecules so that separate images of them can be seen in the electron density map (this effect is indistinguishable from a markedly anharmonic motion of these side chains). Conformational variability, which in some structures extends to more extensive regions than individual side chains,[14,15] cannot be modeled adequately by allocating to each atom a single position and an apparent amplitude of vibration. It is necessary rather to refine the atomic positions, apparent amplitudes of vibration, and occupancies corresponding to each distinguishable conformation in the way that has now been achieved for some side chains in a small number of high-resolution refinements.[16,17] Clearly the identification of each member of a range of alternative conformations is very difficult if the atomic positions overlap, and the complete description of a molecule in which the atoms move within and between complex potential wells presents a considerable challenge. Present experience suggests that the limit to the number of substantially occupied alternative conformations that can be detected in a Fourier map is two or three.

Effect of Experimental Errors and Experimental Limitations

Consideration of relatively simple structures has amply demonstrated that errors in the observed structure amplitudes have a more serious effect upon derived thermal parameters than upon atomic positions. Systematic errors, such as those arising from absorption of X rays within the crystal or its mounting, have particularly serious effects which are difficult to distinguish from those of structural "mobility." By careful use of modern methods of intensity measurement and correction, many convincing analyses of thermal motion have now been made.[12,18] The experimental data used in protein structure analysis are, however, generally less

[13] J. C. Kendrew, H. C. Watson, B. E. Strandberg, R. E. Dickerson, D. C. Phillips, and V. C. Shore, *Nature (London)* **190,** 666 (1961).

[14] I. T. Weber, L. N. Johnson, K. S. Wilson, D. G. R. Yeats, D. L. Wild, and J. A. Jenkins, *Nature (London)* **274,** 433 (1978).

[15] J. Walter, W. Steigmann, T. P. Singh, H. Bartunik, W. Bode, and R. Huber, *Acta Crystallogr., Sect. B* **B38,** 1462 (1982).

[16] W. A. Hendrickson and M. M. Teeter, *Nature (London)* **290,** 107 (1981).

[17] N. Sakabe, K. Sakabe, and K. Sasaki, *in* "Structural Studies on Molecules of Biological Interest" (G. Dodson, J. P. Glusker, and D. Sayre, eds.), p. 509. Oxford Univ. Press, London and New York, 1981.

[18] V. Schomaker and K. N. Trueblood, *Acta Crystallogr., Sect. B* **B24,** 63 (1968).

satisfactory from this point of view since absorption errors[19] and the effects of irradiation damage[20] are more difficult to correct and lead to precisely the kind of systematic variations in structure factor amplitude that are taken up in apparent temperature factors. It is hoped that the interest in studies of "motion" in protein crystals might lead to a renewed interest in improving experimental methods. In the meantime it would seem that the position is similar to that obtaining generally some years ago when Lonsdale and Milledge[21] recommended that structures should be refined against two or more independent sets of experimental data, preferably collected at different temperatures. The justification for this recommendation can be seen from the results, published in 1970, of the International Union of Crystallography Single Crystal Intensity Project[22,23]: Zachariasen's particularly bleak assessment was, "Looking at the many structures which have been published in *Acta Crystallographica* during the last few years it is evident . . . [that] the thermal parameters are all nonsense and must all be done again in a sensible way."[24]

Despite these serious reservations recent studies have shown that much can be learned within the limitations of our present experimental methods. It is true that the atomic displacements are generally much larger in proteins than in small molecules, so that rather larger errors can be tolerated. Unfortunately the increase in "motion" causes a weakening of the diffraction at higher angle so that the errors increase in parallel. It has, indeed, become clear that major systematic errors in the atomic temperature factors arise if the data do not extend to sufficiently high resolution, and it is generally accepted that unless the data extend to 2.0 Å resolution or better there may be a serious underestimation of these temperature factors. This effect is presumably due, in part, to the omission of the disordered structure from the unit cell. The disordered solvent, which usually occupies a substantial fraction of the total crystal volume, will, by Babinet's principle, tend to scatter out of phase with the protein component, thereby reducing the magnitude of the observed diffraction. Since the scattering from the solvent falls off quickly with resolution, a major effect is to render the usual atomic scattering factors inexpedient at low resolution and hence distort the refinement of temperature factors. It has

[19] A. C. T. North, D. C. Phillips, and F. S. Mathews, *Acta Crystallogr., Sect. A* **A24,** 351 (1968).

[20] C. C. F. Blake and D. C. Phillips, *in* "Biological Effects of Ionising Radiation at the Molecular Level," IAEA Symp., p. 183. IAEA, Vienna, 1962.

[21] K. Lonsdale and J. Milledge, *Acta Crystallogr.* **14,** 59 (1961).

[22] S. C. Abrahams, W. C. Hamilton, and A. McL. Mathieson, *Acta Crystallogr., Sect. A* **A26,** 1 (1970).

[23] W. C. Hamilton and S. C. Abrahams, *Acta Crystallogr., Sect. A* **A26,** 18 (1970).

[24] W. H. Zachariasen, *Acta Crystallogr., Sect. A* **A25,** 276 (1969).

also been clear for some time that excluding weak data from the refinement process causes a systematic underestimation of the atomic displacement parameters,[25] and there is now little excuse for such practices, since French and Wilson have shown how such data can be correctly handled.[26] A further minor error will be introduced if, in the course of applying an absorption correction, account is not taken of the fact that the "equivalent sphere," to which the methods of both North *et al.*[19] and Kopfmann and Huber[27] correct the data, does not absorb equally at all Bragg angles. In fact the absorption is less at higher angles, producing an effect that is almost indistinguishable from diminished atomic displacements.[27]

Finally, we will consider the effect of applying restraints during least-squares refinement of the model. [These methods are described in detail elsewhere in the volume.] We have already seen that the vibrational motion of atoms produces marked shortening of the apparent bond lengths with the harmonic approximation. In these cases the restraints on bond lengths often applied during refinement will be inappropriate to the X-ray observations and will consequently further invalidate the thermal parameters obtained for these atoms by moving the atom position away from the centroid of its probability distribution.

Many restrained refinements now incorporate restraints on the variation in thermal motion between covalently bonded atoms. If we consider the atoms as vibrating independently, then it is indeed sensible to restrain their thermal motion, so that the implied instantaneous fluctuations in bond length are close to those observed by other techniques.[28,29] However these fluctuations in bond length are very small (for C—C and C=O bonds they are 0.0025 and 0.0012 Å^2, respectively), and so we would expect much of the intramolecular motion in macromolecules to occur by relatively large fluctuations in rather "softer" variables than bond lengths, such as rotations about bonds. If such motion is modeled by isotropic temperature factors, the implied fluctuations in bond length will tend to be grossly overestimated and attempts to impose sensible restrictions on them will produce a ridiculously "stiff" molecule.

This may be regarded as a particular example of a more general difficulty. The aims of a protein structure analysis are, first, to provide an objective image of the structure in the form of an electron density map and, second, to interpret this image in terms of a stereochemically reasonable structure, though in practice the two stages of the analysis are combined, as the current model of the structure is used in the generation of an

[25] F. L. Hirshfeld and D. Rabinovich, *Acta Crystallogr., Sect. A* **A29,** 510 (1973).

[26] S. French and K. S. Wilson, *Acta Crystallogr., Sect. A* **A34,** 517 (1978).

[27] G. Kopfmann and R. Huber, *Acta Crystallogr., Sect. A* **A24,** 348 (1968).

[28] F. L. Hirshfeld, *Acta Crystallogr., Sect. A* **A32,** 239 (1976).

[29] J. H. Konnert and W. A. Hendrickson, *Acta Crystallogr., Sect. A* **A36,** 344 (1980).

improved image. The difficulty is that the electron density map represents the space and time average of all the participating structures, and a single structure that gives the best fit to such a map cannot be expected to satisfy normal stereochemical criteria. Thus, as a further example, the electron densities corresponding to side chains that happen to move in concert may well approach one another more closely than normal van der Waals contacts would allow. Failure to allow properly for these effects clearly invalidates the estimates both of the positions of the atoms and of their apparent motion.

Molecular Motion from Apparent Thermal Motion

Given a set of $\langle u^2 \rangle$ values for the individual atoms of a macromolecule it may be instructive to attempt to explain them in terms of concerted molecular motions. We shall consider below some of the simpler models that might be investigated. From consideration of a number of well-refined protein structures it has become clear that the scale of motions (i.e., $\langle u^2 \rangle$ values) occurring in proteins runs from a little less than 0.1 Å^2 (crystalline, small organic molecules usually have motions of about half of this) to something over 1 Å^2, at which point the atoms become lost in the noise of the electron density maps. It has further become clear that the protein side chains usually have considerably greater motion than the corresponding main chain. Hence investigations of concerted molecular motion initially only consider the main-chain atoms, and in some cases the greater motion of the carbonyl oxygens has lead to them being omitted from the analysis.[11]

Sternberg *et al.*[30] have derived some useful measures of agreement for testing such models:

$$\Delta = \sum_{j=1}^{n} (\langle u_j^2 \rangle - \langle u_j^2 \rangle_{\text{calc}})^2 \tag{13}$$

and

$$R^2 = \Delta \Big/ \sum_{j=1}^{n} (\langle u_j^2 \rangle - \langle u^2 \rangle_{\text{mean}})^2 \tag{14}$$

where

$$\langle u^2 \rangle_{\text{mean}}) = \Big(\sum_{j=1}^{n} \langle u_j^2 \rangle \Big) \Big/ n \tag{15}$$

[30] M. J. E. Sternberg, D. E. P. Grace, and D. C. Phillips, *J. Mol. Biol.* **130,** 231 (1979).

Here the sums are from 1 to n where n is the number of residues in the molecule. R^2 indicates the extent to which the model produces a better fit to the observed $\langle u^2 \rangle$ values than the simple model that each $\langle u^2 \rangle$ value is equal to the mean. Formally, R^2 is the fraction of the variance of the $\langle u^2 \rangle$ values that is not accounted for by the model. Thus the closer the value of R^2 is to zero, the better the model fits the observations.

Simple Rigid-Body Motion

In general, the values of $\langle u^2 \rangle$ tend to be smallest for the residues that are buried and increase toward the surface of the molecule. Figure 3 shows a plot of $\langle u^2 \rangle$ against the square of the distance from the molecular centroid for hen egg lysozyme.[30] There is considerable variation from a linear relationship. After fitting a least-squares best line through these points the resultant values for Δ and R^2 were 0.78 Å^4 and 0.64, respectively, suggesting that a considerable amount of the variation had not been explained. This model corresponded to the molecule moving as a rigid body with a translational vibration of mean square amplitude 0.13 Å^2 and uniform libration about axes through the centroid. Alternatively it may be the result of intramolecular motion, for example, a periodic expansion and contraction in which the atoms farthest from the centroid

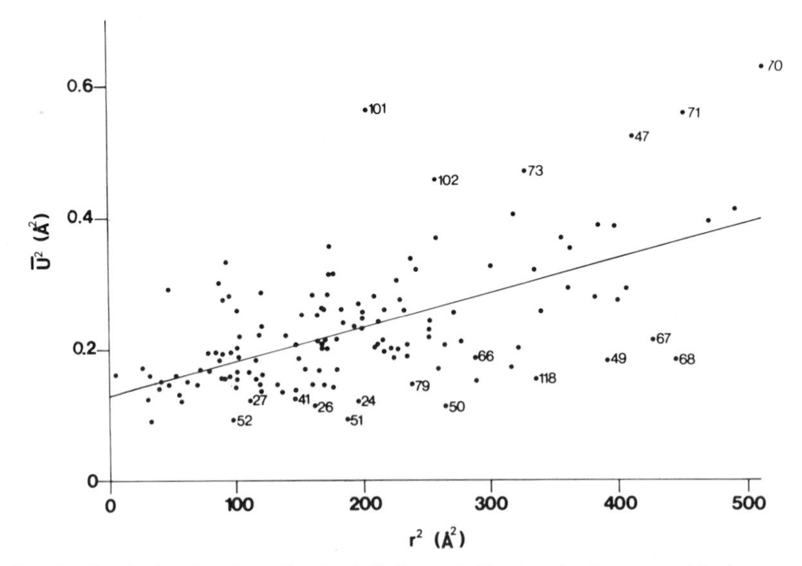

FIG. 3. Graph showing the value for $\langle u^2 \rangle$ for each C_α atom for hen egg-white lysozyme plotted against the square of the distance of the atom from the molecular centroid. The least-squares best straight-line fit to these points is shown.

move most. It probably indicates that the translational disorder in the crystal does not exceed about 0.15 Å^2.

General Rigid-Body Motion

Schomaker and Trueblood have analyzed the relationship between the anisotropic vibration parameter ($\mathbf{U}°$) of an atom and the rigid-body motion of the molecule.[18] For an atom the mean square amplitude of vibration in the direction of the unit vector $\mathbf{l}$ ($=l_1, l_2, l_3$) is given from the symmetric tensor $\mathbf{U}°$ by

$$\sum_{i=1}^{3}\sum_{j=1}^{3} U_{ij}^{\circ} l_i l_j \tag{16}$$

The rigid-body motion of the molecule is represented by two symmetric tensors $\mathbf{T}$ and $\mathbf{L}$, and by an unsymmetric tensor $\mathbf{S}$. The mean square amplitude of translational vibration in the direction $\mathbf{l}$ is given by

$$\sum_{i=1}^{3}\sum_{j=1}^{3} T_{ij} l_i l_j \tag{17}$$

and the mean square amplitude of libration about an axis $\mathbf{l}$ is given by

$$\sum_{i=1}^{3}\sum_{j=1}^{3} L_{ij} l_i l_j \tag{18}$$

The tensor $\mathbf{S}$ is introduced here as the libration axes need not necessarily intersect and so there can be a mean correlation between libration about $\mathbf{l}$ and translation parallel to $\mathbf{l}$ given by

$$\sum_{i=1}^{3}\sum_{j=1}^{3} S_{ij} l_i l_j \tag{19}$$

For any atom,

$$U_{ij}^{\circ} = G_{ijkl} L_{kl} + T_{ij} + H_{ijkl} S_{kl} \tag{20}$$

where the values of the arrays **G** and **H** involve the coordinates of the atom (p_1, p_2, p_3) and are given in Table 2 of Schomaker and Trueblood.[18]

In the case of isotropic refinement $\langle u^2 \rangle$ values are available. As we have seen above [Eq. (12)],

$$\langle u_2 \rangle = \tfrac{1}{3}[(U_{13}^{\circ})^2 + (U_{22}^{\circ})^2 + (U_{33}^{\circ})^2]$$

By finding the trace of the right-hand side of Eq. (20) one obtains

$$\begin{aligned}\langle u^2\rangle = {} & \tfrac{1}{3}[(p_2^2 + p_3^2)L_{11} + (p_1^2 + p_3^2)L_{22} + (p_1^2 + p_2^2)L_{33}] \\ & - 2p_2p_3L_{23} - 2p_1p_3L_{13} - 2p_1p_2L_{12} + \langle t^2\rangle \\ & + \tfrac{2}{3}[p_1(S_{32} - S_{23}) + p_2(S_{13} - S_{31}) + p_3(S_{21} - S_{12})] \qquad (21)\end{aligned}$$

where $\langle t^2\rangle = \frac{1}{3}(T_{11} + T_{22} + T_{33})$ and represents the isotropic mean square amplitude of translational vibration.

Equation (21) holds for each $\langle u^2\rangle$ value, giving a set of observations of $\langle u^2\rangle$ and p_1, p_2, p_3 from which the unknowns can be estimated by linear least squares. The n_{obs} observational equations may be expressed as

$$\sum_{b=1}^{n_{\text{obs}}} A_{ab}v_b = \langle u_a^2\rangle$$

where each v_b is one unknown and a_{ab} is the observation corresponding to a $\langle u_a^2\rangle$. For example, for residue a at $\mathbf{p}(p_1, p_2, p_3)$,

$$A_{a1} = \tfrac{1}{3}[(p_2^a)^2 + (p_3^a)^2]$$

and $v_1 = L_{11}$.

The linear least-squares estimate of the unknowns is

$$v_b = \sum_{c=1}^{n_{\text{par}}} \sum_{a=1}^{n_{\text{obs}}} N_{bc}A_{ca}^{\text{T}} \langle u_a^2\rangle$$

where $\mathbf{N} = [\mathbf{A}^{\text{T}}\,\mathbf{A}]^{-1}$. The accuracy of the determination of p is estimated by

$$\sigma^2(v_b) = N_{bb}\sigma^2(\Delta\langle u^2\rangle)$$

where $\sigma^2(v_b)$ is the variance of v_b, N_{bb} is the appropriate diagonal element of $\mathbf{N}$, and

$$\sigma^2(\Delta\langle u^2\rangle) = 1/(n_{\text{obs}} - n_{\text{par}}) \sum_{a=1}^{n_{\text{obs}}} (\langle u_a^2\rangle_{\text{obs}} - \langle u_a^2\rangle_{\text{calc}})$$

where the subscripts obs and calc denote the value of $\langle u^2\rangle$ that was observed and was calculated from Eq. (21), n_{obs} the number of observations, and n_{par} the number of parameters (10).

The values of the six independent components of **L** can be found and the magnitudes and directions of its principal axes determined. In the **TLS** analysis, the values of **L** are independent of, and the values of **S** and **T** dependent on, the choice of coordinate origin (the molecular centroid).

When applied to lysozyme this analysis gave encouraging results, $\Delta = 0.55$ Å^4, $R^2 = 0.45$. The vibration about the major libration axis was $7^{\circ 2}$.[30]

TL *Rigid-Body Motion*

Prior to the development of the **TLS** model by Schomaker and Trueblood,[18] Cruickshank[31] analyzed rigid-body motion in terms of the **T** and **L** tensors, by assuming that the three librational axes intersect at a point. Schomaker and Trueblood's analysis can be reduced to Cruickshank's by setting **S** to zero in Eq. (21). This leaves seven unknowns to be determined. The principal axes of **L** can be found but both the values of **L** and $\langle t^2 \rangle$ depend on the choice of origin. When applied to lysozyme this analysis was rather less satisfactory than the **TLS** model ($\Delta = 0.65$ Å^4, $R^2 = 0.54$), but the librational parameters were in gross agreement.[30]

Hinge-Bending Motion

It is now believed that a number of protein structures can undergo marked intramolecular motion that can be described quite well in terms of the relative movement of two lobes or domains so that the separation of the lobes (that is the cleft between them) varies. It has been proposed that such motions are of importance to the biological function of several enzymes.[32–34]

Hinge bending may be modeled by a libration about a suitably chosen instantaneous axis (Diamond shows how this axis may be defined[35]). If the libration of the first lobe has mean square amplitude h_1^2 such that the motion is along unit vector **n** then, following Dunitz and White,[36] for a given residue in this lobe

$$U_{ij} = h_1^2 d_1^2 n_i n_j$$

where d_1 is the distance of the residue from the first instantaneous axis. The value of $\langle u^2 \rangle$ for the residue is obtained by

$$\langle u^2 \rangle = \tfrac{1}{3} h_1^2 d_1^2$$

Similarly for residues in the second lobe

$$\langle u^2 \rangle = \tfrac{1}{3} h_2^2 d_2^2$$

The values for the hinge parameters h_1 and h_2 can be found by a least-squares fit to the observed $\langle u^2 \rangle$ values. Sternberg *et al.* in their analysis of

[31] D. W. J. Cruickshank, *Acta Crystallogr.* **9,** 754 (1956).
[32] W. S. Bennett and T. A. Steitz, *J. Mol. Biol.* **140,** 183 (1980).
[33] W. S. Bennett and T. A. Steitz, *J. Mol. Biol.* **140,** 211 (1980).
[34] R. D. Banks, C. C. F. Blake, P. R. Evans, R. Haser, D. W. Rice, G. W. Hardy, M. Merrett, and A. W. Phillips, *Nature (London)* **279,** 773 (1979).
[35] R. Diamond, *J. Mol. Biol.* **130,** 252 (1979).
[36] J. D. Dunitz and D. N. J. White, *Acta Crystallogr., Sect. A* **A29,** 93 (1973).

the motions of lysozyme assumed that $h_1 = h_2$, but this is not an implicit assumption of the method.[30]

General Breathing Motion

There is some evidence that molecules may be subject to expansion and contraction. This breathing motion can be considered by an analysis similar to that of Schomaker and Trueblood.[18] The instantaneous displacement of any atom at position $\mathbf{p}$ (p_1, p_2, p_3) from the centroid in the direction $i(i = 1, 2, \text{ or } 3)$ is given by

$$U_i = \sum_{j=1}^{3} G_{ij} p_j \tag{22}$$

where $\mathbf{G}$ is the displacement gradient tensor. The required mean square displacement U_{ij} is found from $U_i U_j$. Taking the time average,

$$U_{ij} = \langle U_i U_j \rangle$$

It can be shown that

$$\langle u^2 \rangle = E_{11}p_1^2 + E_{22}p_2^2 + E_{33}p_3^2 + E_{23}p_2p_3 + E_{13}p_1p_3 + E_{12}p_1p_2 \tag{23}$$

where the $\mathbf{E}$ terms are constants that depend on the type of breathing motion. The six independent components of the symmetric $\mathbf{E}$ tensor may be found by least squares.

General Breathing with Uniform Translation

A uniform translation term $\langle t^2 \rangle$ can be added to Eq. (23) to give

$$\langle u^2 \rangle = E_{11}p_1^2 + E_{22}p_2^2 + E_{33}p_3^2 + E_{23}p_2p_3 + E_{13}p_1p_3 + E_{12}p_1p_2 + \langle t^2 \rangle \tag{24}$$

This equation may be directly related to the equation for the **TL** model [Eq. (21) minus the **S** terms]. If we compare the coefficients we get

$$E_{33} = \tfrac{1}{3}(L_{11} + L_{22}), \qquad E_{22} = -\tfrac{2}{3}L_{23}, \qquad \text{etc.}$$

Thus one would get the same fit to the observed $\langle u^2 \rangle$ values as from the **TL** model. This analysis shows how the results from the rigid-body model could also be interpreted as intramolecular motions.

Analysis with Anisotropic Atomic Temperature Factors

The above methods can all be extended directly to the case in which anisotropic B factors are available (see Cruickshank[31] and Schomaker and Trueblood[18] for the **TL** and **TLS** cases, respectively). The incorporation of anisotropic information would distinguish between librational and breathing motions.

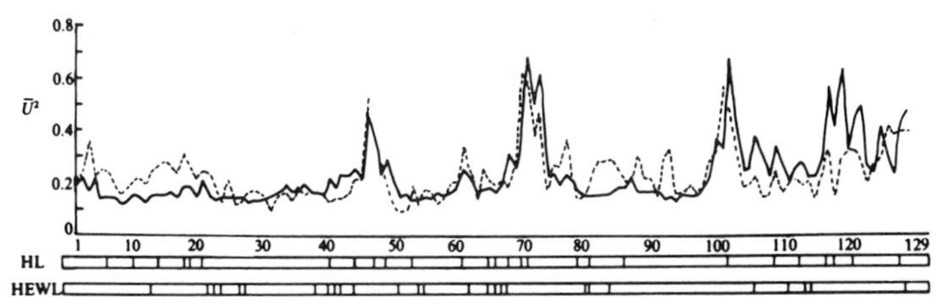

FIG. 4. A plot of mean main-chain $\langle u^2 \rangle$ values against residue number for human lysozyme (full line) and hen egg-white lysozyme (broken line). The numbering scheme is for hen lysozyme. Residues involved in intermolecular contacts in the crystals of the two molecules are indicated in the lower rectangles (HL, human lysozyme; HEWL, hen lysozyme).

Intramolecular Motion

The range of potential intramolecular motion open to a protein is enormous, from simple atomic vibrations through librational motions of side chains to displacements involving stretches of chain, sometimes leading to the disordering of whole domains. Examples of all these forms of motion have been inferred from X-ray diffraction analyses; however at present the descriptions have been largely qualitative. In some cases this is very proper, such as when disorder causes a piece of structure to become "invisible" to the X rays. Even in such a case, however, this observation may provide valuable insights into the molecule's function.[15,37] It is necessary, before we can consider quantitative assessments of intramolecular motion, to address the question of to what extent a protein molecule possesses a characteristic set of motions and, further, to consider to what extent these are affected by the errors in the experimental data and the crystal packing constraints. In an attempt to answer these questions, refinements of the structure of lysozyme from two species were undertaken.[38] In this case the molecules were quite differently arrayed in their respective crystals, the data collection and correction procedures were different, and finally the refinements were by different techniques.

The values of $\langle u^2 \rangle$ of the main-chain atoms in the two molecules are shown, averaged over individual residues, in Fig. 4. Clearly to a first approximation at least, the two plots have the same characteristics, a fairly constant background on which are superimposed several peaks corresponding to residues 47, 70–73, and 101–102. Although there are some differences the general agreement is good, as shown by the correlation

[37] T. Alber, D. W. Banner, A. C. Bloomer, G. A. Petsko, D. C. Phillips, P. S. Rivers, and I. A. Wilson, *Philos. Trans. R. Soc. London, Ser. B* **293,** 159 (1981).

[38] P. J. Artymiuk, C. C. F. Blake, D. E. P. Grace, S. J. Oatley, D. C. Phillips, and M. J. E. Sternberg, *Nature (London)* **280,** 563 (1979).

coefficient of 0.61. The distribution of $\langle u^2 \rangle$ is much smoother, however, for human lysozyme, almost certainly because this structure has been refined more completely ($R = 0.18$) at higher resolution (1.5 Å) than has that of hen egg-white lysozyme ($R = 0.22$ at 2.0 Å resolution). Nevertheless the comparison is valuable because it suggests strongly that the observed variations in $\langle u^2 \rangle$ cannot be due either to the effects of systematic errors in the data (which must be very different in the two analyses) or to interactions between molecules in the crystals (which are different in the two structures). It seems, rather, that the parallel variation of the $\langle u^2 \rangle$ values obtained in the two studies represents some property of the lysozyme molecules themselves.

As we have seen above, to some extent the variation in $\langle u^2 \rangle$ values along the polypeptide chain arises from the motion of the whole of the lysozyme molecule as a rigid body. Even when these rigid-body motions are taken into account, however, the $\langle u^2 \rangle$ values shown in Fig. 4 are not fully explained, and the residual variations imply the existence of intramolecular motion (or conformational variability). Close examination of these variations shows that they correlate well with the three-dimensional structure of the lysozyme molecule, a finding reinforcing the conclusion that they represent a characteristic molecular property of lysozyme and not some artifact of the crystallography.[38] Furthermore this study revealed that the behavior at 1.5 Å resolution of the unrestrained atomic displacements of the side-chain atoms was physically sensible. The tendency, confirmed by subsequent studies, is for side chains to be rather more mobile than the backbone to which they are attached, with the displacements increasing with the number of bonds about which free rotation is possible and dependent on the local environment.

From a graph such as Fig. 4, the next stage in the analysis is simply to locate the regions of large displacement on the molecule. This is shown for lysozyme in Fig. 5 and reveals that they form a continuous strip on the enzyme surface which includes both lips of the active site and most of the main-chain segments that undergo conformational change when inhibitors are bound.[39] This observation, suggesting as it does that the apparent motion may play a part in the activity of the enzyme, emphasizes the importance of distinguishing between true motion and conformational variability. The elegant work of James *et al.*[40] has produced similar results; the residues around the active site have larger than average dis-

[39] C. C. F. Blake, L. N. Johnson, G. A. Mair, A. C. T. North, D. C. Phillips, and V. R. Sarma, *Proc. R. Soc. London, Ser. B* **167,** 378 (1967).

[40] M. N. G. James, A. R. Sielecki, G. D. Brayer, and L. T. J. Delbraere, *J. Mol. Biol.* **144,** 43 (1980).

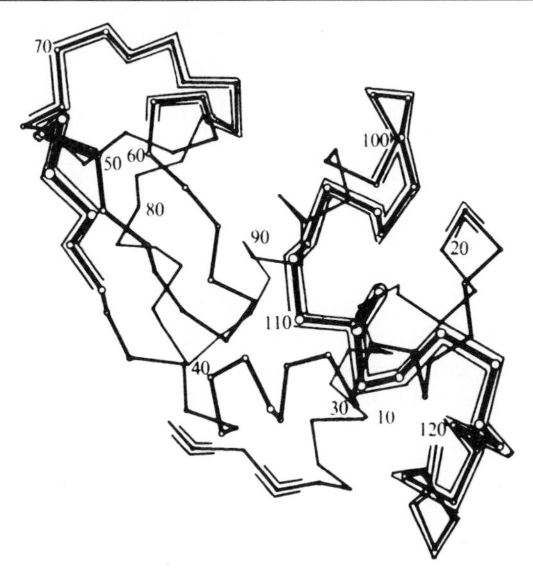

FIG. 5. A perspective drawing of the lysozyme molecule with residues having displacement >0.2 Å^2, in human lysozyme outlined by parallel lines. The active site cleft runs almost vertically down the front of the drawing.

placements in the native enzyme which are reduced on complexing with inhibitors.

A way ahead in crystallographic studies of macromolecular mobility is indicated by the work of Frauenfelder *et al.*[41] and Hartmann *et al.*[42] on myoglobin. By studying the structure of this protein over a range of temperatures from 80 to 300 K these authors have attempted to show how true motion and static disorder can be distinguished. Unfortunately, it has been shown by Willis[43] that the analysis of the displacement potentials by Frauenfelder *et al.*[41] is incorrect, since the isotropic temperature factors on which their analysis is based incorporate the assumption of a simple harmonic displacement. Walter *et al.*[15] have performed a careful analysis of trypsinogen at 103 and 173 K in an attempt to establish whether the so-called activation domain of trypsinogen, which is invisible at room temperature, is subject to static or vibrational disorder. This work, which suggested that the domain was suffering static disorder, also indicates the difficulties of interpretation that arise from such experiments.

[41] H. Frauenfelder, G. A. Petsko, and D. Tsernoglou, *Nature* (*London*) **280,** 558 (1979).

[42] H. Hartmann, F. Parak, W. Steigmann, G. A. Petsko, D. Ringe Ponzi, and H. Frauenfelder, *Proc. Natl. Acad. Sci. U.S.A.* **79,** 4967 (1982).

[43] B. T. M. Willis, *in* "Computational Crystallography" (D. Sayre, ed.), p. 479. Oxford Univ. Press (Clarendon), London and New York, 1982.

Anisotropic Refinements

In a very few high-resolution analyses it has been possible to attempt to refine protein atoms with anisotropic thermal factors. In these cases an enormous wealth of information concerning the atomic displacements is produced. Fortunately, if during the refinement no restraints were placed on the elements of the displacement tensor, it is possible to apply Hirshfeld's "rigid bond" test or its generalization.[28,44] The former test is based upon the postulate that the mean square vibrational amplitudes of two bonded atoms of comparable mass should be very nearly equal in the direction of the bond between them. The generalization states that in a rigid body the test can also be applied to all pairs of nonbonded atoms. Although there is no published measure of the extent to which the anisotropic refinements obey these restrictions, it is encouraging to see that there is indeed evidence for such concerted movement in the anisotropic ellipsoids shown in Fig. 6.[38,45,46]

A Torsion Angle Constrained Model for Thermal Motion

Until now all refinements of individual atomic displacements against X-ray diffraction observations for proteins appear to have been based upon the statistical model in which each atom is considered as an independent harmonic oscillator. In those cases in which restrictions on the atomic displacements have been introduced in order to ameliorate the problems caused by a paucity of X-ray observations, this has been achieved either by the use of restraints on the amplitudes of displacement[29] or by simple averaging.[47] Use of a statistical model has been inherited directly from small-molecule practice where there has been a tendency (fostered by the comparative abundance of observations) to postfit simplified mechanistic models to the results of a statistical analysis, although there has been some lively discussion of this (see, for instance, the report from the summer school of the International Union of Crystallography held in 1969).[48]

We feel that in the more difficult case of proteins it is well worth reassessing the mechanistic approach, whereby one can define a model

[44] R. E. Rosenfeld, K. N. Trueblood, and J. D. Dunitz, *Acta Crystallogr., Sect. A* **A34,** 828 (1978).

[45] K. D. Watenpaugh, L. C. Sieker, and L. H. Jensen, *J. Mol. Biol.* **138,** 615 (1980).

[46] I. Glover, I. Haneff, J. Pitts, S. Wood, D. Moss, I. Tickle, and T. Blundell, *Biopolymers* **22,** 293 (1983).

[47] J. H. Sussman, S. R. Holbrook, G. M. Church, and S.-H. Kim, *Acta Crystallogr., Sect. A* **A33,** 800 (1977).

[48] Discussion *in* "Crystallographic Computing" (F. R. Ahmed, S. R. Hall, and C. P. Huber, eds.), p. 250. Munksgaard, Copenhagen, 1970.

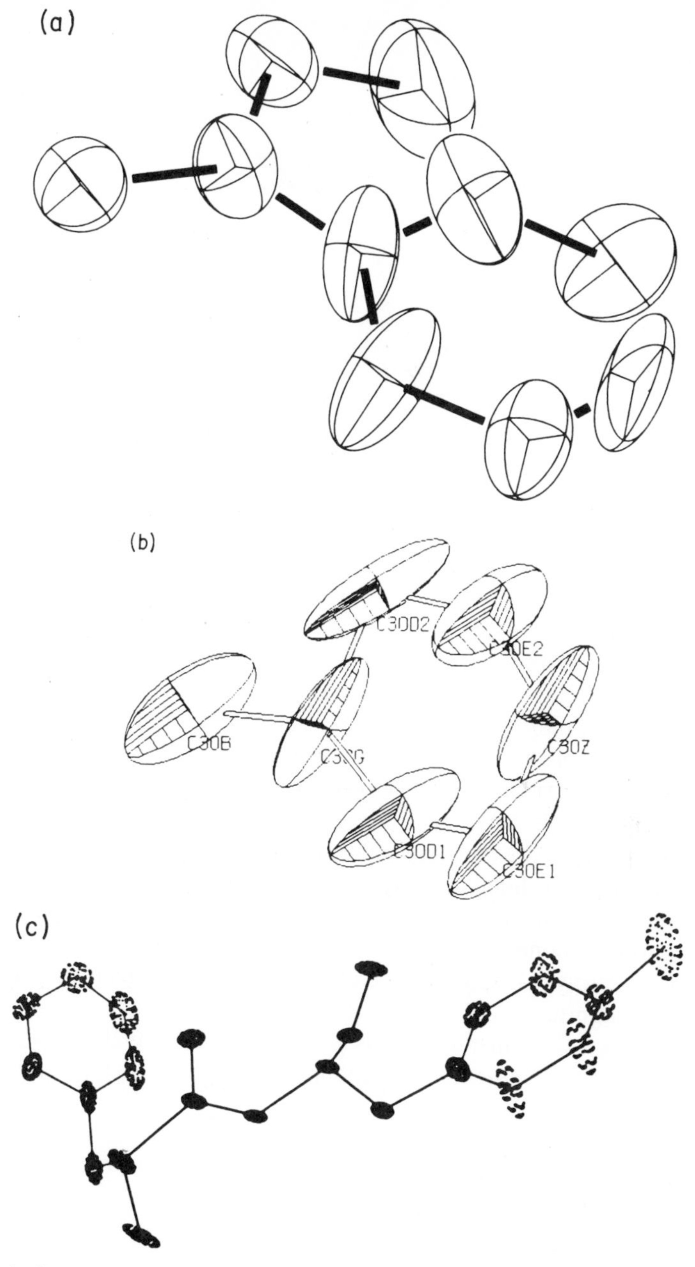

FIG. 6. Conventional drawings of the displacement ellipsoids of (a) Trp 63 of human lysozyme, resolution 1.5 Å, (b) Phe 30 of rubredoxin, resolution 1.2 Å, (c) Phe 20 and Tyr 21 of avian pancreatic polypeptide, resolution 1.0 Å. Note that as the resolution increases there is greater evidence for rigid-body motion of the side chains.

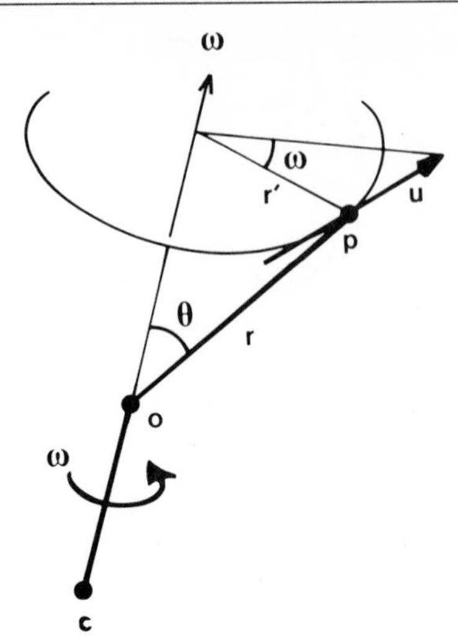

FIG. 7. An illustration of the approximation (**u**) to the motion of atom p due to rotation about the bond between atoms c and o.

with far fewer parameters than are required by the statistical approach. Although the analysis below is independently derived[49] it is related to that of Dunitz and White[36] who proposed a similar model for intramolecular motion and showed how the model parameters could be extracted from the anisotropic thermal ellipsoids of individual atoms. In spite of its limitations, this model has provided some interesting results[1] and it would be instructive to apply it to proteins. We will now show how a similar model can be defined and refined directly against the X-ray observations (it is not difficult to formulate a more exact model; however, the mathematical manipulations become less clear).[49]

Figure 7 shows an atom at p undergoing a displacement due to a rotation of magnitude ω about the bond joining atoms at c and o (defined as direction $\boldsymbol{\omega}$). The instantaneous displacement is in direction $\mathbf{u}$. For small angular displacements we may approximate the displacement by[50]

$$\mathbf{u} = \boldsymbol{\omega} \times \mathbf{r} \tag{25}$$

where $\mathbf{r}$ is the vector to the atom p from the atom at o to which it is bonded.

[49] D. Stuart, M. Levine, and N. Walker, unpublished results.

[50] M. L. Boas, "Mathematical Methods in the Physical Sciences," p. 208. Wiley, New York, 1966.

The vector product can be represented as[51]

$$\boldsymbol{\omega} \times \mathbf{r} = \underset{\mathbf{A}}{\begin{bmatrix} 0 & r_3 & -r_2 \\ -r_3 & 0 & r_1 \\ r_2 & -r_1 & 0 \end{bmatrix}} \underset{\boldsymbol{\omega}}{\begin{bmatrix} \omega_1 \\ \omega_2 \\ \omega_3 \end{bmatrix}} \tag{26}$$

Note that the matrix **A** is unrelated to the matrix of the same name used above (we retain this notation for consistency with the literature). Let **l** be the direction cosines of **ω**, then

$$\boldsymbol{\omega} \times \mathbf{r} = \omega \underset{\mathbf{A}}{\begin{bmatrix} 0 & r_3 & -r_2 \\ -r_3 & 0 & r_1 \\ r_2 & -r_1 & 0 \end{bmatrix}} \underset{\mathbf{l}}{\begin{bmatrix} l_1 \\ l_2 \\ l_3 \end{bmatrix}} \tag{27}$$

To calculate the temperature factor we need to find the projection of the displacement **u** on the reciprocal lattice vector **d***, i.e., **u** · **d***. Let us for simplicity assume an orthogonal unit cell. We may then rewrite Eq. (3) as

$$T^{hkl} = \exp(-2\pi^2 \mathbf{d}^{*\mathrm{T}} \mathbf{U}^\circ \mathbf{d}^*) \tag{28}$$

where

$$\mathbf{U}^\circ = \langle \mathbf{u}\mathbf{u}^\mathrm{T} \rangle$$

from Eq. (26), $\mathbf{u} = \mathbf{A}\boldsymbol{\omega}$, and therefore

$$\begin{aligned} \mathbf{U}^\circ &= \langle \mathbf{A}\boldsymbol{\omega}\boldsymbol{\omega}^\mathrm{T}\mathbf{A}^\mathrm{T} \rangle \\ &= \mathbf{A}\langle \boldsymbol{\omega}\boldsymbol{\omega}^\mathrm{T} \rangle \mathbf{A}^\mathrm{T} \\ &= \mathbf{A}\mathbf{L}\mathbf{A}^\mathrm{T} \end{aligned} \tag{29}$$

(cf. International Tables for Crystallography, Vol. IV, p. 320[52]). Here

$$\mathbf{L} = \begin{bmatrix} \langle \omega_1\omega_1 \rangle & \langle \omega_1\omega_2 \rangle & \langle \omega_1\omega_3 \rangle \\ \langle \omega_2\omega_1 \rangle & \langle \omega_2\omega_2 \rangle & \langle \omega_2\omega_3 \rangle \\ \langle \omega_3\omega_1 \rangle & \langle \omega_3\omega_2 \rangle & \langle \omega_3\omega_3 \rangle \end{bmatrix}$$

Let us express ω_i in terms of ω and the direction cosines of ω,

$$\omega_i = \omega l_i$$

$$\mathbf{L} = \langle \omega^2 \rangle \begin{bmatrix} (l_1 l_1) & (l_1 l_2) & (l_1 l_3) \\ (l_2 l_1) & (l_2 l_2) & (l_2 l_3) \\ (l_3 l_1) & (l_3 l_2) & (l_3 l_3) \end{bmatrix}$$

$$\mathbf{L} = \langle \omega^2 \rangle \mathbf{L}' \tag{30}$$

[51] J. Heading, "Matrix Theory for Physicists," p. 15. Longmans, Green, New York, 1958.

[52] C. K. Johnson and H. A. Levy, *in* "International Tables for X-ray Crystallography" (J. A. Ibers and W. C. Hamilton, eds.), Vol. 4, p. 320. Kynoch Press, Birmingham, England, 1974.

We can now rewrite the temperature factor [Eq. (28)] as

$$T^{hkl} = \exp[-2\pi^2\langle\omega^2\rangle\mathbf{d}^{*\mathrm{T}}\mathbf{A}\mathbf{L}'\mathbf{A}^{\mathrm{T}}\mathbf{d}^*] \quad (31)$$

which may be easily evaluated since $\mathbf{d}^*$, $\mathbf{A}$, and $\mathbf{L}'$ are fixed quantities (i.e., functions of the reciprocal lattice and atomic positions).

To be able to refine the value $\langle\omega^2\rangle$, we may write the structure factor, $\mathbf{F}(\mathbf{h})$, as

$$\mathbf{F}(\mathbf{h}) = \sum_j f_j \exp(2\pi i\mathbf{h}\cdot\mathbf{p}_j)\exp[-2\pi^2\langle(\mathbf{u}_j\cdot\mathbf{d}^*)^2\rangle] \quad (32)$$

where $\mathbf{h}$ is the Miller index triplet vector and the sum is over all the atoms (atom j being at position $\mathbf{p}_j$ with displacement $\mathbf{u}_j$). The temperature factor exponent can be replaced by Eq. (31) and differentiating the resultant expression leads to (omitting the atom subscripts for clarity)

$$\frac{\partial\mathbf{F}(\mathbf{h})}{\partial\langle\omega^2\rangle} = \sum f\exp(2\pi i\mathbf{h}\cdot\mathbf{p})(-2\pi^2)\mathbf{d}^{*\mathrm{T}}\mathbf{A}\mathbf{L}'\mathbf{A}^{\mathrm{T}}\mathbf{d}^* \times \exp[-2\pi^2\langle\omega^2\rangle\mathbf{d}^{*\mathrm{T}}\mathbf{A}\mathbf{L}'\mathbf{A}^{\mathrm{T}}\mathbf{d}^*] \quad (33)$$

In practice we may have to consider the effect of several successive libertions. Let us make the simplifying approximation that the total displacement of atom p by two rotations, i and j, can be taken as the sum of the displacements due to each of these alone:

$$\mathbf{u} = \mathbf{u}_i + \mathbf{u}_j$$

This sum can simply be extended to include any librations that displace atom p. Examining Eq. (31) we see that the temperature factor expression for one libration simply consists of the product of the magnitude ω^2 multiplied by a scalar independent of ω. If we call this scalar a and invoke the central limit theorem for independent random variables, we may represent the temperature factor exponent as

$$\exp(a_i\langle\omega_i^2\rangle + a_j\langle\omega_j^2\rangle + \cdots)$$

or

$$\exp\left[-2\pi^2\sum(\langle\omega^2\rangle\mathbf{d}^{*\mathrm{T}}\mathbf{A}\mathbf{L}'\mathbf{A}^{\mathrm{T}}\mathbf{d}^*)\right]$$

Differentiation of this with respect to $\langle\omega^2\rangle$ is straightforward.

Naturally there may be a tendency for some of the librations to be correlated in an unpredictable way, and this can be accounted for after a fashion in the above scheme. Although such effects will cause unsatisfying negative amplitudes of libration, Johnson has pointed out that this "seasick sailor's librating lighthouse" syndrome is physically reasonable.[53]

Once values have been obtained for the mean square libration it is straightforward to correct the bond lengths (as discussed above) and apply the usual restraints to these corrected values.

Clearly work must be done to see the extent to which such a simplified scheme, perhaps used in conjunction with a **TLS** tensor to account for the rigid-body component of the displacements, can provide an adequate description of protein motion in the crystal.

Dynamic Simulations and Diffraction Experiments

Recently there has been much activity in the field of the computer simulation of protein dynamics.[54] This is of interest to us here only inasmuch as the results bear upon the interpretation of X-ray diffraction experiments.

Van Gunsteren *et al.* have attempted to link the two approaches.[55] These workers performed a 12-psec dynamic simulation of the motions of a complete unit cell (including intermolecular solvent) of the crystal form of trypsin inhibitor, whose structure has been elucidated by X-ray diffraction. From the instantaneous configurations during their simulation, a set of electron density maps was calculated and summed to mimic the diffraction situation. Comparison of the resultant averaged (and hence thermally smeared) structure with the X-ray observations, however, revealed that the correlation was little better than random when taken over the complete set of X-ray observations. A better model was obtained by using the atomic positions from the diffraction study with the vibrational parameters from the dynamical simulation, yet this model was still a markedly worse fit to the experimental data than the simple isotropic model derived from conventional refinement. Clearly these studies are of interest but are as yet of little practical help to the crystallographer.

An alternative approach to the derivation of a theoretical model for the motions of a macromolecule is to attempt to calculate the normal modes of vibration for a complete molecule by consideration of the restoring force on each atom. This represents a fearsome computational problem and also returns to the simplifying assumption of harmonic motion, but

[53] C. K. Johnson, *in* "Computing in Crystallography" (R. Diamond, S. Ramaseshan, and K. Venkatesan, eds.), p. 14.01. Indian Acad. Sci. Bangalore, India, 1980.

[54] M. Karplus and J. A. McCammon, *Annu. Rev. Biochem.* **53,** 263 (1983).

[55] W. F. van Gunsteren, H. J. C. Berendsen, J. Hermans, W. G. J. Hol, and J. P. M. Postma, *Proc. Natl. Acad. Sci. U.S.A.* **80,** 4315 (1983).

there is some hope that the major displacements of the polypeptide backbone may be accounted for by a small number of the lowest frequency modes.[56] If this is the case it will be straightforward, and fascinating, to calculate "dynamic" displacement matrices for every atom and test them against the X-ray observations.

[56] B. Brooks and M. Karplus, *Proc. Natl. Acad. Sci. U.S.A.* **80,** 6571 (1983).

Section II

Modeling

[10] Optical Matching of Physical Models and Electron Density Maps: Early Developments

By FREDERIC M. RICHARDS

Introduction

Even the best electron density maps of globular proteins available today rarely provide atomic resolution at the time of initial interpretation. An essential step has been the construction of an atomic model which is thought to provide a reasonable "fit" to the electron density. From such a model a tentative set of atomic coordinates can be obtained. These coordinates form the basis for subsequent refinement procedures based on machine computation. The model itself can be "made" by computation and displayed two-dimensionally or stereographically[1] (for more recent information and procedures, see this volume [12]). However, it is frequently useful to build a physical model from accurately formed skeleton components. The "rod and clip" method used in the first interpretation of the myoglobin structure by Kendrew *et al.*[2] was a successful attempt to build a model directly into a Fourier map. As a code for the electron density, clips of different colors were put at the proper vertical positions on a "forest" of steel rods. The brass wire model was then built in among the rods. Mechanical interference made adjustment difficult, and the model was hard to see because of the very large number of supporting rods.

In current work on various proteins, the maps are displayed as a stack of contoured sections supported on Perspex sheets. Appropriately scaled coordinate systems are set up for a given model and map. Measurements are transferred back and forth between the model and the map as the construction of the model proceeds. Although faster than the "rod and clip" method, the process is time-consuming and tedious. Ideally, the model would be seen suspended within the contour map so that the "fit" could be evaluated directly. This cannot be done by direct superposition of the two objects, but it can be simulated optically as reported by Richards.[3]

The device for holding and displaying the map and the model is shown schematically in Fig. 1. The half-silvered mirror is set at exactly 45°, so

[1] C. Levinthal, *Sci. Am.* **214,** 42 (1966).

[2] J. C. Kendrew, R. E. Dickerson, B. E. Strandberg, R. G. Hart, D. R. Davies, D. C. Phillips, and V. C. Shore, *Nature (London)* **185,** 422 (1960).

[3] F. M. Richards, *J. Mol. Biol.* **37,** 225 (1968).

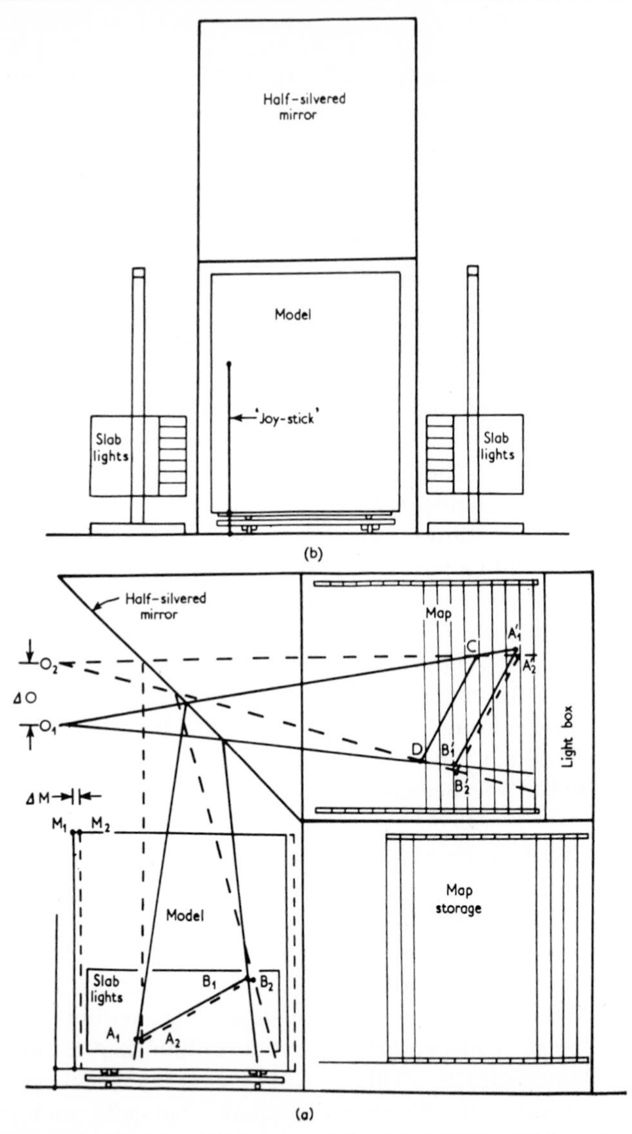

FIG. 1. Schematic diagram of the mirror and frame assembly used for optical superposition of a skeleton model and a three-dimensional electron density map. (a) Side view; (b) front view. The main frame is 7 ft high and 3 ft wide. The model frame is shown mounted on tracks to permit horizontal motion in two dimensions. Such movement is required when the model and map have different scales (see text). When the scales are identical, the model frame is held fixed with respect to the map and mirror for all viewing positions. Except for the first prototype, all subsequent instruments have used only the matched scales and fixed frames.

that the reflection of a vertical axis in the model is exactly collinear with the perpendicular to the map sections. Looking into the mirror from the position O_1, one sees a mirror image of the model and superimposed a direct view of the map as mounted in front of the light box. In order to be in register, the map must be stacked as a mirror image. This is easily done when the map is handled and stored as single sections. Each section is inserted upside down without changing its position in the sequence of sections. Such a stack is then the proper mirror image of its normal counterpart.

If the scale of the model is identical with the scale of the map, then corresponding positions in the map and in the model will appear superimposed and will not move with respect to each other regardless of the position of viewing. This is by far the most effective way of using the device. The most commonly used models are the Kendrew brass wire units supplied by Cambridge Repetition Engineers (Green's Road, Cambridge, England) at a scale of 2 cm = 1 Å. For use with these units, the electron density maps should be drawn to a scale of exactly 2 cm = 1 Å. Less expensive and lighter plastic units can also be used as long as the model and map scales are matched (however, see below). In the future, most maps will be machine drawn either directly on an x–y plotter or from recorded film of cathode ray tube displays. The latter are easily enlarged to the appropriate scale and traced from the back of a translucent glass sheet used as a screen.

When building a model, one manipulates the brass pieces by rotation about single bonds while observing the process in the mirror. The mirror image produces no more difficulty than learning to work under a microscope. The fit to the map can be seen directly and no measurements are required. Horizontal positioning is clearly as accurate as either map or model allow. Vertical positioning depends initially on one's stereoscopic vision. For most people this is remarkably good with a little practice. Errors of 1/4 in. in vertical position are readily visible. If one magnifies the stereoscopic effect by moving one's head from side to side, then higher accuracy is easily obtained. For those without stereoscopic vision this latter procedure would be used in any event.

A thickness of 8–12 sections is usually the maximum thickness of the map that can be profitably used at any one time. When the model is essentially complete, it is convenient to illuminate only that section of the model which corresponds to the portion of the map being viewed. Lights of adjustable height which illuminate only a horizontal slice of the model are very useful. The nonilluminated part of the skeleton model above the lighted section does not interfere with the viewing. The occasional overlaps can always be eliminated by a slight change in viewing position.

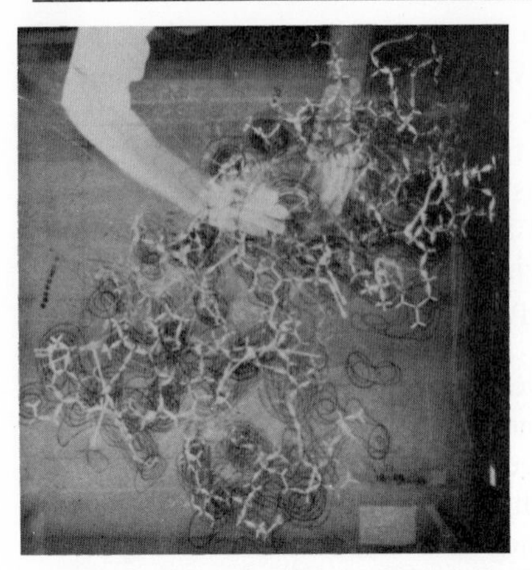
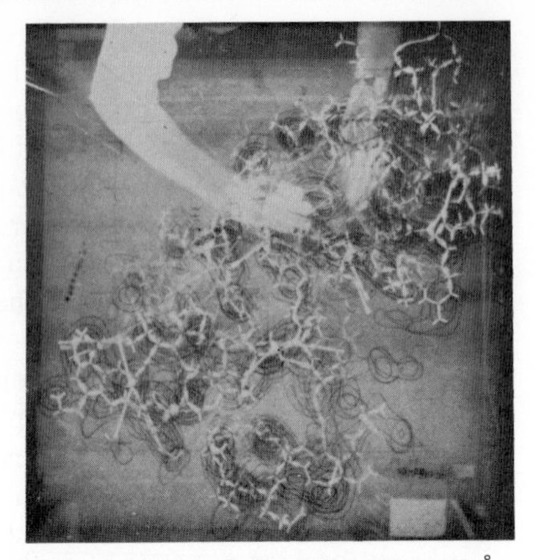

FIG. 2. Stereophotographs taken during the fitting of a model of ribonuclease S to a 3 Å electron density map. The 10 map sections represent the center part of the molecule. The model is brightly lit in that portion corresponding to these sections. Parts of the model above and below these sections can be dimly seen but do not interfere with the matching in progress. The ring of Phe 46 is being adjusted. The high-density peak of the sulfur atom in Met 29 can be seen toward the top of the photographs and the ring of His 12 near the center. Stereographic photographs are easily made with the camera in the position normally used for viewing. To obtain the correct, nonmirror, image, the negatives are turned upside down, and the right and left partners are interchanged.

(Note, however, that with some of the smaller scale plastic parts the fraction of space occupied is much larger than for the wire units. This severely limits the depth of the molecule through which one can see effectively.) All parts of the mounting near the model, including the base, should be painted matt black so as not to show in the image. For molecules the size of lysozyme or ribonuclease, one can look through an entire wire model without undue interference, but much larger molecules may have to be built in sections even with the wire units. A stereographic pair of photographs of some sections of the 3 Å ribonuclease S map[4] and the superimposed model fitted to it are shown in Fig. 2. In some parts the fit can clearly be improved by further adjustment of the model. Positioning of the ring of Phe 46 is under way. It is suggested that diagrams of this sort may be useful in allowing readers without knowledge of crystallography to form their own opinions of the quality of a map and the significance of the model and of the derived list of coordinates.

[4] H. W. Wyckoff, K. D. Hardman, N. M. Allewell, T. Inagami, L. N. Johnson, and F. M. Richards, *J. Biol. Chem.* **242,** 3984 (1967).

Details of Construction and Use

Each unit that has been built is slightly different in detail and materials used. All have the imprint of the "owner." The original description follows but the subsequent chapter in this volume should be consulted before actually building an instrument.

The frame of the map and the mirror is conveniently made from $1\frac{1}{2}$-in. "slotted angle." In the present model, the map, drawn on thin acetate sheet, is mounted on 36 in. × 36 in. Perspex sheets which slide into and out of the rack or storage areas. The sheets slide on plywood to which deal spacers have been tacked to give the proper spacing for the map scale. (The spacings can be compressed in the storage area to provide room for more than one map.) A lighter, cheaper, and more convenient mount could probably be made with the aluminum frame system used by W. N. Lipscomb (personal communication). The light box for the map consists of nine 100-W incandescent bulbs operated from a variable transformer, and has a 36 in. × 36 in. flashed opal glass sheet as a diffuser.

The half-silvered mirror is 36 in. × 42 in., made of 1/4-in. plate glass, and backed with 1/4-in. plate. The silver surface need not be perfectly uniform, and local glass suppliers can usually provide such a mirror quite inexpensively. The high-quality plate glass is required to avoid distortion. Satellite reflections from the front and back surfaces occur but are not normally intense enough to provide serious interference.

The frame for the model is made from 1-in. "speed frame" (Dexion Ltd., Wembley, England) and should be open on all four sides and the top. The present mounting system consists of 16-gauge piano wire. The vertical units (on a 6-in. grid) are spring loaded at the bottom and stretched between wire supports attached to the frame top and bottom. Painted black, these wire supports and springs are almost invisible in the viewer. The tensioned wire resists horizontal flexing as well as the usual 1/8-in. steel or brass rods used in the conventional mounts. It is not as good with respect to torsion. Auxiliary clamping, as in the conventional system, provides considerable improvement. However, no completely satisfactory support system has yet been devised.

The side lights for the model each have three 100-W bulbs fed through a single Variac. The light passes through a macro "Soller slit" system composed of horizontal sheets of thin metal about 4 in. × 24 in. with a vertical spacing of about 1/4 to 3/8 in. stacked to a total height of about 6 in. The metal is painted matt black before assembly. These two side lights are mounted on movable floor-standing frames of "slotted angle" which permit height adjustment. Commercial focusing studio lights would do as well or better, but are much more expensive.

For visualization or photography, the intensities of the map and model lights are adjusted to get the required relative contrast between the superimposed images of the map and model. This will vary slightly for different levels of the model and for different viewing positions, due in part to nonuniformity of the silvering on the mirror.

Use of Nonidentical Scales

The original paper described the process of matching with nonidentical scales. Since this has been of no practical importance, the description is omitted.

Measurement of Coordinates

With the model frame open at the top as well as the sides, the optical measurement of the coordinates of positions in the model becomes very simple. Two speed-frame bars are soldered together at right angles and each is equipped with a scale and a slider holding a small length of speed frame with cross hairs at both ends. The third scale is attached to the appropriate horizontal part of the model frame. One bar and slider lie across the frame with its cross hair piece vertical while the other bar hangs down beside the model with its cross hairs horizontal. When a given atom in the model is centered in both sets of cross hairs, the three rectangular coordinates are read off the scales. With a little practice, an atom can be centered and the coordinates obtained in about 30 sec. With no lenses in the system at all, average errors in position are less than 2 mm, corresponding to 0.1 Å. This is more than adequate for the inherent accuracy represented by models of this type. Viewing through the vertical cross hairs is incidentally made very easy by the 45° mirror conveniently at hand. For these measurements a piece of white cardboard for background is slipped under the model.

If a coordinate grid is inserted in a defined place in the map stack and a vertical scale mounted beside the model, coordinate measurements can also be readily obtained from stereographic photographs. Height measurements are made with a parallax bar. The grid position and height of a given atom are easily converted to the coordinates based on the desired orthogonal Cartesian axes. Description of the procedures can be found in standard texts on photogrammetry (see, for example, Trorey[5]). Simple equipment for this purpose is available from Abrams Instrument Co., Lansing, Michigan, and sold through its agents. More modern and rapid procedures are described in the following chapter.

[5] L. G. Trorey, "Handbook of Aerial Mapping and Photogrammetry," 2nd ed. Cambridge Univ. Press, London and New York, 1952.

Correction to the Original Bibliographic Citations

It is with considerable embarrassment that the author now reports a serious omission in the original literature search conducted as part of this study and the original 1968 report. All of the description reported above with the exception of certain construction details had appeared in much earlier publications. The only excuse is that this material is not located in sources normally consulted by protein crystallographers and was not retrievable through computer search systems in 1968. Even now the original discoverers of the procedure are unknown and reference is limited to a review volume, Hopkins,[6] and not to the original literature. In view of the time interval involved this oversight might be overlooked. However, the incredible correspondence between the illustrations appears to demand the publication of this addendum.

In Fig. 3 is seen the diagram of the 45° mirror and the reflected image superimposed on the background provided in this case by the canvas T–T. The description in the text accompanying this figure is as follows:

> "Amphitrite, come forth!" exclaims the person in charge of the show. All at once, a woman in the costume of an opera nymph rises from the sea without anything being visible to support her in space, in which she turns round and round, gracefully moving her legs and arms, now in one direction, and then in another. When the exhibition is at an end, she straightens out in the position of a swimmer about to make a dive, and plunges behind the curtain representing the ocean.
>
> The illusion that we have just described may be performed as follows: Amphitrite is an image—a specter analogous to those of Robin. If we imagine that a transparent glass, M–M., in our diagram, is inclined 45° with respect to the stage, a person clad in light clothing, lying horizontally upon a black background beneath the stage, and well illuminated, will exhibit an upright image behind the glass.
>
> This image will appear to be formed in front of the back canvas, T–T. Now, as Amphitrite is lying upon a table, P–P, she will be able to go through her evolutions and bend herself in a circle; and if, during this time, the table, movable upon its axis, A, is revolved, her image will turn in all directions. Finally, to cause Amphitrite to appear or disappear, it will suffice to slide the table upon rails, thus bringing it in front of or behind the glass. Amphitrite should be placed upon an absolutely black background. Her costume should be of a light color with metallic spangles, and she should be illuminated by a power electric light.

In Fig. 4 is seen a version corresponding to the more recent models in which the glass plate is vertical. This particular illusion emphasizes the

[6] A. A. Hopkins, "Magic: Stage Illusions and Scientific Diversions," Chapter II. Benjamin Blom, Inc., New York, 1897.

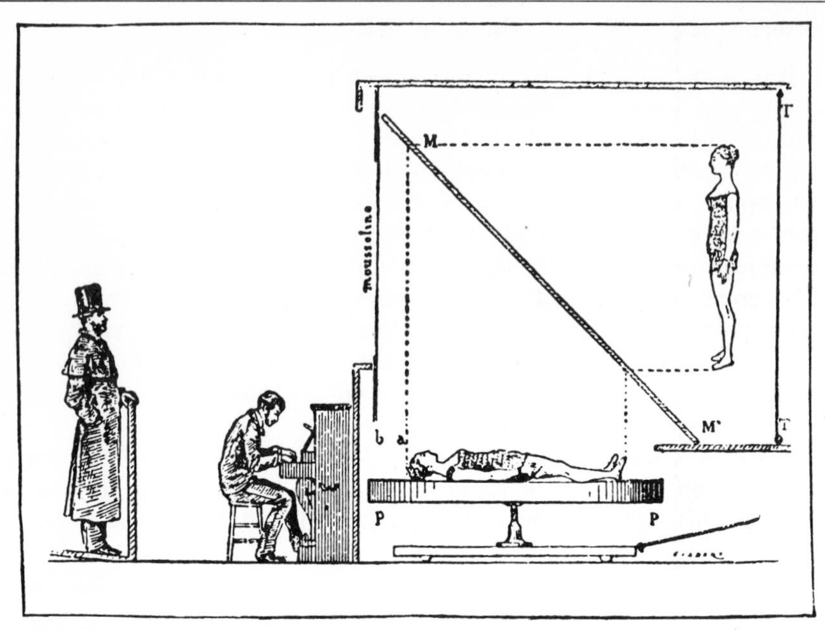

FIG. 3. "Diagram Explaining the Amphitrite Illusion." See the discussion in the text. (Reproduced from Hopkins[6] with permission of the publisher.)

use of the variable light sources to control the image mixing. The accompanying text is as follows:

> At the end of this second chamber, at the back of a stage, is seen a coffin standing upright, in which one of the audience is requested to place himself. Entering the stage by the side door he is conducted by an attendant to the coffin and placed in it. Blocks of wood are placed for him to stand on in quantity sufficient to bring his head to the right height so that the top of it just presses against the top of the coffin, and the attendant with great care adjusts his height according to the predetermined position. Two rows of Argand burners illuminate his figure, which is then wrapped in a white sheet. Now, as the spectators watch him, he gradually dissolves or fades away and in his place appears a skeleton in the coffin. Again, at the word of command the skeleton in its turn slowly disappears, and the draped figure of the spectator appears again. The illusion is perfect to the outer audience. Directly in front of the coffin, crossing the stage obliquely, is a large sheet of the clearest plate glass, which offers no impediment to the view of the coffin with its occupant, when the latter is fully illuminated. At one side of the stage, in the back of the picture, is a painting of a skeleton in a coffin with its own set of Argand burners. It is screened from view. When strongly illuminated, and when the lights of the real coffin are turned down, the spectators see reflected from the glass a brilliant image of the pictured coffin and skeleton. By turning up one set of burners as the others are turned down a perfect dissolving effect is obtained, skeleton replacing spectator and *vice versa* at the will of the exhibitor.

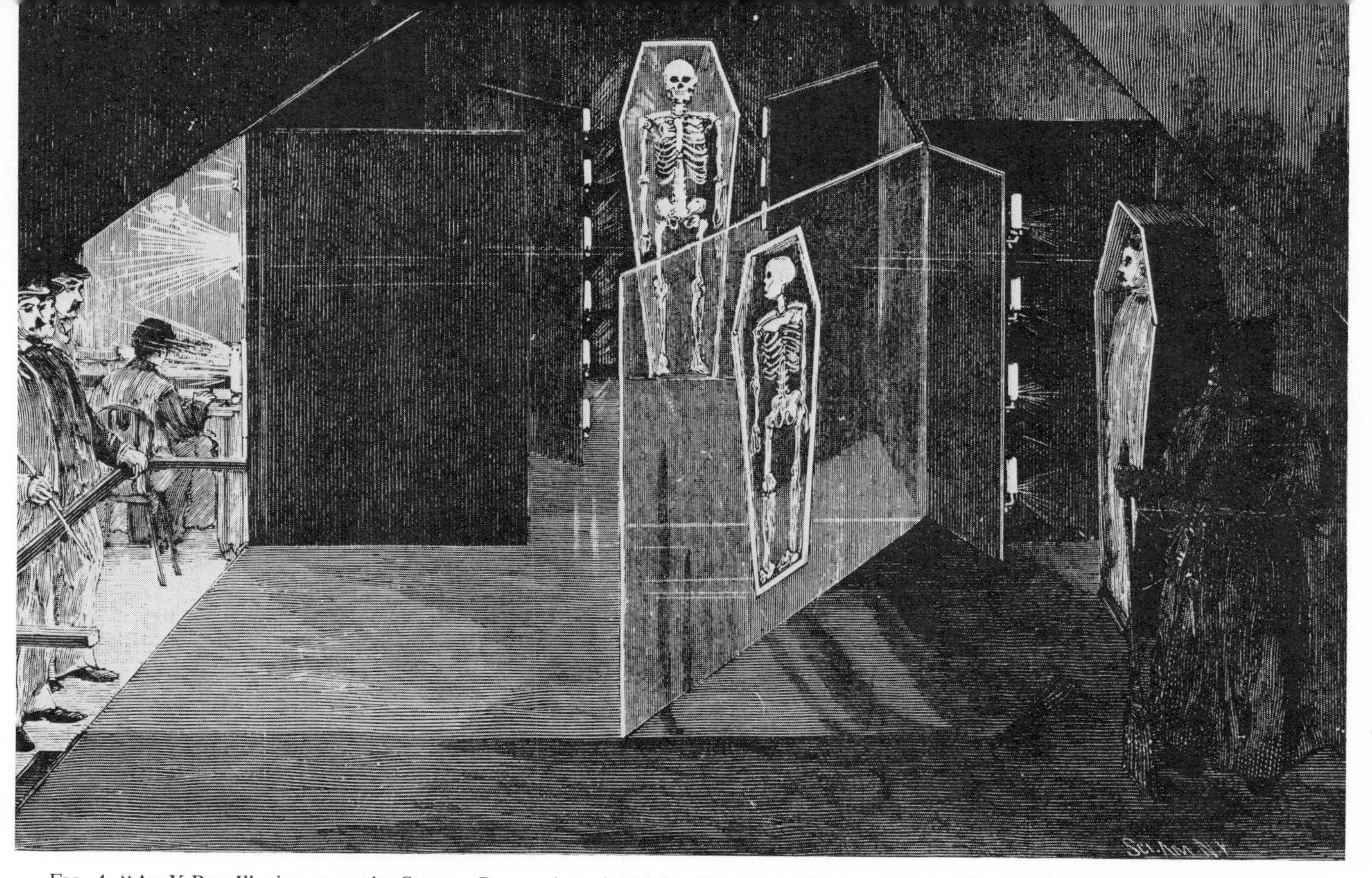

FIG. 4. "An X-Ray Illusion upon the Stage—Conversion of a Living Man into a Skeleton." See discussion in the text. (Reproduced from Hopkins[6] with permission of the publisher.)

The magic lantern operator always realizes that to secure a good dissolving effect perfect registration is essential.

Had this reference been known to the author in 1968 no further description of the "folly" would have been required. I wish to thank Mr. Howard Wellman of Cornell University for bringing the Hopkins' text to my attention.

[11] Some Minor Refinements on the Richards Optical Comparator and Methods for Model Coordinate Measurement

By F. R. SALEMME

Despite the long-standing, virtually universal adoption of the Richards Optical Comparator (this volume [10]) for skeletal model fitting in protein electron density maps, the basic design has remained essentially intact since its first description.[1] The only substantial design modification[2] has been to alter the orientation of the half-silvered mirror, so that it lies in a plane parallel to the electron density sheets rather than at 45° to them. The parallel mirror arrangement has the advantages that it is somewhat easier to construct, saves floor space, and allows the model to be constructed in closer proximity to the map.

An important aspect of fitting in the Richards comparator is the ability of the model builder to alter his or her position relative to the mirror so as to view the model-to-map fit from various aspects. However, occasionally it is desirable to view the map–model superposition from an angle where the map is only partially illuminated by the light box behind it. Owing to both this and the limited transparency of the map sheets, it is useful to construct the light box so that it can slide into the same slots the map sheets are usually mounted in. In this way the position of the light box illuminating the sheets can be adjusted to lie directly behind the last map sheet being examined. This makes the map easier to view, and the device more readily adaptable to maps of different depth.

Minor technical considerations, which nevertheless materially affect the ease of use of the comparator, involve how the map sheets are mounted and the materials used for plotting them. A basic objective is to

[1] F. M. Richards, *J. Mol. Biol.* **37,** 225 (1968).

[2] P. M. Colman, J. N. Jansonius, and B. W. Matthews, *J. Mol. Biol.* **20,** 701 (1972).

make the maps as transparent as possible to allow easy viewing through multiple sections. Perhaps the most cost-effective approach is to stretch thin (0.0005–0.001 in.) Mylar or drawing acetate (0.002–0.05 in.) over aluminum screen frames, which can be easily constructed in the laboratory and will work without undue deformation up to about 1.5 m^2. Although drawing acetate is significantly more transparent than Mylar, it changes dimensions with variations in humidity, which eventually results in the maps either sagging or springing from their frames. Whatever material is finally used, considerable eye strain is avoided by plotting the map with opaque drawing inks designed for drafting on plastic films. This produces a map with more consistent and higher contrast contours than can be obtained with generally available marking pens. Given the time and fatique involved in actually building a skeletal model, it is usually worthwhile to invest the initial effort in optimizing the clarity of the map to be fitted.

Once having obtained a skeletal model of a protein, the next task is to get accurate atomic coordinates as required for structural analysis or refinement. A wide variety of approaches have been used, which can be grouped into methods in which coordinates are obtained either directly from the wire model or in association with the Richards optical comparator.

Many of the most commonly used methods for measuring coordinates from a wire model basically involve positioning the model on a frame arranged as an orthogonal x–y coordinate system. Both the x and y axes of the frame, which might be made of optical bench rails, serve as translational mounts for a cathetometer, telescope, or cross hair sight. Coordinate measurement involves sighting successive atoms through each telescope to give x, z and y, z atom coordinates in the two projected directions. These can then be combined to produce three-dimensional coordinates. Alternatively, the whole process can be repeated twice with a single axis arrangement, by simply rotating the model 90° between successive measurements, while taking appropriate care to keep everything absolutely square and level.

An alternative method involves the use of a surveyor's transit plus a cathetometer.[3] Sighting each atom through the transit gives angular coordinates for each atom relative to an external origin. Combining these values with an atomic z coordinate measured with the cathetometer produces polar coordinates which can be subsequently transformed into Cartesian coordinates. More recently, this approach has been extended to use two transits which are relatively fixed in known positions. Here,

[3] M. A. Frentrup and A. Tulinsky, *J. Appl. Crystallogr.* **14**, 439 (1981).

measurements of angular relationships of the sighted atoms are used to compute Cartesian coordinates. An integrated, ready-to-use system incorporating electronic transits interfaced to a minicomputer is marketed as the Analytical Industrial Measuring System (AIMS) by Keuffel and Esser Co., and can be purchased or rented. This method has been recently proved effective in determining accurate model coordinates for KDPG adolase by A. Tulinksy and co-workers.

Although the optical methods described generally produce coordinates whose accuracy substantially exceeds the map-fitting precision, sighting the individual atoms through a telescope amid the forest of surrounding atoms nevertheless remains a tedious process. Protein crystallographers at Oxford University have overcome these difficulties by constructing a system in which the operator touches a wand to the atom in question and coordinates are then read out directly. The device is constructed as an array of three linear microphones defining a Cartesian axis system external to the model box. Correspondingly, the atom-designating wand has an acoustic transducer which the operator causes to emit a click when contacted with the atom to be measured. Electronic measurement of the acoustic flight time of the sonic pulse to the strip microphones, which are interfaced to a minicomputer, allows direct computation and readout of the atom Cartesian coordinates.

Some alternative methods of model coordinate measurement have utilized the Richards comparator as an integral part. The simplest application to coordinate measurement involves inserting a pair of grid sheets into the slots usually holding map section in the comparator. This allows projected x, z coordinates of each atom to be estimated by viewing the superposition of the model and the grids. The depth coordinate can then be obtained by rotating the model 90° and remeasuring. Coordinates measured in this way generally have a precision of about 0.2–0.3 Å on the 2 cm/Å scale, which may be adequate in some applications.

A more accurate measuring technique utilizing the optical comparator was first devised by H. Wyckoff at Yale. Here a traveling Cartesian coordinate system mounting a small light source was mounted inside the space usually occupied by the electron density map sections. The light could be moved around mechanically so that its transmitted image in the map space superimposed on the reflected image of each model atom in the half-silvered mirror. Atomic positions were read directly to punched paper tape by reading the values of potentiometers connected to the mechanical mechanisms driving the light translation along the three Cartesian coordinate axes. A more sophisticated version using electric motors to drive the light source, but retaining analog output, was subsequently constructed at the California Institute of Technology by M. Ross and

R. Stroud. A wholly digital version of the device, which allowed coordinates to be both read out and entered through a control console (e.g., for model building from coordinates), was constructed at the University of California at San Diego, and subsequently at the University of Arizona.[4] The motorized versions of this superposition device produce coordinates of comparable precision to the optical and acoustic devices described previously (at least ±1 mm), and similarly allow for very rapid measurements (one atomic position per 30 sec). However, construction of the former device is more complicated than either the AIMS (which can simply be rented) or acoustic coordinate measuring devices.

With the advent of computer graphics and associated software for both map fitting and coordinate readout (this volume [12]), it appears only a matter of time before graphics obviates either building physical models of proteins in optical comparators or devising means for measuring their coordinates. Nevertheless, the techniques described here will undoubtedly continue to prove effective aids in determining crystal structures for some time to come. In addition, many modeling studies of protein conformation and interaction continue to be most readily investigated in their initial stages by the construction of physical models. Subsequent model development and analysis by computer necessitates having accurate coordinates, so that the technical feasibility of such studies depends on having an accurate and easy way of measuring model coordinates.

[4] F. R. Salemme and D. G. Fehr, *J. Mol. Biol.* **70,** 697 (1972).

[12] Interactive Computer Graphics: FRODO

By T. ALWYN JONES

Introduction

Computer graphics provides an elegant method of controlling the protein crystallographer's interaction with his model. The graphics display allows the model to "show" the crystallographer a part of its electron density such that he can decide how a molecular fragment best fits. Once the crystallographer has made his decision, the computer merely does the bookkeeping and minor improvements. The aim of the molecular fitting program, therefore, is to create the necessary environment to allow the crystallographer to decide what atoms he wants in what piece of density.

A great deal of effort has been spent to develop the necessary control software and, in some cases, to build the hardware. At present there are approximately 10 density-fitting program systems in active use. Most laboratories are not able to build the necessary hardware, but high-performance equipment is available from a number of vendors. The program FRODO[1,2,3,4] is presently implemented on the Vector General 3400 (with DEC, VAX, or PDP-11 computers), MMS-X, Evans & Sutherland PS 2 (PDP-11), MPS (VAX or PDP-11), and PS300.

The man–machine interface in FRODO varies with the available equipment. All systems use a data tablet to pick menu items and to identify atoms shown on the screen. The VG3400 and Evans & Sutherland PS300 systems use analog-to-digital converters to define the view direction, and other picture-related functions such as clipping, zooming, and intensities. Commands that invoke dihedral angle rotations, single atom, and fragment shifts are also coupled to the A-Ds. On the Evans & Sutherland PS2 and MPS equipment, pseudo A-Ds are drawn on the screen and can be activated by the data tablet pen. The MMS-X version uses the standard user control panel.

Program Flow

The following is strictly applicable to only the VG3400 version of FRODO. Other versions have some minor differences.

After starting the program the user must specify a control data set. This data set contains all of the important parameters needed by FRODO. It includes, for example, the data set name for the user's coordinate file and the regularization zone. It is a source file and can be edited (at one's own risk). The user then gets to the CHAT interface, which has a large number of menu items activated from a terminal keyboard. Every exit from CHAT causes an update of the control file. Control normally then passes to the display loop.

To make the system easier to use an effort has been made to keep the display menu options to a minimum, and to a single "page." The data tablet pen position is marked on the screen by a cursor, and a menu item is activated by moving the pen so that the cursor is positioned over the

[1] T. A. Jones, *J. Appl. Crystallogr.* **11,** 268 (1978).

[2] T. A. Jones, *in* "Computational Crystallography" (D. Sayre, ed.), p. 303. Oxford Univ. Press, London and New York, 1982.

[3] B. L. Bush, *in* "Computers & Chemistry," vol. 8, p. 1. Pergamon, Oxford, 1984.

[4] J. W. Pflugrath, M. A. Saper, and F. A. Quicho, *in* "Methods and Applications in Crystallographic Computing" (S. Hall and T. Ashida, eds.), p. 404. Oxford Univ. Press, London and New York, 1984.

item and then pushing the pen into contact with the tablet. In principle, any number of commands can be activated and each is polled in turn. This means, for example, that an atom can be moved and have its contacts updated at the same time. In practice some care has to be taken, since there are a number of mixed options which may be disagreeable to the user. For example, if a group of atoms is being moved and the user activates the SAVE command, then the current fragment coordinates will get written to disk. The active commands can be seen at a glance because each has a star drawn next to it. To exit from the display loop, the user must pick a suitable menu item to either terminate the session or enter one of the utilities. Reentering the display loop causes a new loading of coordinate information from the disk.

The vectors drawn on the display are constructed from three data sets and normally show some of the atoms in the coordinate data set superimposed on a "chicken wire" representation of some sort of electron density.

The coordinate data set contains more than just atomic coordinates. The atoms are grouped together to form a residue. There are residue records to describe the type of residue (e.g., PHE, MPD), the name of the residue (e.g., A2, 10G), the position in the data set of the atom records for this residue, the center of gravity, and the radius of the residue. The residues are grouped together to form a sequence. As far as the display loop is concerned, the sequence is only important when defining viewing zones, i.e., it does not necessarily force any chemical connectivity between residues (although it may exist). The data set can also contain extra information such as lattice type (P, I, R, F, A, B, C), unit cell constants, and crystal symmetry information. This information is optional but may be required for certain commands. The user must decide at the CHAT interface how he wishes to access this data set. There are three possibilities: (1) Define the start and end residues of a zone. The program then displays all the atoms in the residues within the zone as defined by the sequence. (2) Define a point in space and a radius, and then display all of the atoms in the data set which are within the volume. (3) Define a mixture of 10 display zones plus a sphere.

In the sphere mode the user can choose an option to display any symmetry-related atoms which may fall within the volume. Both the sphere and symmetry options make use of the residue center of gravity information to decide what appears in the volume. Another option allows one to define by name which atoms are to be displayed; e.g., one can define just C_α to see the fold of a protein.

After picking which atoms are to be displayed, one must decide on a connectivity, i.e., which atoms should have a line drawn between them.

The usual connectivity scheme in FRODO is based on distance criteria so that if atoms are closer than a certain distance they are joined. In sphere and mix modes all atoms are tested together. In zone mode the connectivity is built up a residue at a time, and a specific link is made between residues. An atom with no connections appears on the display as a three-dimensional cross. If one is displaying just C_α atoms, for example, there is an option to connect the first atom to the second, the second to the third, etc. The initial connectivity does not necessarily represent chemically correct bonds. It is simply there as an initial framework for the crystallographer to decide how he is to change his structure to fit the density.

The second data set consists of linked vectors. It may represent density contoured at a number of different levels, or skeletonized electron density, or guide points, or a vectorized library of molecular data sets. The vectors are arranged in three-dimensional volume elements and in what are called contour commands (C-COMs). If the data set is a vectorized map, each C-COM corresponds to a contour level. In the vectorized molecular library each molecule is equivalent to a C-COM. The user can decide in the CHAT interface which (if any) C-COMs are to be chosen.

The third data set is an electron density map. It is also arranged in three-dimensional volume elements with each density value packed into one byte. This data set is much smaller than an equivalent vectorized map, and has the advantage that the contour levels can be changed at any time. It is, however, slower to work with than the vectorized data set. The map can also be used to automatically fit molecular fragments to the density. One often uses both a map and vector data set where the contour level has been chosen after a brief inspection of the map.

Building an Initial Model

The crystallographer usually knows the rough fold of his molecule before starting work on the display. This is best determined by extensive study of minimaps plotted on stacked plastic sheets. The structure solution of retinol binding protein by Newcomer *et al.*[5] is one of the few exceptions to this rule. There are then four different ways of building the model on the display.

The first method is a relic of working with Kendrew wire models in a Richards box. FRODO has extensive model-making features to produce coordinates from a given sequence which have standard bond lengths and angles and preferred torsion angles. A zone of residues can be made and

[5] M. E. Newcomer, T. A. Jones, J. Åqvist, J. Sundelin, U. Eriksson, L. Rask, and P. A. Peterson, *The EMBO J.* **3:7,** 1451 (1984).

moved to the place in the map where one wishes to begin building (not necessarily the N terminus). The fragment can be translated and rotated by the display menu command FBRT so that the first residue sits close to its density. Up to six consecutive dihedral angles can be varied at a time, and by repeated FBRTs and TORs the fragment may be made to fit the density. These coordinates are written to disk when the user is satisfied by using the SAVE command. The model-making option can be used to extend the zone of residues and an attempt made to fit these. However, it rapidly gets more difficult to do this while maintaining the constrained, rigid geometry.

In method two, by judicious choice of display zones, the user fits a few residues as described above but introduces a discontinuity in a peptide linkage by separately fitting the next zone of residues.

In method three the user introduces discontinuities directly on the screen. The screen connectivity is used to decide which atoms are affected by dihedral rotations and by fragment rotation/translation. Suppose a zone of residues fits the density but the user sees that a side chain in the center of the zone would fit much better if he could change ϕ (around the N–C_α bond) for the residue. If one changes this angle directly, all the other atoms to the end of the zone would be moved out of density. This is prevented by breaking the C_α–C bond of the residue using the menu BOBR command and then rotating around the N–C_α bond. This, of course, distorts the bond angles around the C_α atom. More commonly, the user disconnects the side chain from the main chain and moves the small fragment straight into the desired density. This is illustrated in Fig. 1a (which is drawn on a plotter directly from the picture on our display), where a growing chain has clear density for the phenylalanine side chain. The C_α–C_β bond is broken and the ring moved into the density using FBRT (Fig. 1b). The same coordinates are shown from a different view in Fig. 1c, where one can clearly see density for the carbonyl oxygen. In Fig. 1d the oxygen has been moved into this density and we now have a very distorted residue.

It should be clear that to simplify the fitting process we must introduce errors in bond lengths, angles, and fixed torsion angles. These can be removed by model regularization. To prevent the buildup of errors in particular variables, the regularization should have no built-in rigid constraints (such as fixed bond lengths, for example). FRODO uses the method described by Hermans and McQueen[6], which they call the method of local change. In this method each atom is shifted to minimize a weighted sum of terms representing the shift from its starting positions,

[6] J. Hermans and J. E. McQueen, *Acta. Crystallogr., Sect. A.* **30,** 730 (1974).

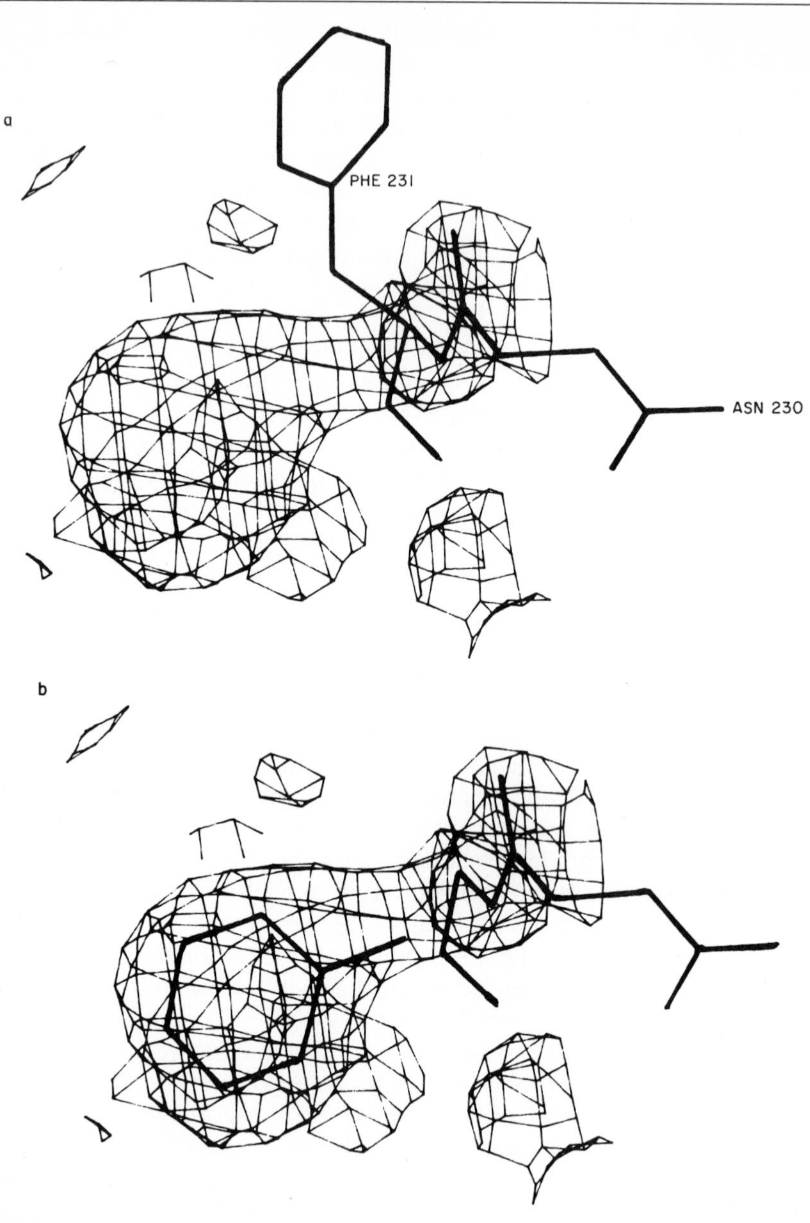

FIG. 1. Fitting a growing protein chain using method 3. (a) The growing chain with the Phe residue out of density. (b) The Phe ring is moved as a rigid group to fit the density. (c) A different view showing density for the carbonyl oxygen. (d) The carbonyl oxygen is moved into density. (e) The result of regularization with certain atoms fixed. (f) The ring stays in density after regularization.

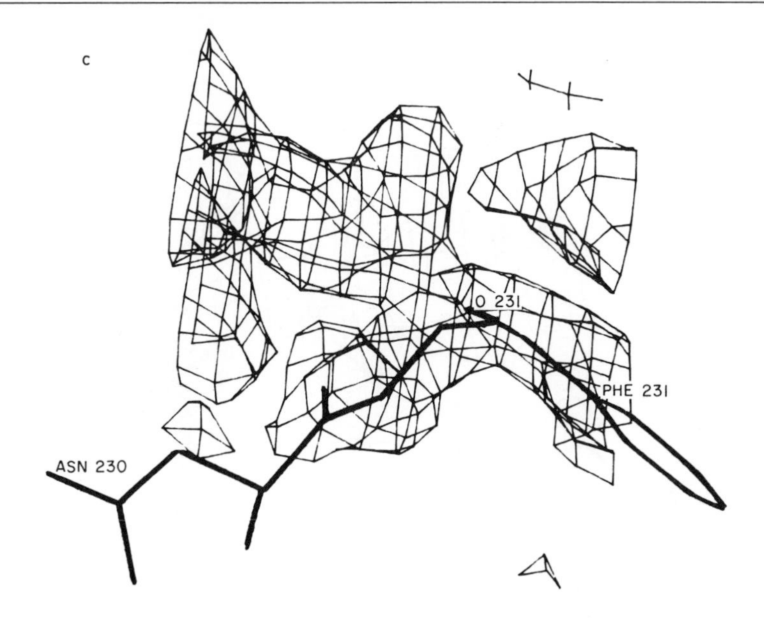

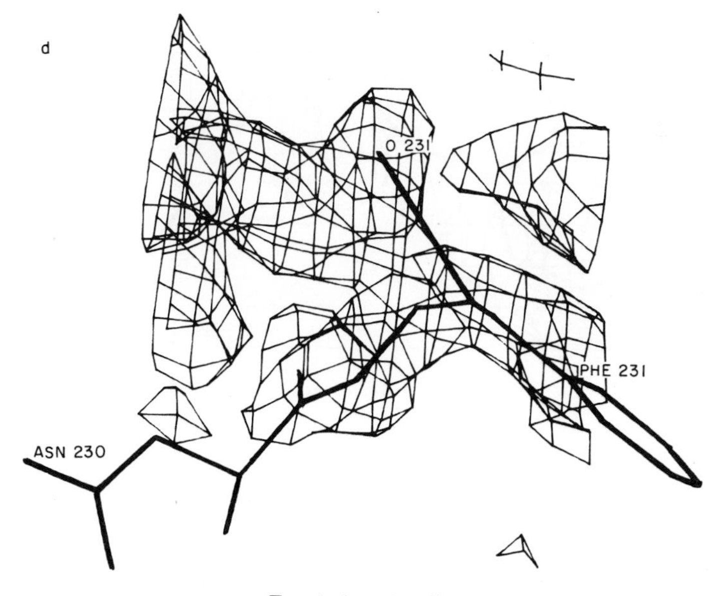

FIG. 1. (*continued*)

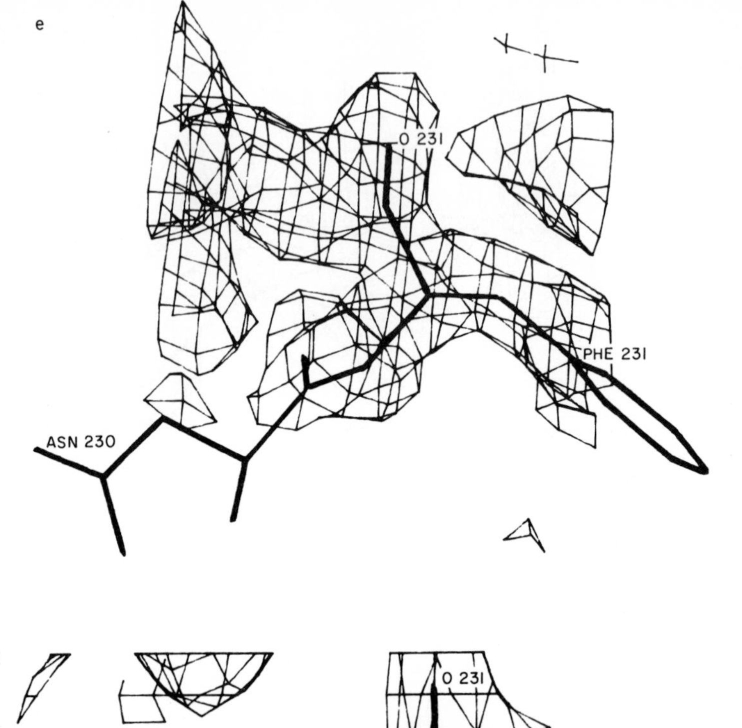

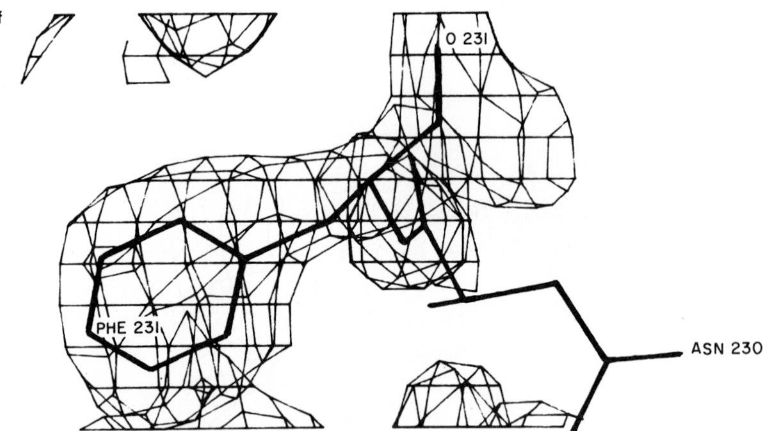

FIG. 1. (*continued*)

errors in bond lengths, bond angles, and fixed torsion angles. Extra terms can easily be added to maintain preferred values for normally variable torsion angles (e.g., for an α helix) and for distances between atoms (e.g., to maintain hydrogen bond distances).

To keep the atoms in density FRODO has a menu option FIX. Any atom hit with the FIX option will not move during regularization. Before regularizing the atoms shown in Fig. 1d, the peptide oxygens of residues 230 and 231 were fixed together with C_β of residues 231. The result of regularization is shown in Figs. 1e and 1f. Sometimes one carries out a second regularization with nothing fixed to remove some buildup of errors (normally in the CO bond). The skills to be learned in using FRODO are to decide how to distort the model to fit the density, and then to decide which atoms to fix so that the main group of atoms stay in density and the rest are pulled into it. Often the fit is a combination of actions such as a bond break to delimit a dihedral rotation, the dihedral rotation to line up some chain with some density, another bond break to separate the fragment, and finally a rotation/translation shift into the density.

The fourth method of fitting attempts to place atoms near their density before actually sitting in front of the display. Guide coordinates are read directly from the minimap. At most one picks (for a protein) a guess for the C_α, O, and one side-chain atom. For regions of the molecule with secondary structure, FRODO's model-making options are used to generate the desired conformation. Another option is then used to find the best rigid body fit to the guide coordinates. The gaps between secondary structure are filled by first inserting guide atom coordinates into the data set and making these atoms fixed. The coordinates for the remaining atoms get added (rather roughly) by the regularizer option, provided coordinates already exist for the first residue in the regularization zone. Because the guide atoms are fixed, the newly added atoms are forced to move close to their density. A second cycle without fixes is normally required to produce good stereochemistry. The last residue of the regularization zone is used as the first of the next zone. Once the complete protein has been made, the map can be inspected and improvements made as described earlier.

After making the first pass through the molecule in zone mode, the user must make a second pass in sphere mode. This is not just to ensure that chains do not interpenetrate. Since one rarely works at atomic resolution, the interpretation must be guided by contacts with neighboring residues. At this stage one can also decide how well the atoms fit the density. Atoms not fitting density are not used in later phase calculations.

The advantages of building a model with FRODO rather than in a Richards box are as follows:

1. *Speed*. One can normally build a model more quickly in a display. This gain can be lost by spending more time on details such as making better hydrogen bonds and trivial changes to improve the fit to the density.

2. *Accuracy*. All parts of the map are equally accessible in the display, but not in the box. Their crystallographic R factors are noticeably better,

often below 40%. The chicken wire representation of the density in the display is usually clearer than stacked sections in a box, and if the crystallographer can correctly interpret the density, he can build the model more accurately.

3. *Control*. Once a newly positioned conformation has been SAVEed with FRODO, the coordinates are safely on disk. Backup and restore features are available on the display menu, so that accidental overwriting of coordinate information is not a disaster. The wire model is easily degraded by gravitational forces prior to measuring coordinates. These coordinates must be regularized and their positions checked by plotting on density sections.

4. *Volume*. For large enzymes and viruses the actual space occupied by a physical model becomes a problem, especially in mature laboratories. A Kendrew model of the plant virus satellite tobacco necrosis virus (STNV) would have a volume of 18 m^3.

5. If the crystallographic asymmetric unit consists of more than one molecule, the model of the first molecule built in the display can be used as a good starting model for the second. This technique was used on southern bean mosaic virus, in which there are three different protein chains.[7] One can also use coordinates from related conformations as was done by Eklund *et al.*[8] in their study of holo-alcohol dehydrogenase, ADH. In this example the coordinates for the two domains of the apo-ADH were rotated into the map, and then rebuilt to fit the density. The second chain of the molecule was then built using the transformed first chain as the initial model.

The main disadvantage of building a model in the display is that there are usually so many people wanting to use it that one cannot just go and check out an idea on the spur of the moment.

A Tool in Crystallographic Refinement

FRODO originated in the groups of Drs. Robert Huber and John Gassman in 1976. By that time it was known that protein molecules could be refined, provided the crystallographer had the necessary patience and computer power. At that time Huber's group used Diamond's constrained real space refinement program[9,10] to improve the model fit in maps calcu-

[7] M. G. Rossmann, C. Abad-Zapatero, M. A. Hermodson, and J. W. Erickson, *J. Mol. Biol.* **166,** 37 (1983).

[8] H. Eklund, J-P. Samama, L. Wallén, C-I. Bränden, Å. Åkeson, and T. A. Jones, *J. Mol. Biol.* **146,** 561 (1981).

[9] R. Diamond, *Acta. Crystallogr., Sect. A* **27,** 436 (1971).

[10] J. Deisenhofer and W. Steigemann, *Acta. Crystallogr., Sect. B* **31,** 238 (1975).

lated with $2|F_o| - |F_c|$ amplitudes and model phases. After some cycles the refinement would stop and it was then necessary to find out why. By studying various sorts of maps (with model or model/isomorphous combined phases) it was usually possible to locate errors in the model but very difficult to correct them. Since then, refinement methods have improved, and mostly gone into reciprocal space.[11,12,13] These programs are very good at removing small random errors from the model, but they cannot correct major nonrandom errors.[2] During refinement one hopes that the improvement in the phase angles is such that the nonrandom errors can be identified and corrected on the display.

When building the first model in the MIR map, I find it useful to divide the atoms into three groups: the good, the bad, and the ugly. During refinement one hopes that the distribution is pushed in the direction of the good atoms. In Munich and Uppsala an atom flag in the FRODO coordinate data set is used to decide if the atom is used in phase calculation. Normally during the course of the refinement more atoms are used, although one frequently removes atoms for a few cycles if one distrusts their behavior. We often use chopped Fouriers, in which a portion of the structure is removed from the phase calculation and a Fourier calculation (normally $2|F_o| - |F_c|$ amplitudes) made around the region of interest.

The bookkeeping qualities inherent in the computer system are an important aid in refinement. With FRODO the crystallographer can maintain a library of past models in the vector data set and check his present model against any past version. It is also helpful to keep a map copy of the MIR map since it is the only density unbiased by the model.

The main disadvantage of working with FRODO is that one is often faced with one's own inadequacy, in that one can see everything perfectly and still not know what to do.

FRODO Crystallographic Extensions

An important development in computer graphics in the last few years has been the use of 32-bit computers to control the display. The increased computer power means that crystallographically meaningful calculations can be made on part of the structure when the user sits at the display. The first to be introduced is a real space refinement option.[14] This option is an

[11] W. A. Hendrickson and J. H. Konnert, *in* "Computing in Crystallography," p. 13.01. Indian Acad. Science, Bangalore, 1980.

[12] J. L. Sussman, S. E. Holbrock, G. M. Church, and S. H. Kim, *Acta Crystallogr., Sect. A* **33,** 800 (1977).

[13] A. Jack and M. Levitt, *Acta. Crystallogr., Sect. A* **34,** 931 (1978).

[14] T. A. Jones and L. Liljas, *Acta Crystallogr., Sect. A* **40,** 50 (1984).

extension of the fragment rotate/translate menu option and is called fingertip refinement. The fragment moves as a rigid body to maximize the grid sum convolution $\Sigma\ \rho_{calc}\rho_{obs}$, where ρ_{calc} is the calculated density obtained from a Gaussian function and ρ_{obs} the observed grid density. The user controls the number of grid points over which to sum and the rotation/translation search range. The technique is illustrated in Fig. 2. The side chain C_β–C_γ bond has been severed and the two fragments have been arbitrarily moved out of the density, as shown in Figs. 2a and b. The results of volume fitting each fragment are shown with the same views in Figs. 2c and d. A noninteractive version can refine a zone of residues. Each residue is split up into smaller fragments, each of which is refined as above. The result can be shown on the display so that the user can decide if there should be any fixes before regularization. These options are not seen as replacing reciprocal space refinement methods but as an aid to the FRODO user. They have, however, been used by Jones and Liljas[15] to carry out the first crystallographic refinement of a virus (satellite tobacco necrosis virus) to a resolution of 2.5 Å.

FRODO as a General Molecular Modeling System

FRODO was designed as a tool for the practicing protein crystallographer. However some of its features make it of more general use:

1. The flexible coordinate data set and the absence of any connectivity dictionary mean that any sort of molecular fragment can be displayed.
2. There are a number of different options available for choosing which part of the coordinate data set should be displayed. However, on a 16-bit computer there are space limitations as to how many atoms can be manipulated (200 atoms on a PDP 11/VG3400 system).
3. The MOL option can be used to create quite complicated display lists. These can be made up of any combination of atoms, zones, spheres, residue types (i.e., all Phes, say) and molecular surfaces. Display lists are sets of vectors and as such take the place of the electron density normally displayed with FRODO.
4. The fragment rotation/translation with neighbor calculations can be used to dock substrates into active sites. This has recently been enhanced to include local energy minimization (Cambillau *et al.* [16]).
5. The edit features in the SAM option allow one to change residue types and delete or insert new residues. If there are sequence similarities between two molecules, and if the X-ray structure has been solved for one

[15] T. A. Jones and L. Liljas, *J. Mol. Biol.* **177,** 735 (1984).

[16] C. Cambillau, E. Horjales, and T. A. Jones, *J. Molecular Graphics* **2,** 53 (1984).

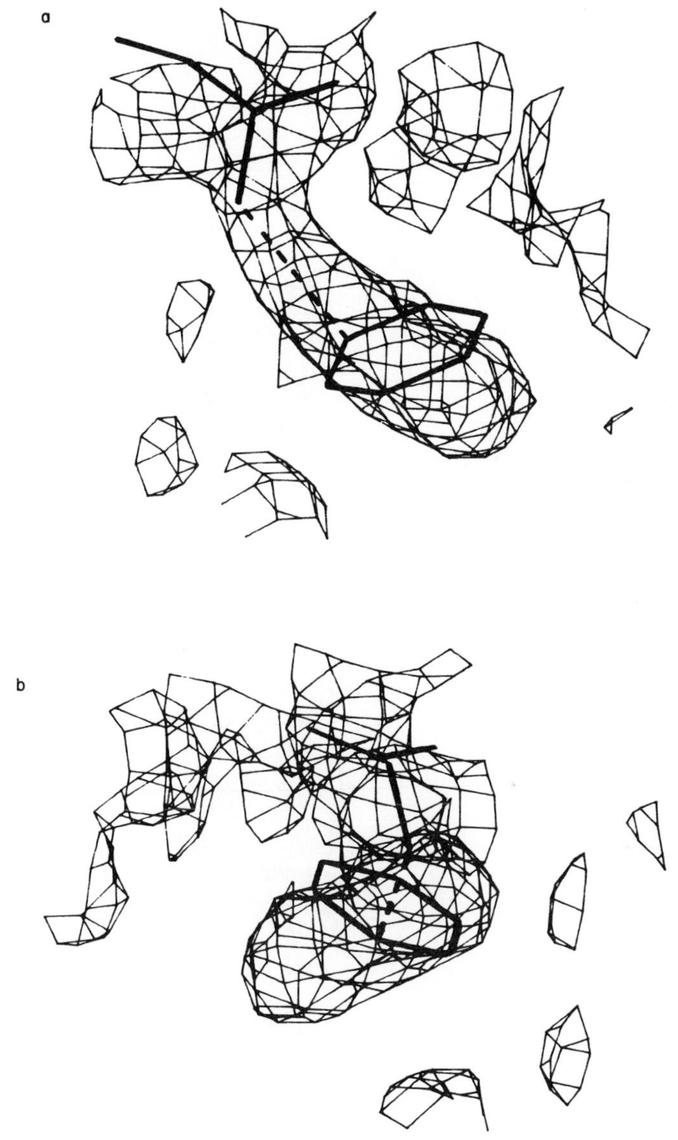

FIG. 2. Fitting a fragment automatically using real-space volume fitting. (a) and (b) are different views of the starting coordinates. The dashed line represents the C_β–C_γ bond, which is not drawn on the display because the atoms are too far apart. (c) and (d) are the results after refinement of each fragment.

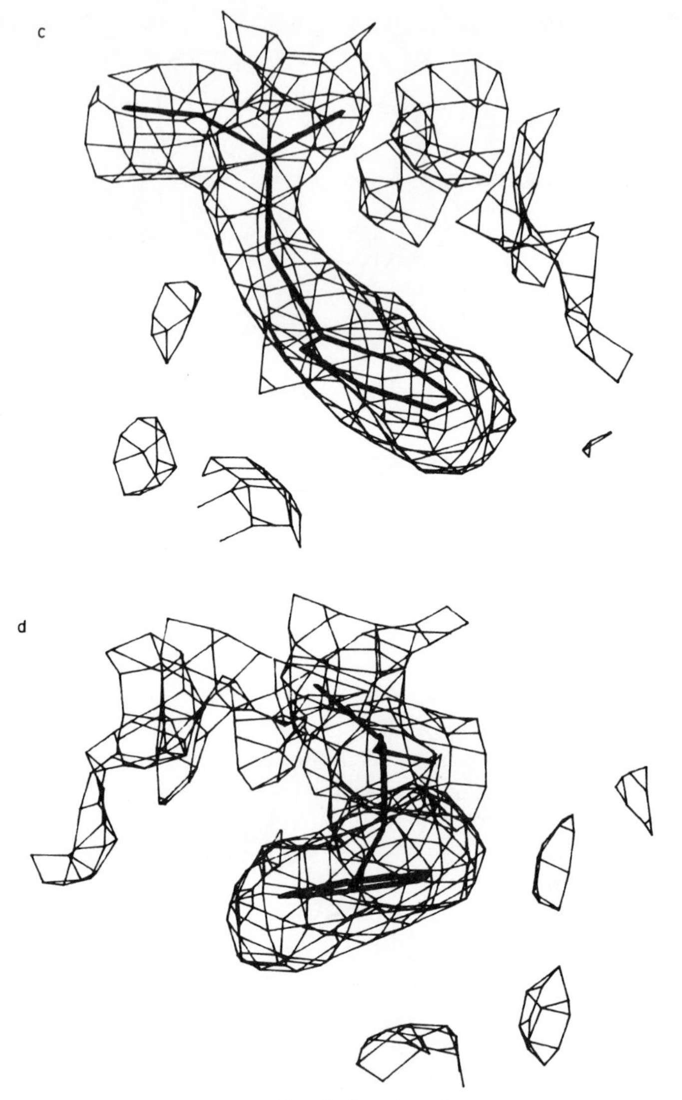

FIG. 2. (*continued*)

of them, then its coordinates can be used as a skeleton to hang on the other sequence.

The program is cumbersome when one tries to compare two related molecules, because only one coordinate file can be active at a time. However, the main drawback of the program is that there is no entry to a complete protein data base system except via a single coordinate data set.

Acknowledgments

Drs. J. Gassmann and R. Huber encouraged me to develop a density-fitting program. Professor T. Blundell invited me to get FRODO running on an Evans & Sutherland PS 2, and Dr. I. Tickle improved my efforts. Dr. L. Bush made the MPS and Drs. J. W. Pflugrath & M. A. Saper made the PS300 conversions. Drs. C. Cambillau and E. Horjales are responsible for the energy-minimization options. Dr. J. Hermans developed the regularization method.

[13] Rationalization of Molecular Models

By JAN HERMANS

Introduction

It is normal to represent the "solution" of a crystal structure, which itself approaches the electron density as closely as the accuracy of the structure factors permits, in terms of a molecular model, i.e., a collection of atoms. There are two principal reasons to do this: chemists and biochemists would pay little or no attention to a structure that was presented in any other form, and the model is an indispensable part of the principal crystallographic refinement technique applicable to proteins. Available chemical information is invariably incorporated in the model; of a protein it is preferable to have the entire amino acid sequence. If the amino acid sequence is not known, a good deal of it may be inferred from the shape of the calculated electron density; the larger side chains are easily recognized. However, individual C, N, and O atoms are not resolved in the first maps. Atomic resolution may be reached at the later stages of refinement, if the experimental data are of sufficiently high resolution.

The fact that the map does not clearly indicate individual positions for the majority of atoms causes considerable uncertainty, which can be much reduced by making the model conform to standard stereochemistry, according to measured bond lengths and bond angles of small molecules such as amino acids and peptides. These were first established in the now classic work of Pauling and co-workers. Bond lengths and bond angles have been found to be very similar for each type of chemical group, even in somewhat different physical (i.e., nonbonded) or chemical environments. In addition, such groups as aromatic rings and the peptide group have been found to have close to planar geometry in these model compounds. A complete set of generally agreed-upon, so-called ideal, or canonical, parameters describing the bonded geometry of proteins exists.

Despite these constraints, the conformation of a protein remains extremely variable as a result of freedom of internal rotations about single bonds; each residue contains two single bonds in the main chain, between α-carbon and peptide carbon and nitrogen, respectively, and a variable number of single bonds in the side chain.

When molecular models were first fitted to electron density maps of proteins, so-called Kendrew models were built from rigid brass parts made precisely to a scale of 2 cm to the angstrom according to Pauling geometry. The parts were joined at each single bond by a connector permitting complete freedom of internal rotation about the bond, and, as much as possible, care was taken not to deform the bond angles or the planarity of the parts during construction of the model.

The first model-building computer program for proteins was developed by Diamond[1]; this program has been used, for example, to correct the often imprecisely measured coordinates of the brass model. Just like the brass model, the computed model consists of parts of unvarying ideal geometry, connected by single bonds, rotation about which provides the necessary conformational freedom. Diamond has subsequently developed his program so that it improves a computed fit of a model to an electron density map, i.e., in real space, by criteria similar to the ones used in least-squares crystallographic refinement in terms of structure factors.

Use of a model of perfectly ideal geometry in protein crystallography is thus both practical and traditional. However, it has turned out that strict adherence to perfect geometry is unnecessary and that it is often more practical to use models of approximately perfect geometry instead. Such less perfect models had suggested themselves from studies of molecular geometry in terms of molecular energy. In these studies, molecular energy is approximated as a sum of terms that correspond both to deviations from ideal geometry and to so-called nonbonded interatomic repulsions and attractions. In this context, a molecule having perfect, ideal geometry simply has all energy terms for geometric deformation equal to the minimum, i.e., equal to zero. However, the important criterion is that the *total* energy should be as low as possible; the minimum of the total energy frequently corresponds to a lower nonbonded energy obtained at the expense of some distortion of the geometry. Use of a model having slightly imperfect geometry may therefore be preferable, especially when high resolution is attained.

[1] R. Diamond, *Acta Crystallogr.* **21**, 253 (1966).

Computer programs that manipulate protein conformation and lower the conformational energy while allowing for geometric deformation were first developed by Levitt and Lifson.[2] These methods were applied to the problem of model rationalization by Hermans and McQueen.[3] Other methods of model rationalization that have since been developed are variations of the two intrinsically different approaches, and were designed to facilitate specific tasks, which vary from constrained refinement to model manipulation on interactive graphic systems.

The objective of these methods is the calculation of a molecular model of ideal geometry such that the atomic positions deviate minimally from a given set, the so-called target positions. The discussion of the methods is split in two parts: the first part describes how this can be done by letting a model of perfectly ideal geometry gradually approach the target positions, while the second part describes the alternative method in which ideal geometry is gradually approximated as the model slightly moves away from the target positions.

Models Having Perfectly Ideal Geometry

Least-squares fitting of the coordinates **X** of a perfectly ideal model to a set of target coordinates $\mathbf{X}_0$, solely by application of internal rotations about single bonds, can be formulated as minimization of the error[4]

$$D = \sum_i \sum_j (x_{ij} - x_{0,ij})^2 \qquad (1)$$

Or, more generally,

$$D = \sum_i W_i \sum_j (x_{ij} - x_{0,ij})^2 \qquad (2)$$

if it is desired to attach a different weight W to the error term of each atom; this may be done, for example, on the basis of the local quality of the electron density map. This error must be minimized with respect to

[2] M. Levitt and S. Lifson, *J. Mol. Biol.* **46,** 269 (1969).

[3] J. Hermans and J. E. McQueen, *Acta Crystallogr., Sect. A* **A30,** 730 (1974).

[4] **X** represents the entire coordinate set, $\mathbf{x}_i$ represents the coordinates of atom i, and x_{ij} represents the jth component of the coordinates of atom i. When necessary, a superscript in parentheses indicates the coordinate system in which **x** is expressed, e.g., $x_{ij}^{(k)}$; $\mathbf{X}^{(0)}$ represents the model coordinates in the global coordinate system, i.e., the coordinate system of the target coordinates, $\mathbf{X}_0$. **v** represents a vector with three components, denoted v_1, v_2, v_3, while either $|\mathbf{v}|$ or v may denote the vector's length. Scalar and vector products are indicated by $\cdot$ and $\times$, respectively.

the set of variable dihedral angles, $\boldsymbol{\rho}$, where $\mathbf{X}$ is a known function of $\boldsymbol{\rho}$. This problem is sufficiently complex that an initial guess of the model coordinates, $\mathbf{X}$, is required, from which by some standard technique of function minimization one then derives a conformation that corresponds to a minimum (but not necessarily the global minimum) of the error D.

This initial guess must be constructed with some care. When the geometry of the target coordinates is good, dihedral angles calculated from the target coordinates are a good initial set of variables from which to calculate the first model with ideal geometry. Care is needed because of the general problem that variation of a single dihedral angle in the main chain of a model displaces a large number of atoms with respect to the remainder, i.e., results in what may be called a *global* change of the model. Consequently, it is advisable to construct the initial model in a piecemeal fashion, in which the coordinates of the pieces are successively improved by the minimization procedure before the model as a whole is made to fit the target position.

A typical model-building program consists of parts that perform the following functions:

1. Generation of a model from standard geometric parameters and a set of variable dihedral angles.
2. Alteration of the model when the variables are changed.
3. Calculation of the derivatives of the error D, relative to each variable parameter.
4. Minimization of the error by repeated alteration of variable parameters and model, and calculation of derivatives.

Each of these operations can be performed by more than one method. Important features and details of currently known best implementation schemes of each of these methods are described below.

Generation of a Model for a Given Set of Dihedral Angles

Matrix Method. The standard method by which atomic coordinates are computed from a complete set of bond lengths, bond angles, and dihedral angles was first given by Eyring.[5] The method uses the concept of two coordinate systems, one of which is the global system of the molecule, in which the coordinates need be expressed, and the other of which is a local system in which atomic coordinates are given by simple expressions. Coordinates are transformed from one system to the other by axis rotation and translation, of which the former is handled as a matrix multiplication. Figure 1 shows an appropriate local coordinate

[5] H. Eyring, *Phys. Rev.* **39,** 746 (1932).

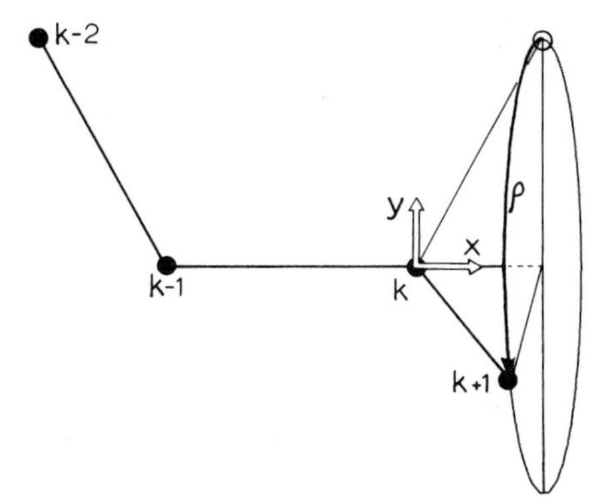

FIG. 1. Definition of an appropriate local coordinate system (x, y, z) and the dihedral angle ρ. (a) The origin lies at atom k and the x axis is parallel to the bond from atom $k - 1$. The y axis is in the plane formed by the x axis and bond $(k - 2, k - 1)$. Displacement of atom $k + 1$ by internal rotation by ρ about bond $(k - 1, k)$ is indicated. The dihedral angle is the angle between two planes, respectively containing atoms $k - 2, k - 1, k$ and $k - 1, k, k + 1$. Sign and magnitude of ρ correspond to conventions which state that ρ is positive if the atom down the spindle is rotated in a clockwise fashion, and that the eclipsed conformation has $\rho = 0$.

system (x, y, z), defined by two chemical bonds. In this kth coordinate system (superscript k) the coordinates of atom $k + 1$ are expressed simply as

$$\mathbf{x}_{k+1}^{(k)} = \begin{pmatrix} -l \cos \theta \\ l \sin \theta \cos \rho \\ l \sin \theta \sin \rho \end{pmatrix} \tag{3}$$

in terms of the appropriate bond length l, bond angle θ, and dihedral angle ρ.

Transformation of coordinates from the local system to the global system (superscript 0) is performed with

$$\mathbf{x}_{k+1}^{(0)} = \mathbf{M}_k \, \mathbf{x}_{k+1}^{(k)} + \mathbf{x}_k^{(0)} \tag{4}$$

Each matrix $\mathbf{M}_k$ may be computed in one of two ways.

1. $\mathbf{M}_k$ follows directly from the coordinates of atoms $k - 2$, $k - 1$, and k, since the components of $\mathbf{M}$ are also the components of the unit vectors $\mathbf{x}_k$, $\mathbf{y}_k$, and $\mathbf{z}_k$, expressed in the global coordinate system

$$\mathbf{M}_k = \begin{pmatrix} x_{k,1}^{(0)} & y_{k,1}^{(0)} & z_{k,1}^{(0)} \\ x_{k,2}^{(0)} & y_{k,2}^{(0)} & z_{k,2}^{(0)} \\ x_{k,3}^{(0)} & y_{k,3}^{(0)} & z_{k,3}^{(0)} \end{pmatrix} \tag{5}$$

Here $\mathbf{x}_k$ is simply the unit vector along the bond $(k-1, k)$, $\mathbf{z}_k$ is the unit vector parallel to the vector product of the bond vector $(k-2, k-1)$, and $\mathbf{x}_k$, and $\mathbf{y}_k$ is obtained as the vector product $\mathbf{z}_k \times \mathbf{x}_k$.

2. Each matrix, $\mathbf{M}_{k+1}$, may be computed from the preceding matrix, $\mathbf{M}_k$, by multiplication by a matrix $\mathbf{T}_{k+1}$,

$$\mathbf{M}_{k+1} = \mathbf{M}_k \mathbf{T}_{k+1} \tag{6}$$

where

$$\mathbf{T}_{k+1} = \begin{pmatrix} -\cos\theta & -\sin\theta & 0 \\ \sin\theta\cos\rho & -\cos\theta\cos\rho & -\sin\rho \\ \sin\theta\sin\rho & -\cos\theta\sin\rho & \cos\rho \end{pmatrix} \tag{7}$$

describes successive rotations about the local x axis by $-\rho$ and the local z axis by $\theta - \pi$.

Use of the matrix method implies selection of a group of three atoms whose positions in the global coordinate system are unrelated to molecular geometry; this is called the root segment of the molecule. Transformation from the overall molecular coordinate system, centered on the root segment, to the global coordinate system is again performed by successive axis rotation and translation, i.e.,

$$\mathbf{X}^{(0)} = \mathbf{M}_1 \mathbf{X}^{(1)} + \mathbf{t} \tag{8}$$

Consequently, the position of the root is formally determined by six independent parameters. Three of these are the components of $\mathbf{t}$, which happen to be equal to the coordinates of a root atom; $\mathbf{M}_1$ depends on three additional parameters, for which one may choose a set of three Euler angles $\boldsymbol{\phi}$.

Regardless of how an axis rotation matrix $\mathbf{R}$ such as $\mathbf{M}$ or $\mathbf{T}$ is formulated, it has several useful properties: the transpose of $\mathbf{R}$ is the inverse of $\mathbf{R}$,

$$\mathbf{R}\,\mathbf{R}^{\mathrm{T}} = \mathbf{I} \tag{9}$$

the columns of R are orthogonal unit vectors, as are the rows; each column represents what was a unit vector along one of the coordinate axes before the rotation; the product of two axis rotation matrices is also an axis rotation matrix.

Piecewise Construction. An alternative to use of the matrix method is construction of the protein from a dictionary of coordinates of amino acid residues with standard geometry, each of which is completely described in a local coordinate system.[1] Each residue is linked to the end of the molecule by simple rotation and translation; the proper conformation of the molecule is subsequently obtained by changing each dihedral angle to the desired value by internal rotation over an angle δ_k equal to the difference of the desired and initial values of ρ_k, according to the second of the two methods described in the following section.

Alteration of the Model

There are three ways by which the coordinates of a model may be changed without a change of bonded geometry.

1. By the matrix method. The first consists of recomputing the model for the new set of ρ values following the same procedure that was used to generate the original model, as has been described in the preceding section, under "The Matrix Method."

2. By successive internal rotations. The second consists of in turn applying each rotation by an amount δ_k about the kth spindle, by moving every affected atom i by an arc of length $R_{ik}\delta_k$, where R_{ik} is the distance from the atom to the spindle (Fig. 2). The atom's displacement is

$$\delta \mathbf{X}_i = R_{ik}[\mathbf{w} \sin \delta_k - \mathbf{v}(1 - \cos \delta_k)] \tag{10}$$

u, **v**, and **w** are orthogonal unit vectors, of which **v** is along $\mathbf{R}_{ik}$ and **u** is down the spindle. (The portion of the molecule that moves is considered to lie on the "down" side of the spindle.)

It is most direct to calculate $R_{ik}\mathbf{w}$ and $R_{ik}\mathbf{v}$ as

$$R_{ik}\mathbf{w} = \mathbf{S}_{ik} \times \mathbf{u} \tag{11a}$$

$$R_{ik}\mathbf{v} = (\mathbf{S}_{ik} \times \mathbf{u}) \times \mathbf{u} \tag{11b}$$

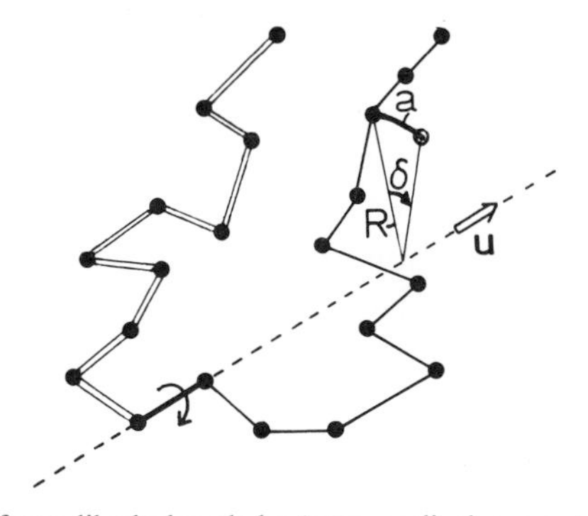

FIG. 2. A change of one dihedral angle by δ causes displacement of one part of the model in relation to the remainder; the two parts are distinguished by the manner in which the bonds are drawn. R is the distance of a selected atom from the axis of rotation, or spindle; the atom moves over an arc a of length $R\delta$. Other symbols are explained in the text.

where $\mathbf{S}_{ik}$ is a vector from atom i to one of the two atoms that define the spindle.

3. Rigid-body rotation and translation. Optimization by rigid-body rotation and translation according to

$$\mathbf{X}_R = \mathbf{R}\mathbf{X} + \mathbf{t}_R \tag{12}$$

is best accomplished by a rotation about the centroid $\mathbf{c}$ of the model, defined by

$$c_j = \sum_i W_i x_{ij} \Big/ \sum_i W_i \qquad (j = 1, 2, 3) \tag{13}$$

followed by a translation that makes centroids of model and target coincide. The entire rigid-body fit may therefore be summarized by

$$\mathbf{X}_R = \mathbf{R}(\mathbf{X} - \mathbf{c}) + \mathbf{c}_0 \tag{14}$$

As was first shown by McLachlan[6] the desired rotation matrix $\mathbf{R}$ is such that it maximizes the sum of the diagonal terms of the product of $\mathbf{R}$ and a 3×3 correlation matrix $\mathbf{C}$, given by

$$C_{lm} = \sum_i W_i[(x_{il} - c_l)(x_{0,im} - c_{0,m})] \tag{15}$$

From this condition, the matrix $\mathbf{R}$ can be computed by matrix algebra or, about equally effectively, by iterative adjustment of the three Euler angles on which $\mathbf{R}$ depends.[6,7] This method has the advantage that $\mathbf{R}$ is computed from $\mathbf{C}$, and that the coordinates are used only twice, first when $\mathbf{t}$ and $\mathbf{C}$ are computed and second when $\mathbf{X}_R$ is obtained from $\mathbf{X}$ with Eq. (14).

Derivatives of the Error D with Respect to the Variables

Various minimization techniques have been successfully applied to this problem. All of these require computation of the derivatives of the error D with respect to the variables $\mathbf{X}$. Since D depends on $\mathbf{X}$ by Eq. (1), and $\mathbf{X}$ again on ρ, the derivatives are

$$\partial D/\partial \rho_k = 2 \sum_i W_i \sum_j (x_{ij} - x_{0,ij})\, \partial x_{ij}/\partial \rho_k \tag{16}$$

[6] A. D. McLachlan, *Acta Crystallogr., Sect. A* **A28,** 656 (1972).
[7] D. R. Ferro and J. Hermans, *Acta Crystallogr., Sect. A* **A33,** 345 (1977).

As was shown by Thompson,[8] the derivatives $\partial x_{ij}/\partial \rho_k$ ($j = 1, 2, 3$) are most simply formulated as the components of a vector tangent to the arc along which atom i moves as a result of rotation about spindle k

$$\partial \mathbf{x}_i/\partial \rho_k = R_{ik}\mathbf{w} \tag{17}$$

An expression for $R_{ik}\mathbf{w}$ has been given above in Eq. (11a) (see also Fig. 2).

As noted, the coordinates of the model depend not only on the variable dihedral angles, but also on an overall rotation and translation of the entire molecule. Optimal rigid-body rotation and translation may be maintained during the minimization of D by application of a rigid-body fit after each change of the conformation. Alternatively, six variables, i.e., three Euler angles, $\boldsymbol{\phi}$, and the three components of a translation $\mathbf{t}$, may be added to the set of variable dihedral angles. One then calculates the corresponding derivatives of D and uses the altered values of $\boldsymbol{\rho}$ and of $\boldsymbol{\phi}$ and $\mathbf{t}$ when computing the new model.

The derivatives with respect to the components of $\mathbf{t}$ are simply

$$\partial D/\partial t_j = 2 \sum_i W_i(x_{ij} - x_{0,ij}) \tag{18}$$

Derivatives with respect to three Euler angles, $\boldsymbol{\phi}$, that describe rotation about the global coordinates axes from the *current* orientation of the root segment to a new position are easily calculated by Thompson's geometric method,[8] as follows. The derivatives of the error can be written in terms of the derivatives of the coordinates, $\partial \mathbf{x}_i/\partial \phi_k$, using Eq. (16) with ρ_k replaced by ϕ_k. The latter derivatives are given by

$$\partial \mathbf{x}_i/\partial \phi_k = R_{i1}\mathbf{w} \qquad (k = 1, 2, 3) \tag{19}$$

with

$$R_{i1}\mathbf{w} = \mathbf{S}_{i1} \times \mathbf{e}_k \tag{20}$$

where $\mathbf{S}_{i1}$ is the vector from atom i to the root atom on which the first molecular coordinate system is centered, and $\mathbf{e}_k$ is the negative unit vector along an axis of the global coordinate system. Rotation about a coordinate axis may thus be treated as if it were a rotation about a bond in the model. This approach is a variation of Diamond's use of dummy atoms in order to provide the root segment with six degrees of freedom.[1]

The analogy can be easily extended to the application of finite rotations about the axes in order to update an existing model during the minimization process, which therefore requires only self-explanatory ex-

[8] H. B. Thompson, *J. Chem. Phys.* **47**, 3407 (1967).

tension of Eqs. (10)–(12). Alternatively, when the matrix method is used to generate a new model, one simply updates $\mathbf{M}_1$ with

$$\mathbf{M}_1' = \begin{pmatrix} 1 & 0 & 0 \\ 0 & \cos\phi_1 & \sin\phi_1 \\ 0 & -\sin\phi_1 & \cos\phi_1 \end{pmatrix} \begin{pmatrix} \cos\phi_2 & 0 & -\sin\phi_2 \\ 0 & 1 & 0 \\ \sin\phi_2 & 0 & \cos\phi_2 \end{pmatrix} \begin{pmatrix} \cos\phi_3 & \sin\phi_3 & 0 \\ -\sin\phi_3 & \cos\phi_3 & 0 \\ 0 & 0 & 1 \end{pmatrix} \mathbf{M}_1 \quad (21)$$

and resets $\boldsymbol{\phi}$ to zero, each time rotation of the root segment takes place. (These updates of $\mathbf{M}$ should be temporary ones for the tentative function evaluations that are performed by some minimization routines.)

Comparison of the Two Methods of Model Alteration. Each of the two methods has advantages over the other. (1) The effort required to compute a new model for a given set of dihedral angles is roughly proportional to the number N of atoms in the molecule, whereas the effort required to adjust the atomic positions for given changes of the dihedral angles may vary as N^2, since it depends both on the number of atoms and the number of variable dihedral angles, which is, itself, proportional to N if all degrees of freedom are varied simultaneously. (2) Repeated adjustment of a conformation by internal rotation may cause some deterioration of geometry as a result of cumulative round off errors. This and the first consideration favor the first method. (3) The second method has the advantage that it is easily used to move either one of the two parts of the molecule that are separated by the spindle, relative to the global coordinate system; this is precluded in the first method by the need to select a unique root segment, whose position is unaffected by changes in any of the dihedral angles.

Minimization

Several function minimization procedures that all require calculation of the function's first derivatives with respect to the independent variables have been successfully applied to this problem. We recommend the use of one of several general-purpose routines developed in recent years for minimization of functions of many variables, the Gauss method as modified by Diamond,[1] the Fletcher–Powell–Davidon method,[9,10] the method of conjugate gradients,[11] and others.[12] A properly written model

[9] P. K. Warme, N. Go, and H. A. Scheraga, *J. Comput. Phys.* **9**, 303 (1972).
[10] R. Fletcher and M. J. D. Powell, *Comput. J.* **6**, 163 (1963).
[11] R. Fletcher and C. M. Reeves, *Comput. J.* **7**, 149 (1964).
[12] S. Fitzwater and H. A. Scheraga, *Acta Crystallogr., Sect. A* **A36**, 211 (1980).

rationalization program will allow substitution of different minimizers with little change; communication between application program and minimization routine should be restricted to the value of the function being minimized (in this case the error *D*), the variable parameters (the dihedral angles), and the partial derivatives of the former with respect to the latter.

The Problem of Ring Closure

The methods to fit models of perfect geometry to a set of coordinates do not give a satisfactory solution to the problem of maintaining perfect closure of flexible annular structures, such as proline rings and the larger rings formed in the presence of disulfide bridges, or when more than one part of the model is kept stationary during error minimization, as when the conformation of a stretch of residues lying between known, and already fit, portions of the molecule, is to be adjusted.

The problem has been extensively analyzed by Go and Scheraga.[13] Each ring closure and/or bridge between separate fixed chain segments reduces the number of independent variables by six. Thus, if a ring contains six or more single bonds that are free to rotate, then closure is generally possible without occurrence of geometric strain; if the number is greater than six, the ring has a continuously variable conformation.

It follows that the ring of proline has strained geometry and that its conformation cannot be changed continuously by alteration of dihedral angles alone. Hence the ring's flexibility cannot be represented in a model having rigid geometry. A choice between possible pseudo-rotational alternatives of proline must therefore be made by specification of the corresponding "ideal" geometry and is not affected by the minimization process.

Larger rings, such as are formed by disulfide bridges, do retain considerable conformational freedom. However, application of the fitting procedure requires formal section of all rings in order to reduce the structure to one that is logically entirely branched or linear; subsequent internal rotations, applied to improve the fit by lowering the error, destroy the model's perfect geometry at the points of section. The geometric errors are reduced if (1) each severed chain end is formally extended with two atoms that correspond to the first atoms of the other chain end, and the error in the position of these atoms is included in the total error *D*, and (2) the weights *W* for the atoms near each point of section are increased [cf. Eq. (1)]. Very similar steps can be taken to improve the result of fitting a bridge of atoms between two fixed positions.

[13] N. Go and H. A. Scheraga, *Macromolecules* **3,** 178 (1970).

Models Having Flexible Geometry

Molecular models may be made to approach ideal geometry by a series of small atomic displacements. The immediate displacement of any one atom is determined by the positions of only those other atoms that are connected to it via, at most, three successive chemical bonds. With this method one may improve a model's geometry by moving a single atom at a time. Even when for convenience all atoms are displaced simultaneously in each step of the procedure, overall improvement of the model in fact proceeds as a collective *local* improvement.

The process can be formulated as the minimization of an error F in the geometry, with respect to the atomic coordinates $\mathbf{X}$. The following discussion treats alternative methods by which the error may be defined and minimized.

Alternative Definitions of Geometric Error

Bond Lengths and Angles; Dihedral Angles. Using these standard descriptors of molecular geometry, one may write a weighted error as

$$F_g = \frac{1}{2}\sum_k K_{l,k}(l_k - l_{k,0})^2 + \frac{1}{2}\sum_k K_{\theta,k}(\theta_k - \theta_{k,0})^2 + \frac{1}{2}\sum_k K_{\rho,k}(\rho_k - \rho_{k,0})^2 \quad (22)$$

where the l, θ, and ρ are the bond lengths, bond angles, and dihedral angles of the model and the subscript 0 indicates the corresponding ideal values. This expression for F_g resembles expressions that are commonly used to approximate the conformational energy of a macromolecule; expressions for the conformational energy also contain terms for nonbonded interactions. By analogy, one may think of the K values as force constants, and thereby form a quasi-mechanical concept of idealization as the result of atomic displacements required to balance the resulting forces. However, the values used for K in the idealization procedure need not be related to actual force constants; one normally uses a single value for all K_l and possibly different values for all K_θ and for all K_ρ.

Interatomic Distances. One may include in the error, F_g, terms that depend on distance, s, between atoms that are not directly connected by a chemical bond,

$$F_s = \frac{1}{2}\sum_i K_{s,i}(s_i - s_{i,0})^2 \quad (23)$$

For example, the value of a bond angle may be restrained indirectly by specification of all distances between the three atoms defining the bond angle. In fact, the entire geometric error may be satisfactorily defined in terms of interatomic distances alone.[14,15]

Target Restraint. In addition to the geometric terms, idealization may require use of a restraint that maintains atoms near specified target positions, $\mathbf{X}_t$, which may, but need not, be the starting coordinates of the model. This term may be formulated in terms of the squares of the distances between atomic and target positions, as

$$F_t = \frac{1}{2} \sum_i K_{t,i}(\mathbf{x}_i - \mathbf{x}_{t,i})^2 \tag{24}$$

In the mechanical analogy, each atom is held near its target position with a spring of force constant $K_{t,i}$.

Derivatives. The derivatives of the various error terms with respect to the atomic positions are given by

$$\partial F_t/\partial x_{ij} = K_{t,i}(x_{ij} - x_{t,ij}) \tag{25}$$

and

$$\partial F_l/\partial x_{ij} = \sum_k K_{l,k}(l_k - l_{k,0})\ \partial l_k/\partial x_{ij} \tag{26}$$

$$\vdots$$

Since the position of any one atom affects only a few bond lengths, etc., the number of terms in each sum is quite limited. The derivatives of the bond lengths, l, and other distances, s, between two atoms, i and m, are very simple, i.e.,

$$\partial l/\partial x_{ij} = l_j/l \tag{27}$$

with $\mathbf{l}$ the bond vector

$$\mathbf{l} = \mathbf{x}_i - \mathbf{x}_m$$

and similarly with s replacing l.

Bond angles and dihedral angles and their derivatives are most succinctly given in terms of interatomic vectors. In the following, **u**, **v**, and **w** are three successive bond vectors connecting atoms 1 through 4 and

[14] E. J. Dodson, N. W. Isaacs, and J. S. Rollett, *Acta Crystallogr., Sect. A* **A32,** 311 (1976).

[15] L. F. Ten Eyck, L. H. Weaver, and B. W. Matthews, *Acta Crystallogr., Sect. A* **A32,** 349 (1976).

pointing in the same direction along the chain. The first bond angle, θ, is given by

$$\cos\theta = -\mathbf{u}\cdot\mathbf{v}/uv \tag{28}$$

For the outlying atoms the derivatives are

$$\partial\theta/\partial x_{1j} = \frac{1}{|\mathbf{u}\times\mathbf{v}|}(\mathbf{u}\cdot\mathbf{v}u_j/u^2 - v_j) \tag{29a}$$

$$\partial\theta/\partial x_{3j} = \frac{1}{|\mathbf{u}\times\mathbf{v}|}(\mathbf{u}\cdot\mathbf{v}v_j/v^2 - u_j) \tag{29b}$$

and for the central atom

$$\partial\theta/\partial x_{2j} = -(\partial\theta/\partial x_{1j} + \partial\theta/\partial x_{3j}) \tag{29c}$$

the dihedral angle ρ is given by

$$\cos\rho = (\mathbf{u}\times\mathbf{v})\cdot(\mathbf{v}\times\mathbf{w})/|\mathbf{v}\times\mathbf{u}|\cdot|\mathbf{v}\times\mathbf{w}| \tag{30a}$$

$$\sin\rho = [(\mathbf{u}\times\mathbf{v})\times(\mathbf{v}\times\mathbf{w})]\cdot\mathbf{v}/|\mathbf{v}\times\mathbf{u}|\cdot|\mathbf{v}\times\mathbf{w}|v \tag{30b}$$

The derivatives are found with

$$\frac{\partial\rho}{\partial\mathbf{x}_i} = \frac{\partial\mathbf{x}_i/\partial\rho}{|\partial\mathbf{x}_i/\partial\rho|^2} \tag{31}$$

For the atoms that do not lie on the central bond, a simple expression for the derivative $\partial\mathbf{x}_1/\partial\rho$ may be obtained by Thompson's method [Eqs. (11a) and (17)]. In this case

$$\partial\mathbf{x}_1/\partial\rho = \mathbf{u}\times\mathbf{v}/v \tag{32a}$$

An analogous expression gives $\partial\mathbf{x}_4/\partial\rho$. For the other two atoms, the derivatives are obtained from what in the mechanical analogy amount to conditions of zero net force and torque, and are

$$\frac{\partial\mathbf{x}_2}{\partial\rho} = \left(\mathbf{S}_{31}\times\frac{\partial\mathbf{x}_1}{\partial\rho} + \mathbf{w}\times\frac{\partial\mathbf{x}_4}{\partial\rho}\right)\times\frac{\mathbf{v}}{v^2} \tag{32b}$$

$$\frac{\partial\mathbf{x}_3}{\partial\rho} = -\left(\frac{\partial\mathbf{x}_1}{\partial\rho} + \frac{\partial\mathbf{x}_2}{\partial\rho} + \frac{\partial\mathbf{x}_4}{\partial\rho}\right) \tag{32c}$$

Selection of Restraints; Maintenance of Planarity and Chirality. There is no unique choice of what local geometric descriptors should be made to approach their ideal values in order to obtain a model that has perfect geometry everywhere. For example, the geometry of an atom and four substituents may be completely described in terms of four bond lengths and a minimum either of five bond angles, or of three bond angles

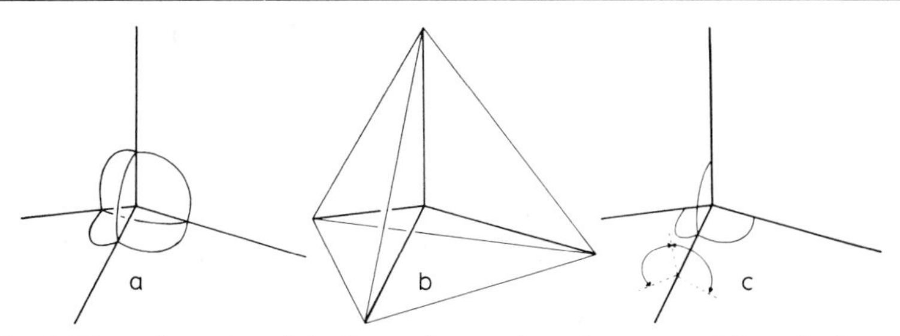

FIG. 3. Three alternative minimal sets of constraints of an atom and four substituents. (In each of parts a and b, one constraint is superfluous.)

and two dihedral angles, or of five of the six distances between substituent atoms (Fig. 3). It would be not even incorrect to use all possible angle and distance constraints simultaneously. However, this is unnecessary, since a model that satisfies a required minimum set of geometric constraints also satisfies all possible alternative constraints.

Simplicity of algebraic form of error term and derivatives favors expression of ideal geometry in terms of interatomic distances, alone. However, this is complicated by the fact that interatomic distances alone do not provide tight restraint of coplanar and of colinear groups of atoms. In terms of algebra, tightness of a restraint corresponds to the second derivative of the error F. When F is expressed as a sum of errors in interatomic distances, its second derivative with respect to a dihedral angle, ρ, defined by a group of four atoms, vanishes when the atoms are coplanar, i.e., when $\rho = 0$ or $\rho = \pi$.

Two alternative strategies exist that provide tight planarity restraint when all restraints are in terms of interatomic distances. The first consists of adding an out-of-plane dummy atom to each planar group and firmly tying it by distance restraints to all atoms in the planar group (Fig. 4). Our mechanical analogy of model and restraints readily makes obvious the effectiveness of such a cantilevered arrangement. According to the second strategy, suggested by Ten Eyck,[16] but never implemented, the planarity restraint may be tightened by imposition of a deliberate imbalance of the distance restraints. This is illustrated by a simple example in Fig. 4: restraint nonbond interatomic distances, b, do not match restraint bond distances, a, i.e., $b > a\sqrt{3}$; the result of this is a proper planarity restraint. As a side effect, the effective restraint bond length has a value between a and $b/\sqrt{3}$. A mechanical analogy of this arrangement is provided by the trampoline.

[16] L. F. Ten Eyck, personal communication (1977).

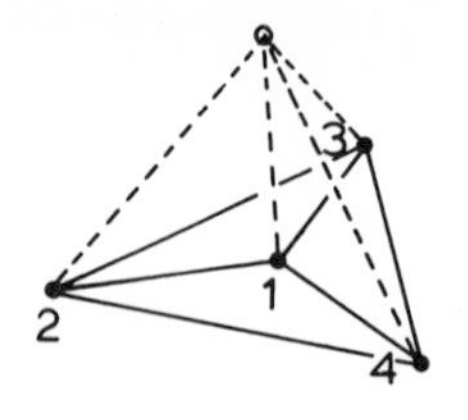

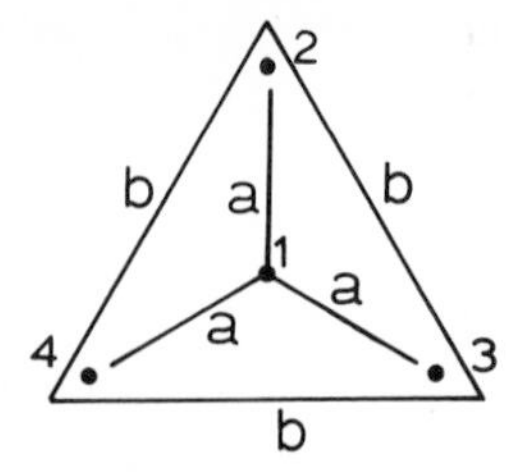

FIG. 4. An illustration of methods to impose planarity on a group of atoms, numbered 1 through 4. (a) The first requires four additional constraints to an out-of-plane dummy atom; these are shown as dashed lines. (b) In the second the constraints are mismatched; the constraint distance a is less than the equilibrium interatomic distance, while the constraint distance b exceeds it.

Correct chirality of asymmetric centers cannot be imposed by constraint of interatomic distances, or of distances and bond angles, alone. Here the problem is not that distances alone provide a restraint that is insufficiently tight, but that two conformations of opposite chirality exist that equally satisfy the constraints. Chirality may be imposed and maintained via constraints expressed in terms of dihedral angles or of chiral volumes, the latter defined as[17]

$$\chi = (\mathbf{u} \times \mathbf{v}) \cdot \mathbf{w} \tag{33}$$

where $\mathbf{u}$, $\mathbf{v}$, and $\mathbf{w}$ are different bond vectors from the asymmetric atom. Dihedral angle and chiral volume both change sign when the asymmetric center is inverted. Since other types of constraints suffice in order to restrain the conformation when chirality is correct, one need include a chirality-dependent term in the error F only when dihedral angle or chiral volume differ considerably from the ideal value. One can accomplish this by defining chiral error as

$$F_\chi = \frac{1}{2} \sum_k K_{\chi,k}(\chi_k - \chi_{k,0}/n)^2, \qquad \chi_k/\chi_{k,0} < 1/n \tag{34}$$
$$F_\chi = 0 \qquad \text{otherwise}$$

with $n > 1$, and $K_{\chi,k}$ sufficiently large to overcome distance or bond angle restraints that oppose inversion when the chirality is incorrect. Thus defined, the chiral contribution to the geometric error is nearly always zero. [If chirality is defined in terms of dihedral angle, ρ replaces χ in Eq. (34).

[17] W. A. Hendrickson and J. H. Konnert, *in* "Biomolecular Structure: Conformation and Evolution" (R. Srinavasan, ed.), p. 43. Pergamon, Oxford, 1979.

This is strictly correct when ρ is near zero; when ρ is near π, the value of $\pi - \rho$ replaces χ.]

Minimization of the Error

In the first application of model rationalization via local geometric improvement, Hermans and McQueen[3] calculated an error term based on bond lengths, bond angles, and dihedral angles. They computed first and second partial derivatives of the error with respect to the coordinates of a single atom and applied an atomic shift obtained from these by the Newton–Raphson method. This was repeated for every atom in turn, for a specified number of cycles over the entire molecule. This algorithm has been incorporated into at least two molecular graphics systems.[18–20]

Minimization may also be rapidly accomplished with use of the conjugate gradient minimizer[11] that changes all atomic coordinates in every minimization step, and requires repeated evaluation of the error and all partial first derivatives of the error with respect to the variable atomic coordinates.[14,15]

Minimization is straightforward if its objective is the attainment of a conformation that strikes a compromise between geometric and target restraints, in which case the function to be minimized is defined as

$$F = F_g + w_t F_t \tag{35}$$

In general, F will possess one or more minima with respect to the atomic coordinates $\mathbf{X}$. Which particular minimum is attained will depend on the starting coordinates. Size and distribution of geometric error in the compromise conformation depend on the values of the weight w_t and of the restraint force constants $K_{l,k}$, $K_{\theta,k}$, and $K_{\rho,k}$ used in the expression for F_g in Eq. (22).

In contrast, F_g does not have one or more unique minima relative to $\mathbf{X}$, but is generally zero for a continuum of $\mathbf{X}$. Nevertheless, application of the two cited methods to minimization of F versus $\mathbf{X}$ produces final coordinates of near-perfect geometry that deviate minimally from the starting coordinates. Best results are obtained if the value of w_t is gradually decreased to zero during the minimization.[3] This is straightforward in the minimization method of Hermans and McQueen, while with the conjugate gradient method, w_t can be decreased automatically if its square root is

[18] The University of North Carolina molecular graphics project, directed by F. P. Brooks, with contributions from W. V. Wright, E. G. Britton, M. E. Pique, J. S. Lipscomb, J. E. McQueen, J. Hermans, and others.

[19] D. Tsernoglou, G. A. Petsko, J. E. McQueen, and J. Hermans, *Science* **197,** 1378 (1977).

[20] T. A. Jones, *J. Appl. Crystallogr.* **11,** 268 (1978).

added to the set of independent variables, the atomic coordinates. To be precise, one may minimize

$$F = F_g + q_t^2 W F_t \tag{36}$$

with W a constant, and q_t a variable; selection of a small value of W serves to retard the decrease of q_t during minimization.

Evaluation

In this chapter two methods have been described by which one can obtain a model of perfect geometry that is an optimal fit to a set of experimental or target coordinates. The first method uses a model of perfect geometry that is adjusted by internal rotation about single valence bonds until the fit is optimal. In the second method, the local geometry of the experimental model is made to gradually approach perfection, as a result of which the model coordinates move away minimally from the initial target coordinates. Use of the second method is preferable for several important reasons.

Speed. For a large molecule, minimization of geometric error by the second method proceeds much more rapidly than does optimization of the fit of two sets of coordinates by the first method. The principal reason for this is that the length of the calculation of the derivatives of the geometric error with respect to the atomic coordinates is linear in the number of atoms, N, while evaluation of the derivatives of the fit with respect to the variable dihedral angles requires a number of steps that varies as the square of N. A second reason is an apparent difference of the ease with which at least one minimization routine, the conjugate gradient program, performs the two minimizations.

Only under special circumstances is it preferable to optimize the fit by internal rotation about single bonds. This is the case if the bonds about which rotation is possible are few in number, and the atoms are many, or if particularly large atomic displacements are required in order to achieve the fit.

Accuracy of ring closure. We have discussed the difficulties of maintaining proper ring geometry that arise when a model of rigid geometry is used. In contrast, when geometry is improved by local adjustment of a flexible model, both inherent geometric strain of small rings and geometry of large rings are appropriately treated, without need of any special measures.

Size. Because routines, such as the conjugate gradient minimizer, require storage for derivatives and auxiliary arrays, minimal program size is obtained with use of the method of Hermans and McQueen,[3] in which only one atom is moved in each step.

Convenience. When program size is not a consideration, model rationalization is most conveniently performed by simultaneous adjustment of the coordinates of all atoms in each step of a minimization of geometric strain with use of the conjugate gradient minimizer.[11,14,15] The constraints may be defined either in terms of bond lengths, bond angles, and dihedral angles,[3] or in terms of interatomic distances.[14,15] In the latter case, special constraints must be provided that will maintain planarity and chirality.

Acknowledgments

This work was supported by the National Science Foundation, Grant PCM81-12234. Computations done with support by the National Institutes of Health, Division of Research Resources, Grant RR-00898.

[14] Interpretation of Electron Density Maps

By JANE S. RICHARDSON and DAVID C. RICHARDSON

In the best of cases no real "interpretation" of an electron density map is necessary—a well-phased 2 Å protein map is a joy to work with and will show what the appearance of various side chains and conformations should be. However, maps are unfortunately the exception; therefore the following section offers indications of what to expect and strategies for interpretation of protein maps at lower resolution or with less accurate phasing.

Low Resolution (5–6 Å)

The resolution level around 5–6 Å is one of the traditional milestones in the solution of a protein structure: it is a low point in most radial distributions of diffraction intensity (see Fig. 1), it is sufficient for location of heavy-atom positions, and it provides information about some overall features of the structure. The simplest information commonly obtained at low resolution is the distinction between protein and solvent density, which shows the molecular shape and packing. Subunit arrangement and symmetry can be seen, as can domains within a subunit if they are well separated. The ease and certainty with which molecular boundaries can be drawn at low resolution vary with crystal packing density (tight packing produces ambiguous contacts), crystallization medium (contrast is

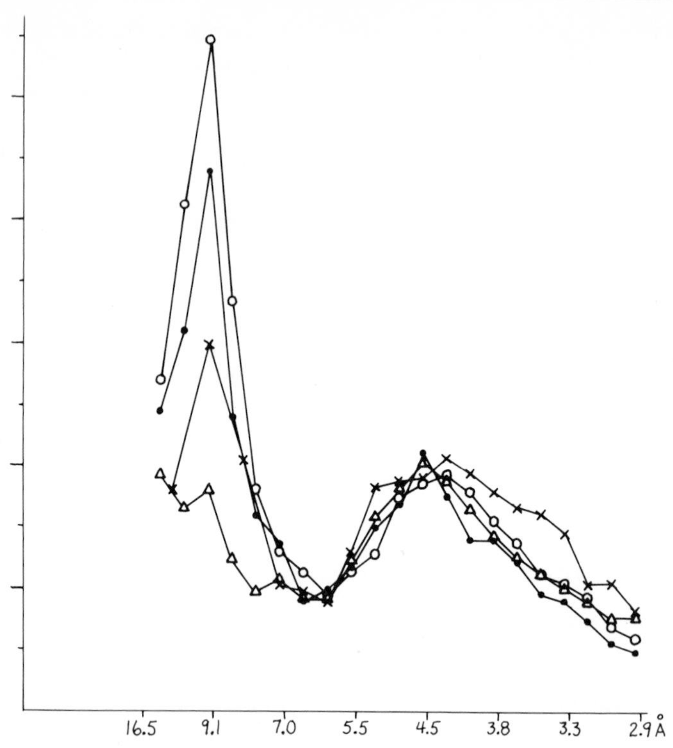

FIG. 1. Radial distribution (as a function of resolution) of diffracted intensity (F^2) for a representative crystal from each of the four major classes of protein tertiary structure. The α and α/β proteins have the highest peaks at 10 Å resolution (mainly from helix–helix contacts) and the antiparallel β protein the lowest, while the small protein has the most relative intensity at high resolution. All types have a trough at about 6 Å and a peak at about 4.5 Å.

better in alcohols or PEG than with concentrated salts), and accuracy of phasing (which can be improved, for instance, by averaging over noncrystallographic symmetry). Figure 2 is a sample slice through a low-resolution map showing the boundary between protein and solvent.

Within the molecule, at low resolution the density is being averaged over a volume greater than that of an individual side chain or backbone. The result is that most side chains disappear altogether (below average solvent density), especially the hydrophobic ones which contain little N or O and a rather high proportion of low-density hydrogens. Large, apparently empty, regions inside the protein as much as 10–15 Å across (as in Fig. 4) usually represent well-packed hydrophobic side chains rather than solvent-accessible holes. Sometimes small peaks appear at the positions of denser side chains such as Cys or Trp, but they are not usually enough

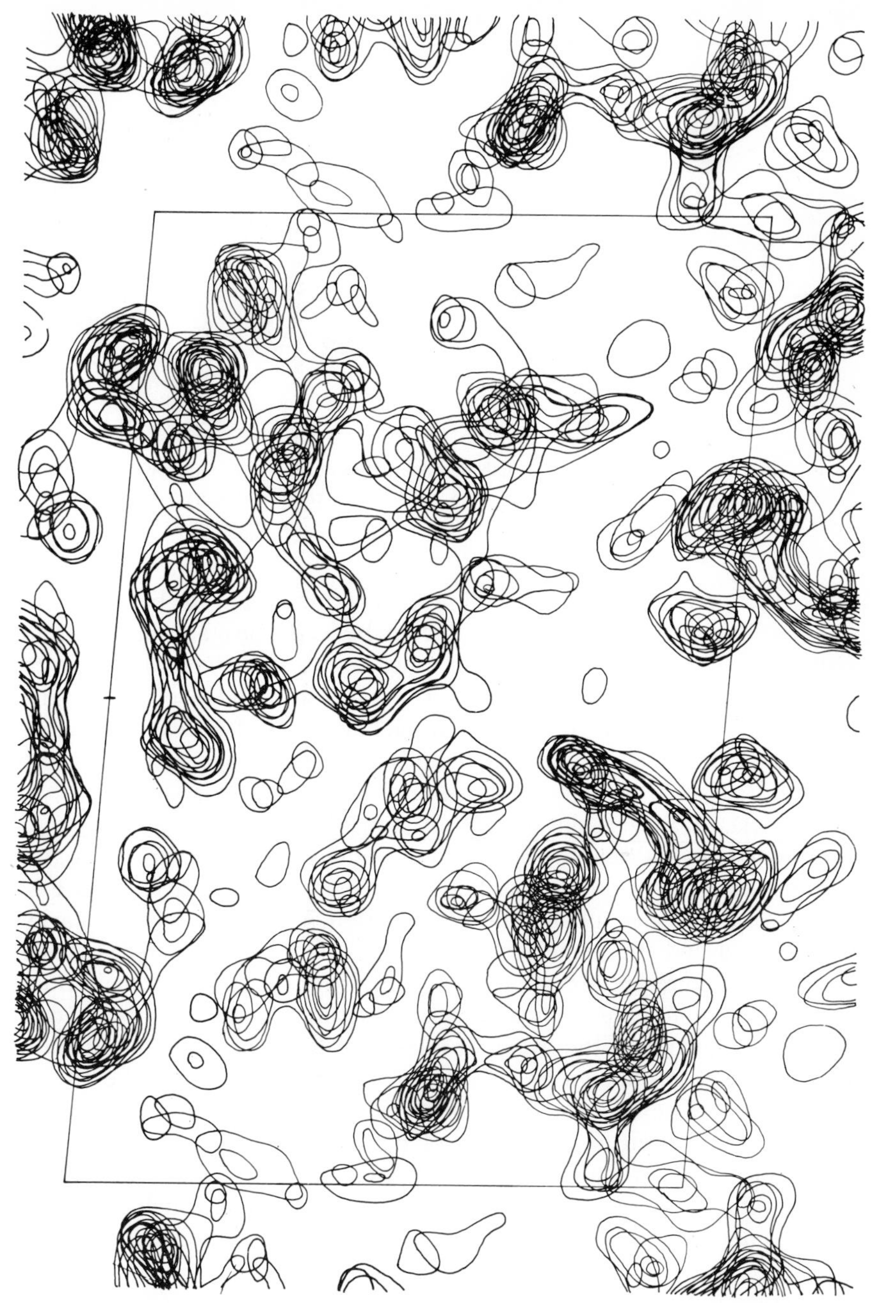

FIG. 2. A slice through the low-resolution (5.5 Å) map of Cu,Zn superoxide dismutase, with dense regions for a pair of subunits in the upper left and lower right of the outlined asymmetric unit, and solvent regions at low density with small, isolated peaks.

FIG. 3. Low-resolution map of a fish hemoglobin subunit,[1] with backbone model from horse hemoglobin superimposed. α-Helices show up as rods of density, sometimes with breaks of main-chain continuity at their ends.

above the random noise level to allow interpretation. For the backbone, at this resolution there is a critical distinction between the main chains within H-bonding distance of one another (such as within helix or between β-strands), which show up as strong positive features, and places where the backbone is separated by side-chain packing or even by van der Waals contact of backbone atoms, in which case the average density is low enough that such "isolated" main chains often disappear.

The classic low-resolution feature is the α-helix, which appears as a high-density cylindrical rod 4–6 Å in diameter (depending, of course, on contour level). Typical spacing between helices is 10 or 12 Å. Figure 3[1] shows a hemoglobin subunit at 6 Å resolution. The globin fold can easily be followed and the heme located, but in the absence of such outside information there are almost always ambiguities that prevent chain tracing at low resolution even in all-helical structures. Connections from one helix to the next are often weak or missing because of series termination effects (which produce a negative ripple in the map around the strong positive rod of helix density); chain termini are very difficult to identify; and there are often extra bridges of density through such features as hemes or metals, groupings of heavy side chains, or even unusually close helix contacts. Myohemerythrin[1a] is one simple case in which an unambiguous and correct chain tracing was possible at low resolution for a

[1] E. D. Getzoff, J. A. Tainer, J. S. Sack, D. Bickar, J. S. Richardson, and D. C. Richardson, *Fed. Proc., Fed. Am. Soc. Exp. Biol.* **39**, 2192 (1980).

[1a] W. A. Hendrickson, G. L. Klippenstein, and K. B. Ward, *Proc. Natl. Acad. Sci., U.S.A.* **72**, 2160 (1975).

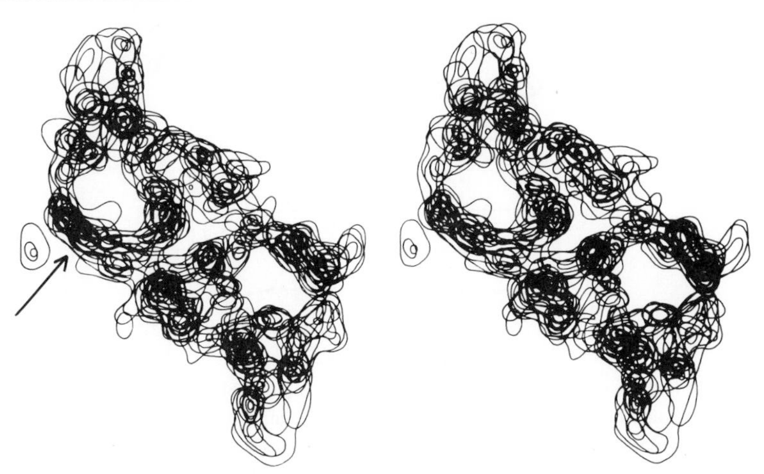

FIG. 4. Low-resolution map of one dimer of Cu,Zn superoxide dismutase, in stereo. The arrow points to a "wall" of density that represents a β-sheet.

novel protein structure (the first four-helix cluster). The number and placement of helices seen at low resolution are sometimes diagnostic of the general category of tertiary structure (see the "mini-atlas" of illustrations in [23], this volume), as is the case for the domain of glycolate oxidase that has been identified as having a triose-P-isomerase-type barrel mainly on the evidence of the outer ring of eight helices.[2]

A region of β-sheet can often be recognized at low resolution, as a smooth isotropic sheet of somewhat lower density than in a helical rod (see Figs. 4 and 13). The sheet looks essentially identical along the strands and across them, so that there must be additional information in order to tell which way the strands run. This can be done in a barrel, or in a nucleotide-binding domain, by matching the overall twist and placement of features with those in known structures. If the sheet density has a clear twist and definite corners, then strand direction can be determined: if upper left and lower right corners twist upward toward you (as in Fig. 5), then the strands run vertically. Also, if one opposite pair of sheet edges looks smooth and parallel while the other pair is jagged or far from parallel, then the smooth pair probably follow the edge strands of the sheet. If such a pair can be identified, then the number of β-strands (n) in the sheet can be estimated from the distance d along a curving line (within the sheet of density) perpendicular to those two edges: $d = 4.8(n - 1) + 3$ Å (approximate) (see Fig. 5). Spacing can vary slightly, but probably not

[2] Y. Lindqvist and C. I. Brändén, *J. Mol. Biol.* **143,** 201 (1980).

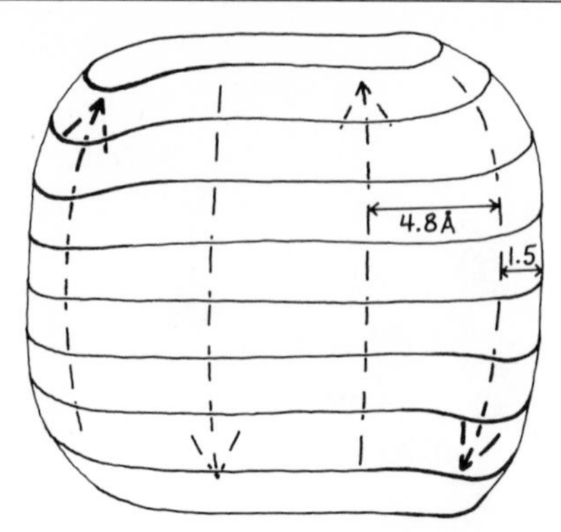

FIG. 5. Illustration of how to determine the probable number and direction of β-strands from the width and twist of the low-resolution electron density for a β-sheet. If the upper left and lower right corners twist forward (as above), then the strands are vertical, while if the other two corners twisted forward the strands would have to be horizontal. The total width of the sheet of density in a direction perpendicular to the strands should be approximately $4.8(n - 1) + 3$ Å.

enough to give the wrong integer for n. If a β-sheet has helical or irregular density next to both its faces, it is almost certainly predominantly parallel, while if it is exposed to solvent on one face, then it is almost surely predominantly antiparallel.

At 5–6 Å resolution, an isotropic sheet of density represents that portion of the β-sheet which has regular hydrogen bonding. An irregularity in the middle of the β-sheet produces a hole in the sheet of density, and where the H bonding stops between neighboring strands at the end of the β-sheet, the continuity of electron density is broken along the strand direction as well as between strands.

Medium Resolution (2.5–3.5 Å)

The region around 4–4.5 Å resolution is not discussed specifically here because it should generally be avoided for protein maps (although not necessarily for nucleic acids). This region is near a maximum in the radial intensity distribution (see Fig. 1) so that it produces particularly severe series termination errors, and the resulting maps are exceptionally difficult to interpret correctly. Helices are rather undistinguished wiggles rather than spirals (see Fig. 11c), and historically have quite frequently been overlooked in this resolution range. If structure factors are for some reason available to 4 Å resolution, it is probably worth calculating and examining an electron density map at 5 or 6 Å as well, as a check on the location of secondary structure features.

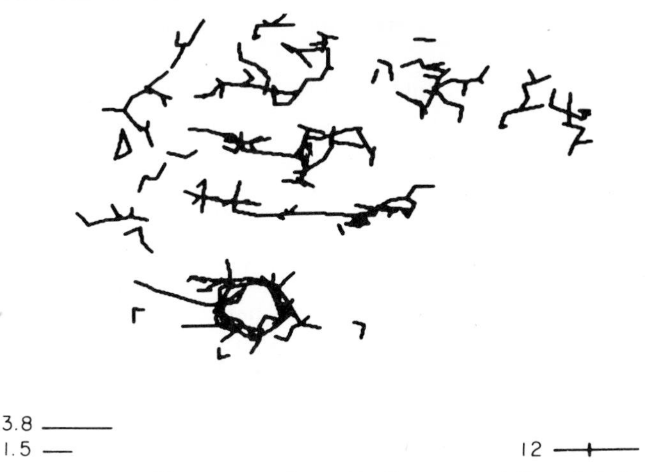

FIG. 6. A computer graphics system using a ridgeline representation of electron density instead of contours, designed to allow interactive initial chain tracing.[3] This is the MIR map of staphylococcal nuclease truncated at 2.5 Å resolution, with a very obvious α-helix seen end on at the bottom and a small β-sheet above it. Reference lines at the bottom show scale at left and contour level, which is variable in real time, at right.

Somewhere in the 2.5–3.5 Å resolution range there is usually a point at which a correct overall chain tracing is first possible. That point depends strongly on whether or not the amino acid sequence is known, because the sequence allows one to resolve a limited number of ambiguities. At better than 2.5 Å resolution the sequence is not necessary for correct chain tracing (although of course it greatly enhances the usefulness of the result), while at something like 3.5 Å the sequence information ceases to help because so few of the side-chain shapes can be reliably recognized in the map. A plausible chain tracing must of course use all parts of the molecular density once and only once, and not make closed loops or tie knots. In order to keep track of such criteria and to see continuity of secondary structures, it is necessary to work with a fairly large three-dimensional portion of the map at once. Initial chain tracings are traditionally done on "minimaps" (stacked plastic sheets at small scale) rather than on computer graphics. However, an interactive graphics system based on ridgeline representation of density rather than contours[3] promises to allow large-scale pattern recognition, with real-time change of scale and contour level (see Fig. 6). The risk of making an incorrect choice somewhere is of course less for small structures with fewer choice points, and is usually less for a structure with clear analogy to one of the

[3] T. V. Williams, Ph.D. Dissertation, University of North Carolina, Chapel Hill (1982).

recognized common folding patterns (see the "mini-atlas," this volume [23]).

In the intermediate resolution range it is crucial that the sequence be actively used if it is available. One simple way to do so is to position independently within the sequence a number of the clearest stretches in the map. For a run of at least six or seven residues, preferably in either a helix or an extended strand (where one can be sure of not inserting or omitting a residue), the shape and location of each side chain in the map are evaluated and a probability assigned (say, on a scale of 1 to 5) for that density representing each of the 20 side-chain types. A well-formed tryptophan shape, for instance, might rate 5 for Trp, 2 or 3 for Tyr or Phe, and only 1 for anything else. An apparent alanine on the surface might rate 5 for Ala, 4 for Ser, 3 for Lys (since it could be disordered, as explained below), etc., but on the inside of the protein an alanine shape would rate only 1 for Lys. For the branched side chains one needs to evaluate whether the branch looks flat or tetrahedral, how far out it is from the C_α, and the hydrophobicity of the environment. If carbonyl oxygens are visible a probability can be estimated for which direction the chain is going, perhaps nearing certainty for a clear α-helix. Then a simple computer program can evaluate the overall probability (by multiplying all relevant numbers) that this stretch in the map represents any given stretch in the sequence. A run of six or seven residues will normally fit at only one place in the sequence, and a possible chain tracing which is tied down to the sequence on each side of each ambiguity is quite sure to be correct.

The possible sources of ambiguity at medium resolution include those mentioned below for high-resolution maps, plus several more. A disulfide (Fig. 7) or a salt link (especially an Arg–Asp or Arg–Glu with a double H bond) can look just as dense and continuous as the polypeptide backbone, and connectivity may also be confusing around metal sites. It is not always easy to distinguish the peaks or side branches in the density that are caused by the side chains from those caused either by carbonyl oxygens or by close solvent molecules. Even α-helical regions can be confusing because the density at this resolution is sometimes continuous along the hydrogen bonds as well as, or instead of, along the main chain (see Fig. 12).

At medium resolution, perhaps even if the phases are very good, there are almost always a few deep breaks in the density along the backbone. These occur most often at either the ends of loops or turns at the surface, or in β-sheets at points of local irregularities in the H bonding or strand spacing (Fig. 13). Unfortunately, for such regions in a β-sheet it is apparently common for the density to be continuous perpendicular to the chains so that the overall connectivity of the density is incorrect. This has

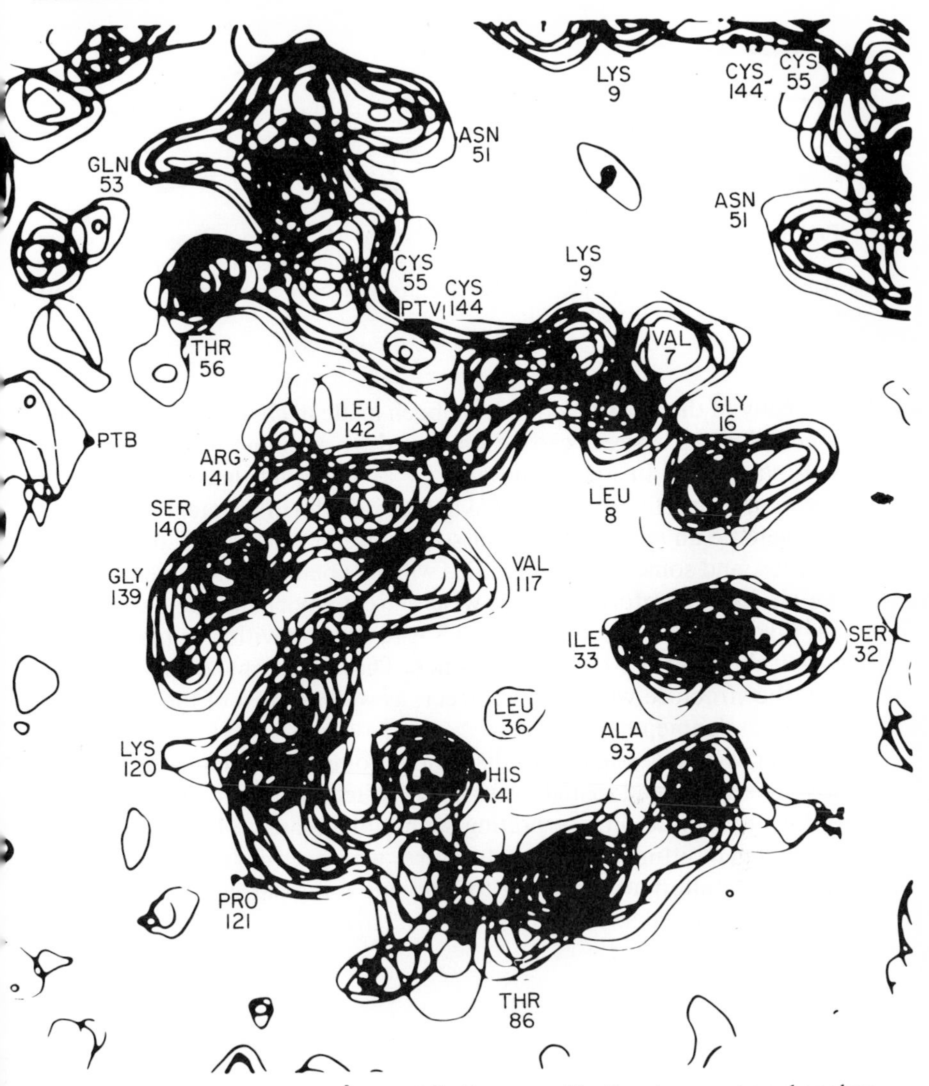

FIG. 7. A slice through the 3 Å map of Cu,Zn superoxide dismutase, contoured on glass sheets with some labels and triangular beads marking the α-carbons. Density for the disulfide, which bridges between a loop at the top and the central β-barrel, is approximately as heavy as the main-chain density.

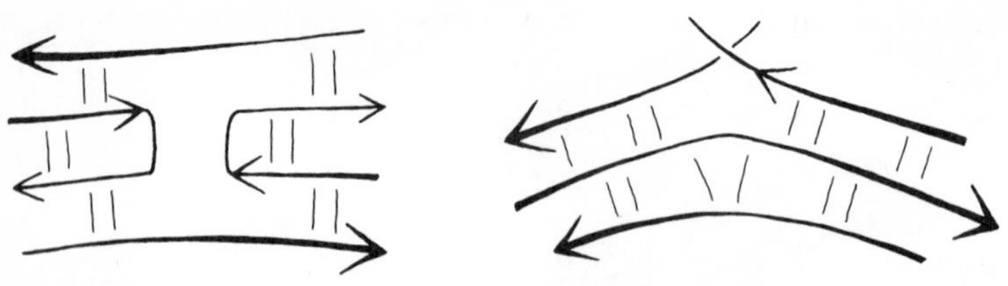

FIG. 8. At the left is a type of interrupted β-strand which almost always turns out to represent an incorrect chain tracing (the middle strands should continue straight across). At the right is a legitimate sort of interrupted strand, in which the two top strand segments twist strongly and are not at all colinear.

led to many of the revisions that have been made in chain tracings.[4–11] The chances of being correct can be improved by avoiding features such as shown on the left in Fig. 8 and also by avoiding left-handed crossover connections[12] and topologies that skip back and forth frequently.[12a] However, even the unusual forms occur sometimes in well-established protein structures, and some map ambiguities will give plausible arrangements for either choice. Therefore, it seems wise to treat any chain tracing as tentative if it was made at much less than 3 Å resolution with a sequence or at less than about 2.5 Å without a sequence. The overall shape and organization of the structure will still be correct, as well as most of the secondary structure assignments.

In helices and β-strands, the distortions common at medium resolution almost invariably lead to underestimation of secondary structures. Therefore it is appropriate in interpreting such initial maps to be prejudiced in the direction of fitting repetitive structures wherever possible, and always trying as one alternative tracing to continue a regular structure through apparent local breaks.

[4] T. N. Bryant, H. C. Watson, and P. L. Wendell, *Nature (London)* **247,** (1974).

[5] R. D. Banks, C. C. F. Blake, P. R. Evans, R. Haser, D. W. Rice, G. W. Hardy, M. Merrett, and A. W. Phillips, *Nature (London)* **279,** 773 (1979).

[6] C. I. Branden, H. Eklund, B. Nordstrom, T. Boiwe, G. Soderlund, E. Zeppezauer, I. Ohlsson, and Å. Åkeson, *Proc. Natl. Acad. Sci., U.S.A.* **70,** 2439 (1973).

[7] H. Eklund, B. Nordstrom, E. Zeppezauer, G. Soderlund, I. Ohlsson, T. Boiwe, B.-O. Soderberg, O. Tapia, C.-I. Bränden, and Å. Åkeson, *J. Mol. Biol.* **102,** 27 (1976).

[8] R. E. Fenna and B. W. Matthews, *Nature (London)* **258,** 573 (1975).

[9] R. E. Fenna and B. W. Matthews, *Brookhaven Symp. Biol.* **28,** 170 (1977).

[10] I. M. Mavridis, and A. Tulinsky, *Biochemistry* **15,** 4410 (1976).

[11] I. M. Mavridis, M. H. Hatada, A. Tulinsky, and L. Lebioda, *J. Mol. Biol.* **162,** 419 (1982).

[12a] J. S. Richardson, *Nature (London)* **268,** 495 (1977).

Medium resolution maps contain many potential pitfalls, but they are also of great importance because (especially if the sequence is known) they can provide chain tracings and initial model fittings for new protein structures.

High Resolution (2–2.3 Å, or better)

At high resolution the chain tracing is normally unambiguous even if the amino acid sequence is unknown. Phase errors produce some breaks in the continuity of the main chain, but such breaks are usually quite local and the correct backbone positioning can be deduced from the shape of the density on either side. Many side chains can be identified from their characteristic shapes, of which a few examples are shown in Fig. 9. Six-membered rings can be oriented quite accurately at high resolution, and even the rounder, fatter histidine rings usually show some flattening.

A resolution of 2 Å is traditionally the highest resolution to which data is collected for heavy-atom derivatives, so that it is normally the highest resolution at which initial interpretation of electron density maps is done.

FIG. 9. Appearance of side chains for an Arg and a Phe in the original MIR map of staphylococcal nuclease at 2 Å resolution, contoured on stacked glass sheets and with brass Kendrew models lying on top of the stack to show the interpretation. The Phe ring has a slight dimple, and the hammerhead shape of the guanidinium can be seen to hydrogen bond both to a main-chain carbonyl and to a dense, triangular shape that is a phosphate of the bound inhibitor.[15]

Turn type	CO(1) orientation	CO(2) orientation	C_α dihedral
I	Nearly 90° down	>90° down	+45°
II	Slightly down	Nearly 90° up	0°
I′	Nearly 90° up	>90° up	−45°
II′	Slightly up	Nearly 90° down	0°

One exception is when resolved anomalous phasing is used,[13] which may become commoner for small proteins and perhaps with synchrotron data, both of which encourage high-resolution data collection. Initial interpretation at, for example, 1.5 Å resolution has real advantages, since in good parts of the map it allows recognition of resolved atoms in such features as aromatic side chains or proline rings.[13]

At 2–2.5 Å resolution individual atoms do not show separate peaks, but they produce clear effects on the shape of the density contours. Probably the most important atom type for accurate fitting is the carbonyl oxygen of the main chain: its orientation directly determines the ϕ and ψ angles on either side, and in a well-phased map at high resolution most of the carbonyl oxygens should be visible as distinct bumps. In an α-helix, their orientation immediately determines the direction of the polypeptide chain, since all the carbonyl oxygens point toward the COOH-terminal end of the helix. They can sometimes be used to determine direction in other parts of the chain by measuring which α-carbon they are closest to. Their position is the major feature which allows assignment of hydrogen bonding, either directly for the O end of a bond or indirectly by determining the orientation of the peptide plane and therefore the NH position.

A specific example of the usefulness of carbonyl oxygens in map interpretation is the assignment of types of tight turns. The types are generally defined in terms of ϕ and ψ angles for the two central residues, exact values for which cannot usually be obtained until later. However, turn types I and II and their mirror images I′ and II′ can be assigned from carbonyl oxygen orientations CO(1) (in the H bond) and CO(2) (in the central peptide) (see Fig. 31 of Ref. 14). Those characteristic orientations are summarized in the table, along with the dihedral angle formed by the four α-carbons, which provides an additional check that can be seen in the map. Although there is considerable variation in the exact CO position within each turn type, genuinely intermediate conformations are quite strongly disfavored: for example, it is almost impossible to form a type I

[13] W. A. Hendrickson and M. M. Teeter, *Nature (London)* **290,** 107 (1981).

[14] J. S. Richardson, *Adv. Protein Chem.* **34,** 167 (1981).

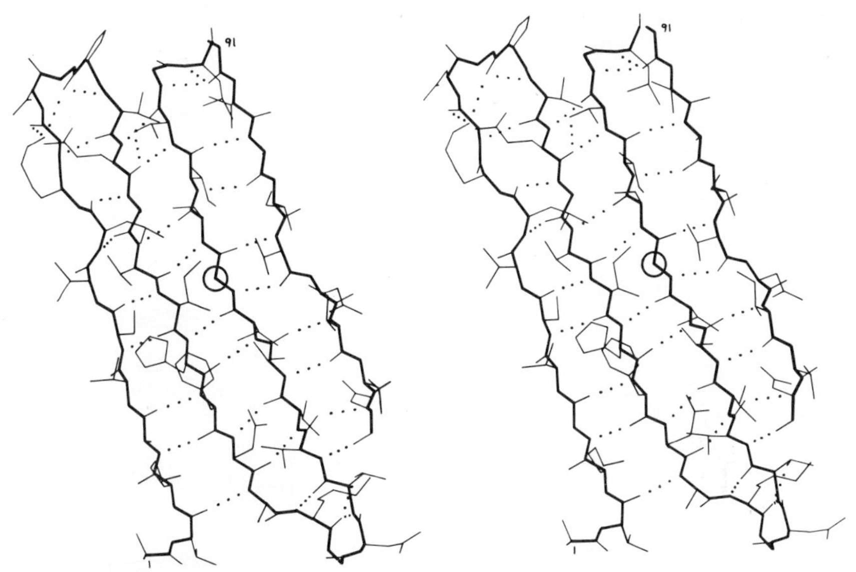

FIG. 10. A four-stranded β-sheet from Cu,Zn superoxide dismutase[17] in stereo, showing Gly 31 (circled) in a left-handed β conformation; it has normal hydrogen bonding but alters the "pleat" and makes the sheet slightly convex.

turn with a flat dihedral angle and a straight hydrogen bond. Type III turns differ from type I by only 30° in one conformational angle, which cannot be distinguished at this initial level unless there is a second overlapping turn making it into a short piece of 3_{10} helix. If the carbonyl oxygens are not visible in the map, then it is possible to identify the region as a turn but not to assign the type.

Although, in general, fitting can proceed quite easily at high resolution, there are still some problems, mostly caused by either inaccurate phasing or actual disorder in the structure. Sometimes a stretch of several residues, including the backbone, can be partially or completely disordered if there are no hydrophobic side chains near a chain terminus or in a loop at the surface (e.g., Ref. 15). These regions either disappear altogether or appear at a low density in the map, and can only be fit approximately if at all. They will usually show up at least a bit more clearly after refinement of the rest of the structure.

Individual side chains at the molecular surface frequently show disorder. About three-fourths of the lysines and one-fourth to one-third of the

[15] A. Arnone, C. J. Bier, F. A. Cotton, V. W. Day, E. E. Hazen, Jr., D. C. Richardson, J. S. Richardson, and A. Yonath, *J. Biol. Chem.* **246,** 2302 (1971).

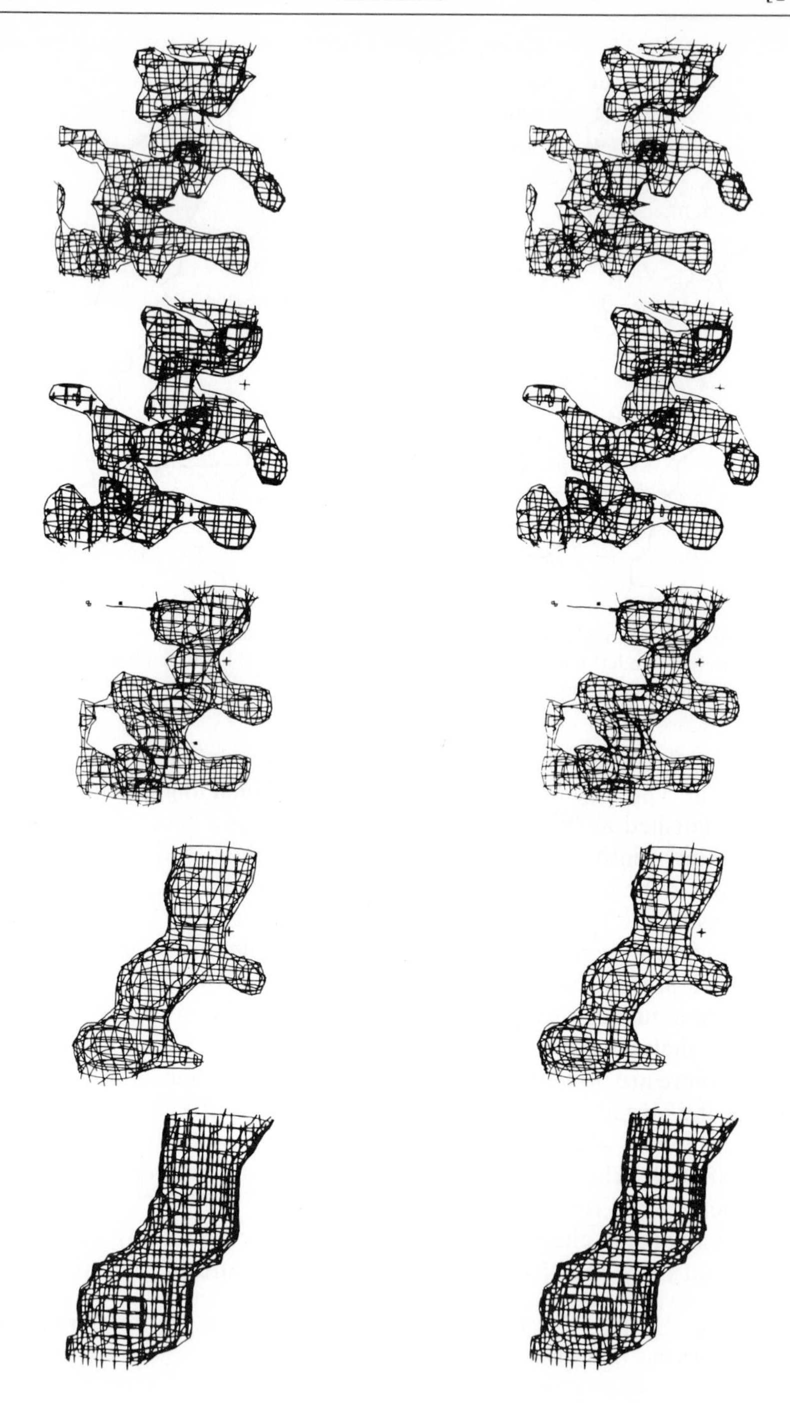

other charged side chains are disordered,[14] most often indicated by density visible only out to C_β or C_γ, but sometimes by having an unconnected peak for the end of the side chain. This disorder may often represent the superposition of two discrete, definite conformations,[13] but that fact can only occasionally be seen in the initial map.

Either phase errors or series termination effects can distort the local appearance of the map. At high resolution the most susceptible place seems to be the β-carbon. Instead of showing the "elbow" that should result from the tetrahedral angle at C_β, it is not uncommon to have the density there rather weak and extending in a straight line from C_α to C_γ, which usually makes the χ_1 and χ_2 angles ambiguous. With a branched C_β, especially for valine, the map may show a straight bar of density joining the C_γ atoms, extended more or less parallel to the backbone but with no indication to which side the C_β should protrude. Such situations illustrate a basic conflict in map interpretation: for instance, valine has a very strong preference for the staggered conformation of χ_1,[16] and choosing that alternative in ambiguous cases greatly improves one's chances of being correct; however, that undermines later attempts to get unbiased statistics for χ_1 frequencies. Perhaps the best strategy for such problems is to use the expected value in initial fittings but make note of the ambiguity as one that should be explicitly checked later during refinement.

Some particular residues are especially prone to fitting difficulties. It is hard to determine the correct ϕ and ψ angles for glycine, since there is no C_β to give a "lever arm" on what are otherwise quite subtle differences in shape along the backbone. For instance, it is quite possible for a glycine in the center of a β-sheet to be in left-handed (i.e., with ϕ positive and ψ less than $-40°$) rather than in normal β conformation, although with good hydrogen bonds on both sides and only a change in the "pleat" of the C_α position[17] (Fig. 10). Disulfides are difficult to fit because they must simultaneously satisfy so many nonorthogonal constraints and also because they are especially likely to show indecisive directionality at C_β; again,

[16] J. Janin, S. Wodak, M. Levitt, and B. Maigret, *J. Mol. Biol.* **125,** 357 (1978).

[17] J. A. Tainer, E. D. Getzoff, K. M. Beem, J. S. Richardson, and D. C. Richardson, *J. Mol. Biol.* **162,** 181 (1982).

FIG. 11. Stereo views of the electron density for an α-helix in staphylococcal nuclease[15] at (from top to bottom) 2, 3, 4, 5, and 6 Å resolution. All maps were made using F_{obs}, the original MIR phases, and the same grid spacing. Viewpoint is the same, and contour levels were adjusted to be approximately equivalent. All carbonyl oxygens are clear at 2 Å, but almost all of them are absent at 3 Å, although side-chain shapes can still be judged. At 4 Å, density has begun to coalesce along the helix axis, and there is a false connection between side chains at the lower left. Figures 3, 11, 12, and 13 were made using the GRIP molecular graphics system developed by Wright, Pique, Britton, Lipscomb, Brooks, and others at the University of North Carolina, Chapel Hill.

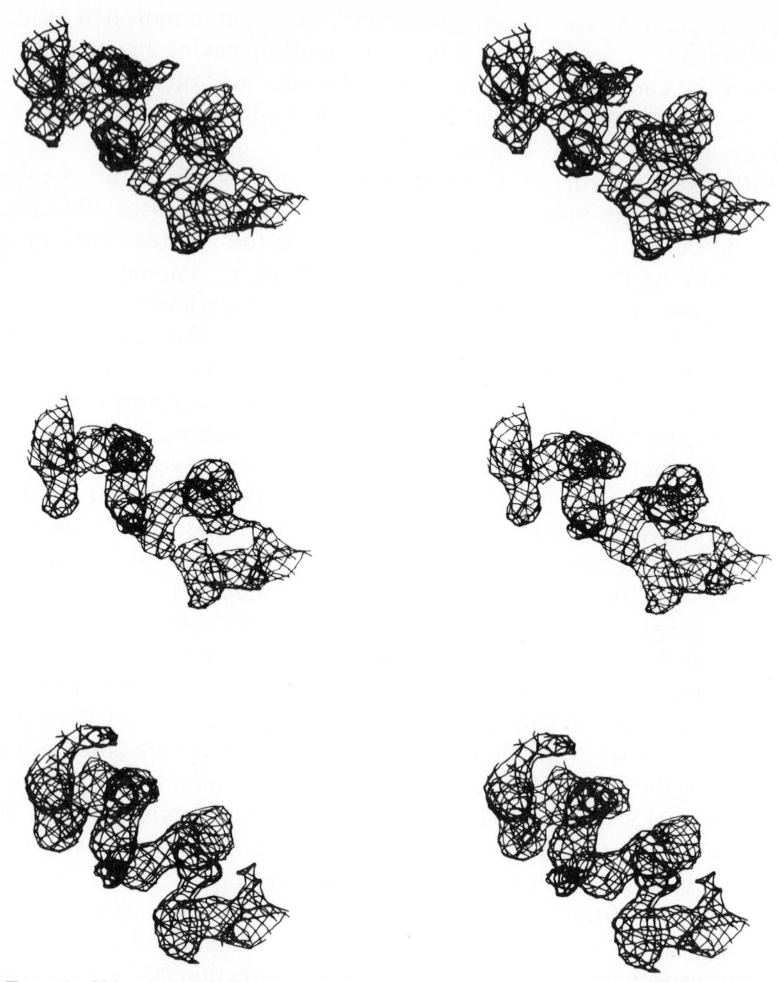

FIG. 12. Side views of the same helix as in Fig. 11, at 2, 3, and 3.5 Å resolution. At intermediate resolution the density connects through a hydrogen bond (lower right) more strongly than through the nearby helical main chain, although the connectivity is correct at both higher and lower resolutions.

→

FIG. 13. Stereo views of the electron density for two strands of antiparallel β-sheet in staphylococcal nuclease at 2, 3, 4, 5, and 6 Å resolution. In this case the strands separate correctly at 4 Å, but that would not always be true. At 5 and 6 Å the density is sheetlike, but with holes in variable locations. At 6 Å the right-hand side extends further out because it is no longer separated from a third strand.

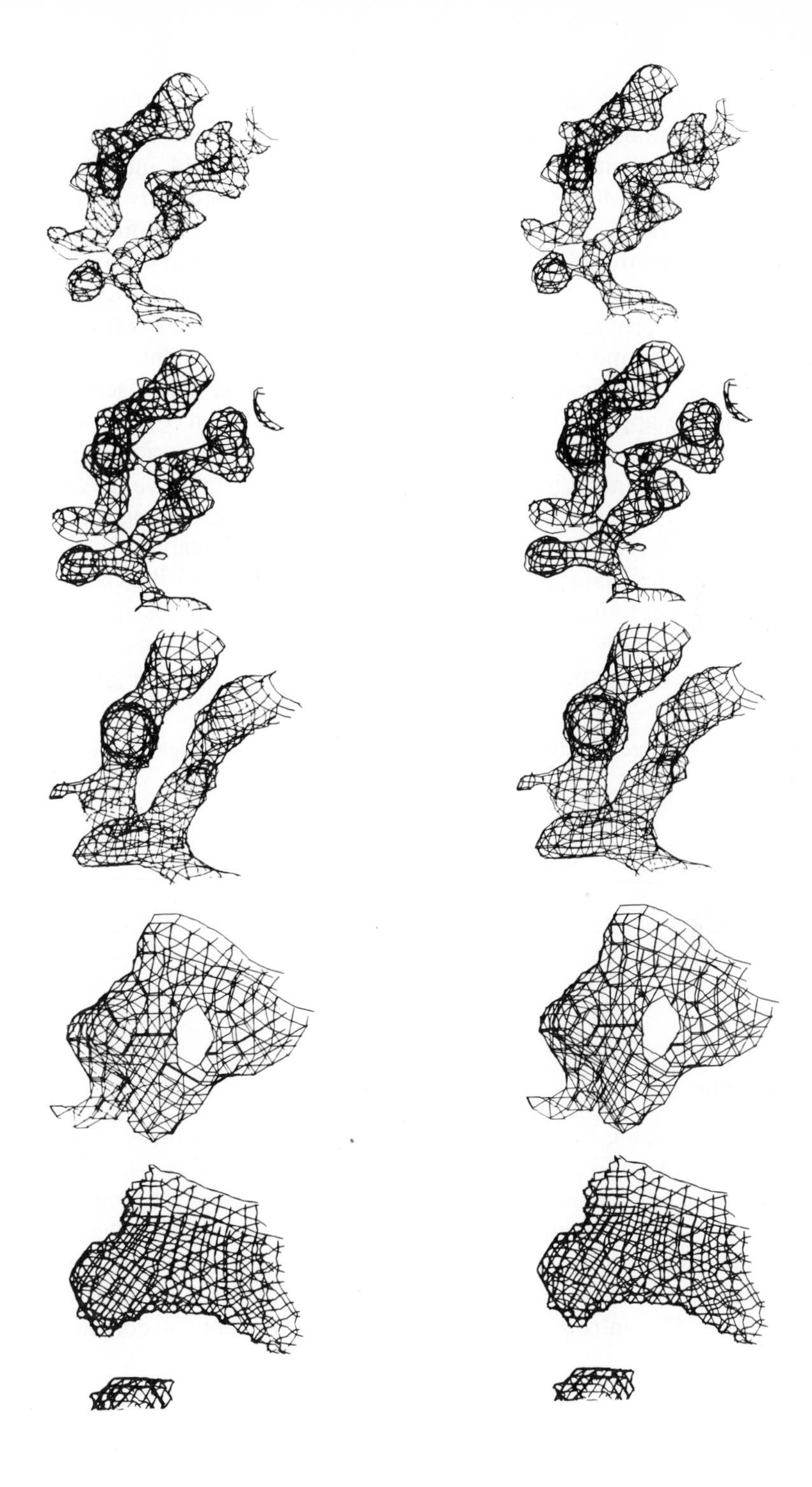

expected conformations[14] can be chosen in initial fitting and reexamined later. For prolines, one should keep in mind that about 15% of them[18] have cis rather than trans peptides (examples in Refs. 14 and 19).

However, all of the above problems are local and only sometimes present. In general, at 2–2.5 Å resolution a quite accurate fitting of the model to the electron density map can be made, even when the amino acid sequence is not known. When it is possible to collect data out to 2 Å, that resolution can provide both a quite respectable structure from initial fitting and also the most suitable starting point for refinement.

The final set of figures (11, 12, and 13) summarize this discussion by displaying the appearance of the electron density of an α-helix and of a piece of β-sheet as a function of resolution, starting from a multiple isomorphous replacement initial map of a real protein structure at 2 Å resolution. It can be seen that there are some particular stages along the way at which the density coalesces in ways that do not quite match the covalent continuity of the chain. Although one should not necessarily assume that incorrect connectivities will always occur in the same ranges of resolution that they do here, these illustrations give some concrete examples of what level of clarity to expect and what features to look for at a given resolution.

[18] There are 11 cis prolines and 63 trans prolines in a sample of 12 highly refined protein structures.

[19] R. Huber and W. Steigemann, *FEBS Lett.* **48,** 235 (1974).

[15] Computer Skeletonization and Automatic Electron Density Map Analysis

By JONATHAN GREER

Introduction

The past decade has seen the three-dimensional structures of proteins solved by X-ray diffraction methods at an ever increasing pace. In addition, the sizes of the molecules studied have also increased dramatically with larger proteins being analyzed all the time. This situation has placed greater pressure on the ability to interpret the electron density maps of these proteins with an atomic model that can then be used for further refinement. The introduction by Richards of the optical comparator[1,2]

[1] F. M. Richards, *J. Mol. Biol.* **37,** 225 (1968).

[2] F. M. Richards, this volume [10].

moved electron density map interpretation out of the dark ages into the modern era. Yet, as the molecules studied became larger and new structures more frequent, sheer size and space limitations, as well as length of time required and accuracy of model building, have increased the need for further advances in map interpretation techniques.

Two approaches were possible to achieve improvement in map interpretation methods: computer graphics-based Richards box systems and an automated approach to map interpretation. Several groups have now developed useful computer graphics map interpreting programs. These allow protein crystallographers to interpret their map with the considerable aid and accuracy of computer graphics which can be directly coupled to a larger computer for immediate coordinate refinement. Such programs are user friendly and crystallographers have rapidly adapted to their use. This chapter deals with the alternative—automated electron density map interpretation[3–6] utilizing basic pattern recognition and other advances in computer science and artificial intelligence. Because this approach is less familiar to the protein crystallographer, this chapter is intended to show the crystallographer how such a system works, give examples of the results it has achieved, and suggest future directions in which it should be expanded. Thereby, the investigator community may become acquainted with the automated system and allow its use to expand.

Method of Analysis

Preparation of the Electron Density Map

The electron density map should be between 2.0 and 3.5 Å in resolution. At this resolution, continuity of the main chain should be optimal for analysis. The map is interpolated, if necessary, into a Cartesian coordinate system of definition 1 Å. The portion of the map to be examined should contain the whole of one molecule with a border of at least 2 Å around the outer edge of the molecule. Ideally, the center of gravity of the molecule should be near the center of the map space being analyzed. In order to illustrate the various steps performed on an electron density map, a simulated two-dimensional electron density map is presented and operated upon by each successive step in the process. This map is shown in Fig. 1a in a typical contour form.

[3] J. Greer, *J. Mol. Biol.* **82,** 279 (1974).
[4] J. Greer, *J. Mol. Biol.* **98,** 649 (1975).
[5] J. Greer, *J. Mol. Biol.* **100,** 427 (1976).
[6] J. Greer, *J. Mol. Biol.* **104,** 371 (1976).

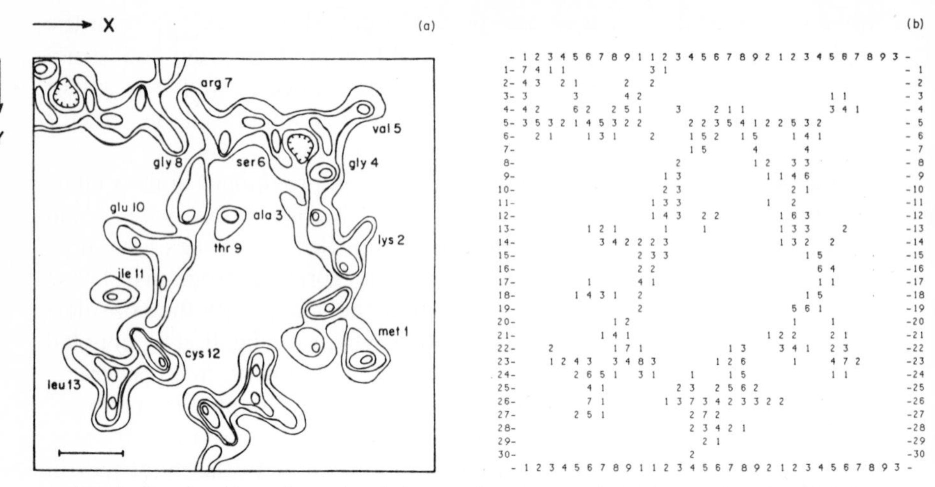

FIG. 1. Simulated two-dimensional electron density map of a tridecapeptide, in which the central molecule is labeled and neighboring molecules are positioned consistent with the symmetry of the planar space group p2. The scale bar, where shown, is 5 Å long. (a) A map presented in contour form, which has been selected to highlight several common features of a protein map. The sequence, starting from the lower right, is N-fMet-Lys-Ala-Gly-Val-Ser-Arg-Gly-Thr-Glu-Ile-Cys-Leu. Notice that very little of the side chains of Lys 2 and Arg 7 appears. Ser 6 hydroxyl forms a hydrogen bond with the main-chain carbonyl oxygen of Ala-3, creating a loop of density. The side chains of Thr 9 and Ile 11 are disconnected from the main chain at this contour level. The first contour is at 0 and the contour interval is 2. Negative contours are suppressed. (b) A digital representation of the contour map. The grid interval is 1 Å. The minimum density that is being considered here is 1, so all points that are 0 or below are replaced by a blank. The symmetry operators are 2-fold axes that lie perpendicular to the plane very close to (12, 5) and (12, 25). Notice that the symmetry-related density is not always identical. This is due to interpolation onto the 1 Å grid. (c) Skeleton map of the simulated two-dimensional map. The connectivity of the structure has been preserved and the side-chain appendages can be seen at intervals along the main chain. Notice that the symmetry-related features are not identical in the skeleton. This is due not only to the interpolation process but also to idiosyncracies in the point selection procedure of the skeletonization process (see text and Greer[3,5] for a more complete discussion). (d) Skeleton map shown as a graph structure with the side chains removed. + marks the positions along the main chain where the side chains were removed. The side-chain branches at the ends of the chain have been retained, since no a priori information is assumed as to which branch is the end of the main chain and which is the terminal side chain. X marks the position of the center of gravity point used for the molecular isolation procedure. (e) Skeleton of the isolated central molecule generated using the isolation procedure described in the text. The center of gravity point used in the calculations is labeled by X. (f) Composite drawing of the contour map of the isolated molecule with the skeleton of this molecule from (e). This demonstrates that the skeleton follows the electron density faithfully. (g) The simplified skeleton of the molecule. The loop around (21, 7) has been removed and the main chain extends as one single segment of length 47.9 Å from (24, 19) all the way to (6, 24). (h) α-Carbon plot of this tridecapeptide. The side-chain branch points are α-carbon positions. Interpolation using the 3.8 Å constant inter-α-carbon distance was used to generate the remaining positions as described in the text. (Adapted from Greer.[3,5])

(c)

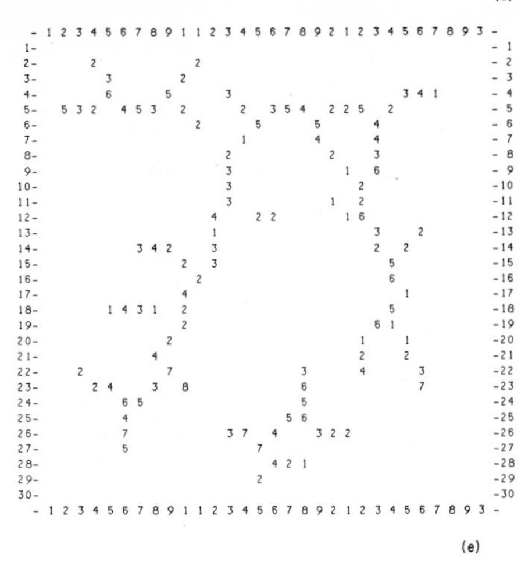

(d)

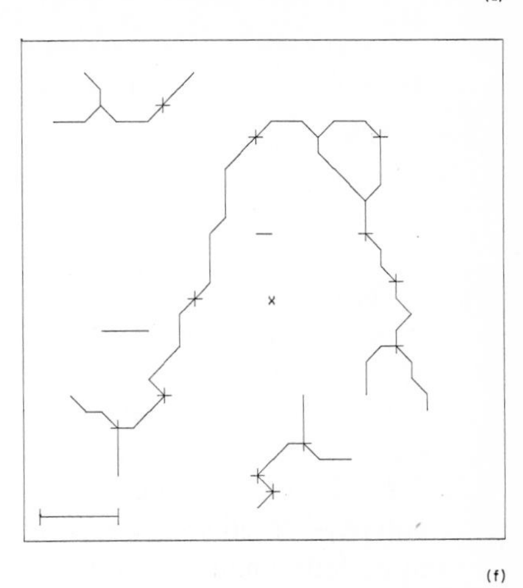

(e)

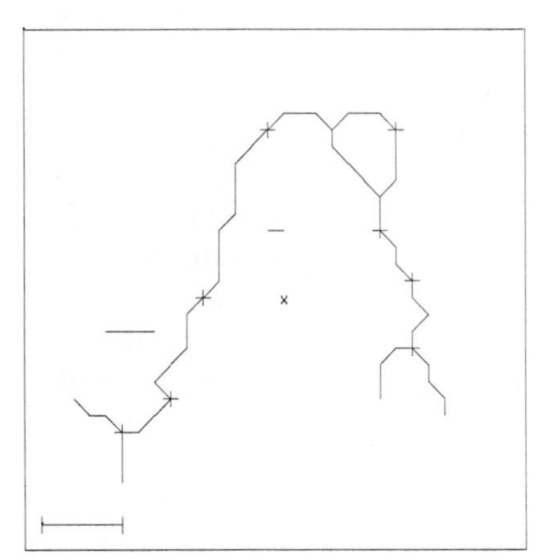

(f)

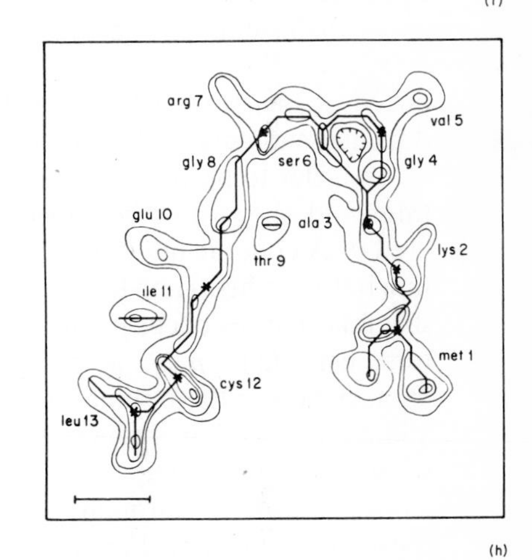
arg 7
val 5
gly 8
ser 6
gly 4
glu 10
ala 3
thr 9
lys 2
ile 11
met 1
cys 12
leu 13

(g)

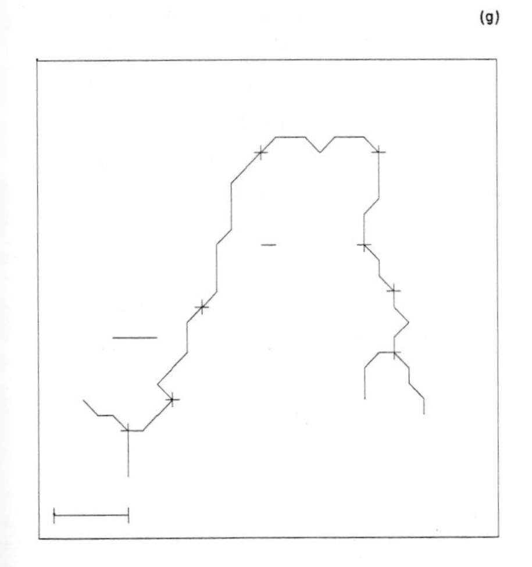

(h)

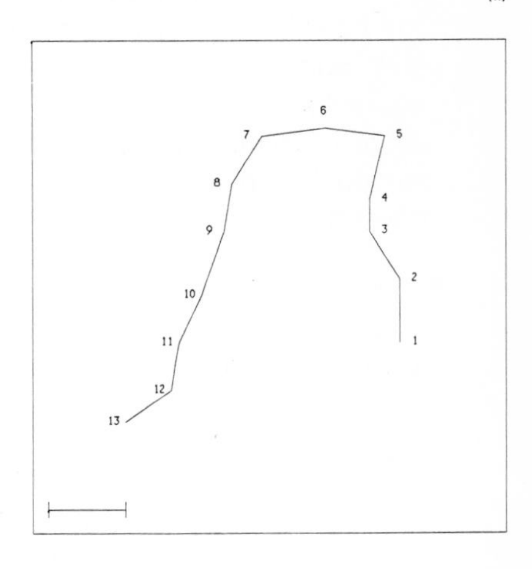

Skeletonization of the Density Map

The electron density map contains a vast number of points which must be reduced to a workable number in order to analyze the map efficiently, yet without causing irrevokable loss of important information. Thus, the first step is to choose a sensible reduced representation of the map.

Protein molecules consist of a small number of polypeptide chains which are reasonably thin in diameter, yet very long, often up to hundreds of angstroms. Branches extrude from this chain at close to regular intervals. Occasionally the chains are cross-linked by disulfide bridges or very electron-dense hydrogen bonds. One possible formalism would be to reduce the electron density to idealized thin lines which follow the long polypeptide chains preserving at all times the connectivity of the structure. Side chains would be represented by thin lines that come off the main chain and cross-bridges would become thin lines connecting the main chain lines. Therefore, three-dimensional skeletonization was introduced at this stage using a method that was generalized from the two-dimensional work of Hilditch.[7] The method is summarized here.

In order to prepare the map for this procedure, a density significance threshold, ρ_{min}, must be chosen; below this value, the density is considered 0. A contour interval, $\Delta\rho$, is also selected which retains the distinction between high- and low-density features of the map yet is not so grainy that false curves occur in the skeleton. The selection of these two parameters, especially ρ_{min}, is critical, and may require examining several choices before the optimal ρ_{min} and $\Delta\rho$ can be found. The values of ρ_{min} and $\Delta\rho$ are used to reduce the density values of the map to a series of integers corresponding to the contour level above the threshold (Fig. 1b).

A pass is made through the map for each contour interval starting with the lowest level first. For each pass, the grid points are divided operationally into three groups. Points higher than the current density interval are in I; points at zero or previously removed are in N; and the current interval is R. The treatment of any grid point in R depends on the status of its 26 immediate neighbors. During each pass, grid points in R are removed (i.e., placed in N at the end of the pass) if they satisfy the following rules in sequence:

1. The grid point must be in R, i.e., at the current contour interval.
2. No "hole" is created by removing this point. This test is performed as three orthogonal two-dimensional tests. At least one of the four points

[7] C. J. Hilditch, *Mach. Intell.* **4,** 403 (1969).

in each of these three sets must be in N for the grid point x, y, z to be removed:

$$x + 1, y, z; x - 1, y, z; x, y + 1, z; x, y - 1, z$$
$$x + 1, y, z; x - 1, y, z; x, y, z + 1; x, y, z - 1$$
$$x, y + 1, z; x, y - 1, z; x, y, z + 1; x, y, z - 1$$

3. The point cannot be a "tip" or single point in the skeleton. This means a grid point cannot be removed if it has only one neighbor (out of the 26) or no neighbors that lie in I.

4. The removal of the central grid point cannot disconnect those points among the 26 immediate neighbors that are in I.

These tests are performed on each point at the contour interval of that pass, i.e., that is in R, repetitively until no further points are removed. The contour levels are scanned from the lowest to the highest contour density in the map.

When the process is completed, a skeleton representation of the map remains (see Fig. 1c). This skeleton, however, is not unique. The precise position of the skeletal lines varies slightly depending upon the order in which the points are considered. In all cases, this variation is not more than one grid interval. Thus, as can be seen, Fig. 1f, the skeleton follows the contour density quite faithfully.

Tracing the Chain

At this stage, the skeleton is composed of grid points in a three-dimensional map. Typically, only 1–3% of the total number of grid points of the original map are retained in the skeleton. In order to work with this skeleton efficiently it must be represented in a useful way.

Since proteins consist of main chain with side chains branching off, the first step is to categorize possible side chains and store them on a list separately from the rest of the skeleton. A side chain on the skeleton is detected as a fragment of the skeleton which begins at a tip and ends at a branch point on the skeleton. These are removed [except at the very ends of the skeletal chain, e.g. (6, 24) and (24, 19)], and the branch point on the skeleton where the fragment was removed is flagged (see Fig. 1d).

For purposes of efficient access and use, the skeleton is then described as a graph or treelike structure. The nodes of the graph are the branch points and the tips. A segment is defined as that portion of the skeleton that lies between two nodes. Thus, a full description of a skeleton consists of a list of segments and a table of how they are connected (see table for a description corresponding to the two-dimensional skeleton of Fig. 1d).

SEGMENT LIST AND CONNECTIONS TABLE FOR THE TWO-DIMENSIONAL SIMULATED MAP

Segment	No. of points	Length	End-to-end distance	Beginning to . . . end[a]								Segment connections			Segment connections		
				First		Second		$(n-1)$th		nth		No.	Beginning[b]		No.	End[b]	
				X	Y	X	Y	X	Y	X	Y						
1	5	4.8284	4.4721	26	23	26	22	25	20	24	19	0			2	2	3
2	10	10.6569	9.2195	24	19	24	18	22	11	22	10	2	1	−3	2	4	5
3	5	4.4142	3.6056	24	19	23	19	22	21	22	22	2	1	−2	0		
4	5	5.2426	5.0000	22	10	21	9	19	7	19	6	2	2	−5	2	−5	6
5	9	9.2426	5.0000	22	10	23	9	20	5	19	6	2	2	−4	2	−4	6
6	24	27.9705	22.2036	19	6	18	5	7	24	6	24	2	4	5	2	7	8
7	4	3.8284	3.6056	6	24	5	23	4	23	3	22	2	6	−8	0		
8	4	3.0000	3.0000	6	24	6	25	6	26	6	27	2	6	−7	0		
9	4	3.4142	3.1623	21	26	20	26	19	26	18	25	0			2	10	11
10	4	3.0000	3.0000	18	25	18	24	18	23	18	22	2	9	−11	0		
11	6	6.6569	5.0000	18	25	17	25	16	28	15	29	2	9	−10	0		
12	2	1.0000	1.0000	5	4	5	3	5	4	5	3	2	−13	−14	0		
13	3	2.4142	2.2361	5	4	4	5	4	5	3	5	2	−12	−14	0		
14	6	6.2426	5.0990	5	4	6	5	9	4	10	3	2	−12	−13	0		
15	2	1.0000	1.0000	15	12	16	12	15	12	16	12	0			0		
16	4	3.0000	3.0000	5	18	6	18	7	18	8	18	0			0		

[a] The coordinates for the second and penultimate points of each segment are recorded here in order to specify the segment uniquely since several segments will have the same endpoints.

[b] The sign of the segment indicates the direction of the connection. If it is positive, the segment and its neighbor can be connected without changing the direction of either. If the entry is negative, then the direction of one must be changed to connect them.

This formulation allows tracing and parsing of the skeleton to be performed with great facility.

Isolating the Molecule

Since the electron density map is that of a crystal of the protein, it is likely that in addition to the molecule encompassed by this section of the map, parts of other neighboring molecules are included. Thus, the next step is isolating the central molecule from its neighbors. Success at this step depends upon the extent to which neighboring molecules are intertwined. Therefore, a conservative procedure was designed which errs on the side of retention of segments of the skeleton.

Symmetry-related segments of the skeleton are identified using the symmetry operations for the space group of the respective crystal. Clearly, only one of the symmetry-related segments is in the molecule being isolated. The following criteria were used to choose which segment should be retained and which deleted:

1. If one of the symmetry-related segments is entirely within the map space defining the molecule, while part of the other falls outside the map space, the latter segment is deleted.
2. A center of gravity point is defined either for the molecule, or if more appropriate, for each lobe of the molecule. The distance from the center of gravity of each of the two related segments to its closest molecular center of gravity point is computed. If one of the two segments lies more than 20% further away from the molecular center of gravity point(s) than the average distance of the two segments, it is deleted.
3. Unique segments, i.e., those with no symmetry-related mates, are labeled as being in the molecule.
4. If all the neighboring segments (see table) of one of the segments have been deleted as being in neighboring molecules, then that segment is deleted as well. Similarly, if all the neighboring segments of one of the segments either were deleted or are indeterminate, but at least one of the neighbors of the second segment is clearly in the molecule, the former is deleted and the latter is retained.
5. Isolated segments and single points which are more than 4 Å from any other segment in the skeleton are removed.
6. Upon occasion, the results of criteria 2 and 4 will conflict, i.e., the segment that is more distant from the center of gravity is connected to other segments in the molecule, while the related segment that is closer to the center of gravity is connected to segments that have already been deleted as being in neighboring molecules. In such a case, the connectiv-

ity criterion 4 is given preference and the more distant segment is retained.

The above rules were applied to the map in Fig. 1d. One center of gravity point is used at (16, 16) represented by an X in this figure. The resulting map appears in Fig. 1e and shows the skeleton for a single, isolated molecule. This skeleton is presented, superimposed upon the original contour map, Fig. 1f, to show how faithfully the skeleton is following the density and how the skeleton retains many of the topographical features of the original map.

In previously published applications of the molecular isolating program, criterion 6 was not used, which led to the result that in case of a conflict between criteria 2 and 4, both segments were deleted. As a result, important parts of the molecule could be deleted inadvertantly as occurred for the two-dimensional simulated map in Fig. 1d of Greer[5] and for the COOH terminus of the Bence-Jones Rhe protein[6] (see Application section). Criterion 6 was therefore added to solve this problem.

Simplification of the Skeleton

Since the main chain is linear, if any segment is connected to another segment by one or more segments (see for example segments 4 and 5 that connect segments 2 and 6 in Fig. 1 and the table) then only one of these paths can be the main chain. The following criteria are used to distinguish which of the two should be retained: (1) If only one of the two connecting segments has side-chain branches, then that one is retained and the other is deleted (see Fig. 1g). If both have side chains, both are retained. (2) When neither segment has side chains, if the longer one is four times larger than the shorter one and is longer than 8 Å, it is retained and the other deleted. More complex cage-like structures can appear, usually at α-helices. These can often be resolved by using a finer definition map.[4]

Selecting the Main Chain

Previous steps have left main-chain and inter-main-chain bridges primarily. Thus, the next step is to attempt to identify the main chain and connect it together to trace the molecule in the map. Two criteria are used to identify main chain:

1. Segments which are larger than 10 Å are main chain. This is usually true because hydrogen bonds and disulfide bridges are 4 and 9 Å long, respectively. Occasionally, two side chains may interact to form a longer bridge. This occurs once in ribonuclease S (see section c of Greer[5]) and

once in the Bence-Jones Rhe structure.[6,8] Thus, this 10 Å limit may have to be increased somewhat.

2. A segment which contains side chains must be main chain.

The longest segment of the skeleton is taken first and each end is pursued, determining which of the segments at each node to assign as main chain using the above criteria. This process continues at both ends until a tip is reached or until the program cannot decide between the possibilities. The next largest unassigned segment is then taken and the process continued at its ends until all the long segments (>10 Å) have been examined.

Typically, at the end of this step several long sections of the main chain have been connected. The complete main chain cannot usually be connected because of low-density discontinuities in the electron density map. Connecting these fragments has in the past been left to intervention of the human map interpeter, often by using maps or skeletons calculated with a somewhat lower ρ_{min} value. This aspect will be discussed further in the section on "Future Prospects."

Generating Atomic Coordinates

The next stage is to convert the skeletal lines into atomic coordinates of the protein. The ability to perform this task depends upon how well the skeleton preserves the distances and angles of the original map. The first step is to determine the α-carbon coordinates of the molecule. These are assigned as follows:

1. The juncture of the main chain with a side chain, which was previously removed in the step described in the section on "Tracing the Chain" (Fig. 1c), is tentatively assigned as an α-carbon. The β-carbon is set 1.54 Å along the side-chain branch.

2. If two side-chain branches are less than 2 A apart, one is probably a carbonyl oxygen and not a real side chain. Which is designated side chain and which carbonyl oxygen is determined by the distances to the neighboring side chains consistent with the 3.8 Å inter-α-carbon distance (see below).

3. Since adjacent α-carbons are always 3.8 Å apart, regardless of the conformation of the chain, this value is used to interpolate α-carbons along the skeletal chain when a stretch appears without side chain branches (Fig. 1g).

[8] B. C. Wang, C. S. Yoo, W. Furey, and M. Sax, *J. Mol. Biol.* **135,** 305 (1979).

Application of these rules to the two-dimensional skeleton is shown in Fig. 1h. However, this procedure is not foolproof. It sometimes adds or misses α-carbons. This will be discussed in more detail in the Applications section. However, those α-carbons that it does find are usually good trial coordinates for the actual α-carbon positions.

The occasional occurrence of a branch corresponding to the carbonyl oxygen allows the determination of the directionality of the main chain. This is because a carbonyl oxygen branch should be 1.5 Å COOH-terminal to its α-carbon and 2.5 Å NH_2-terminal to its adjacent α-carbon. This distinction in distances can usually be made accurately enough in the skeletal representation to assign directionality.

More sophisticated model building is possible in the secondary structure regions. The various α-helices and β-strands can be identified from their relative α-carbon positions along the main chain as has been previously described in detail.[5,9] Once assigned, a standard α-helix or double-stranded parallel or antiparallel β-sheet can be fit to the α-carbon coordinates. In this way, a considerable part of the structure can be constructed in preparation for coordinate refinement (see appropriate chapters, this volume, on refinement).

Application to Electron Density Maps

The automated interpretation system was first applied to the 2.0 Å electron density map of ribonuclease S,[10] and was subsequently used on a number of other protein electron density maps of different quality and magnitude. The most complete analysis was performed on the 3 Å map of the Bence-Jones protein Rhe kindly provided by B. C. Wang and his co-workers.[11] The results of the various steps in the analysis of the ribonuclease S (RNase S) and the Bence-Jones Rhe maps have been described in detail.[3–6,8,11] Here, we will review the models obtained from these two maps and outline the problems that were present in these interpretations.

Ribonuclease S

Figure 2 presents a plot of the α-carbon coordinates determined by the automated interpretation system together with a plot of the α-carbon coordinates of Richards and Wyckoff.[12] The former set of coordinates are

[9] M. Levitt and J. Greer, *J. Mol. Biol.* **114,** 181 (1977).

[10] H. W. Wyckoff, D. Tsernoglou, A. W. Hanson, J. R. Knox, B. Lee, and F. M. Richards, *J. Biol. Chem.* **245,** 305 (1970).

[11] B. C. Wang, C. S. Yoo, and M. Sax, *J. Mol. Biol.* **129,** 657 (1979).

[12] F. M. Richards and H. W. Wyckoff, *in* "Atlas of Protein Structures" (D. C. Phillips and F. M. Richards, eds), Vol. 1, p. 9. Oxford Univ. Press, London and New York, 1973.

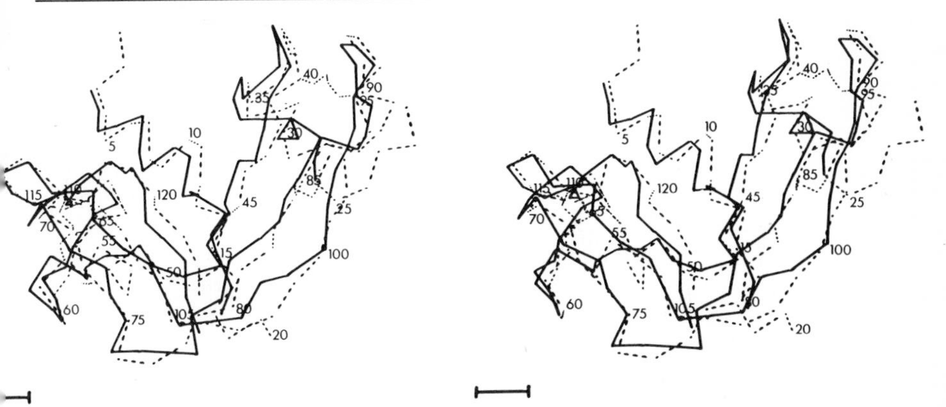

FIG. 2. α-Carbon line drawing of ribonuclease S produced using the coordinates derived by the automated procedure (solid lines) superimposed upon the α-carbon drawing using the coordinates of Richards and Wyckoff[12] (broken lines). The residue labels refer to the actual protein sequence and structure; every fifth residue is labeled. (From Greer.[5])

the result of the step-by-step analysis described in the previous section. Human intervention has occurred solely to determine how to connect several of the main-chain segments together and to suggest a directionality for the S peptide based upon the proximity of its two ends to the NH_2 terminus of the S protein.[5] Figure 3 shows a residue-by-residue comparison of the deviations between the program-derived coordinates and those of Richards and Wyckoff.[12] The arrows mark the places where the program has predicted an additional α-carbon and the dashed lines represent α-carbons that have been missed by the program.

Comparison of the model produced by the automated system with the Richards–Wyckoff model shows that the overall topography of the molecule has been reproduced remarkably faithfully from the map by the automated system. Thus, the automated system can produce an excellent likeness of the molecular structure. In particular, when the models are compared residue by residue (Fig. 3) for 112 α-carbons, the mean absolute deviation is 1.472 Å before secondary structure modeling and 1.613 Å afterwards. This small increase is due to the fact that the ideal α-helices and β-sheet used for modeling do not fit the structure perfectly. In order to describe the strengths and the weaknesses of the program system, a brief tour through the molecule is useful.

The program begins its interpretation with residue 3 of the molecule. Residue 1 is missed because of weak density. The density for residue 2 main chain is there but cannot be distinguished by the program from the density for the side chain of Thr 3. Missing the chain termini is quite

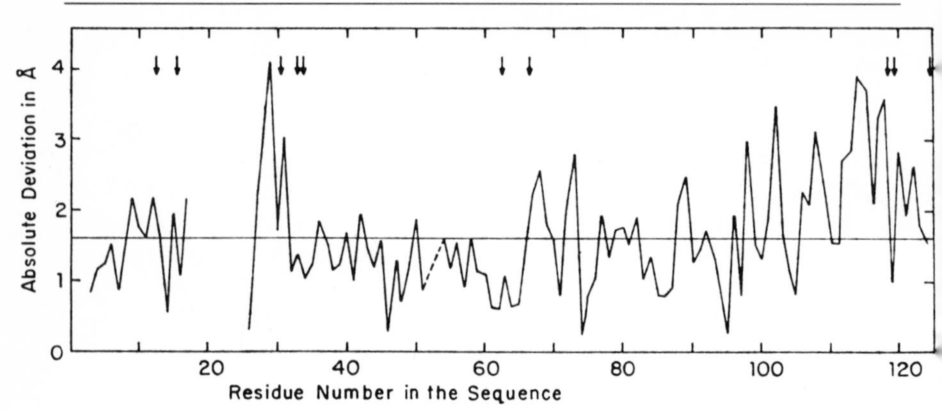

FIG. 3. A plot of the absolute deviation between the α-carbon coordinates predicted by the program and those of Richards and Wyckoff[12] for the residues of ribonuclease S. The horizontal line at 1.613 Å shows the mean absolute deviation for the whole molecule. The broken line crosses over those residues missed by the program (52 and 53). Arrows mark the places where the program has erroneously added a residue. (Adapted from Greer.[5])

expected as they tend to be relatively disordered in the crystal and appear at weaker density. Reexamination of this area with a lower $\rho_{\min}$ value may be quite fruitful.

Residues 3–13 of the S peptide are α-helical and they are recognized as such by the program and built nicely. Residues 13–17 have more difficulties; two additional α-carbons have been predicted, between 12 and 13 and between 15 and 16. The density for the S peptide COOH terminus, 18–20, is very weak in the map and thus missed by the program.

The NH_2 terminus of the S protein, 21–23, is also missing in the skeleton because of weak density. The main chain from 24 to 26 exists in the skeleton but, once again, the program does not know how to assign it as main chain relative to the large side chain of Tyr 25. Residues 26–34 form an α-helix in RNase. The program has a very difficult time with this region (as well as with the helix at 50–58) because some of the main chain appears with weaker density while the α-helical hydrogen bonds have much stronger density. As a result, the chain determined by the program does not look like an α-helix at all (see Fig. 2). This problem can be solved by using a finer grid interval in the map[4] as well as by more sophisticated methods that would detect this α-helical cagelike density and process it correctly.

The map between residues 34 and 50 is quite well interpreted and coordinate refinement would probably produce an excellent structure in this region. Residues 50–58 are α-helical and suffer from the problems noted above for helical regions. Residues 52 and 53 are missed entirely. However, using a 0.5 Å grid map for this area improved dramatically the interpretation of this helix.[4]

Residues 58–108 are interpreted fairly well with two extra residues inserted between 62 and 63 and between 66 and 67. The region includes an external loop from 65 to 72 which has relatively weak density. Much of this portion of the molecule is β structure which is interpreted well by the program.

The structure from residues 109 to 117 forms a large loop (Fig. 2). The predicted coordinates here correspond less well to those of Richards and Wyckoff[12] (see Fig. 3). In particular, the ends of the loop approach too closely and some confusion occurs because of the high-density disulfide bridge between residues 58 and 110. The last seven residues of RNase S, 118–124, appear with strong density because they are a β-strand forming antiparallel β interactions to residues 105–110. This strand is clear in the skeleton and is interpreted with two additional residues, between 118 and 120 and one after residue 124.

Bence-Jones Rhe

When this map was originally interpreted by the automated system,[6] Wang and co-workers had not yet interpreted their map in detail.[11] The program's current interpretation differs somewhat from that reported previously due to a correction in the automated system as is described below.

The α-carbon coordinates of the automated interpretation are shown superimposed upon the interpretation of Wang and co-workers in Fig. 4. Once again, the overall representation of the structure by the program's model is excellent. The two β-sheets and the short α-helix are clearly present and detectable. The α-carbon coordinates in the two models are compared residue by residue in Fig. 5. The mean absolute deviation for the 111 α-carbons which correspond was calculated to be 1.34 Å. A short comparison of the two models, which differs slightly from that of Wang *et al.*,[8] follows:

Residue 1 of the protein does not appear in the automated system's model because the program cannot choose between the main chain for residue 1, which is present in the skeleton, and the side chain of residue 2. The first extended chain is interpreted with two extra residues, between 5 and 6 and between 7 and 8. The α-helix at 25–30 is interpreted very well, with none of the problems found in two of the three α-helices in RNase S. Similarly, the rest of the β-sheet is interpreted quite nicely with additional residues inserted erroneously between 47 and 48, 63 and 64, and 73 and 74. The only real error in interpretation occurs in the region of residues

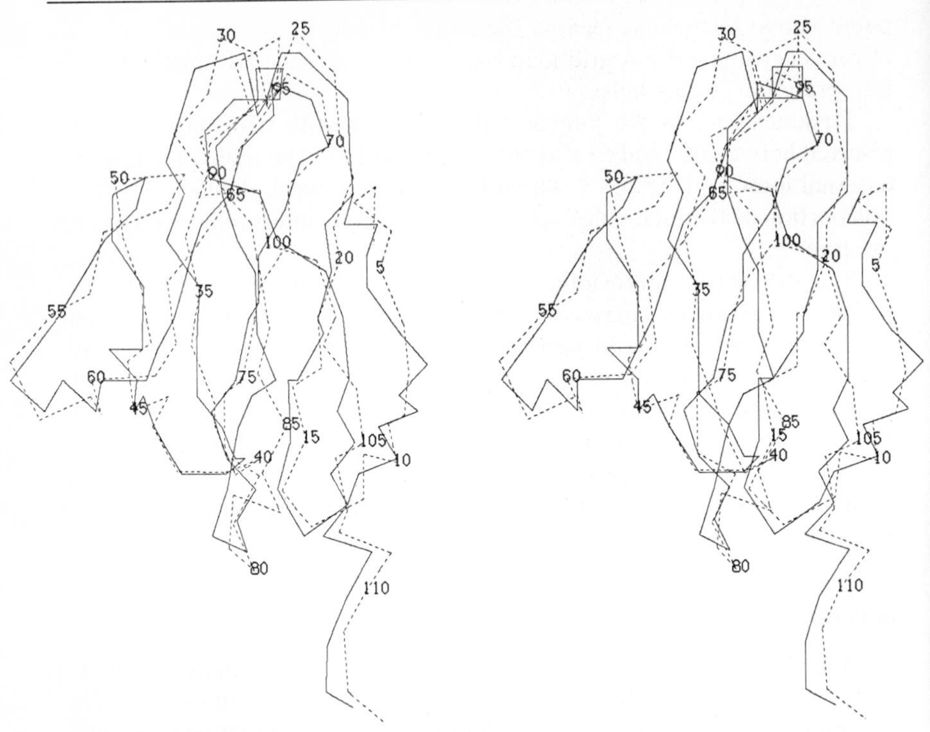

FIG. 4. α-Carbon plot of the program produced coordinates of Bence-Jones Rhe (solid lines) superimposed upon those of Wang *et al.*[11] (broken lines). The residue labels refer to the Wang *et al.* structure; every fifth residue is labeled. Notice how closely the two interpretations agree.

83–88 of the structure. Here the density for what would appear to be the main chain of an extended β-strand is actually the density for the interacting side chains of Asp 84 and Tyr 88. The true density for the main chain itself deviates from a simple straight β-strand and forms a little excursion about the side chain density before continuing on as a β-strand. The program interprets the side-chain density as the main chain and in the process misses one residue. This error would be difficult to correct without human intervention.

In the original automated interpretation of this molecule an error existed in the routine which isolated the molecule. In particular, rule 6 in the section, "Isolating the Molecule," was not applied, as previously described. As a result, all the skeletal segments that corresponded to the true density for residues 107–113 were deleted. After we received the model from Wang and co-workers, we went back, found this error in the

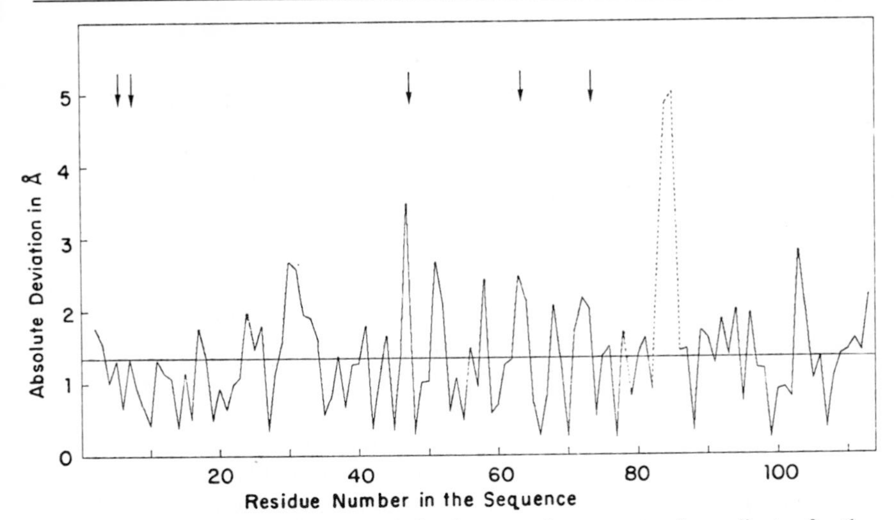

FIG. 5. A plot of the absolute deviation between the two sets of coordinates for the Bence-Jones Rhe molecule. The horizontal line marks the mean absolute deviation for the whole molecule, 1.34 Å. The broken line marks the misinterpreted region of 83–86 including one missed residue (see text). The arrows mark the points where residues were added erroneously by the program system.

isolation program, and corrected it. As a result, the skeleton for the COOH terminus is clearly present and the program's α-carbon coordinates for these residues fit the Wang model very well as can be seen in Figs. 4 and 5.

Critique of the Method

The interpretation of the electron density map of a protein at less than atomic resolution, i.e., 2.0–3.5 Å, is not a simple pattern recognition problem. This is particularly true because of the quite variable density for the main chain that is due to protein dynamics, crystal disorder, and phase and measurement errors. Further complications are introduced by disulfide bridges, strong hydrogen bonds, and prosthetic groups. Nevertheless, a certain degree of success has been achieved, as can be seen in Figs. 2 through 5. The experience obtained in interpreting these and other maps has indicated some of the problems that are encountered in map interpretation and has suggested future directions to pursue in improving and expanding the usefulness of the automated interpretation system. In this section, we discuss some of these problems, and in the next section, possible future directions are described.

One of the most difficult tasks undertaken by the program is isolation of a single molecule. It is easy to imagine that a strangely shaped subunit, as occurs with one loop in lactate dehydrogenase, could make this task very difficult. In some cases, interpretation by the standard Richards box methods have erred in assignment of molecular boundary. In both RNase S and Bence-Jones Rhe, a satisfactory solution was found for this problem. This was accomplished by assuming a basic compactness to the structure and judging protrusions, not by their distance from the center of the molecule, but primarily by their connectivity to the rest of the structure. This technique will probably not always work. Inclusion of information about the sequence would be very valuable in solving this problem (see next section). Ultimately, human intervention may be necessary to resolve the most difficult cases of molecule isolation.

The second problem which is most commonly encountered is designation of the main chain. This may be due to difficulty in distinguishing skeletal segments which are main chain from those which are side chains, disulfide bridges, or hydrogen bonds. Another major difficulty is that lower density regions of the main chain cause discontinuities in the skeleton which must be bridged by some method. Both of the above have been known to confound the human interpreter as well. One solution to the first problem is to detect more side chains, which are the best indicator of main chain. There is the need for a more sophisticated analysis of the possible connections between the designated main-chain fragments. When necessary, these decisions were performed manually in the case of RNase S and Bence-Jones Rhe. In larger proteins, though, so many fragments occur that a program is needed to consider all the possible connections and exclude those which are topologically unreasonable.

In some cases, α-helical regions are difficult to interpret, as in RNase S (see above). The presence of strong hydrogen bonds causes cagelike structures, and weak density for the main chain causes the wrong density to be followed. Earlier studies suggested using a finer grid definition, such as 0.5 Å, to solve this problem.[4] Some evidence indicates that a somewhat lower resolution map, around 3.0 Å, is easier to interpret than a 2.0 Å map for these α-helical regions. More attention must be paid to this problem and more experience gained with other proteins which contain α-helical regions.

Another common problem with the automated system is in the α-carbon assignment routines. The program adds extra α-carbons where none occur in the structure or, less frequently, may miss an α-carbon. Such errors do occur when humans interpret electron density maps in the absence of sequence data although not as frequently as it does with the program system. Several solutions exist to help this problem. If more side chains could be located, then α-carbon positions would be assigned more accurately. Even more useful would be the incorporation of sequence information, when available, to help in locating the main chain and keeping the residue count correct (see next section).

Last, it is worth pointing out that unusual situations will always occur in which the sophistication of human pattern recognition will be needed to interpret a portion of the map. The region about residues 83–88 of Bence-Jones Rhe is an illustration of such a difficult area. It is not worth the large programming effort to include complex decisions and strategies for that small percentage of cases which are obscure. It is probably worth considering criteria for detecting these situations and calling them to the attention of the human interpreter so that it will be apparent when human intervention is necessary. Such criteria are often easy to describe and include in the programming system.

Future Directions

The last few years have seen the successful introduction of several computer graphics Richards box programs that have proved very valuable in aiding the crystallographer to interpret the map (see [12], this volume). The ability to use computerized graphics to manipulate map and model simultaneously at the will of the user is clearly an invaluable aid to the map interpreter. Nevertheless, certain problems remain in such systems that can be improved. Much of the modeling is now routine, but it is still time consuming on an expensive computer graphics system. Furthermore, a contoured electron density map is hard to visualize on a graphics terminal in other than small pieces containing two to three residues. An overall view of the map or of larger features of the map is very hard to present, but it is just in such aspects that an automated map interpretation can be most powerful.

The most important first step for future development is to marry the basic routines of the automated system to one of the good computer graphics Richards box programs. The ability to produce the skeleton of the isolated molecule will be invaluable in aiding the earliest parts of map interpretation, when it is often difficult to obtain a toehold in the map to begin the interpretation. The skeleton of the map allows the visualization of large parts of the map in a useful and efficient way.

Further steps in the automated interpretation process, such as α-carbon assignment, chain direction assignment, and construction of the full main chain for secondary structure elements, can help with much of the more routine parts of the map interpretation. This can greatly speed up the procedure, and remove much of the drudgery, yet leave the more exciting and difficult parts of the map interpretation to the crystallographer. Naturally, the environment of a good computer graphics program will allow the crystallographer to intervene as necessary to correct errors and choose from among several alternatives that may be offered by the automated system.

Certain directions are indicated for the further development of the automated interpretation which are worthwhile mentioning here briefly. A valuable next step to the interpretation process would be a systematic analysis of the side chains of the molecule. The program system should examine the density for each identified side chain (see section on "Tracing the Chain" above) and classify it by size, length, total integrated density, and highest density. Such a list would be a useful first step in assigning possible residue types to the side chains. If the sequence of the protein were available, this list of side-chain attributes would be invaluable for corresponding regions of the interpreted molecule to the sequence data. Any available sequence information should be incorporated into the automated system to help in the interpretation and, in particular, to avoid the erroneous addition and deletion of residues.

Many electron density maps are difficult to interpret because of discontinuities in the chain. This is a major problem for the human model builder as well as for the program system. The program system can be designed to analyze the electron density map with a variable, local ρ_{min} value. Such an analysis, which has been studied in great detail in several fields of computer science including interpretation of photographs, could help considerably in resolving some of the discontinuities found in electron density maps. The larger the molecule being built, the more difficult it becomes to bridge these gaps in the chain by visual inspection. The automated system should be programmed to try the various possibilities, including tests for chain end proximity and the important constraint of passing through each fragment of the chain only once and in the correct direction. Such a routine would be a valuable aid to the interpreting crystallographer.

By combining the automated system with a computer graphics map interpretation package and by expanding the programs in these directions described above while leaving to the human operator to interpret the unusual and difficult anomalies that occur in each map, electron density map interpretation can be transformed into a maximally efficient and enjoyable procedure for the protein crystallographer.

Section III

Combined Model and Phase Refinement

[16] Overview of Refinement in Macromolecular Structure Analysis

By LYLE H. JENSEN

Introduction

Three-dimensional models of macromolecules based on X-ray diffraction data are derived initially from electron density maps. The atomic coordinates based on such models are not very accurate, in part because the maps are of insufficient resolution to show the constituent atoms individually and in part because the phases used to calculate the electron densities are not very accurate. Even for models derived from the best electron density maps, the accuracy of the atomic coordinates is likely to be no better than about 0.5 Å, and localized regions of models derived from average quality maps may suffer more serious errors. Such models, however, are very useful in showing the overall features of macromolecules and the pattern of chain folding, but to understand much of the chemistry of these molecules and how they function requires more accurate parameters. It is important, therefore, to improve the models of macromolecules by refining them to the fullest extent possible.

Meaning of Refinement

Refining molecular models derived from single-crystal X-ray diffraction data has long been considered an essential part of molecular structure analysis. In its traditional usage, the term *refinement* refers to the process of adjusting the parameters of the models, mainly the positions (x_j, y_j, z_j) and the thermal parameters (B_j) of the j atoms within the unit cell of the structure, in order to improve the agreement between the amplitudes (or intensities) of the observed reflections and the values calculated from the model parameters.[1] The process is an iterative one, and as the parameters

[1] The observed amplitudes are designated by $|F_o|$ and are equal to $\sqrt{I_o}$, I_o being the intensities of the observed reflections. The calculated amplitudes are designated by $|F_c|$ and are equal to $\sqrt{A_c^2 + B_c^2}$, where

$$A_c = \sum f_j \cos 2\pi(hx_j + ky_j + lz_j),$$
$$B_c = \sum f_j \sin 2\pi(hx_j + ky_j + lz_j),$$

and f_j is the scattering factor of atom j. The phase $\alpha_c = \tan^{-1} B_c/A_c$. The coordinates x_j, y_j, z_j are fractional; the indices of the X-ray reflections are hkl.

change from one cycle to the next, the calculated phases also change. This is refinement in the crystallographic sense of the term, long established by use among the community of X-ray crystallographers.

The process of refinement is said to have *converged* when the changes in parameters (and thus the phases) are no longer considered significant. If the initial parameters are too far from their true values, the refinement may converge to a false minimum. In such cases, the initial parameters of the model (or part of it) are said to lie outside the *convergence range*.

Refinement in Macromolecular Crystallography

In the field of macromolecular crystallography, the term refinement is used not only in the traditional sense just described, but also in other ways. Thus, in the case of Diamond's "real space refinement program"[1a] as it was initially used, only the model parameters (torsion angles and selected interbond angles) were adjusted to best fit the electron density map (see this volume [18]). The phases were usually experimental, based on heavy-atom derivatives, and remained unchanged in the course of the refinement. However, Huber and his associates have applied Diamond's program to refine models in the crystallographic sense by recomputing the electron density maps, using calculated phases after each cycle of fitting[2] (see also this volume [21]).

Another use of the term refinement involves what is known as phase correction by density modification[3,4] (see also this volume [6]). For example, if the X-ray data are properly scaled, the electron density map should nowhere be negative. If it is, the negative density is set to zero or if in the solvent continuum of a macromolecular crystal, it is set to the expected value of the solution. The modified map is then Fourier inverted, i.e., the amplitudes and phases of all reflections *hkl* are calculated, and used to compute a new electron density map. The process is repeated until there are no significant changes in the map from one cycle to the next. It is unnecessary to assume a molecular model of the structure; the electron density itself is refined to provide a better basis for fitting the model.

Density modification has proved particularly effective in improving electron density maps when multiple copies of a subunit or a major part of a structure are repeated within the crystallographically unique volume,

[1a] R. Diamond, *Acta Crystallogr., Sect. A* **A27** 435 (1971).

[2] R. Huber, D. Kukla, W. Bode, P. Schwager, K. Bartles, J. Deisenhoefer, and W. Steigemann, *J. Mol. Biol.* **89,** 73 (1974).

[3] W. Hoppe and J. Gassman, *Acta Crystallogr., Sect. B* **B24,** 97 (1969).

[4] D. M. Collins, *Acta Crystallogr., Sect. A* **A31,** 388 (1975).

the *asymmetric unit*.[5–7] Transforming the electron densities of the individual copies onto one of them, averaging, and then backtransforming the average to the positions of the individual copies has led to dramatically improved phases as judged from the improved electron density maps.[8]

Limit of Resolution

Whatever the method of refinement chosen for a particular macromolecular problem, we encounter limitations not ordinarily met in structures of small molecules. The root of the problem is inherent in the nature of macromolecules in the crystalline state: the constituent atoms of the molecules are not well localized. They often suffer both static and dynamic disorder,[9] an especially acute problem for surface groups such as amino acid side chains projecting into the solvent continuum surrounding protein molecules in a crystal. As a consequence, the intensities of reflections from crystals of proteins and other macromolecules decrease rapidly with increasing diffraction angle, thus limiting the extent of the data that can be observed.

Because of the sheer size of most macromolecular structures, the investigator may, for practical reasons in the early stages of a study, restrict the data set to something less than the limit to which reflections can be observed. By whatever means a data set is limited, whether by the inherent falloff of intensity or by terminating at a particular diffraction angle, that limit determines the resolution in the optical sense[10,11] and will set the ultimate limit of detail that can be observed in the electron density map. It can be shown that the resolution achieved with a given data set is $\sim 0.92 d_{min}$, the minimum interplanar spacing of the reflections.[12] In an electron density map based on such data, structures separated by $0.92 d_{min}$ should appear separately. This limit may not be reached in practice, how-

[5] M. G. Rossmann and D. M. Blow, *Acta Crystallogr.* **16,** 39 (1963).

[6] M. G. Rossmann and D. M. Blow, *Acta Crystallogr.* **17,** 1474 (1964).

[7] G. Bricogne, *Acta Crystallogr., Sect. A* **A32,** 832 (1976).

[8] A. C. Bloomer, J. Grahm, S. Hovmöller, P. J. G. Butler, and A. Klug, *in* "Structural Aspects of Recognition and Assembly in Biological Macromolecules" (M. Balaban, ed.), Vol. 2, p. 851. BALABAN Int. Sci. Serv., Rehovot, Israel, and Philadelphia, Pennsylvania, 1981.

[9] A. S. Brill, *in* "Tunneling in Biological Systems" (B. Chance *et al.*, eds.), p. 561. Academic Press, New York, 1979.

[10] J. Strong, "Concepts of Classical Optics," p. 210. Freeman, San Francisco, California, 1958.

[11] R. W. James, "Optical Principles of the Diffraction of X-rays," p. 390. Cornell Univ. Press, Ithaca, New York, 1965.

[12] R. E. Stenkamp and L. H. Jensen, *Acta Crystallogr., Sect. A* **A40,** 251 (1984).

ever, so that it is customary in the field of macromolecular structure analysis to take the limit of resolution simply as d_{min} of the data set. From Bragg's law in the form

$$d_{min} = \lambda/2 \sin \theta_{max} \quad (1)$$

it is evident that the maximum diffraction angle, θ_{max}, determines d_{min} and thus the limit of resolution.

Number of Unique Reflections

In any structure determination the number of observations must at least equal the number of parameters to be evaluated, and in practice should greatly exceed it. In general, the greater the ratio of the number of observations to the number of parameters, the more reliable the derived parameters will be. Thus we strive to maximize the number of observations or conversely to minimize the number of parameters used to describe the model.

It is enlightening to consider the number of unique reflections accessible at a given resolution. For Cu K_α radiation the approximate number for a structure is given by the equation

$$\text{number} \cong (4.2)V/nd^3_{min} \quad (2)$$

where V is the volume of the unit cell in Å^3 and n = 2,4,6,8,12,24, or 48 depending on the symmetry of the structure (see Table 3.2 in Ref. 13). If the crystal lattice is not simple primitive (one lattice point per unit cell), the number is reduced by a subintegral fraction: one-half for a body-centered or side-centered lattice, one-third for a rhombohedral lattice referred to hexagonal axes, and one-quarter for a face-centered lattice.

As an example, assume a protein with 1000 nonhydrogen atoms in the unit cell of a crystal with 45% solvent. If the lattice is primitive and if n is taken as 2 in Eq. (2), a typical unit cell volume would be 36,000 Å^3. We calculate approximately 2800 unique reflections to 3 Å resolution. Since three positional parameters (x, y, z) are required for each atom, at least 3000 observations are required to fix the positions of all 1000 atoms. It is clear, therefore, that at 3 Å resolution the X-ray data alone are insufficient to define the positions of all atoms in this example.

At 2 Å resolution approximately 9450 reflections are accessible in the example; the problem is now overdetermined because the ratio of the number of observations to the number of parameters exceeds unity, and

[13] G. H. Stout and L. H. Jensen, "X-Ray Structure Determination," p. 54. Macmillan, New York, 1968.

refinement solely against the X-ray data would be possible. One should note, however, that the number of accessible reflections is the maximum that can be observed at any resolution; the actual number will be less, possibly much less, because of the relatively rapid falloff of intensities from macromolecular crystals with increasing sin θ.

Other Information

Although a data set from a macromolecular crystal is severely limited compared to that from typical crystals of smaller molecules, we have much additional information such as bond lengths, bond angles, certain torsion angles, and van der Waals radii that can be incorporated in the refinement. The added information can be imposed on the model either by fixing the geometry of known structural features[14-16] such as peptide groups or aromatic side chains of proteins, or by including the additional information as "observations"[17-19] along with the X-ray data against which the model is to be refined. Both methods are discussed in this volume, [19] and [20], and if either one is used, the added information will allow refinement at lower resolution, even at 3 Å resolution or lower if the problem is sufficiently overdetermined, i.e., if the ratio of the number of observations to the number of parameters is sufficiently great. In any case, the more information that can be included and the greater the overdetermination of the problem, the better the refinement should behave and the more reliable the derived parameters are expected to be.

Monitoring Refinement

The progress of a refinement is usually monitored by the conventional crystallographic R index defined by the equation

$$R = \sum ||F_o| - |F_c|| \Big/ \sum |F_o| \qquad (3)$$

Note that this is a kind of relative error, the sums being over all observed reflections in the data set. Clearly, the better the agreement between the observed amplitudes $|F_o|$ and the calculated values $|F_c|$, the lower R will

[14] C. Scheringer, *Acta Crystallogr.* **16,** 546 (1963).

[15] J. L. Sussman, S. R. Holbrook, G. M. Church, and S. H. Kim, *Acta Crystallogr., Sect. A* **A33,** 800 (1977).

[16] L. G. Hoard and C. E. Nordman, *Acta Crystallogr., Sect. A* **A35,** 1010 (1979).

[17] J. Waser, *Acta Crystallogr.* **16,** 1091 (1963).

[18] J. H. Konnert, *Acta Crystallogr., Sect. A* **A32,** 614 (1976).

[19] W. A. Hendrickson, *in* "Proceedings of the Daresbury Study Weekend," p. 1. Science and Engineering Council, Daresbury Laboratory, Warrington, England, 1981.

be and presumably the better the model. For comparative purposes we note that if atoms in a given structure were to be randomly distributed in the unit cell with the symmetry of the structure, the expected value of R would be 0.59 for a noncentrosymmetric structure, 0.83 for a centrosymmetric one.[20] In view of the fact that when biological macromolecules crystallize, the structures must be noncentrosymmetric, we conclude that R values in the vicinity of 0.6 indicate a structure that does not fit the data any better than does a random one. Protein models derived from readily interpretable electron density maps based on heavy-atom derivatives typically lead to models with R values in the range of 0.4–0.5, not very much below what would be expected for a random structure.

Refinements of macromolecular models often converge at R values in the vicinity of 0.20[2,21–24]; but for the most extensively refined models based on data sets with resolution better than 2 Å, R values approaching 0.12 have been achieved.[25–27] This is in contrast to small organic and biological molecules for which extensive data can be collected, and models typically refine to R values in the range 0.03–0.05.

Reliability of Models

Since refinement of macromolecular models is being widely applied and the implication is that the models have been improved, we need to question the extent to which this is achieved and to assess the reliability of atomic coordinates. On the basis of refinements to date, it is evident that reliability varies not only from one structure to another but also from one part of the same structure to another part. In general, we can say that atoms with lower B values, those well tethered by hydrogen bonds or tightly packed within the molecule, will be better determined than those with higher B values. In other words, atoms with small B values tend to be more precisely located as indicated by smaller estimated standard deviations (σ) in their positional parameters. In the case of the small protein rubredoxin from *Clostridium pasteurianum* (54 amino acids), the σ values of the positional parameters for the C_α and C_β atoms were found to vary

[20] A. J. C. Wilson, *Acta Crystallogr.* **2,** 318 (1949).
[21] S. T. Freer, R. A. Alden, C. W. Carter, Jr., and J. Kraut, *J. Biol. Chem.* **250,** 46 (1975).
[22] J. Deisenhoefer and W. Steigemann, *Acta Crystallogr., Sect. B* **B31,** 238 (1975).
[23] E. T. Adman, L. C. Sieker, and L. H. Jensen, *J. Biol. Chem.* **251,** 3801 (1976).
[24] R. E. Stenkamp, L. C. Sieker, and L. H. Jensen, *Acta Crystallogr., Sect. B* **B38,** 784 (1982).
[25] K. D. Watenpaugh, L. C. Sieker, J. R. Herriott, and L. H. Jensen, *Acta Crystallogr., Sect. B* **B29,** 943 (1973).
[26] A. R. Sieleki, W. A. Hendrickson, C. G. Broughton, L. T. J. Delbaere, G. D. Brayer, and M. N. G. James, *J. Mol. Biol.* **134,** 781 (1979).
[27] W. A. Hendrickson and M. M. Teeter, *Nature* (*London*) **290,** 107 (1981).

approximately linearly from 0.03 to 0.06 Å for atoms with B values ranging from 7 to 21 Å^2.[28] The relatively low values of σ for this protein are not to be taken as representative of protein models generally, in part because the B values for rubredoxin are relatively low and in part because the model was highly refined and based on what for proteins is regarded as very high-resolution data (1.2 Å).

The conventional R defined above is usually taken as a measure of the reliability of a model based on X-ray data. As a single number, it is an overall indicator, and for large molecules it is insensitive even to serious errors in restricted regions of the structure, such as misoriented peptide groups or misplaced side chains for a few amino acids.

While R is useful for comparative purposes and as a rough guide to the reliability of a model, it is the σ values of the atoms that are of immediate value and general interest to investigators in other fields. An overall σ for a set of coordinates can be estimated from a plot of the relationship between R as a function of $(\sin \theta)/\lambda$ for various assumed errors as deduced by Luzzati.[29] Figure 1 is such a plot (solid lines) with results from rubredoxin superimposed (broken lines).[30] For the lowest error in coordinates (0.05 Å), the plot is essentially linear over the range of $(\sin \theta)/\lambda$ shown. For larger errors the plots become progressively more nonlinear, and for an error of 0.5 Å the plot in the figure approaches 0.59, the value of R corresponding to a random structure.

The results for rubredoxin in Fig. 1 are plotted for the initial model, after four refinement cycles against 1.5 Å resolution data, and after eight cycles.[25] Data in $(\sin \theta)/\lambda$ ranging from 0.1–0.33 (d spacings 5–1.5 Å) were divided into three groups and R for each group plotted as shown. After the eighth refinement cycle, the distribution of R follows the theoretical plots quite well, indicating an error of 0.11–0.13 Å. This is close to the value of ~0.14 Å estimated for the C_α and C_β atoms by two other methods.[31] Note that the σ values for rubredoxin refined against the 1.5 Å data are two to four times greater than the range of values for the best determined C_α and C_β atoms from the refinement against the 1.2 Å data quoted earlier.

The Luzzati plot emphasizes the importance of resolution in assessing R. Thus for a model with error 0.5 Å, R would be ~0.15 for data in the $(\sin \theta)/\lambda$ range 0.05–0.10 (d from 10–5 Å), but for the same error, R would be ~0.3 for data in the range 0.10–0.20 (d from 5–2.5 Å).

Finally, one should remember that if the estimated σ values for a set of coordinates are realistic and if the errors are random, then two-thirds of

[28] K. D. Watenpaugh, L. C. Sieker, and L. H. Jensen, *J. Mol. Biol.* **138,** 615 (1980).

[29] V. Luzzati, *Acta Crystallogr.* **5,** 802 (1952).

[30] L. H. Jensen, *Abstr., Am. Cryst. Assoc.* p. 75 (1974).

[31] L. H. Jensen, *in* "Crystallographic Computing Techniques" (F. R. Ahmed, K. Huml, and Sedláček, ed.), p. 314. Munksgaard, Copenhagen, 1976.

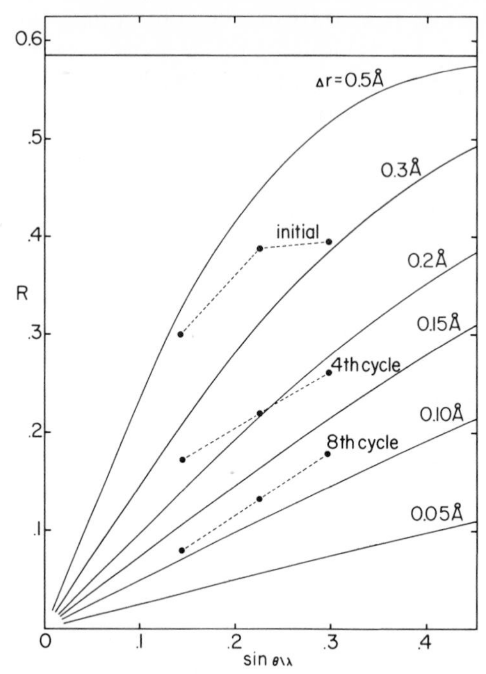

FIG. 1. Luzzati plot (solid lines) of R as a function of $(\sin\theta)/\lambda$ for errors in the range 0.05–0.5 Å. R values for rubredoxin (broken lines) in the $(\sin\theta)/\lambda$ range 0.1–0.33 for the initial model (overall $R = 0.372$), after four refinement cycles (overall $R = 0.224$), and after eight refinement cycles (overall $R = 0.126$). Refinement was against the 1.5 Å resolution data set.

the atoms in the model should be within σ of their true positions, but at the same time, approximately five atoms out of 100 will be more than 2σ from their true positions.

In the chapters that follow, the present status of refining macromolecules is set forth in detail, covering the range of refinement techniques that have been developed. The information in these chapters should enable investigators in other fields to understand the nature of the refinement process and to make reasonable assessments of the reliability of macromolecular models encountered in the literature.

Acknowledgments

I wish to thank R. E. Stenkamp for a number of helpful comments and to acknowledge support of studies involving some of the examples cited in this account under USPHS Grant GM-13366 from the National Institutes of Health.

[17] Classic ($F_o - F_c$) Fourier Refinement

by STEPHAN T. FREER

The use of the ($F_o - F_c$) synthesis for crystal structure refinement first suggested by Cochran in 1951[1] was employed by Jensen and his co-workers in refinement of rubredoxin,[2] and was the first demonstration that protein crystal structures can be crystallographically refined by the same techniques that were successful for small structures. $F_o - F_c$ Fourier refinement is mathematically equivalent to least-squares refinement in which the residuals are weighted by the reciprocal of the unitary atomic scattering factor,[3] and may be carried out on a small computer.[4] Indeed, a portion of the rubredoxin refinement was done by hand. Although $F_o - F_c$ Fourier refinement is economical and provides direct visual representation of the correction of the trial structure, it has been largely supplanted by least-squares refinement which can be more easily automated, has more realistic weighting, can easily accommodate stereochemical constraints directly into the refinement process, and can provide a direct estimate of errors in the atomic parameters of the trial structure.

A Fourier synthesis computed with coefficients $(F_o - F_c)e^{i\alpha_c}$, where F_o is the observed structure amplitude and F_c and α_c are the structure amplitude and phase angle calculated from the current trial structure, reveals any gross errors in the trial structure and also indicates the magnitude and direction of slight parameter adjustments. For example, if the trial structure contains an atom in an incorrect position, the $F_o - F_c$ map will have a negative peak of about one-half theoretical height[5] in the incorrect position and a corresponding positive peak of about one-half theoretical height in the correct position, provided these two locations are separated by at least an atomic diameter and that the phase angles are reasonably accurate. In practice, the peaks are often partially obscured by noise, although the noise is progressively reduced as increasingly accurate phases are obtained during the course of the refinement. If an atom is only slightly mislocated, the resulting peak and hole in the $F_o - F_c$ map

[1] W. Cochran, *Acta Crystallogr.* **4,** 408 (1951).

[2] K. D. Watenpaugh, L. C. Sieker, J. R. Herriott, and L. H. Jensen, *Acta Crystallogr., Sect. B* **B29,** 943 (1973).

[3] H. Lipson and W. Cochran, *in* "The Determination of Crystal Structures" (L. Bragg, ed.), Vol. III, p. 299. Bell, London, 1957.

[4] J. L. Chambers and R. M. Stroud, *Acta Crystallogr., Sect. B* **B33,** 1824 (1977).

[5] R. Henderson and J. K. Moffat, *Acta Crystallogr., Sect. B* **B27,** 1414 (1971).

coalesce and are reduced in magnitude. In this case, a correction to the atomic coordinates can be determined from the slopes of the $F_o - F_c$ map in three orthogonal directions evaluated at the assumed atomic center: the atom is shifted up the slope (toward the positive peak) by an amount proportional to the magnitude of the slope. If an atom is correctly located, the peak and hole coincide and the $F_o - F_c$ map is flat at the atomic center, unless the temperature factor is an error. In this case, the $F_o - F_c$ synthesis provides a means of estimating corrections to the temperature factor. The effect of an increase in temperature factor is a reduction in the magnitude of the structure factor. For instance, if an isotropically vibrating atom has been given too large a temperature factor, its contribution to the F_c will be too small and, therefore, at this atom's location in the $F_o - F_c$ map there will be a peak whose height is proportional to the amount by which the temperature factor should be reduced. Anisotropic temperature factor corrections can also be estimated from $F_o - F_c$ maps, albeit with difficulty.

The input to any cycle of $F_o - F_c$ Fourier refinement is a trial structure (atomic coordinates and thermal parameters) together with the structure factors computed from it, and the output is a new trial structure and new calculated structure factors. A cycle is composed of up to five distinct steps: (1) calculation of an $F_o - F_c$ map, typically with a grid spacing slightly less than one-half the resolution of the intensity data; (2) determination of parameter shifts from the map; (3) computer-aided deletion, addition, or gross repositioning of atoms; (4) application of stereochemical constraints; and (5) calculation of new structure factors. Steps 3 and 4 are not included in every cycle. Steps 1 and 5 can be performed with standard crystallographic Fourier and structure factor programs. Step 4, the application of constraints, can be done by a number of methods, some of which are described in this volume. Step 3, editing the trial structure, is most conveniently done with the aid of computer graphics and may be assisted by computer programs to locate atoms in adjacent molecules which come too close together, locate close, nonbonded contacts within a single molecule, locate atoms which do not lie in electron density in the F_o map, locate potential solvent atoms, etc. Shifts in positional parameters may be calculated from

$$\Delta x_{ij} = -(\text{slope})_{ij}/(\text{curvature})_j = -[\Delta\rho(x+\varepsilon) - \Delta\rho(x-\varepsilon)/2\varepsilon/]C_j,$$

where $i = 1, 2, 3$ for x, y, z (the map is evaluated on an orthogonal grid); index j specifies a particular atom; $\Delta\rho$ is the density of the $F_o - F_c$ map linearly interpolated from the surrounding grid points; $\varepsilon = 0.5$ times the grid interval; and the curvature $C_j = -0.2$ times the atomic number of atom j, an empirically determined value that will produce reasonable parameter shifts. Alternate, but similar, values for curvatures have been used.[2,6] Shifts in temperature factor B may be calculated from $\Delta B_j = -[\Delta\rho(\mathbf{x}_j)/\overline{\Delta\rho}]/C_j$, where $\overline{\Delta\rho}$ is the mean absolute value of the difference electron density at all atomic centers of the $F_o - F_c$ map from the previous cycle. Making the temperature factor correction inversely proportional to $\overline{\Delta\rho}$ has the effect of proportionally increasing shifts in B as the atomic coordinates become more accurate during refinement. Different empirical schemes for temperature factor refinement, which nevertheless are similar in principle, are given, for example, by Moews and Kretsinger[6] and Chambers and Stroud.[4]

[6] P. C. Moews and R. H. Kretsinger, *J. Mol. Biol.* **91,** 201 (1975).

[18] Real Space Refinement

By R. DIAMOND

Early attempts to refine protein structures in the 1960s met with some serious difficulties which arose both from the formidable size of the task (by the computing standards of the time) and from the lack of resolution. Blow and Crick[1] had by then shown how to obtain optimal phases from the multiple isomorphous replacement (MIR) method, and, while it was realized that such phases may be of rather poor accuracy, it was by no means clear at that time how such phases should be handled in any attempt to refine the parameters of an atomic model in relation to such data. The real space refinement method[2] (referred to here as RSR) was conceived against that background, and in a defensive frame of mind in the sense that much emphasis was placed on ensuring that beneficial adjustments would be made and that the canons of stereochemistry would be observed in most respects even if the process stopped short of the ultimate minimum residual. The lack of atomic resolution, which makes free refinement of atomic coordinates impossible, was combated by reducing the number of positional parameters from three per atom to less than one per atom by using dihedral angles as independent variables and by incorporating filtering techniques to combat ill-conditioning. More modern

[1] D. M. Blow and F. H. C. Crick, *Acta Crystallogr.* **12,** 794 (1959).
[2] R. Diamond, *Acta Crystallogr., Sect. A* **A27,** 436 (1971).

techniques, such as that of Hendrickson and Konnert,[3,4] aim to solve these same problems by effectively augmenting the number of observations (by reference to a notional potential energy) rather than by reducing the number of parameters. Such procedures require larger machines than were generally available in the 1960s.

There are two main characteristics of the old method which are logically independent and which may therefore be considered separately. These are, first, the fact that the process aims to minimize $\int(\rho_o - \rho_m)^2\, dv$ for observed and model densities ρ_o and ρ_m, respectively, and second, the manner of handling the stereochemical aspects of the problem through flexible chain techniques. Some of the properties flowing from these characteristics have already been reviewed.[5]

The first of these characteristics gives the RSR method a number of properties, some of which are quite distinctive and not shared by some of the alternatives. Chief among these is that, by refining against ρ_o, it is equivalent to a reciprocal space refinement against the complex F rather than $|F|$. Selecting this method therefore implies that credence is to be attached to the currently available phases. This means, I think, that the method is appropriate to the early stages when a rough model is being brought into line with an MIR map, but that once that stage has passed, processes which treat the phases as disposable rather than observed may be preferable. Such a process might either refine against $|F_o|$, or against $|F_o|\alpha_c$ with α_c redetermined at each cycle, but the former should have faster convergence. RSR has, however, been widely used in a cyclic fashion using a fresh $|F_o|\alpha_c$ map on each cycle and used in this way it has been very effective, as illustrated by Fig. 1.[5a] Watenpaugh *et al.*,[6] Huber *et al.*,[7] and others since then have reported that phases derived from an initial model are closer to those derived from a final refined model than are the MIR phases. This could, of course, be for either or both of two reasons: (1) MIR phases are so poor that calculated phases, even from an incompletely refined model, are likely to be better; and (2), refinement methods are only partially effective in modifying phases so that initial and

[3] J. H. Konnert, *Acta Crystallogr., Sect. A* **A32,** 614 (1976).

[4] W. A. Hendrickson and J. H. Konnert, *in* "Biomolecular Structure, Conformation, Function and Evolution" (R. Srinivasan, E. Subramanian, and N. Yathindra, eds.), p. 43. Pergamon, Oxford, 1981.

[5] R. Diamond, *in* "Crystallographic Computing Techniques" (F. R. Ahmed, K. Huml, and B. Sedláček, eds.), p. 291. Munksgaard, Copenhagen, 1976.

[5a] J. Deisenhofer and W. Steigemann, *Acta Crystallogr., Sect. B* **B31,** 238 (1975).

[6] K. D. Watenpaugh, L. C. Sieker, J. R. Herriott, and L. H. Jensen, *Acta Crystallogr. Sect. B* **B29,** 943 (1973).

[7] R. Huber, D. Kukla, W. Bode, P. Schwager, K. Bartels, J. Deisenhofer, and W. Steigemann, *J. Mol. Biol.* **89,** 73 (1974).

final calculated phases are bound to be similar. Calculated phases usually more further away from the MIR phases during refinement, suggesting that the first reason given above is the main one.

A second property arising from the minimization of $\int(\rho_o - \rho_m)^2\, dv$ is concerned with weighting, for this residual is equivalent to $V^{-1} \Sigma |\Delta F|^2$, so that the reflections are uniformly weighted and all of the density within the molecule is involved in the refinement. This contrasts with methods which seek to place each atomic center at a maximum of ρ_o without regard

(*a*)

FIG. 1. Stereo pairs showing sections around (a) Phe 45 and (b) Pro 8 of bovine pancreatic trypsin inhibitor as refined by J. Deisenhofer and W. Steigemann using RSR and phase relaxation. The quality of the map (phased from the refined model) and the fit of the refined model (which is also shown) are evident. The map is at 1.5 Å resolution contoured at intervals of 0.5 e Å^{-3}, beginning at 0.5 e Å^{-3}. (Reproduced with permission from Deisenhofer and Steigemann.[5a])

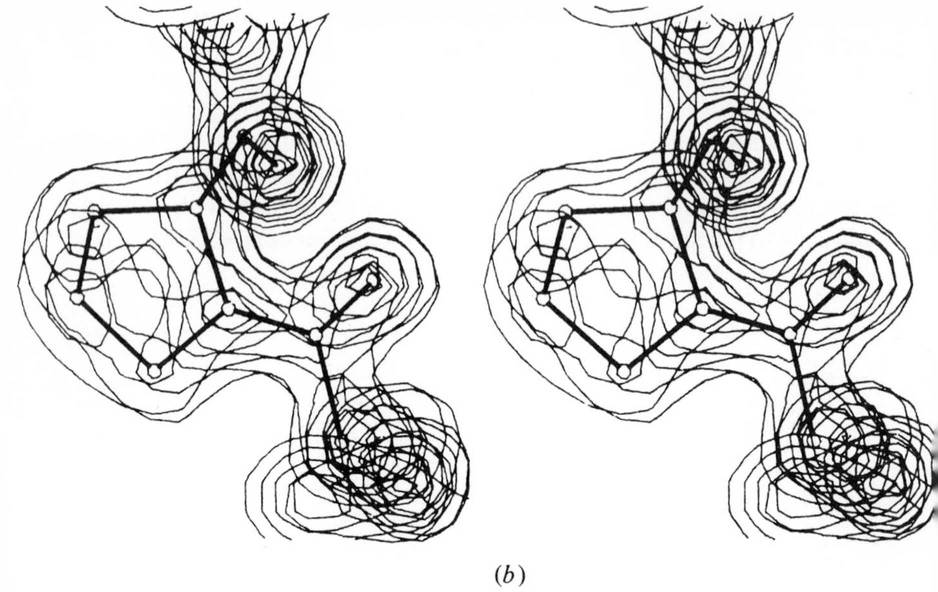

(*b*)

FIG. 1. (*continued*)

to the form of ρ_o in the vicinity of each maximum. This latter is equivalent to minimizing $V^{-1} \Sigma f^{-1}|\Delta F|^2$, which attaches increasing weight to reflections at high angles.[8,9]

A third feature arising from the choice of residual is a wide range of convergence. Provided that ρ_m overlaps ρ_o at least partially, convergence should ensue, whereas a method sensitive to grad ρ_o at the center of each model atom may detect no driving signal in the same circumstances. This may be the case when a model atom is displaced between one and two atomic radii from its true position.

A fourth property is that, once ρ_o has been determined, the size of the computational task rises linearly with the number of atoms, whereas a quadratic characteristic arises in refinements against reciprocal space data because every atom is mathematically linked to every reflection, the number of which is also proportional to the number of atoms.

Finally, in common with any procedure operating in real space, it is trivially easy to refine only a selected portion of the structure against data relating only to that portion independently of the effects of any uninterpreted regions.

The second main characteristic concerns the stereochemical aspects

[8] W. Cochran, *Acta Crystallogr.* **4,** 408 (1951).

[9] H. Lipson and W. Cochran, "The Determination of Crystal Structures," 3rd ed., p. 332. Bell, London, 1966.

of the problem. The RSR method treats the protein or other polymer as a flexible chain using only internal rotations as independent variables, the atomic coordinates being dependent on these. As already mentioned, the objectives in so doing were the reduction of the number of parameters in order that the ratio of the number of observations to the number of parameters should be sufficiently high, and making the provision of a routine refinement procedure for proteins possible on a 32K machine (IBM 360/44). The more modern techniques of Hendrickson and Konnert[4] and of Jack and Levitt[10] take advantage of the larger machines available today in order to add an estimate of the mechanical potential energy to the X-ray residual and minimize this sum using the atomic coordinates as independent variables. This has a number of advantages, perhaps the chief of which is that, should an apparently strained structure be encountered, the strain is spread over all features of the structure in the appropriate proportions rather than being channeled exclusively into those features of the chain which have been nominated as angular variables. The enlarged number of positional parameters (3 per atom compared to 0.75 per atom) has also led to lower residuals being achieved, as is to be expected. The older method, however, retains some advantages in relation to the handling of planar groups for which, in the newer methods, restoring forces for out-of-plane displacements may be vaninshingly small, and angular variables are also convenient whenever it is necessary to reinitialize a side chain, for example, by a large rotation about a bond, as was exploited by Mandel *et al.*[11] using the version of the program (RS4H) with model-adjusting capability.

These workers also encountered a wide scatter of values of the main-chain bond angle at C_α (τ_α) which they countered by introducing flexibility in the other two main-chain bond angles τ_C and τ_N, giving six degrees of freedom per peptide plus the usual side-chain freedoms. They appear to be the only workers to have parameterized the chain in this way, though the method has always had the capability to vary any angles (with some restrictions within rings). They write, "The best fitting with the model-building routine ended with τ_α values that deviated an average 9° from ideal tetrahedral values; but iteration with the model-adjusting routine spread these distortions among all angles, with an average deviation of only 1.5°, and without major changes in chain conformation. The occasional serious distortions found earlier in particular angles were eliminated in a most satisfactory manner."

In some procedures, notably those of Freer *et al.*[12] and of Agarwal and

[10] A. Jack and M. Levitt, *Acta Crystallogr., Sect. A* **A34,** 931 (1978).

[11] N. Mandel, G. Mandel, B. L. Trus, J. Rosenberg, G. Carlson, and R. E. Dickerson, *J. Biol. Chem.* **252,** 4619 (1977).

[12] S. T. Freer, R. A. Alden, C. W. Carter, and J. Kraut, *J. Biol. Chem.* **250,** 46 (1975).

Isaacs,[13,14] the stereochemical problems are handled independently of the refinement steps, being performed by a quite separate idealization routine. This is equally possible with the RSR procedure, which does in any case proceed in two stages. The first stage determines by reference to the map what translational atomic shifts are wanted and the second determines the angular changes which most nearly yield these translations. The procedure has the option of omitting the second stage, in which case the wanted translations are applied directly. There are, however, theoretical grounds for preferring to handle the X-ray and energy considerations simultaneously rather than in alternation, because the maximum joint probability (over X-ray and energy data) is not at a position in parameter space which is a linear combination of the positions found by application of these criteria independently.[15,16] If energy minimization alone leads to a solution x_1 in parameter space with $M_1 = (1/kT)(\delta^2 E/\delta x_i\ \delta x_j)$ and if the X-ray data, taken alone, lead to a solution at x_2 with normal matrix M_2 for properly weighted reflections, then the joint solution using both criteria is at $x = (M_1 + M_2)^{-1} (M_1 x_1 + M_2 x_2)$, which is analogous to a weighted mean of x_1 and x_2 but is not colinear with them. While the Konnert–Hendrickson and Jack–Levitt methods give results of this form, I am not aware that the results of separate idealization and free X-ray refinement have ever been merged in this way.

Although the RSR method is not integrated with energy refinement Levitt[17] has made an intensive study of the relationships between these procedures. He has taken a sequence of six coordinate data sets for hen egg-white lysozyme as follows:

1. W2, hand-measured coordinates from a wire model.[18]
2. M2, coordinates derived from W2 by the "model-building" procedure of Diamond.[19]
3. RS5D, coordinates derived from M2 by RSR refinement against MIR map[20] allowing rotation in N–C peptide bonds.

[13] N. W. Isaacs and R. C. Agarwal, *Acta Crystallogr., Sect. A* **A34,** 782 (1978).

[14] R. C. Agarwal, *Acta Crystallogr., Sect. A* **A34,** 791 (1978).

[15] R. Diamond, *in* "Refinement of Protein Structures" (P. A. Machin, J. W. Campbell, and M. Elder, eds.), p. 47. Science and Engineering Research Council, Daresbury Laboratory, Daresbury, Warrington, England (DL/SCI/R16), 1981.

[16] R. Diamond, *in* "Structural Aspects of Biomolecules" (R. Srinivasan and V. Pattabhi, eds.), p. 81. Macmillan India Ltd., Delhi, 1981.

[17] M. Levitt, Ph.D. Dissertation, University of Cambridge (1971).

[18] C. C. F. Blake, G. A. Mair, A. C. T. North, D. C. Phillips, and V. R. Sarma, *Proc. R. Soc. London, Ser. B* **167,** 365 (1967).

[19] R. Diamond, *Acta Crystallogr.* **21,** 253 (1966).

[20] R. Diamond, *J. Mol. Biol.* **82,** 371 (1974).

4. ER5D, coordinates derived from RS5D by 30 cycles of steepest descent energy minimization using simple repulsive nonbonded forces and a penalty term on deviations from RS5D coordinates.

5. RS7A, coordinates derived from ER5D in the same manner as RS5D using the same map.

6. ER7A, coordinates derived from RS7A in the manner of ER5D but with a penalty term operating on deviations from RS5D.

This sequence showed several interesting features:

1. In the five steps taken between the six data sets in this sequence, statistical orthogonality of the shifts arose between the first and second steps and between the second and third, and thereafter negative correlations arose increasingly. The second step (M2–RS5D) was the largest, after which the step lengths decreased monotonically. See Fig. 2. This reflects the fact that the first three steps are concerned with different aspects of the coordinate set and that each may be said to leave the work of previous steps undisturbed until alternation of the two procedures produces a measure of alternation in the shifts produced.

2. Levitt[17] gives the agreement with X-ray data in terms of the conventional R factor and of the correlation coefficient between F_o and F_c in Table I (omitting water contributions). This shows that although the second and fourth steps are the only ones designed to improve the X-ray agreement, this is in fact improved by every step except the fourth,[21] although the effects may be insignificantly small after the third step.

3. Levitt[17] gives the energies of the six coordinate sets as in Table II. Here the large figures in the last column are due to a small number of nonbonded close contacts which are removed by the first energy refinement, RSR and model building being insensitive to their existence. Once removed, however, they do not recur. Nevertheless, RSR, by reference to the map, does reduce the nonbonded energy term and the model-building step reduces the bond and angle energies so that the net effect of both the first two steps is beneficial to the energy. This is not so of the fourth step (RSR), however, as the easy gains have already been achieved and competition between the procedures is beginning.

4. Levitt[17] also gives data relating to the stereochemistry of these coordinate sets in Table III. Levitt defines his criteria for including or excluding a contact in the count of close contacts or for counting H bonds. The details are not important for present purposes. What is important is that those criteria were not varied, so that the last three columns

[21] It is possible to encounter a small increase in R factor when $\int(\rho_o - \rho_m)^2\, dv$ (the function minimized) is reduced. See, for example, Table III of Ref. 5.

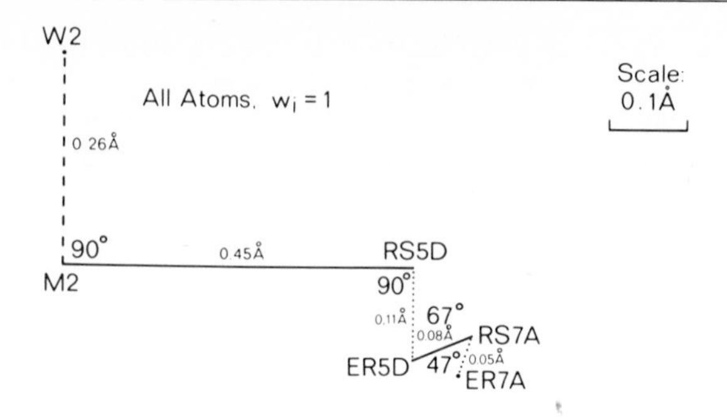

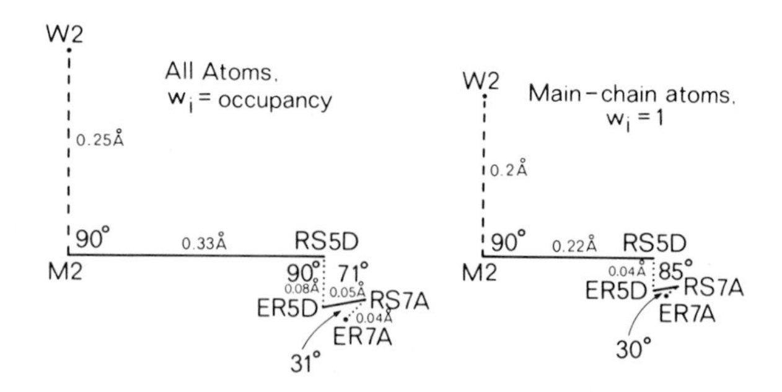

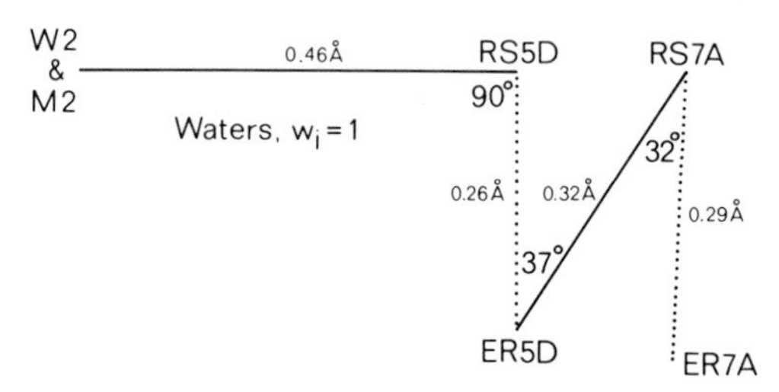

FIG. 2. Shift correlation diagrams for the five sequential steps relating the six coordinate sets for hen egg-white lysozyme. The lengths of the lines are $[\Sigma\, w(\Delta r)^2/\Sigma\, w]^{1/2}$ for vector shifts $\Delta\mathbf{r}$ between coordinates, and the weights w are either unity or the apparent occupancy as measured by RSR, as indicated. The cosines of the indicated angles when negated are the similarly weighted correlation coefficients on consecutive steps. Dashed lines indicate computer model building, solid lines RSR, dotted lines energy minimization. For a fuller description of shift correlation diagrams see Diamond.[20] (Figure calculated by M. Levitt and reproduced with permission from his Ph.D. Dissertation.[17])

TABLE I

X-RAY RESIDUALS AND CORRELATIONS OF THE SIX SETS OF COORDINATES[17]

	W2	M2	RS5D	ER5D	RS7A	ER7A
R^a	0.430	0.411	0.352	0.350	0.352	0.350
C^b	0.498	0.518	0.661	0.668	0.664	0.664

[a] $R = \Sigma||F_o| - |F_c||/\Sigma|F_o|$

[b] $C = \Sigma(|F_o| - \langle|F_o|\rangle)(|F_c| - \langle|F_c|\rangle)[\Sigma(|F_o| - \langle|F_o|\rangle)^2\ \Sigma(|F_c| - \langle|F_c|\rangle)^2]^{-1/2}$

TABLE II

ENERGIES OF THE SIX DATA SETS[17]

Coordinates	Energy (kcal/mol) Total	Bonds	Angles	Torsions	Nonbonded
W2	61006	13774	3972	87	43173
M2	48348	346	333	0	47669
RS5D	42638	351	308	43	41935
ER5D	558	12	91	17	438
RS7A	849	14	97	20	718
ER7A	508	10	85	15	398

TABLE III

SOME STATISTICS OF STEREOCHEMICAL FEATURES[17]

Coordinates	$\tau_\alpha{}^a$ (°) Mean	$\tau_\alpha{}^a$ (°) SD	ω^b (°) Mean	ω^b (°) SD	Close contacts[c]	No. of H bonds between: Atoms of the protein	No. of H bonds between: Water[d]
W2	111.5	10.9	180.9	15.0	67	125	144
M2	112.7	8.4	180.0	0.0	61	132	145
RS5D	112.6	8.3	179.6	4.9	45	132	155
ER5D	112.9	4.2	179.4	4.0	4	139	157
RS7A	112.9	4.2	179.3	4.6	20	138	151
ER7A	112.9	4.1	179.4	4.0	2	139	159

[a] τ_α is the bond angle N–C_α–C.

[b] ω is the dihedral angle in the peptide bond C_α–C–N–C_α. $\omega = 180°$, is planar trans.

[c] Contacts between atoms separated by more than three covalent bonds for which the repulsive force exceeds 20 kcal/mol-Å.

[d] Among water or between water and protein.

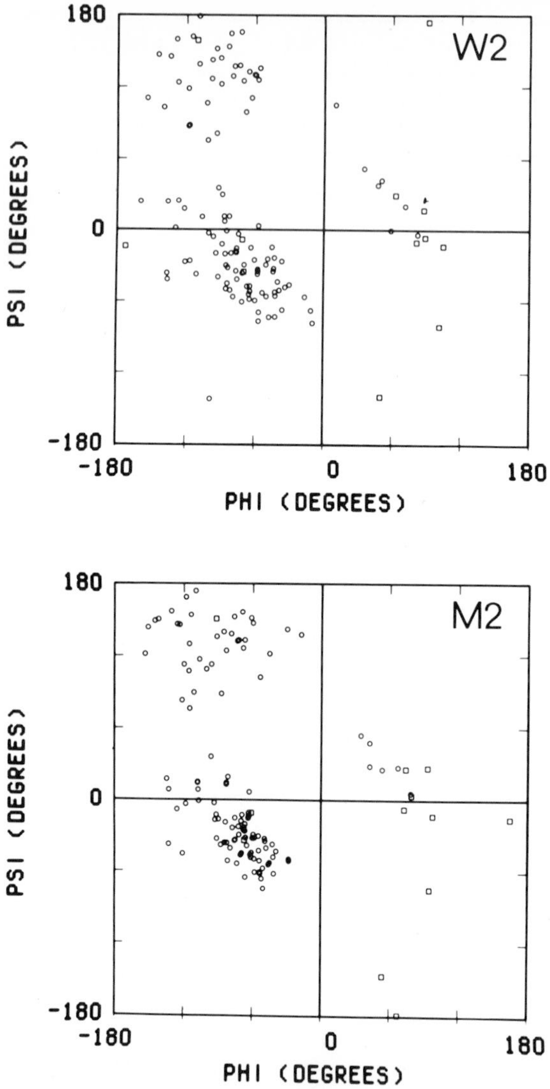

FIG. 3. Distributions of ϕ and ψ for the six coordinate data sets of hen egg-white lysozyme. Glycine residues are marked with squares, all others with circles. (ϕ is the dihedral angle C_{n-1}–N_n–$C_{\alpha n}$–C_n. ψ is the dihedral angle N_n–$C_{\alpha n}$–C_n–N_{n+1}.) The steps W2→M2 and M2→RS5D, which are not controlled by energies, show a bunching of residues into the allowed regions, especially the α-helical region in the vicinity of $\phi = \psi = -50°$. The energy-driven step RS5D→ER5D continues this trend, and the result is not disturbed by further X-ray or energy refinement. (From Levitt.[17])

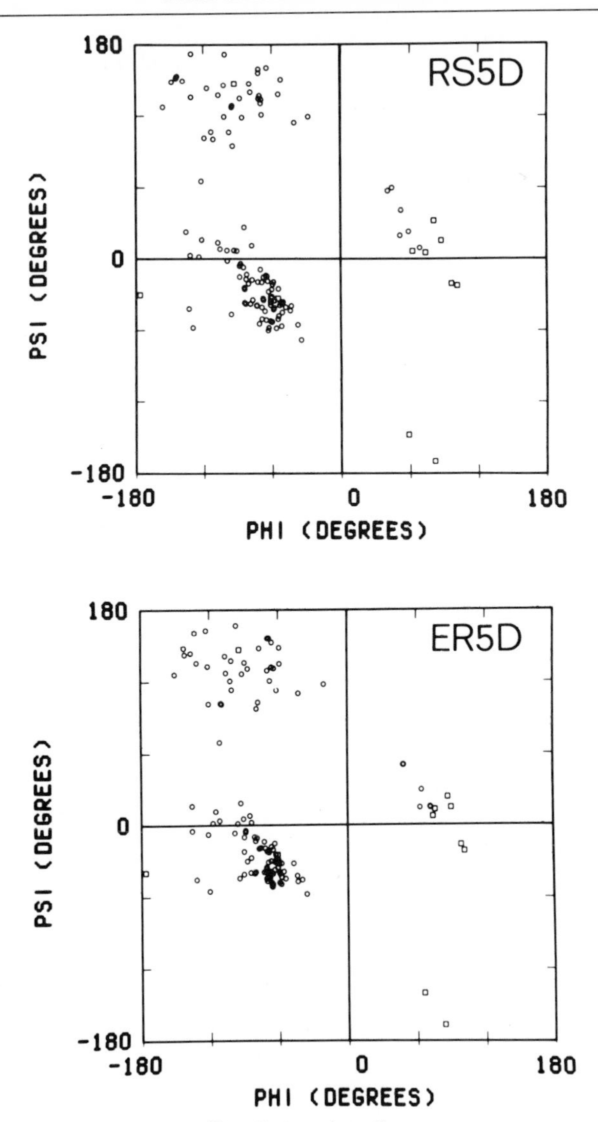

FIG. 3. (*continued*)

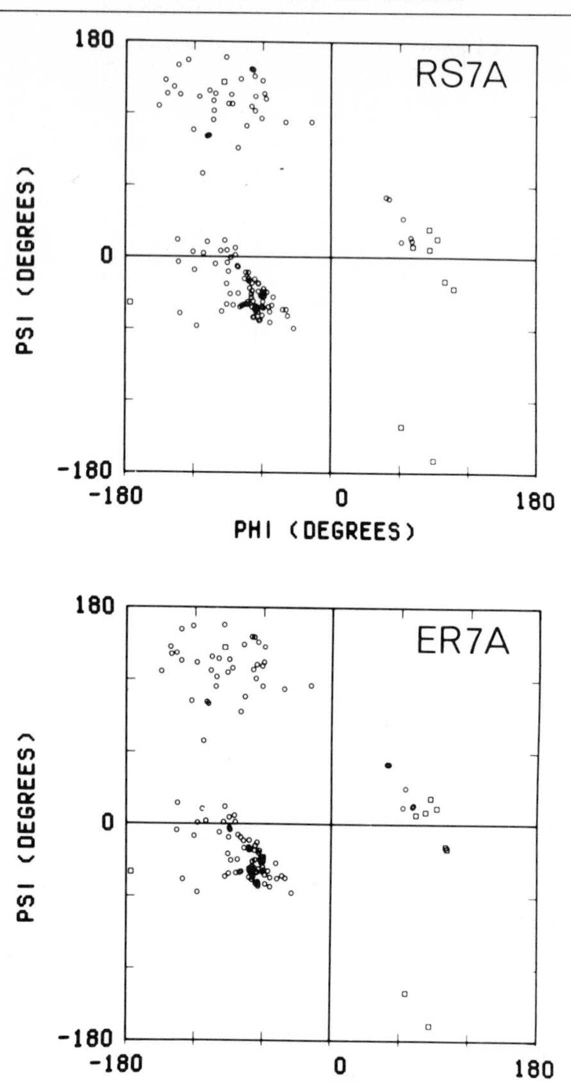

FIG. 3. (*continued*)

have comparative value. Note that the standard deviations on τ_α, for which the RSR method has been justly criticized, in fact improves on every step including the first two even though the first two processes are insensitive to its value. The major gain, of course, comes on the third step and is maintained by RSR on the fourth. Similarly the standard deviation of ω, which is brought to zero by the use of planar peptides in the first step, subsequently behaves entirely reasonably with little suggestion that RSR introduces more fluctuation into ω values than is energetically reasonable.

Similarly the close contacts show that there are initially some 22 of these which can be resolved by reference to the X-ray data, but that the majority require reference to energies. RSR unfortunately reintroduces some close contacts on the fourth step, which are in areas of the map where the density is weak. The H bond counts tell a similar story.

5. Finally, Fig. 3, also from Levitt[17] shows the ϕ–ψ distributions for these six coordinate sets. These, too, show that a bunching of the distributions into the allowed regions is achieved by the first two steps as well as by the third, with no evidence of opposition between the processes.

The above experiences may be summarized by saying that both RSR and energy refinement are beneficial by both criteria, and that they complement one another with little conflict. These conclusions are based on applications of RSR to a single MIR map. However, it is clear from the work of Deisenhofer and Steigemann[5a] and others since then that a more satisfactory result is obtained by abandoning the MIR phases in favor of phases calculated at each stage from current estimates of atomic coordinates and by recalculating the observed electron density map as the transform of $[(n+1)|F_o| - n|F_c|]e^{i\alpha_c}$ with $n = 0$ or 1 with each cycle of RSR. Used in this way some 15 protein and nucleic acid structure refinements have been done, as detailed in Table IV. These have an average R factor of 24.7 ± 3.7%. In addition, nine refinements using RSR without changing the map have been published, and these have an average R factor of 41.7 ± 6.1%, although in some of these cases further work was done on the coordinates before regarding them as final. Deisenhofer and Steigemann[5a] also report that the mean phase difference between the final calculated phases and the MIR phases, ~40°, is only about 10° greater than the expected error on the MIR phases as derived from figures of merit. This shows that the phase relaxation involved in the Deisenhofer and Steigemann procedure is amply justified. This is also confirmed by the fact that final difference maps calculated with coefficients $(|F_o| - |F_c|)e^{i\alpha_c}$ are essentially featureless, whereas no maps of the form $|F_o|e^{i\alpha_{MIR}} - |F_c|e^{i\alpha}$ are featureless regardless of what phases α are used on the second term.

TABLE IV

REPRESENTATIVE *R* FACTORS OBTAINED WITH REAL SPACE REFINEMENT

Crystal	*R* factor	Resolution (Å)	Phases used	Notes
Bovine pancreatic trypsin inhibitor[a] (BPTI)	0.197	1.5	Calculated from coordinates	15 cycles, 5 maps. Established significance of small fluctuations in ω
Bovine β-trypsin[b]	0.23	1.8		20 cycles, 5 maps
Bovine trypsin + BPTI[c]	0.23	1.9		11 maps, 20+ cycles, proline conformation and nonplanar peptides determined
Bovine trypsinogen[d]	0.225	1.8		13 cycles, 6 maps
Bovine trypsinogen + BPTI[e]	0.25	1.9		5 cycles, 3 maps (?)
Bence-Jones protein REI[f]	0.24	2.0		~30 cycles, 11 maps
Bence-Jones protein Au[g]	0.31	2.5		3 cycles from *f*
Aquomet erythrocruorin[h] (plus four related structures)	0.183	1.4		20 cycles, 6 difference maps
Metmyoglobin[i]	0.235	2.0		10 maps, 10 cycles, Fe positions by RSR and conventional least-squares differ by 0.04 Å.
Deoxymyoglobin[j]	0.233	2.0		8 maps, 8 cycles starting from cycle 6 of *i*
Deoxyhemoglobin[k]	0.276	2.5		*R* = 0.288 attained with MIR phases in four to eight passes
Methemoglobin[l]	0.231	2.0		9 cycles. Shows oscillatory behavior with $3F_o - 2F_c$ maps. Detected sequence errors
Yeast tRNA[phe m]	0.31	2.5–3.0		13 cycles, 8 maps
Yeast tRNA[phe n]	0.266	2.5		*R* then taken to 0.21 by Jack–Levitt method
Yeast tRNA[phe o]	0.29	2.5		*R* then taken to 0.21 by Hendrickson–Konnert method
Hen egg-white lysozyme[p]	0.35	2.0	MIR	RS5D, 2 cycles, 1 map
Porcine pancreatic elastase[q]	0.44	2.5	MIR	4 cycles, one map, later reached 0.32
Bovine pancreatic phospholipase A2[r]	0.41	1.7	MIR	Later reached 0.17
Dihydrofolate reductase from *Lactobacillus casei*[s]	0.45	2.5	MIR	1 cycle

Fab fragment from human IgG[t]	0.46	2.0	MIR	5 cycles
Metmyoglobin[u]	0.32	2.0	Calculated from X-ray structure	Neutron diffraction H and D included with suitable rotations. H and D automatically distinguished. 4 cycles, 1 map
CO-myoglobin[v]	0.37	2.0		Same as above but various numbers of cycles
Apo dogfish M_4 lactate dehydrogenase[w]	0.43	2.0	MIR	1 cycle only
LDHase: NAD-pyruvate[w]	0.52	3.0	MIR	1 cycle only

[a] J. Deisenhofer and W. Steigemann, *Acta Crystallogr., Sect. B* **B31,** 238 (1975).
[b] W. Bode and P. Schwager, *J. Mol. Biol.* **98,** 693 (1975).
[c] R. Huber, D. Kukla, W. Bode, P. Schwager, K. Bartels, J. Deisenhofer, and W. Steigemann, *J. Mol. Biol.* **89,** 73 (1974).
[d] H. Fehlhammer, W. Bode, and R. Huber, *J. Mol. Biol.* **111,** 415 (1977).
[e] W. Bode, P. Schwager, and R. Huber, *J. Mol. Biol.* **118,** 99 (1978).
[f] O. Epp, E. E. Lattman, M. Schiffer, R. Huber, and W. Palm, *Biochemistry* **14,** 4943 (1975).
[g] H. Fehlhammer, M. Schiffer, O. Epp, P. M. Colman, E. E. Lattman, P. Schwager, W. Steigemann, and H. J. Schramm, *Biophys. Struct. Mech.* **1,** 139 (1975).
[h] W. Steigemann and E. Weber, *J. Mol. Biol.* **127,** 309 (1979).
[i] T. Takano, *J. Mol. Biol.* **110,** 537 (1977).
[j] T. Takano, *J. Mol. Biol.* **110,** 569 (1977).
[k] G. Fermi, *J. Mol. Biol.* **97,** 237 (1975).
[l] R. C. Ladner, E. J. Heidner, and M. F. Perutz, *J. Mol. Biol.* **114,** 385 (1977).
[m] A. Jack, J. E. Ladner, and A. Klug, *J. Mol. Biol.* **108,** 619 (1976).
[n] B. Hingerty, R. S. Brown, and A. Jack, *J. Mol. Biol.* **124,** 523 (1978).
[o] G. J. Quigley and A. Rich, *Science* **194,** 794 (1976).
[p] R. Diamond, *J. Mol. Biol.* **82,** 371 (1974).
[q] L. Sawyer, D. M. Shotton, J. W. Campbell, P. L. Wendell, H. Muirhead, H. C. Watson, R. Diamond, and R. C. Ladner, *J. Mol. Biol.* **118,** 137 (1978).
[r] B. W. Dijkstra, K. H. Kalk, W. G. J. Hol, and J. Drenth, *J. Mol. Biol.* **147,** 97 (1981).
[s] D. A. Matthews, R. A. Alden, J. T. Bolin, D. J. Filman, S. T. Freer, R. Hamlin, W. G. J. Hol, R. L. Kisliuk, E. J. Pastore, L. T. Plante, N. Xuong, and J. Kraut, *J. Biol. Chem.* **253,** 6946 (1978).
[t] F. A. Saul, L. M. Amzel, and R. J. Poljak, *J. Biol. Chem.* **253,** 585 (1978).
[u] B. P. Schoenborn and R. Diamond, *Brookhaven Symp. Biol.* **27,** II-3 (1975).
[v] J. C. Norvell and B. P. Schoenborn, *Brookhaven Symp. Biol.* **27,** II-12 (1975).
[w] J. L. White, M. L. Hackert, M. Buehner, M. J. Adams, G. C. Ford, P. J. Lentz, I. E. Smiley, S. J. Steindl, and M. G. Rossmann, *J. Mol. Biol.* **102,** 759 (1976).

Although the improvement in R factor achieved by phase relaxation is very striking and the quality of final difference maps is usually excellent, it should be remembered that the number of observations per parameter is effectively halved when the phases used to calculate the map are allowed to drift to suit the current model.

[19] Stereochemically Restrained Refinement of Macromolecular Structures

By WAYNE A. HENDRICKSON

Introduction

The theory that relates the geometric parameters of an atomic model to its crystallographic diffraction pattern is exceptionally sound. Thus, in favorable cases it is possible to refine an atomic model to a match of high fidelity with the experimental diffraction data. While rigorous structure refinement has long been a routine procedure in small-molecule crystallography, several factors frustrate the straightforward extension to crystals of macromolecules. First, the sheer size of the computational problem is daunting. Second, most macromolecular crystals diffract relatively weakly. Hence, typically there is a paucity of observable data and this severely restricts the degree of overdetermination in the problem. Third, initial models of macromolecular structures are usually quite inaccurate. This poor start coupled with the nonlinear character of the equations generally results in convergence to a false minimum.

Despite the difficulties, there is great incentive for thorough refinement of crystal structures. Accurate and chemically reasonable atomic models are essential for a detailed understanding of biochemical activity. Fortunately, there is a wealth of prior knowledge about the stereochemistry of macromolecular structures. This knowledge can supplement the limited diffraction data and thereby make the refinement problem better conditioned. At the same time, models that conform with known stereochemical features have inherent chemical reasonableness.

Several different methods have been used to incorporate stereochemical knowledge into the refinement process. These have been reviewed elsewhere[1] and some are elaborated upon in other chapters of this vol-

[1] W. A. Hendrickson and J. H. Konnert, *in* "Computing in Crystallography" (R. Diamond, S. Rameseshan, and K. Venkatesan, eds.), p. 13.01. Indian Acad. Sci., Bangalore, India, 1980.

ume. Here attention is specifically directed at describing a refinement procedure that treats stereochemical knowledge as additional observations, as suggested by Waser[2] and first implemented by Konnert.[3] These subsidiary conditions on the refinement against diffraction data serve to restrict model features to a realistic range of possibilities. Consequently they are termed *restraints* in contrast to constraints, which confine features to specific values. The earlier uses of restraints in refinement focused on bonding distances as subsidiary conditions.[2,3] In the case of macromolecules, many other kinds of restraints are needed.[1,4]

My emphasis in this chapter is on the practical application of stereochemically restrained refinement to macromolecular crystals. Details of computational procedures and minimization algorithms are treated elsewhere[1,4] and need not be considered in routine applications. However, it is important to understand the nature of the function being minimized. Thus, I begin with some description of the several kinds of terms that are included. This draws very heavily from an account prepared for a school on crystallographic computing held in Bangalore.[1] I then turn to a description of the program design, weighting considerations, and refinement strategy. These sections are taken with little change from a report for a Daresbury workshop on refinement of protein structures.[5] The concluding section describes some extensions of the current export versions of the programs that have been implemented or are envisioned.

Observational Functions

Stereochemically restrained refinement, as we have implemented it, is based on the principle of least squares. Each piece of information is treated as an observational equation,

$$g^{\text{obs}} = g^{\text{calc}}(\mathbf{x}) + \varepsilon \tag{1}$$

in which a residual ε describes the discrepancy between an observation g^{obs} and the corresponding theoretical value g^{calc} computed from the parameters $\mathbf{x}$ of a model. According to the principle of least squares, the "best" set of parameters are those which minimize the weighted sum,

[2] J. Waser, *Acta Crystallogr.* **16,** 1091 (1963).

[3] J. H. Konnert, *Acta Crystallogr., Sect. A* **A32,** 614 (1976).

[4] W. A. Hendrickson and J. H. Konnert, *in* "Biomolecular Structure, Function, Conformation and Evolution" (R. Srinivasan, ed.), Vol. 1, p. 43. Pergamon, Oxford, 1981.

[5] W. A. Hendrickson, *in* "Refinement of Protein Structures" (P. A. Machin, J. W. Campbell, and M. Elder, eds.), p. 1. Daresbury Laboratory, Daresbury, Warrington, England, 1981.

over all observations, of the squared residuals:

$$\phi(\mathbf{x}) = \sum_h w_h[g_h^{\text{obs}} - g_h^{\text{calc}}(\mathbf{x})]^2 \tag{2}$$

The appropriate weights w_h are the inverses of the variances of the observations. In this case there are several qualitatively different classes of observations. It is convenient to treat the sum of squared residuals from each class of observations as a separate observational function in the form of Eq. (2). These in turn become terms in a grand function for minimization:

$$\Phi = \sum \phi_i \tag{3}$$

It is not essential that each observational function be of the least-squares form. Other functions that are also optimized at minima, such as potential energy functions, can also be included.

Crystallographic Structure Factors

The theoretical relationship between the parameters of an atomic model and its expected crystallographic X-ray diffraction pattern is given by the structure factor equation:

$$F_{hkl}^{\text{calc}} = K \sum_j f_j(s_{hkl}) \exp(-B_j s_{hkl}^2) \exp[2\pi i(hx_j + ky_j + lz_j)] \tag{4}$$

Here F_{hkl} is the structure factor of the X-ray reflection with indices hkl; K is a scale factor; f_j is the atomic scattering factor of atom j; B_j is an isotropic temperature parameter related by $B = 8\pi\overline{u^2}$ to the mean square displacement of the atom from its average position; this atomic position is given, in fractions of a cell edge, by the coordinates x_j, y_j, and z_j; and $s_{hkl} = (\sin\theta)/\lambda$, where θ is the Bragg scattering angle and λ is the radiation wavelength. The structure factor equation may also be generalized to include occupancy factors for certain atoms, particularly for solvent molecules, and to replace the isotropic thermal factor with an anisotropic representation or even an extension to higher cumulants.

The observational function related to structure factors is

$$\phi_1 = \sum_h^{\text{reflections}} \frac{1}{\sigma_F^2(h)} (|F_h^{\text{obs}}| - |F_h^{\text{calc}}|)^2 \tag{5}$$

where the weighting factor is determined by σ_F (h substitutes here for hkl). The standard deviation of F^{obs} is the value that should be given to σ_F at the completion of refinement. However, other weighting schemes may be advantageous during earlier stages. We have found it to be expeditious to relate σ_F to the current average residuals as discussed in the weighting section.

The preceding analysis relates to X-ray diffraction. Obviously a parallel development pertains in the case of neutron diffraction wherein the X-ray atomic scattering factor $f_j(s)$ is replaced in Eq. (4) by the neutron scattering length b_j, which is independent of scattering angle. If both X-ray and neutron data are available for the same crystalline species, then a joint refinement that includes an additional set of observational functions can be carried out simultaneously against the neutron and X-ray data.[6]

Stereochemistry of Atomic Positions

Bonding Distances. A most restrictive class of stereochemical knowledge concerns the interatomic distances related to chemical bonding. At least three types of distances can usefully be distinguished: actual bond distances, the next-nearest-neighbor distances from the triples of atoms that define bond angles, the first- to fourth-atom distances that relate to prescribed dihedral angles as within planar groups, and possibly hydrogen bond lengths. The calculated distance between two atoms, A and B, in the model is

$$d_{\mathrm{AB}}^{\mathrm{model}} = |\mathbf{r}_{\mathrm{A}} - \mathbf{r}_{\mathrm{B}}| \tag{6}$$

where $\mathbf{r}$ is an atomic position vector in a homogeneously dimensioned Cartesian frame (i.e., $\mathbf{r}$ is a function of both lattice parameters and the fractional atomic coordinates). The resulting observational function is

$$\phi_2 = \sum_{j}^{\mathrm{distances}} \frac{1}{\sigma_{\mathrm{D}}^2(j)} (d_j^{\mathrm{ideal}} - d_j^{\mathrm{model}})^2 \tag{7}$$

where σ_{D} is the standard deviation for distribution of values expected for the distances of a particular type.

Planar Groups. The coplanarity of the atoms within certain groups is another important facet of the stereochemistry of macromolecules. Various techniques have been used to impose planarity. We have used the very direct and effective method of restraining the deviations of atoms from the least-squares plane of the group. This gives the observational function

$$\phi_3 = \sum_{k}^{\mathrm{planes}} \sum_{i}^{\substack{\mathrm{coplanar} \\ \mathrm{atoms}}} \frac{1}{\sigma_{\mathrm{P}}^2(i,k)} (\mathbf{m}_k \cdot \mathbf{r}_{i,k} - d_k)^2 \tag{8}$$

[6] A. Wlodawer and W. A. Hendrickson, *Acta Crystallogr., Sect. A* **A38,** 239 (1982).

where $\mathbf{m}_k$ and d_k are parameters of the current least-squares plane[7] and σ_P is the standard deviation to be permitted.

Chiral Centers. The stereoconfiguration at chiral centers (e.g., asymmetric carbon atoms) is another known aspect of macromolecular structure. The set of interatomic distances is insensitive to handedness so additional restraints must be imposed to assure the preservation of chirality. We introduce a chiral volume equal to the triple scalar product of the vectors from a central atom to three attached atoms to quantify chirality. For example, the chiral volume of the α-carbon of an amino acid is

$$V_{C_\alpha} = (\mathbf{r}_N - \mathbf{r}_{C_\alpha}) \cdot [(\mathbf{r}_C - \mathbf{r}_{C_\alpha}) \times (\mathbf{r}_{C_\beta} - \mathbf{r}_{C_\alpha})] \tag{9}$$

The sign of the chiral volume depends on the handedness of the group and its magnitude equals the volume of the parallelepiped formed by the three vectors. The relevant observational function is

$$\phi_4 = \sum_{l}^{\substack{\text{chiral}\\ \text{centers}}} \frac{1}{\sigma_C^2(l)} (V_l^{\text{ideal}} - V_l^{\text{model}})^2 \tag{10}$$

Nonbonded Contacts. Contacts between nonbonded atoms are important determinants of macromolecular conformation. These contacts are characterized by potential energy functions $U(d)$ that feature a steep repulsive barrier against close contact and a shallow attractive well. For our purposes it is sufficient to restrict close contacts. The repulsive part of the functional form of a Lennard-Jones or Buckingham potential can be approximated by

$$U(d) - U(d^{\min}) \simeq \frac{1}{\sigma^{2n}} (d - d^{\min})^{2n}, \qquad d < d^{\min} \tag{11}$$

Best fittings in the range between $d^{\min} - 2$ Å and $d^{\min}$ have $n = 2$ for contacts among H, C, N, O, and S. Thus a useful observational function is

$$\phi_5 = \sum_{m}^{\substack{\text{nonbonded}\\ \text{contacts}}} \frac{1}{\sigma_N^4(m)} (d_m^{\min} - d_m^{\text{model}})^4 \tag{12}$$

when taken only over "repulsive" contacts, i.e., $d^{\text{model}} < d^{\min}$. The value of $d^{\min}$ depends on the atomic elements in contact and on the type of contact: single-torsion separated atom pairs, multiple-torsion separated atoms, or possibly hydrogen-bonded atom pairs. In principle, σ_N also varies with atom atom types, but in practice the fitted values are quite uniform.

[7] V. Schomaker, J. Waser, R. E. Marsh, and G. Bergman, *Acta Crystallogr.* **12,** 600 (1959).

Torsion Angles. It is useful to distinguish conformational torsion angles, in which certain flexibility resides, from the strictly fixed dihedral angles within planar groups. Conformational torsion angles are among the least restricted of stereochemical features. Indeed, these are the free parameters in many constrained refinements. Yet it is clear that certain restrictions, related to nonbonded contacts, do apply. The nature of restrictions on the main-chain conformation in proteins has long been known and those for side chains have also been analyzed. More sophisticated torsional potentials can be devised, but for our purposes a simple quadratic form suffices. Thus, the observational function to be used is

$$\phi_6 = \sum_{t}^{\substack{\text{conformational} \\ \text{torsion angles}}} \frac{1}{\sigma_{\mathrm{T}}^2(t)} (\chi_t^{\text{ideal}} - \chi_t^{\text{model}})^2 \tag{13}$$

where χ^{ideal} is the angle of the torsional potential minimum to which χ^{model} is currently nearest. Several different cases arise. These include quasi-planar torsions such as in the peptide bond, staggered potentials as in aliphatic side chains, transverse preferred conformations as in aromatic side chains, and possibly targeted main-chain conformation angles such as for α-helices. Here again, σ_{T} is the standard deviation of the expected distribution.

Stereochemistry of Atomic Motion

Stereochemistry has implications for thermal parameters as well as for positional parameters. The rms displacements from mean atomic positions in macromolecules are high, usually several tenths of an angstrom, whereas the variation with time in covalently bonded distances is generally not greater than a few hundredths of an angstrom. Variation in distances associated with bond angles is also quite limited. Thus atomic motions must be highly correlated. Qualitatively, if an atom undergoes large displacements from its mean positions then certain other atoms stereochemically related to it must undergo similarly large, concerted displacements. The thermal parameters that describe atomic displacements (whether due to thermal vibration, dynamic conformational disorder, or static variations within the lattice) should be made to agree with known stereochemistry. We have imposed this stereochemistry by restraining the variances of time-dependent interatomic distance distributions to suitably small values.[8,9]

[8] J. H. Konnert and W. A. Hendrickson, *Acta Crystallogr., Sect. A* **A36,** 344 (1980).

[9] W. A. Hendrickson and J. H. Konnert, *Biophys. J.* **32,** 645 (1980).

Isotropic Temperature Factors. If the correlation of atomic motion is described in a manner analogous to "riding motion," the variances for isotropic thermal motion take on a particularly simple form. The variance is in the magnitude of the differences in mean square displacements $\overline{u^2}$ of the two atoms joined by a bonding distance. Since these displacements are directly related to the thermal parameters by $B = 8\pi\overline{u^2}$, the pertinent observational function is

$$\phi_7 = \sum_j^{\text{distances}} \frac{1}{\sigma_B^2(j)} (B_j^{\text{origin}} - B_j^{\text{target}})^2 \tag{14}$$

where B^{origin} and B^{target} denote the thermal parameters of the atom pair related by a certain bonding distance.

Anisotropic Temperature Factors. Restraints for anisotropic thermal parameters are somewhat more complicated and in general must be separately imposed along selected, orthogonal directions rather than in aggregate as in the isotropic case. The appropriate observational function is then

$$\phi_8 = \sum_j^{\text{distances}} \frac{1}{\sigma_U^2(j)} \left[\Delta_v^2 \cos^2\theta + \frac{\Delta_v^4(\sin^4\theta - 3\cos^2\theta\sin^2\theta)}{d_j^2}\right]^2 \tag{15}$$

where

$$\Delta_v^2 = \pm(\overline{u_{v,j}^{2}}^{\text{origin}} - \overline{u_{v,j}^{2}}^{\text{target}}) \tag{16}$$

which is positive. The angle between a given bond direction and a specific direction of motion v is $\theta(v, j)$ and the mean square displacement along the direction specified by v is $\overline{u_v^2}$. Standard deviations here must relate to mean square displacements rather than to B values as in Eq. (14). It is possible to specify principal axis directions for thermal ellipsoids and thereby reduce the formal parameters to three per atom, but it appears that all six parameters will generally be needed. It is also possible to impose restraints related to group motion.

Noncrystallographic Symmetry

It frequently happens that macromolecules crystallize with multiple copies of basically similar, if not identical, molecules in the crystallographic asymmetric unit. If the internal structure within these subunits (or within parts of them) is the same within certain limits of detectability, then very substantial reduction in numbers of parameters can be had. However, it is also important to permit real deviations from noncrystallographic symmetry to be expressed if they exist. Restraints permit a very useful and flexible compromise between these two objectives.

Positions. Symmetrically related objects can be brought into approximate coincidence with orthogonal transformations such as

$$\mathbf{r}'_{i,k} = \mathbf{R}_k \mathbf{r}_{i,k} + \mathbf{t}_k \tag{17}$$

where $\mathbf{R}$ is a rotation matrix and $\mathbf{t}$ is a translation vector. Transformations must be found for each object k such as to bring them all into common coincidence. It is then possible to define an average structure $\overline{\mathbf{r}'_{i,k}}$ and to minimize deviations from the average structure. The resulting observational function is simply

$$\phi_9 = \sum_i^{\text{unique atoms}} \frac{1}{\sigma^2_{\text{SP}}(i)} \sum_k^{\text{copies}} (\mathbf{r}'_{i,k} - \overline{\mathbf{r}'_{i,k}})^2 \tag{18}$$

The standard deviations σ_{SP} can be varied so as, for example, to restrain weakly those atoms that might be expected to deviate from symmetry (such as those at subunit interfaces) while maintaining strict symmetry in the heart of the structure.

Two different conditions arise with respect to the required transformation parameters of $\mathbf{R}$ and $\mathbf{t}$. In some cases these are known exactly from the alignment of noncrystallographic symmetry axes with crystallographic axes. In other cases they must be refined in successive cycles by finding the transformations that optimize the superposition. Direct methods have been described for doing this.[10]

Thermal Factors. If the atomic positions of multiple copies of a structure are essentially superimposable, it is to be expected that the thermal parameters for these similar local environments will also be similar. An obvious observational function with which to impose this restraint in the isotropic case is

$$\phi_{10} = \sum_i^{\text{unique atoms}} \frac{1}{\sigma^2_{\text{SB}}(i)} \sum_k^{\text{copies}} (B_{i,k} - \overline{B_{i,k}})^2 \tag{19}$$

Other Factors

Resistance to Excessive Shifts. It sometimes happens that parts of a structure are poorly determined by the crystallographic data even in combination with stereochemical restraints. This can lead to instability in the minimization process. An effective technique for circumventing this ill-conditioning is to impose restraints against excessive shifts. This leads to

[10] W. A. Hendrickson, *Acta Crystallogr., Sect. A* **A35,** 158 (1979).

the observational function

$$\phi_{11} = \sum_{i}^{\text{atoms}} \frac{1}{\sigma_{\text{EP}}^2} (\mathbf{r}_i - \mathbf{r}_i^{\text{o}})^2 + \frac{1}{\sigma_{\text{EB}}^2} (B_i - B_i^{\text{o}})^2 \tag{20}$$

Here σ_{EP} and σ_{EB} are the permissible breadths for the distributions of shifts from the current values, $\mathbf{r}_i^{\text{o}}$ and B_i^{o}, for positional and thermal parameters, respectively. The mathematical effect is to augment the diagonal elements of the normal matrix by $1/\sigma^2$. Restraints against excessive shifts also make possible refinement without structure factors, i.e., model idealization.

Occupancy Factors. It sometimes happens that a group of atoms which have variable occupancy factors can all be expected to have the same occupancy. Examples include solvent molecules such as sulfate, ethanol, or water (if hydrogen positions are included) and noncovalently bound substrates, inhibitors, or cofactors. It may prove to be computationally expedient to impose the desired constraint through tight restraints rather than by elimination of variables. The desired observational function is then

$$\phi_{12} = \sum_{k}^{\text{groups}} \frac{1}{\sigma_Q^2(k)} \sum_{i}^{\substack{\text{atoms in} \\ \text{group}}} (Q_{i,k} - \overline{Q_{i,k}})_2 \tag{21}$$

where $Q_{i,k}$ is the occupancy factor of atom i in group k, $\overline{Q_{i,k}}$ is the average of occupancy factors in the group, and σ_Q is the expected standard deviation for the distribution.

Program Design

The initial application of restrained refinement was not to a protein but to the mineral tridymite (a silica structure that has 240 atoms per asymmetric unit and is twinned four ways). Konnert[3] adapted ORFLS,[11] a standard crystallographic refinement program, to include twinning, distance restraints, and conjugate gradients for use in the tridymite refinement. When we decided to test this new refinement method on proteins, we found adaptation of the refinement program itself to be quite straightforward but soon realized the need for a general program to identify restraint distances and specify ideal values. It was also evident that restraints on features other than distances would eventually be needed in protein refinements. Hence we designed a set of protein refinement pro-

[11] W. R. Busing, K. O. Martin, and H. A. Levy, "ORFLS," ORNL-TM-305. Oak Ridge Nat. Lab., Oak Ridge, Tennessee, 1962.

grams that we could quickly implement for our specific test problem, parvalbumin, but that could readily be extended to include other restraints and generalized to apply to other molecules.

The parvalbumin test clearly showed the need for planarity restraints. This capability was then added to the rudimentary program set. Experience gained in further applications (both at the Naval Research Laboratory and elsewhere) has fed back other enhancements. These have been incorporated in turn until the exported programs now have the overall structure shown in Fig. 1. The evolutionary process continues; a new set of programs will soon supplant the present export versions.

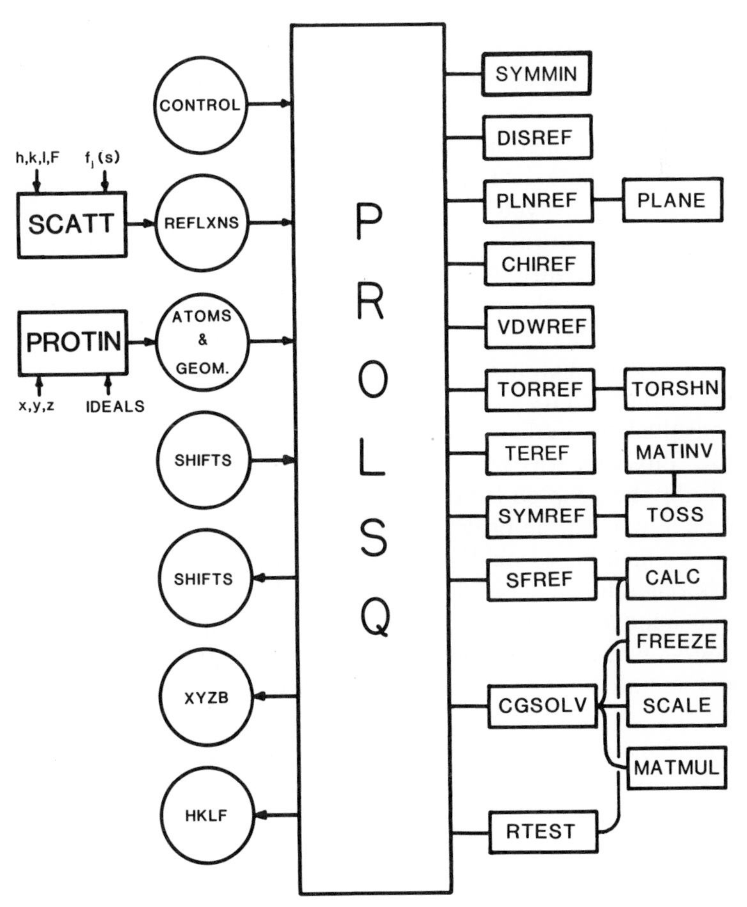

FIG. 1. Schematic structure of the PROLSQ (PROtein Least SQuares) refinement programs.

PROLSQ (PRotein Least SQuares) is the actual refinement program. It reads diffraction data and scattering factors prepared by SCATT (SCATTering data), initial atomic coordinates and restraint specifications prepared by PROTIN (PROTein model INput), parameter shifts from previous refinement cycles, and control card images. It then augments the normal-equation elements pertinent to each of the stereochemical restraints and the structure factor observations. Fractional atomic coordinates are used in order to speed the rate-limiting calculations concerning structure factors. For the same reason, a highly optimized space-group-specific routine, CALC, is used for computing structure factors and their derivatives. Elements of the resulting sparse normal matrix are stored in a singly dimensioned array that is indexed by pointers. Next, PROLSQ uses a conjugate gradients procedure to solve the new parameter shifts. Finally, it tests the expected impact of the new shifts on the R value. An optimal shift damping factor is searched for in trials against a sample of the data.

PROTIN is run once before a series of refinement cycles to prepare the atomic coordinate data needed by PROLSQ and to identify the atoms and ideal values involved in the individual stereochemical restraints. It incidentally also performs a useful verification function for initial models. Ideal values for the various stereochemical features are taken from those in particular small-molecule crystal structures of constituent parts of the macromolecule. This is patterned after Diamond's early model-building program.[12] Group dictionaries specifying ideal values and the atoms involved are compiled for each category of restraint. These dictionaries are then consulted to produce the specifications for a particular polymeric structure. PROTIN was written with proteins specifically in mind, but it could readily be adapted for other polymers.

We have written the programs in FORTRAN for the TI-ASC (a vectorizing machine based on 32-bit words) at the Naval Research Laboratory. Although the code is optimized for the ASC, care has been taken to avoid nonstandard features that might seriously impede transportability. The programs have in fact been implemented on quite a variety of computers. Execution times of course depend on the problem and on the machine. Isotropic refinement of crambin at 1.5 Å resolution (414 atoms and 5638 reflections) consumes 35 sec for each PROLSQ cycle on the ASC, whereas a cycle of refinement on β_4 hemoglobin at 2.5 Å resolution (4664 atoms and 16918 reflections) takes 751 sec. Nearly 94% of the latter time is spent in SFREF and CALC (see Fig. 1). The same refinement of β_4 hemoglobin (which includes individual temperature factors and noncry-

[12] R. Diamond, *Acta Crystallogr.* **21**, 253 (1966).

stallographic symmetry restraints) uses 4.8 hr of CPU time per cycle on a DEC VAX 11/780 that has a floating-point accelerator. This is with a version modified by Pat Briley of Iowa to minimize page faults on the VAX; the standard export version would have taken much longer (A. Arnone, personal communication). Briley's modification splits PROLSQ into two parts. Gerson Cohen of the National Institutes of Health (personal communication) has achieved similar page-faulting economy on the VAX without dividing the program.

Weighting

In practical terms, the optimal weighting strategy for a nonlinear least-squares problem is that which gives the fastest convergence to the deepest accessible local minimum if not to the absolute minimum. There is no established theory to govern this weighting, yet experience shows that progress in protein refinements is quite sensitive to weighting. Thus it may be useful here to recount a weighting scheme that we find to be reasonably effective.

In keeping with the theoretical basis to be approached as a refinement nears completion we cast the weights in terms of variances ($w = 1/\sigma^2$). Each class of observation is assigned a "standard deviation" σ, related to the desired breadth of the distribution for features of this class. Generally there is little difficulty in driving the local geometry of a model to an excellent fit with the ideal. What proves to be extremely troublesome is bringing about a match between calculated and observed diffraction data. The essential problem in weighting reduces to balancing the weights for diffraction terms against those for stereochemical terms.

It is convenient to maintain an internal balance among restraint weights by placing these on a quasi-absolute scale. But in the early stages of a protein refinement the discrepancy between $|F_{obs}|$ and $|F_{calc}|$ is generally much greater than the counting-statistics error in $|F_{obs}|$. Sigmas (σ values) from counting statistics will then vastly overweight the structure factor terms and thereby cause ill-conditioning and deterioration of model ideality. However, good behavior generally follows if structure factor σ values are set at a uniform value related to the average discrepancy between $|F_{obs}|$ and $|F_{calc}|$. As a rule of thumb, $\sigma_F \simeq \frac{1}{2}\langle||F_{obs}| - |F_{calc}||\rangle$ is a good starting point. The value of σ_F is then used to control the progress of refinement whereas the restraint σ values are varied primarily to fine tune the model. Second moments of the distributions in each of the restraint classes are computed during each PROLSQ cycle and compared with the weighting σ values to monitor progress. A set of typical weighting parameters for the current version of PROLSQ are given in the table.

TYPICAL WEIGHTING PARAMETERS FOR VARIOUS OBSERVATIONAL CLASSES

Bonding distances		
Bond length (1–2 neighbor)		σ_D = 0.02 Å
Angle-related distance (1–3 neighbor)		0.03
Intraplanar distance (1–4 neighbor)		0.05
Hydrogen bond or metal coordination		0.05
Planar groups		
Deviation from plane		σ_P = 0.02 Å
Chiral centers		
Chiral volume		σ_C = 0.15 Å
Nonbonded contacts		
Single torsion[a]	σ_N = 0.50 Å	d_{VDW} = −0.30 Å
Multiple torsion	0.50	0.00
Possible hydrogen bond (X····Y)	0.50	−0.20
Possible hydrogen bond (X–H··Y)	0.50	−0.90
Torsion angles		
Specified (e.g., helix ϕ and ψ)		σ_T = 15°
Planar (e.g., peptide ω)		3
Staggered (e.g., aliphatic χ)		15
Transverse (e.g., aromatic χ_2)		20
Thermal factors	Anisotropic	Isotropic
Main-chain bond (1–2 neighbor)	σ_v = 0.05 Å	σ_B = 1.0 Å^2
Main-chain angle (1–3 neighbor)	0.10	1.5
Side-chain bond	0.05	1.5
Side-chain angle	0.10	2.0
Noncrystallographic symmetry	Positional	Thermal
Tight class	σ_{SP} = 0.05 Å	σ_{SB} = 0.5 Å^2
Medium class	0.50	2.0
Weak class	5.00	10.0
Restraints against excessive shifts	Idealize	Refine
Positional parameter	σ_{EP} = 0.1 Å	σ_{EP} = 0.3Å
Thermal parameter		σ_{EB} = 3.0 Å^2
Occupancy parameter		σ_{EQ} = 0.2
Diffraction data		
Structure factor modulus	$\sigma_F \simeq \frac{1}{2}\langle\|\|F_{obs}\| - \|F_{calc}\|\|\rangle$	

[a] σ values are as defined in Eqs. (5) −(21); d_{VDW}, increment to van der Waals radius to yield d^{min} for Eq. (12).

Refinement Strategy

There can be no single prescription for the refinement of protein structures—each problem is idiosyncratic. The range of starting situations includes questionable models based on poorly phased maps, models developed by molecular repositioning from a related by appreciably different structure, and excellent models based on several good derivatives. Obviously, different courses will be followed in the refinement and revision of such disparate models. Nonetheless there are many common considerations in devising a strategy for conducting a refinement.

Resolution Limits. One general area of concern relates to resolution limits. A typical starting model might be based on a 2.5 Å resolution map phased with several derivatives to 3 or 4 Å resolution but only one that goes to 2.5 Å. There might also be a full 2 Å set of native data. What data should be included at the start of refinement and when should others be added? Clearly the radius of convergence is greater at lower resolution; discrepancies between model and true positions must be within a quarter-wavelength of the Fourier wave correspondent to a given reflection for shift indications even to have the right sign. On the other hand, if high-angle terms used in the map are excluded from the refinement one then runs the risk of squandering properly interpreted details. As a rule it is wise to start refinement with data corresponding to the resolution of the strong features in the map to which the model was fitted. At the other end of the scale, very low-order terms (spacings greater than 10 Å) are greatly influenced by the solvent continuum and those in the 5–10 Å shell are usually heavily affected by partially ordered solvent molecules that are omitted in early models. It is often wise to exclude these data until solvent is properly treated.

One typical refinement problem might start with a few cycles of refinement against the 10–3 Å shell of data. One might then exclude the 10–5 Å shell as the R value in this shell becomes relatively worse than in the 5–4 Å shell. When progress slows at 3 Å resolution, higher angle data might be included in a few shell-wise expansions (e.g., 2.5, 2.2, and then 2.0 Å) with three additional cycles of refinement per expansion.

Manual Revision. Except in circumstances so rare that they have yet to be encountered, an initial atomic model of a protein cannot be brought directly by automated refinement to a satisfactory match with the diffraction data. The refinement invariably becomes stuck in one of the many false minima that abound in such marginally overdetermined problems. Manual intervention is then essential to further progress. Refinement usually proceeds in stages of automated cycles followed by gross revision motivated by inspection of Fourier syntheses based on model phases.

Especially in the early stages and at moderate resolution (2–3 Å), the $(2|F_{obs}| - |F_{calc}|) \exp(\alpha_{calc})$ synthesis (or one of its generalizations) is particularly convenient for identifying major imperfections and incompleteness in a model. The $(|F_{obs}| - |F_{calc}|) \exp(\alpha_{calc})$ synthesis often provides definitive clues for a revision, especially at higher resolution. Another useful synthesis has the coefficients $(|F_{obs}| - |F'_{calc}|) \exp(\alpha'_{calc})$ where F'_{calc} is the structure factor of a partial model from which a fragment has been deleted. We have found it advantageous to examine a systematic series of such fragment ΔF maps showing segments about 10 residues in length.

The initial refinement usually greatly improves the quality of maps based on model phases, often to the point of making them superior to the initial map based on experimental phases. However, there may sometimes, particularly for incomplete models, be considerable advantage in combining the model phase information with that from the multiple isomorphous replacement or other phasing experiment. One then uses a synthesis based on coefficients of $m|F_{obs}| \exp(\alpha_{combine})$, where the figure of merit m and centroid phase $\alpha_{combine}$ are computed from the combined phase probability distribution for the two kinds of information.

It is important to be able to superimpose the model on an image of an electron density distribution. This can be done on paper sheets or transparency stacks. However, computer graphics systems for molecular modeling not only solve the display problem exceptionally well but also greatly facilitate the rebuilding process. Graphics becomes an indispensable aid in the refinement process.

Thermal Parameters. Strategic decisions for the course of a refinement must also be made regarding the treatment of thermal parameters. There usually is substantial and meaningful variation in vibrational amplitudes within protein structures. This must be expressed to effect a good match with the diffraction data. Yet on the other hand, one must worry that at moderate resolution freely varying thermal parameters can take on meaningless values as they simply absorb errors in a problem that approaches underdetermination. The high correlations between thermal parameters and scale or occupancy factors must also be taken into account.

The approach to thermal parameters that we have commonly taken in refinements at the Naval Research Laboratory is as follows. Intensity statistics from all available data are used to place the data on an approximately absolute scale and to estimate an overall thermal parameter (B value). In the early, low-resolution cycles of refinement only the scale factor is varied and the overall B is held fixed. (If B is allowed to vary at 3 Å resolution it typically takes on unreasonably low values.) After refinement has been extended to somewhat higher resolution and has reduced R to the vicinity of 0.30, then individual isotropic thermal parameters are varied but with relatively tight restraints. The intent here is to permit a smoothly varying expression of differences in thermal factors. Later, if there are sufficient data, we temporarily release restraints on B values. This permits the effective elimination of atoms that are grossly misplaced, it often identifies wrongly interpreted regions, and at higher resolution it can be used to discriminate between atom types (e.g., N vs O in amide side chains). Finally, highly restrained anisotropic thermal parameters might be used.

Of course, nearly endless variations are possible in the approach to thermal parameter refinement. As with weighting, facilitation of the process and avoidance of false minima are guiding principles during intermediate stages whereas conformity with stereochemical rules is the criterion for a final model.

Solvent Structure. Although much of the large solvent fraction in protein crystals is essentially fluid, usually sufficient of it is well enough ordered to make inclusion of the solvent structure necessary in an advanced refinement model. Properties of the solvent structure are often of interest in their own right. In addition, frequently, correct modeling of the protein structure can only be completed after water and other solvent molecules have been included.

The most tightly bound of the solvent molecules can be readily identified. However, much of the solvent structure cannot be located until after initial refinement has reduced the R value to 0.25 or less. It is often helpful to include the solvent affected 10–5 Å shell of data in refinement cycles leading to difference maps for solvent interpretation. One should be mindful of hydrogen-bonding preferences when ascribing density features to solvent molecules. A sorted list of interpolated peaks in a difference density map and a listing of distances to nearby atoms in the current model are useful aids for solvent interpretation.

Many solvent sites are only partially occupied. Thus occupancy parameters are essential variables in the solvent model. However, occupancy factors are highly correlated with thermal parameters. In lieu of a full-matrix treatment of these covariations, for those atoms with variable occupancy factors we simply alternate the application of occupancy and thermal parameter shifts in successive cycles. In order to minimize the inclusion of meaningless variables we generally eliminate solvent sites that refine to very low occupancy factors (e.g., $Q < 0.3$) or very high thermal values (e.g., $B > 50$ Å^2).

Conformational Restraints. It is desirable in the early stages of some refinement problems to reduce the conformational freedom in a model. If the model is sufficiently restricted, meaningful refinement is then possible against low-resolution data and this can yield a large radius of convergence. Rigid-body refinement is a particularly effective procedure for models derived from molecular repositioning of a related structure. Refinement of a model composed of linked rigid groups[13] affords another direct and fruitful approach. It is also possible to effect a reduction of conformational freedom by imposing certain tight restraints.

One option is to restrain the torsion angles of the model to remain very close to those in the initial model (e.g., $\sigma_T = 2°$). This is particularly sensitive in the case of a model derived from molecular replacement of a

[13] J. L. Sussman, S. R. Holbrook, G. M. Church, and S.-H. Kim, *Acta Crystallogr., Sect. A* **A33,** 800–804 (1977).

related and well-known structure. Another option is to restrain the backbone torsions in elements of known secondary structure to be those expected in ideal helices, β-sheets, etc. The inclusion of special distance restraints related to the hydrogen bonding within structural elements also rigidifies these units. Of course, as the refinement proceeds to higher resolution it is usually wise to relax special conformational restraints.

A special conformational restraint that we have found generally useful concerns peptide planarity. In the initial stages we usually restrain five atoms of a peptide unit (C_α^i, C^i, O^i, N^{i+1}, C_α^{i+1}) to be coplanar. Later, after high-angle data have been included and R is below 0.20, we drop to four-atom peptide planes (C_α^i, C^i, O^i, and N^{i+1}) and rely on torsion restraints to maintain reasonable ω angles.

Convergence. Refinements at low to moderate resolution, particularly in the early stages, are prone to ill-conditioning. That is, the marginal state of overdetermination leads to near-singularity in the normal matrix and consequent instability in the solution for parameter shifts. The conjugate gradients (CG) method is relatively insensitive to ill-conditioning, but such as exists manifests itself by slow and irregular convergence, or even divergence, of the (CG) iterations. PROLSQ monitors the (CG) shifts for a sample of parameters. If these are not monotonically convergent, a change in the relative weighting between diffraction and stereochemistry observation might be needed or restraints against excessive shifts (preconditioning) should be invoked.

Restraints against excessive shifts are simply tethers to current parameters. In the absence of significant pressure for movement, parameters are then held near their starting values. A suitable choice of the weighting σ for these restraints permits model idealization by PROLSQ even without diffraction data. These restraints also control the behavior of groups that are poorly specified by the data (e.g., some lysine side chains).

Generally, continued refinement after the first few cycles (5–10) on a given model leads to diminishing returns and manual revisions are soon in order. Sometimes, though, the progress can be rejuvenated by a sudden relaxation of stereochemical or thermal restraints or by dropping back to lower resolution. This can relax barriers between local minima and increase the radius of convergence. However, such a process also runs the risk that properly fitted features might escape and be trapped into wrong positions when tight restraints and high resolution are reimposed.

Extensions

Thorough structural refinement has become an integral part of macromolecular crystallography during the past few years. Availability of effective programs and access to suitable computers have been instrumental in this increased emphasis on detail. However, refinement strategies and computer programs have certainly not reached maturity and generality. Many useful extensions can be envisioned and several problem areas remain. In this section I turn to some extensions that pertain to the restrained refinement procedure embodied in our programs.

One failing in generality of the programs is the specialization for proteins. PROTIN in particular was written with proteins specifically in mind. However, the extension to other bipolymers is relatively straightforward. Indeed, Gary Quigley has written a corresponding NUCLIN for nucleic acid refinement (personal communication), and Eric Westhof has developed a refinement system for nucleic acids that introduces appropriate restraints for sugar pucker in nucleic acid structures (personal communication). Amendments are also needed to handle glycosylation of proteins. Marcel Knossow has made modifications for such polysaccharide groups (personal communication). Very often the restraints for cofactors, metal centers, or other special groups must be generated. Clearly, extensions to generalize for these cases are important. In particular, a general program is needed for convenient refinement of protein : nucleic acid complexes.

Atomic motion and conformational heterogeneity (or disorder) are major impediments to successful refinement. Vibrational amplitudes and structural variability tend to be large in protein crystals. This causes diffraction intensities to diminish rapidly with scattering angle and greatly limits the extent of measurable data. The distribution functions required to model the highly anisotropic and harmonic characteristics of such large displacements must have many variables. Yet it is just in the case of large displacements that data are few. Hence these interesting and necessary parameters tend to be indeterminate. The challenge is great to produce adequate but economical models for atomic displacements in proteins. Our initial attempts at anisotropic refinement with fixed ellipsoid orientations suggest that this three-parameter thermal representation is inadequate. Thus the extension to full anisotropic refinement, perhaps with librational restraints, is a high priority. Extensions to handle disorder are in place but they should be made more convenient. Eventually, the handling of dynamic properties should be aided by recourse to results from theoretical simulations.

Computational economy is always a paramount concern. It is clear that the use of Fourier transformations to compute structure factors and gradient vectors might greatly improve speed for large problems.[14,15] Such a procedure has been incorporated into the present programs by Mitchell

[14] R. C. Agarwal, *Acta Crystallogr., Sect. A* **A34,** 791 (1978).
[15] A. Jack and M. Levitt, *Acta Crystallogr., Sect. A* **A34,** 931 (1978).

Lewis (personal communication). On vectorizing machines, such as the ASC or Cray, speeds for smaller proteins are already so good as to provide little incentive for FFT acceleration. The use of array processors on other machines can also greatly increase the speed of structure-factor-related computations.[16] Experience in our own laboratory corroborates the utility of this approach. Refinements of lamprey hemoglobin at 2 Å resolution (1199 atoms and 8447 reflections) that require 261 min of CPU time on a VAX 11/780 can be run in 49 min of total elapsed time when the CALC portion is executed in standard calls to a Floating Point Systems AP 120B array processor.[17] Residual calculations on an otherwise empty VAX take 27 min of CPU time and the remaining 22 min are divided between AP execution time and data transfers between units.

There are numerous other improvements that can be envisioned. These include a method for modeling the fluid solvent, an appropriate treatment of the correlation between occupancy and thermal parameters of discrete solvent molecules, restraints for nonbonded contacts from crystal packing, inclusion of attractive potentials for nonbonded contacts, provision for refining partial structures, and proper estimation of standard deviations. Extensions such as these are expected to be important in realizing the goal of producing refined structural models that reproduce the diffraction patterns to within the accuracy of the measured data and which are compatible with prior stereochemical knowledge of macromolecules.

Acknowledgments

As noted in the introduction, this chapter draws heavily from two articles previously published in relatively inaccessible meeting reports.[1,5] I thank the publishers of those articles for the permission to adapt and reproduce portions of those articles here.

The refinement procedures described here have evolved through stimulating interactions with users of the programs and my colleagues at the Naval Research Laboratory. I especially thank John Konnert for his seminal contributions to the method. I am also particularly thankful to Janet Smith, Steven Sheriff, and Richard Honzatko for their roles in developing our standard refinement strategies. Finally, although I cannot single each one out, I wish to thank collectively the several users whose feedback has led to improved capabilities in the programs.

[16] W. Furey, Jr., B. C. Wang, and M. Sax, *J. Appl. Crystallogr.* **15,** 160 (1982).

[17] R. B. Honzatko, unpublished results (1983).

[20] Constrained–Restrained Least-Squares (CORELS) Refinement of Proteins and Nucleic Acids

By JOEL L. SUSSMAN

Introduction

In order to study in detail the three-dimensional structure and flexibility of biological macromolecules, it is crucial to refine the atomic model. During the last few years there has been great progress of refinement methods, which has resulted in much more accurate structure determinations.

Approximate models are usually derived either by the heavy-atom isomorphous replacement method[1] or by the molecular replacement method.[2] From such models it is possible to calculate structure factors and compare them to the observed ones by the conventional agreement factor R summed over all measured reflections h:

$$R = \sum_h \left| |F_{o,h}| - |F_{c,h}| \right| \Big/ \sum_h |F_{o,h}| \tag{1}$$

The R factor for good initial models is in the range of 0.4–0.5 where a random structure results in an R factor of ~0.6. For low molecular weight structures the R factor often begins in this same range but it is possible by conventional least-squares techniques[3] to refine the structure to a final R factor of ~0.05 or less.

The Difficulty in Refining Biological Macromolecules

The difficulty in refining proteins or nucleic acids lies both in their enormous size and in the limited amount of available X-ray data. With a few exceptions, X-ray diffraction data from protein and nucleic acid crystals do not warrant the conventional structure factor least-squares approach, wherein individual atomic positions are refined independently.

[1] K. D. Watenpaugh, this volume [1].

[2] M. G. Rossmann, ed. "The Molecular Replacement Method," Int. Sci. Rev. Ser. No. 13. Gordon & Breach, New York, 1972.

[3] G. H. Stout and L. H. Jensen, "X-ray Structure Determination—A Practical Guide." Macmillan, New York, 1968.

Meaningless shifts of atomic positions result from such refinement, primarily as a result of the low ratio of data to parameters.

The resolution at which protein data sets are obtained governs the ratio of data to parameters. It is necessary that the ratio be greater than one in order that the structure be solvable. For example, if 100 is the number of reflections per atom that is observed at the limit of Cu K_α radiation for a representative structure, then for a set of diffraction data extending to 1.54 Å, the problem is well determined: there are 12 observations per atom or 4 observations per translational degree of freedom. If the diffraction can be measured to only 2.5 Å resolution, however,[4] then there is less than one observation per degree of freedom, and the problem is underdetermined without additional assumptions or data.

If, as in the former case, one has data of sufficient resolution, then it is possible to refine a protein structure by a least-squares method on the individual atomic parameters. This was demonstrated by the pioneering work in structure refinement of Jensen and colleagues for the protein rubredoxin (55 amino acids) that diffracts to better than 1.2 Å resolution.[5] Unfortunately this method is often not applicable, since many reported protein structures diffract to at best 2.5 Å resolution. Therefore it is often necessary to seek methods that will increase the ratio of data to parameters, and thereby make the problem well determined.

Restraints and Constraints in Refinement

One can improve the ratio of observations to parameters by increasing the number of observations and/or reducing the number of parameters. The former can be accomplished for macromolecules by adding information from low molecular weight crystal structures in the form of restraints on bond lengths, bond angles, and torsion angles. The latter can be realized by imposing constraints on the same quantities, reducing the number of independent parameters. Whereas restraints introduce more data by supplying ideal values for relevant parameters, constraints specify strict relationships between the parameters, thereby reducing the number of independent variables. Restraints can be thought of as the addition of ideal configurations with springs pulling the real atoms toward them; constraints can be thought of as the imposition of rigid rods between atoms, reducing their degrees of freedom.

[4] L. H. Jensen, *in* "Crystallographic Computing Techniques" (F. R. Ahmed, K. Huml, and B. Sedláček, eds.), p. 308. Munksgaard, Copenhagen, 1976.

[5] K. D. Watenpaugh, L. C. Sieker, J. R. Herriot, and L. H. Jensen, *Acta Crystallogr., Sect. B* **B29**, 943 (1973).

In 1976 Konnert and Hendrickson[6] introduced, in a way first described by Waser,[7] distance *restraints* to maintain proper stereochemistry throughout refinement and to ensure convergence with a limited set of data. In this procedure the Cartesian coordinates of each atom are varied independently, while the stereochemistry is *restrained* to standard values by springlike forces between atoms. This method of using restraints to improve the ratio of observations to parameters is being used in several different model building and refinement programs,[8–10] and is illustrated in Fig. 1a. Alternatively, the procedure of Diamond[11] is to *constrain* the geometry of all atoms to fixed bond lengths and bond angles and allow only backbone and side-chain torsion angles as degrees of freedom. This approach is illustrated in Fig. 1b.

A somewhat more flexible approach to increase the ratio of observations to parameters is to *constrain* only certain groups of atoms to a particular local stereochemistry and then refine these sets of atoms as rigid groups[12,13] (see Fig. 1c and Fig. 2).[14] This kind of approach was attempted soon after the myoglobin structure was determined[15] and in refinement of models using fiber diffraction data.[16] Such approaches are justifiable because proteins and nucleic acids can be considered as composed of many "rigid" groups, such as certain amino acid side chains (phenyl, tyrosyl, tryptophanyl, prolyl, valyl, etc.), bases, phosphates, and riboses.

The CORELS (COnstrained–REstrained Least-Squares) program[12] was developed for proteins and nucleic acids in order to take advantage of the intrinsic rigid groups found in these molecules and to overcome the relatively low resolution of the X-ray data from their crystals. CORELS combines Scheringer's rigid groups constraints,[17] extended to allow for

[6] J. H. Konnert, *Acta Crystallogr., Sect. A* **A32,** 614 (1976); W. A. Hendrickson and J. H. Konnert, *in* "Computing in Crystallography" (R. Diamond, S. Ramaseshan, and K. Venkatesan, eds.), p. 13.01. Indian Acad. Sci., Bangalore, India, 1980; W. A. Hendrickson, this volume [19].

[7] J. Waser, *Acta Crystallogr.,* **16,** 1091 (1963).

[8] J. Hermans and J. McQueen, *Acta Crystallogr., Sect. A* **A30,** 730 (1974).

[9] A. Jack and M. Levitt, *Acta Crystallogr., Sect. A* **A34,** 931 (1978).

[10] E. J. Dodson, N. W. Isaacs, and J. S. Rollett, *Acta Crystallogr., Sect. A* **A32,** 311 (1976).

[11] R. Diamond, this volume [18].

[12] J. L. Sussman, S. R. Holbrook, G. M. Church, and S. H. Kim, *Acta Crystallogr., Sect. A* **A33,** 800 (1977).

[13] L. G. Hoard and C. E. Nordman, *Acta Crystallogr., Sect. A* **A35,** 1010 (1979).

[14] M. Marquart, J. Deisenhofer, R. Huber, and W. Palm, *J. Mol. Biol.* **141,** 363 (1980).

[15] C.-I. Branden, K. C. Holmes, and J. C. Kendrew, *Acta Crystallogr.* **16,** A175 (1963).

[16] S. Arnott, S. D. Dover, and A. J. Wonacott, *Acta Crystallogr., Sect. B* **B25,** 2192 (1969).

[17] C. Scheringer, *Acta Crystallogr.* **16,** 546 (1963).

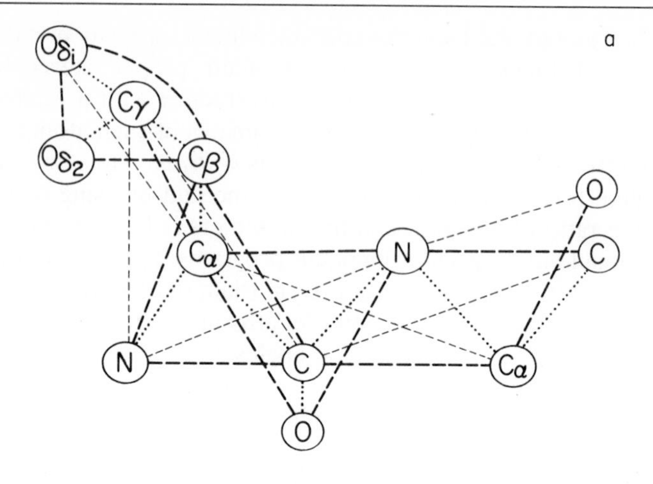

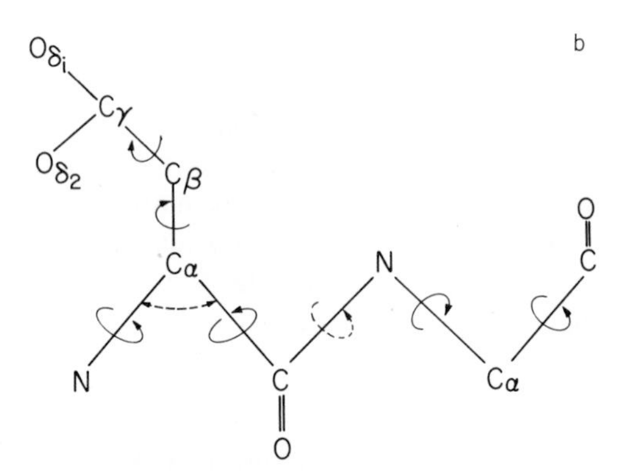

FIG. 1. Schematic illustration of the ways in which structural parameters can be varied while maintaining stereochemistry in three different kinds of model building/refinement programs. (a) A *restrained refinement* procedure.[6,8–10] The Cartesian coordinates of each atom are the variables. The stereochemistry is maintained by specific restraints that correspond to springlike connections, of specific bond lengths (dotted lines), bond angles (heavy dashed lines), torsion angles (light dashed lines), or nonbonded contacts (not shown). (b) A *constrained refinement* procedure.[11] Here the variables consist of only the backbone torsion angles and selected bond angles, while all bond lengths are strictly constrained. (The torsion or bond angles indicated with dashes are optional degrees of freedom.) (c) A *constrained–restrained refinement* procedure.[12,13] Here two constrained groups are illustrated. Each is free to move with 6 degrees of freedom (translation and rotation) as well as any number of internal torsion or bond angles. The bond lengths within any one group are strictly constrained. The stereochemistry between groups is restrained as in (a). The planar peptide bond is maintained by attaching a dummy atom at about 10 Å above the peptide plan[10] and restraining its distance to the N and C_α of the next amino acid.

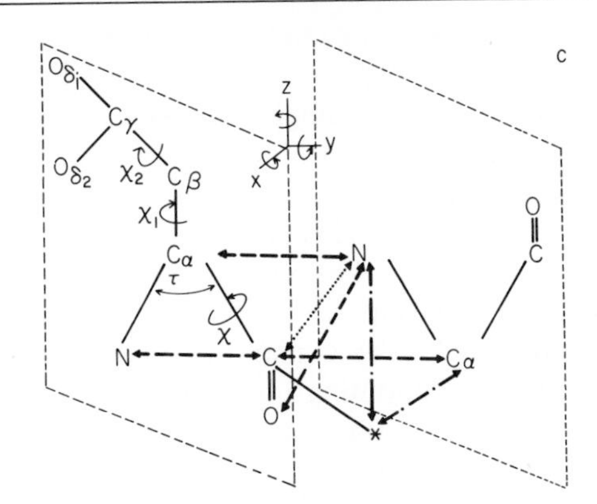

FIG. 1. (*continued*)

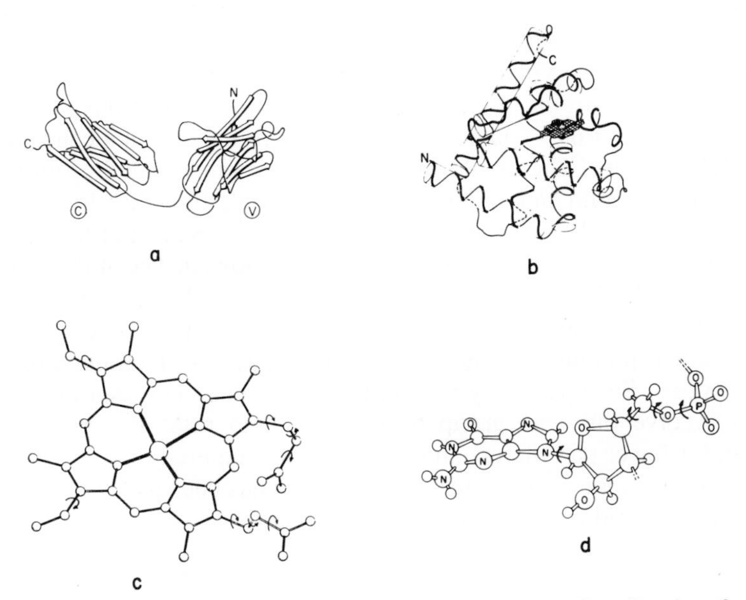

FIG. 2. Examples of different kinds of constrained groups. (a) Two domains of an immunoglobulin structure (see, e.g., Ref. 14). (b) Six major α-helices and heme group of the myoglobin structure. (c) An enlarged view of the heme group showing the various torsion angles that permit additional degrees of freedom. (d) A nucleotide unit showing the variable torsion angles.

variable torsion angles,[18] with distance restraints[7] to maintain stereochemistry between groups within a specified error limit. Even though allowing variable internal dihedral angles introduces torsional degrees of freedom within the otherwise constrained group, it can reduce the total number of structural parameters in the structure by decreasing the number of groups. The advantages of this approach are (1) a large increase in the data-to-parameter ratio over the restrained refinement methods; (2) automatic maintenance of group stereochemistry: within the group, all bond lengths and bond angles are constrained; (3) an increased range of convergence; (4) reduced computing time; and (5) applicability to low-, intermediate-, and high-resolution data.

In order to maintain the appropriate stereochemistry, distance restraints are applied in a way analogous to the method developed by Konnert,[6] i.e., maintaining proper bond lengths and angles at connections both within groups and between rigid groups. Moreover, distance restraints are used to minimize unacceptable nonbonded contacts. A list of minimum nonbonded contact radii is prepared for each atom type, and whenever a group is shifted by refinement such that any of its member atoms comes into close contact with any other nonbonded atoms,[19] appropriately weighted restraint terms shift the group in such a way that the unacceptable contact is alleviated. This approach is analogous to applying a repulsive van der Waals potential, and is a restraint in contrast to the hard-sphere constraint often employed in energy calculations.

Distance restraints can also be used for stereochemical idealization of a model structure. The usual procedure in other model-building programs has been to restrain *all* bond distances, bond angles, and any fixed torsion angles to specified "ideal" values.[8,10,20] In CORELS an approach consistent with the structure factor refinement was followed. The atomic coordinates of standard constrained groups are fitted onto the coordinates of the model by a least-squares minimization; simultaneously, intergroup distances corresponding to bond lengths and angles are restrained to standard values. Since the only restraints are the links between groups, all shifts involve only the group translational and rotational parameters; hence, far fewer parameters and restraints are required than when allowing all atoms to move separately. This method has been used for idealizing both nucleic acid and protein models.

[18] C. Scheringer, *Acta Crystallogr.* **16,** A175 (1963).

[19] G. N. Ramachandran and V. Sasisekharan, *Adv. Protein Chem.* **23,** 283 (1969).

[20] M. Levitt and S. Lifson, *J. Mol. Biol.* **46,** 269 (1969).

Mathematical Description of CORELS

A detailed mathematical derivation of the equations and derivatives used in CORELS has been given in Ref. 12. Only a brief summary is presented here.

The quantity Q to be minimized in the least-squares procedure consists of the sum of four separate observational functions:

$$Q = w_F \mathrm{DF} + w_D \mathrm{DD} + w_V \mathrm{DV} + w_T \mathrm{DT} \tag{2}$$

where w_F, w_D, w_V, and w_T are overall weights for each term.

The first term, DF, is the usual sum of structure factor differences summed over all or part of the reflections h:

$$\mathrm{DF} = \sum_h \frac{1}{\sigma_F} (|F_{o,h}| - k|F_{c,h}|)^2 \tag{3}$$

Here σ_F represents the standard deviation of F_o at the completion of refinement. However, other weighting schemes may be advantageous during earlier stages of refinement; specifically, it has been useful to relate σ_F to the current average residuals ($\sigma_F \simeq \frac{1}{2}\langle|F_o - F_c|\rangle$).[6] The variable k represents one or more scale factors.

The second term restrains the stereochemistry and is the sum over all distance restraints d:

$$\mathrm{DD} = \sum_d \frac{1}{\sigma_d} (D_{o,d} - D_{c,d})^2 \tag{4}$$

where $D_{o,d}$ is the "ideal" distance between a specified pair of atoms, and may correspond to a bond length, or arise from bond angle or torsion angle restraints, and where $D_{c,d}$ is the distance calculated from the model. Here σ_d is the standard deviation in distances of type d observed in small molecule structures.

The third term restrains the nonbonded close contacts and is the sum over all nonbonded distance restraints v:

$$\mathrm{DV} = \sum_v \frac{1}{\sigma_v} (D_{o,v} - D_{c,v})^2 \tag{5}$$

where $D_{o,v}$ is the "close contact" distance between a specified pair of atoms and $D_{c,v}$ is the distance calculated from the model. As this term refers to only the repulsive van der Waals-like potential, the sum is only over distances where $D_{c,v} < D_{o,v}$. Here σ_v is the standard deviation of values for the nonbonded distances of a particular type.

The fourth term restrains the structure from moving away from the input coordinates and is used primarily for model building. Here the sum is over all atoms i and over the three positional coordinates j of each atom:

$$DT = \sum_{i} w_i \sum_{j=1}^{3} (X_{T,i,j} - X_{i,j})^2 \tag{6}$$

where the $X_{T,i,j}$ are the Cartesian coordinates in angstroms of the target (input) atoms, while the $X_{i,j}$ are the corresponding coordinates of the working model.

For constrained–restrained structure factor least-squares refinement, usually we set $w_T = 0$ in Eq. (2), while for distance–target idealization (model building), $w_F = 0$. The relative magnitudes of w_F, w_D, and w_T have been discussed earlier.[6,7]

Each constrained group can be described fully by specifying its orientation with respect to an arbitrarily chosen origin, by means of a translation vector and a rotation vector, in addition to its internal dihedral angles, if any, and temperature factor(s). (Not all parameters within each group are independent.) Its parameterization is relatively simple because its only internal degrees of freedom are thermal parameters and torsional angles, if present. The quantity Q which is to be minimized from Eq. (2) depends upon the parameters of each of the groups in the structure as well as the scale factor relating F_o and F_c:

$$Q = Q(\mathbf{X}_i, \mathbf{R}_i, \psi_{i1}, \psi_{i2}, \ldots, B_{i1}, B_{i2}, \ldots, k) \tag{7}$$

where $\mathbf{X}_i$ and $\mathbf{R}_i$ are the translation and rotation vectors, respectively, of group i; ψ_{in} are the n internal dihedral angles, and B_{im} are the m temperature factors of group i; and k is the scale factor of the structure.

The minimization of Q with respect to the orientational parameters, $\mathbf{X}_i$ and $\mathbf{R}_i$, of the groups is approached by finding the appropriate derivatives of Q from Eq. (2).[12,21] Furthermore, if each group contains subgroups of atoms which are constrained to the same temperature factor, then the derivatives of each subgroup can be calculated and depend only upon F_c. From these sets of orientational and thermal derivatives, the least-squares normal equations follow.[12] The CORELS program solves the matrix of normal equations by the conjugate gradient iterative method.[6,22] Only a sparse matrix (of about 1–5% of the full matrix) is constructed by omitting terms whenever two parameters occur in groups with no distance re-

[21] R. Doedens, *in* "Crystallographic Computing" (F. R. Ahmed, S. R. Hall, and C. P. Huber, eds.), p. 198. Munksgaard, Copenhagen, 1970.

[22] M. R. Hestenes and E. Steifel, *J. Res. Natl. Bur. Stand. (U.S.)* **49,** 409 (1952).

straints between them. This results in an enormous saving of computer time and storage.

In order to reduce computer time farther, the program was originally written with space-group-specific subroutines for the calculation of structure factors and derivatives. Takusagawa[23] has written a space-group-independent subroutine for CORELS to do these calculations, which is especially useful for the refinement of high-symmetry space groups. Recently, using this feature, we have begun the refinement of the cubic form of yeast $tRNA^{Phe}$ in space group $I4_132$.[24] In addition the program permits the refinement of a partial structure, consisting even of groups of atoms not contiguous in the sequence.

Implementation of CORELS

In CORELS the structure is treated as a series of discrete units. Each unit is a constrained (CORELS) group wherein the bond lengths and bond angles are not allowed to vary, the dihedral angles and temperature factors may vary, and the stereochemistry is usually based on a crystal structure of a small molecule similar to the unit. In protein work, for simplicity, we have usually chosen the different amino acids as the building blocks of the structure (see Fig. 1c), while for nucleic acids we often use larger groups as shown in Fig. 3.

A preliminary input program called PRECORels is generally run before using CORELS to prepare most of the input files (see Tables I–IV). The user supplies PRECOR with an approximate set of input GUIDE coordinates as well as with the instructions on how to divide the molecule into any number of constrained groups such as amino acid residues. In addition PRECOR facilitates the explicit definition of inter- and intragroup restraint distances between specific pairs of atoms (see Fig. 1). At present these may be of the following types: PEPTIDE, CISPEPTide, RNA, DNA, SS bond, etc. Finally PRECOR prepares a list of potential close contacts which are used for the repulsive potential in CORELS (see above).

Two dictionaries are supplied with PRECOR: (1) RESGRP.DIC, a dictionary of (a) the default types of input residues and the atom names defining them and (b) a description of currently available CORELS groups with their degrees of freedom; (2) IDEALXYZ.DIC, a dictionary of ideal

[23] F. Takusagawa, private communication.

[24] J. K. Nachmann, J. L. Sussman, R. W. Warrant, and S. H. Kim, *Acta Crystallogr.* **A37,** *Abstr., IUCr Congr., 12th, 1981,* Abstract No. 02.4-02 (1981).

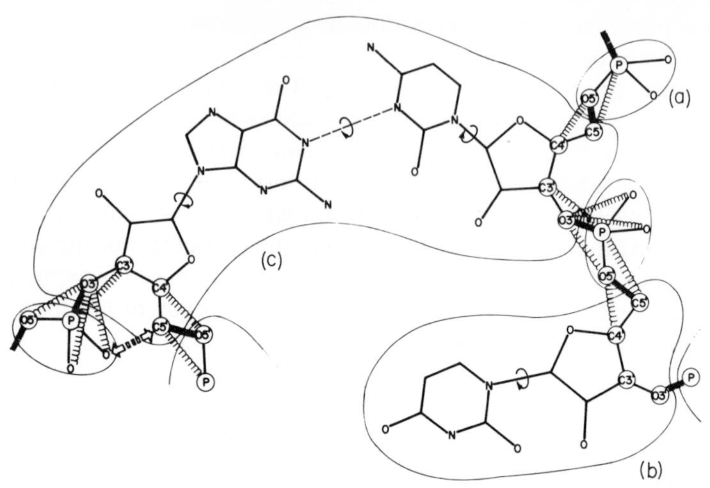

FIG. 3. The three different kinds of constrained groups used in the higher resolution refinement of tRNA are shown. (a) The smallest is a phosphate group with 6 positional degrees of freedom and a single temperature factor (a total of 9 degrees of freedom). (c) The largest group is a constrained base pair (for the double-helical stem regions). In addition to the torsional flexibility of each ribose relative to its respective base, one of the nucleosides is permitted to twist as a unit about a vector shown as a dashed line between the N1 of the purine to the N3 of the pyrimidine (for a total of 13 degrees of freedom). Restrained distances corresponding to bond lengths are shown by dotted lines, distances corresponding to bond angles by dashed lines, and nonbonded contacts with a double-headed arrow.

atomic coordinates (in angstroms and each centered around the origin) of the constrained groups defined in RESGRP.DIC.

One of the additional files that PRECOR produces defines the dimensions for variables in CORELS. This file is automatically included by using the FORTRAN-77 statement: "INCLUDE 'CORELS.DIM' " when CORELS is compiled. PRECOR must be rerun and CORELS recompiled with a fresh CORELS.DIM file if the structure is divided into different groups or the restraints are changed. In addition to the files generated by PRECOR, CORELS requires a dictionary of ideal distances and their standard deviations (Table V) if restraints are used to link atoms between or within groups, and a list of F_{obs} is required for structure factor least-squares. As an example of the way in which PRECOR parameterizes a polypeptide chain, see Table III and Table IV and Fig. 1c.

It must be stressed that the size of CORELS groups is not limited to just amino acids or nucleotides but can be easily adjusted to fit the amount of data available. This allows for meaningful refinement even at extremely low resolution. In order to increase the radius of convergence during the course of the refinement of a structure, it is best to start with large rigid

TABLE I
INPUT/OUTPUT FILES PRECOR AND CORELS

Input		Output	
Unit no.	Name/description	Unit no.	Name/description
	PRECOR		
15	Guide coordinates[a]—user supplies	10	XYZ file
		12	PARAMETERS file
4	RESGRP.DIC—dictionary of possible residues and constrained groups—supplied	11	TARGETS file sorted guide coordinates
9	IDEALXYZ.DIC—dictionary of ideal coordinates for the constrained groups—supplied	(8	RESTRAINTS file)
		2	CORELS.DIM dimension file to be "included" when compiling CORELS
(5	Sequence of constrained groups)[b]		
(5	List of restraints)		
	CORELS		
44,42,43	XYZ, PARAMETERS, TARGETS	34	XYZ file
(8)	(and RESTRAINTS) files all prepared by PRECOR	32	PARAMETERS file
(9	DISSIG.DIC—Dictionary of distances and sigmas—supplied)	33	TARGETS file
(11	Input structure factor file)[c]	(1,2	Calculated structure factor files)

[a] AATYPE, RESNUMB, ATNAME, XF, YF, ZF, B, WT (3X,A3,2X,I3,A4,5F10.5).
[b] Anything enclosed in parentheses is optional.
[c] h, k, l, F, σ (3I4,2F8.3).

bodies and relatively low-resolution data, initially varying the fewest number of degrees of freedom.

Application of CORELS to Specific Problems

In this section several different examples of the use of CORELS for both model building and reciprocal space refinement are described.

TABLE II
TARGETS FILE

Group name and number	Atom name	X	Y	Z	Temperature factor	Weight
TRP	1N	0.18934	−0.19327	0.03861	10.0	1.0
TRP	1CA	0.19883	−0.17843	0.03233	10.0	1.0
TRP	1C	0.22334	−0.18430	0.02786	10.0	1.0
TRP	1O	0.24317	−0.17903	0.03163	10.0	1.0
TRP	1CB	0.20231	−0.16179	0.03890	10.0	1.0
TRP	1CG	0.17918	−0.14985	0.03953	13.4	1.0
TRP	1CD1	0.16765	−0.14385	0.04795	13.4	1.0
TRP	1NE1	0.14717	−0.13315	0.04518	13.4	1.0
TRP	1CE2	0.14640	−0.13221	0.03549	13.4	1.0
TRP	1CZ2	0.12963	−0.12274	0.02958	13.4	1.0
TRP	1CH2	0.13348	−0.12395	0.01908	13.4	1.0
TRP	1CZ3	0.15292	−0.13390	0.01489	13.4	1.0
TRP	1CE3	0.16995	−0.14335	0.02088	13.4	1.0
TRP	1CD2	0.16623	−0.14227	0.03134	13.4	1.0
LYS	2N	0.22209	−0.19519	0.01983	10.0	1.0
LYS	2CA	0.24440	−0.20195	0.01473	10.0	1.0
LYS	2C	0.25068	−0.18908	0.00614	10.0	1.0
LYS	2O	0.23454	−0.18406	−0.00010	10.0	1.0
LYS	2CB	0.23978	−0.22062	0.01018	10.0	1.0
LYS	2CG	0.21253	−0.22438	0.00900	12.8	1.0
LYS	2CD	0.20023	−0.23453	0.01676	12.8	1.0
LYS	2CE	0.21786	−0.24781	0.02185	12.8	1.0
LYS	2NZ	0.23226	−0.25729	0.01400	12.8	1.0

Refinement of Yeast tRNA^{Phe}

CORELS was developed initially for refinement of the crystal structure of yeast phenylalanine tRNA.[25] It was felt that a refinement with constraints as well as restraints was necessary, considering the limited resolution of the data and the resulting poor data-to-parameter ratio.

We generated an idealized starting set of atomic coordinates based on a model of tRNA^{Phe} derived by the multiple isomorphous replacement (MIR) and partial Fourier methods.[26] This was done by linking together stereochemically ideal CORELS groups with specific restraint distances, while simultaneously minimizing the distances between the coordinates of the ideal groups and the MIR target coordinates.

[25] J. L. Sussman, S. R. Holbrook, R. W. Warrant, G. M. Church, and S. H. Kim, *J. Mol. Biol.* **122,** 607 (1978).

[26] J. L. Sussman and S. H. Kim, *Biochem. Biophys. Res. Commun.* **68,** 89 (1976).

TABLE III
PARAMETERS FILE

[a]		[b]			[c]							
TRP	1	13	2	15	0.198831	−0.178427	0.032329					
[d]								[e]				
10.00000								1	6			Temperature factor 1
13.40000								7	15			Temperature factor 2
		[f]										
−153.85760		9.01039	143.79213									Rigid body angles
[g]								[h]				
−78.82773								4	5	2	3	Relative dih ang ψ
0.00000								3	15	1	2	Relative dih ang ϕ
0.00000								1	1	2	6	Relative dih ang τ
30.07171								7	15	2	6	Relative dih ang X_1
−21.71884								8	15	6	7	Relative dih ang X_2
LYS	2	15	2	10	0.244395	−0.201949	0.014730					
10.0000								1	6			Temperature factor 1
12.80000								7	10			Temperature factor 2
32.15706		−13.46437	49.92391									Rigid body angles
−131.50339								4	5	2	3	Relative dih ang ψ
0.00000								3	10	1	2	Relative dih ang ϕ
0.00000								1	1	2	6	Relative dih ang τ
148.53725								7	10	2	6	Relative dih ang X_1
−78.29115								8	10	6	7	Relative dih ang X_2
−118.92149								9	10	7	8	Relative dih ang X_3
−146.58324								10	10	8	9	Relative dih ang X_4

[a] CORELS group name and number.
[b] Number of parameters, temperature factors, and atoms for the group.
[c] X, Y, Z fractional coordinates of the origin of the group (see Table IV).
[d] Subgroup temperature factors.
[e] First through last atom constrained to each subgroup temperature factor.
[f] Rigid-body rotation angles: Φ (about z axis), Θ (about new x axis), ρ (about new y axis).
[g] Torsion angle: the relative change in torsional angle of the current coordinates of the XYZ file (see Table IV).
[h] Torsion angle definition: first two columns correspond to the first through last atom affected by the rotation, while the last two columns represent the two atoms that define the rotation vector.

TABLE IV
XYZ FILE

Group name and number	Atom name	Atomic coordinates[a] (Å)			Atomic number flag[b]
TRP	1N	−1.17244	0.88199	0.00000	2
TRP	1CA	0.00000	0.00000	0.00000	1
TRP	1C	1.23700	0.87200	0.00000	1
TRP	1O	2.37300	0.36400	0.00000	3
TRP	1ODUM	1.69250	1.64050	10.00000	0
TRP	1CB	−0.02113	−0.87777	−1.23596	1
TRP	1CG	−1.27908	−1.75197	−1.34430	1
TRP	1CD1	−2.33212	−1.58312	−2.15885	1
TRP	1NE1	−3.27170	−2.61362	−1.94963	2
TRP	1CE2	−2.80771	−3.40270	−1.06941	1
TRP	1CZ2	−3.39050	−4.56193	−0.58936	1
TRP	1CH2	−2.64063	−5.24194	0.36623	1
TRP	1CZ3	−1.39168	−4.78688	0.79675	1
TRP	1CE3	−0.80004	−3.62085	0.30233	1
TRP	1CD2	−1.54091	−2.93437	−0.64701	1
LYS	2N	−1.17244	0.88199	0.00000	2
LYS	2CA	0.00000	0.00000	0.00000	1
LYS	2C	1.23700	0.87200	0.00000	1
LYS	2O	2.37300	0.36400	0.00000	3
LYS	2ODUM	1.69250	1.64050	10.00000	0
LYS	2CB	0.01725	−0.87257	−1.25432	1
LYS	2CG	1.30320	−1.68186	−1.41133	1
LYS	2CD	1.34948	−2.41669	−2.74554	1
LYS	2CE	2.22252	−3.66536	−2.68066	1
LYS	2NZ	2.06415	−4.44155	−3.91673	2

[a] Coordinates of the atoms of the ideal group relative to the group origin at (0,0,0).
[b] Atomic number flag: 0, dummy atom; 1, carbon; 2, nitrogen; 3, oxygen; 4, magnesium; 5, phosphorus; 6, sulfur; 7, iron; 8, bromine.

TABLE V
DICTIONARY OF DISTANCES AND SIGMAS

Distance	Sigma (σ)	Description of Restraint		
1.3257	0.08	01	C–N	Bond Len
2.4107	0.20	02	CA–N	Ang (CA–C–N)
2.4432	0.20	03	C–CA	Ang (C–N–CA)
2.2612	0.20	04	O–N	Ang (O–C–N)
2.7913	0.08	05	O–CA	Tors Ang OMEGA
2.4095	0.20	06	N–C	Ang (N–CA–C, TAU)
1.5186	0.80	07	N–CD	Bond Len (PRO)
2.2998	0.20	08	N–CG	Ang (N–CD–CG: PRO)
2.4872	0.20	09	CA–CD	Ang (CA–N–CD: PRO)
2.5212	0.20	10	C–CD	Ang (C(i − 1)–N–CD: PRO)
2.040	0.08	11	S–S	Bond Len Disulfide

TABLE V (*continued*)

Distance	Sigma (σ)			Description of Restraint
3.062	0.20	12	C–S	Ang (C–S–S, Disulfide)
3.800	0.00	13	CA–CA	Tors Ang OMEGA
10.037	0.08	14	ODUM–N	Planar Peptide Linkage
10.122	0.08	15	ODUM–CA	Planar Peptide Linkage
3.00	1.00	16	O(N)–O(N)	Bump
3.40	1.00	17	C–N	Bump
3.35	1.00	18	C–O	Bump
3.70	1.00	19	C–C	Bump
2.0	0.20	20		Restraint Len of 2.0
2.1	0.20	21		Restraint Len of 2.1
4.9	0.20	49		Restraint Len of 4.9
5.0	0.20	50		Restraint Len of 5.0
1.4389	0.08	51	O5′–C5′	RNA: Bond Len
2.6079	0.20	52	P–C5′	RNA: Ang (P–O5′–C5′)
2.4297	0.20	53	O5′–C4′	RNA: Ang (O5′–C5′–C4′)
1.5934	0.08	54	O3′–P	RNA: Bond Len
2.6036	0.20	55	C3′–P	RNA: Ang (C3′–O3"–P)
2.4794	0.20	56	O3′–O5′	RNA: Ang (O3′–P–O5′)
2.515	0.20	57	O3′–O1P	RNA: Ang (O3′–P–O1P)
2.515	0.20	58	O3′–O2P	RNA: Ang (O3′–P–O2P)
1.458	0.08	59	O1′–C4′	RNA: Bond Len
2.420	0.20	60	O1′–C5′	RNA: Ang (O1′–C4′–C5′)
2.379	0.20	61	O1′–C3′	RNA: Ang (O1′–C4′–C3′)
2.348	0.20	62	C1′–C4′	RNA: Ang (C1′–O1′–C4′)
3.693	0.0	63		RNA: Sug Pucker Flag
2.90	0.20	64		Tertiary Interaction
2.785	0.20	65	N6–O4	H-Bond Len
3.858	0.20	66	C2–O2	H-Bond Len
2.672	0.20	67	O6–N4	H-Bond Len
3.035	0.20	68	N2–O2	H-Bond Len
6.9	0.20	69		Restraint Len of 6.9
7.0	0.10	70		Restraint Len of 7.0
1.4499	0.08	71	O5′–C5′	DNA: Bond Len
2.6209	0.20	72	P–C5′	DNA: Ang (P–O5′–C5′)
2.4227	0.20	73	O5′–C4′	DNA: Ang (O5′–C5′–C4′)
1.594	0.08	74	O3′–P	DNA: Bond Len
2.6096	0.20	75	C3′–P	DNA: Ang (C3′–O3′–P)
2.4694	0.20	76	O3′–O5′	DNA: Ang (O3′–P–O5′)
2.513	0.20	77	O3′–O1P	DNA: Ang (O3′–P–O1P)
2.510	0.20	78	O3′–O2P	DNA: Ang (O3′–P–O2P)
1.463	0.08	79	O1′–C4′	DNA: Bond Len
2.416	0.20	80	O1′–C5′	DNA: Ang (O1′–C4′–C5′)
2.361	0.20	81	O1′–C3′	DNA: Ang (O1′–C4′–C3′)
2.354	0.20	82	C1′–C4′	DNA: Ang (C1′–O1′–C4′)
3.253	0.0	83		DNA: Sug Pucker Flag
2.7341	0.20	84	CA–CA	Tors Ang OMEGA CIS-PROLINE
3.5902	0.20	85	O–CA	Tors Ang OMEGA CIS-PROLINE

In the early stages of refinement, only 6153 reflections between 3 and 10 Å were used in order to increase the radius of convergence of the least-squares procedure and to save calculation time. The 1652 atoms of tRNA were divided into 132 CORELS groups (Fig. 3). Distance restraints were included which specified the stereochemical connection between phosphate and sugar groups, hydrogen bonds for the assumed secondary and tertiary interactions, and hard-sphere contact distances for nonbonded atom pairs. These restraints, along with maintaining the proper stereochemistry, served to improve effectively the data to parameter ratio. At this point, there were 888 positional parameters representing the entire structure with an overall temperature and scaling factor.

After four cycles of refinement of these parameters, we increased the resolution of the observations to 2.7 Å (8207 reflections with $F_o > 2\sigma_F$ used out of 8579 measured), removed distance restraints for the tertiary hydrogen bonds, and allowed temperature factors to vary for each phosophate, ribose, or base moiety. We then introduced variable sugar pucker by defining three dihedral rotations for each sugar and simultaneously restraining distances between atoms within each ribose. The total number of positional parameters was thus increased to 1121 plus 228 subgroup (phosphates, riboses, and bases) temperature factors. At this point, 88% of the total bond lengths and 84% of the total bond angles in the structure were constrained and the rest were restrained. Three cycles of refinement of 1652 variable atomic thermal parameters concluded the refinement.

The complete refinement lowered the R value from 42% for 6153 reflections (3 Å data) to 23% with isotropic subgroup thermal parameters (or 20% with individual atomic thermal parameters) for 8207 reflections (2.7 Å data). The model included four bound magnesium ions and 60 water molecules. The average change in atomic coordinates from the starting model was 0.83 Å and the average phase shift was 61° from the MIR phases. The overall quality of the difference Fourier and Fourier maps was greatly improved (see Figs. 8 and 11 of Ref. 25).

All bond distances and angles within constrained groups (over 85% of total bond distances and angles in the structure) were held at their canonical values. The average deviation in restrained distances from ideal, for distances corresponding to bond lengths was 0.01 Å, while those corresponding to bond angles and dihedral angles deviated by an average of about 1.1°. The shortest nonbonded contact was 0.1 Å less than the closest allowed contact distance.

The results of the refinement of the thermal parameters of each moiety (base, ribose, and phosphate) to some extent represent the flexibility of the structure. To illustrate this overall "thermal flexibility," the thermal

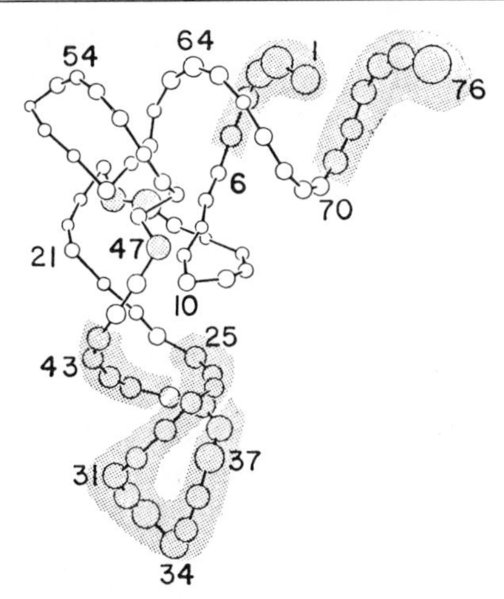

FIG. 4. To provide the impression of overall flexibility of the molecule, an average thermal motion of each residue of yeast $tRNA^{Phe}$ [25] is indicated by a circle proportional to the average mean displacement of each residue. The larger the circle is, the higher the thermal vibration of that residue in the crystal. The term *thermal* is used to cover a wide variety of small motions, such as true thermal vibration of each residue, swaying, bending, or precessing motion of arms, and also the partial opening and closing (unwinding and winding) of the double-helical stems. The overall thermal behavior was found not to be correlated to the lattice contact. Regions of larger thermal motion are shaded.

parameters of the base, ribose, and phosphate of each nucleotide were averaged to give one value per nucleotide, and these are shown in Fig. 4.[27] The radius of each circle is proportional to the mean square displacement of each residue. The most prominent feature of the figure is the unusually high thermal vibration of the anticodon arm and of the acceptor stem, suggesting that those two arms are more flexible than the other parts of the molecule, or that their stems are partially unwinding. This molecular flexibility may have a functional rationale: when tRNA participates in transpeptidation and translocation within the ribosome, it is likely that the acceptor stem and the anticodon arm move slightly.

Yeast Initiator $tRNA_f^{Met}$

The initial model of the crystal structure of yeast initiator $tRNA_f^{Met}$ was based on the interpretation of a rather noisy electron density map at 4.5 Å resolution, prepared by the method of multiple isomorphous re-

[27] S. R. Holbrook, J. L. Sussman, R. W. Warrant, and S. H. Kim, *J. Mol. Biol.* **122,** 631 (1978).

placement (MIR) and augmented by direct methods.[28-30] The interpretation of this map was aided by locating four covalently bound heavy-atom markers (at extreme positions in the structure) by difference Fourier maps. These markers helped to place the structure in the unit cell.[29,31] After the refinement of the entire structure, the poor quality of the MIR map was found to be primarily due to an incorrect assignment of the *z* coordinate of one heavy-atom derivative.[32]

Based on this map an approximate model of $tRNA_f^{Met}$ was constructed using a computer graphics. The model of $tRNA^{Met}$ which was built from its MIR map was similar to the structure of $tRNA^{Phe}$, although due to limitations in the model-building procedure it had somewhat poorer stereochemistry. To correct this, we decided to start the refinement with a stereochemically more reasonable model, so the $tRNA^{Phe}$ structure[25] was fitted by least-squares technique[33] matching up the phosphorus coordinates to those of the $tRNA_f^{Met}$. The rms distance between the 75 phosphorus atoms in common for both coordinate sets was 5.4 Å. The agreement factor between observed and calculated structure factors in the range 12.5–20 Å resolution for the fitted $tRNA^{Phe}$ model was 58.1% for 143 reflections (random is 67% for the space group $P6_422$). Study of packing conditions and of the known location of the covalently bound heavy-atom markers led us to believe that the derived model was approximately correct, although it probably had very large errors, possibly as much as 5 Å (as estimated from model calculations). We had then a rough interpretation of an electron density map, and clearly it was not possible to proceed into the kind of high-resolution refinement that was used in $tRNA^{Phe}$.

As the $tRNA^{Phe}$ molecule has very well-defined double-helical domains with the known stereochemistry,[27,34] it was possible to reduce drastically the number of degrees of freedom by introducing that information. Furthermore, based on test cases, we worked at very low resolution

[28] A. D. Podjarny, R. W. Schevitz, and P. B. Sigler, *Acta Crystallogr., Sect. A* **A37,** 662 (1981).

[29] R. W. Schevitz, A. D. Podjarny, N. Krishnamachari, J. J. Hughes, P. B. Sigler, and J. L. Sussman, *Nature* (*London*) **278,** 188 (1979).

[30] J. L. Sussman, A. D. Podjarny, R. W. Schevitz, and P. B. Sigler, *in* "Structural Aspects of Recognition and Assembly in Biological Macromolecules" (M. Balaban, ed.), p. 597. (BALABAN Int Sci. Serv., Rehovot, Israel, and Philadelphia, Pennsylvania, 1981).

[31] R. W. Schevitz, N. Krishnamachari, J. Hughes, J. Rosa, M. Pasek, G. Cornick, M. A. Navia, and P. B. Sigler, *in* "Structure and Conformation of Nucleic Acids and Protein–Nucleic Acid Interactions" (M. Sundaralingam and S. T. Rao, eds.), p. 85, University Park Press, Baltimore, Maryland, 1975.

[32] R. W. Schevitz, A. D. Podjarny, M. Zwick, J. J. Hughes, and P. B. Sigler, *Acta Crystallogr., Sect. A* **A37,** 669 (1981).

[33] S. C. Nyburg, *Acta Crystallogr., Sect. B* **B30,** 251 (1974).

[34] A. Jack, J. E. Ladner, and A. Klug, *J. Mol. Biol.* **108,** 613 (1976).

(20.0–12.5 Å) to increase the radius of convergence, so that the method would overcome the large errors of the original model. Even at this low resolution the number of degrees of freedom was made small enough by the constraints to maintain a reasonable ratio of observations to parameters.

In order to attempt refinement at low resolution,[35] the CORELS program was modified to handle a constrained group with any number of atoms. In this way it was possible to incorporate the known stereochemistry of large portions of the tRNA structure.

The refinement proceeded in steps, initially with the smallest possible number of degrees of freedom (the entire structure treated as one rigid group) and only the lowest resolution data (12.5–20 Å). After each step of the refinement converged, the number of degrees of freedom was successively increased by dividing the structure into two and four rigid groups, corresponding to the different domains of the structure (see Fig. 5). This procedure caused the *R* factor to drop from 58 to 33% for the 12.5–20 Å data (see Fig. 6). More data were then included and the structure refined as four rigid groups until at 6 Å resolution the *R* factor was 42% (see Fig. 7).

The most striking result of this procedure was a shift of the center of mass of the whole structure by almost 5 Å (see Fig. 8). This is a large movement and could result only because the initial stages of the refinement use low-resolution data (12.5–20 Å). At this stage more data were used (first 4.5 Å and then to 4 Å resolution) and the rigid group constraints were relaxed. For the loop regions the groups were separate phosphates and nucleosides while for the double-helical stem regions they were phosphates and base-paired nucleosides. This is the same kind of scheme that was used in the $tRNA^{Phe}$ refinement[25] and is shown in Fig. 3. Restraints were imposed between the CORELS groups to maintain reasonable stereochemistry as well as to prevent unacceptably close contacts between nonbonded atoms. No restraints were imposed on the tertiary base–base interactions until the very final stages of refinement after the least-squares minimization had converged; i.e., each residue not in a helical stem was refined independently at first.

At virtually all stages of the CORELS refinement we examined difference electron density maps to be certain that the structure had not fallen into a false minimum. Based on these maps we refitted the structure on a static computer graphics system[36,37] as necessary. This was especially important in the single-stranded 3′ end where the largest differences in

[35] J. L. Sussman and A. D. Podjarny, *Acta Crystallogr., Sect. B* **B39,** 495 (1983).

[36] L. Katz and C. Levinthal, *Annu. Rev. Biophys. Bioeng.* **1,** 465 (1972).

[37] A. D. Podjarny, Ph.D. Thesis, Weizmann Institute of Science, Rehovot, Israel (1976).

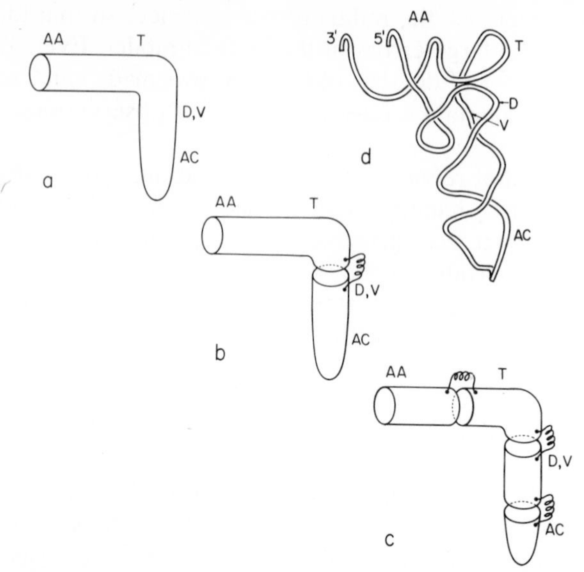

FIG. 5. Schematic representation of how the structure of $tRNA_f^{Met}$ was successively refined as rigid groups at low resolution. (a) The entire L-shaped structure was treated as a *single* rigid group (6 degrees of freedom, i.e., 3 rotational and 3 translational). (b) The structure was divided into the *two* major domains, i.e., one consisting of the AA–T arms and the other consisting of the D–V–AC arms (12 degrees of freedom). The springlike connections between groups represent stereochemical restraints used to maintain reasonable bond lengths and bond angles at the division between groups. (c) The structure was divided into *four* rigid groups, i.e., the AA stem, T arm, AC and V arms, and the D arm (24 degrees of freedom). (d) For comparison, the backbone of the $tRNA^{Phe}$, in approximately the same orientation, is shown.

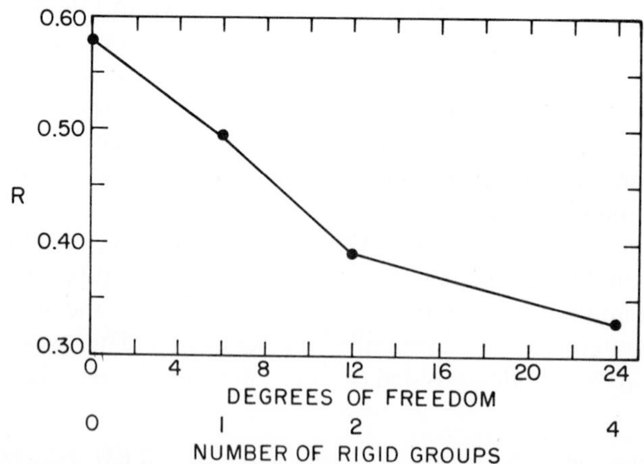

FIG. 6. *R* factor for the low-resolution (12.5–20 Å) domain refinement of $tRNA_f^{Met}$ as the structure was successively divided into one, two, and four rigid groups (see Fig. 5).

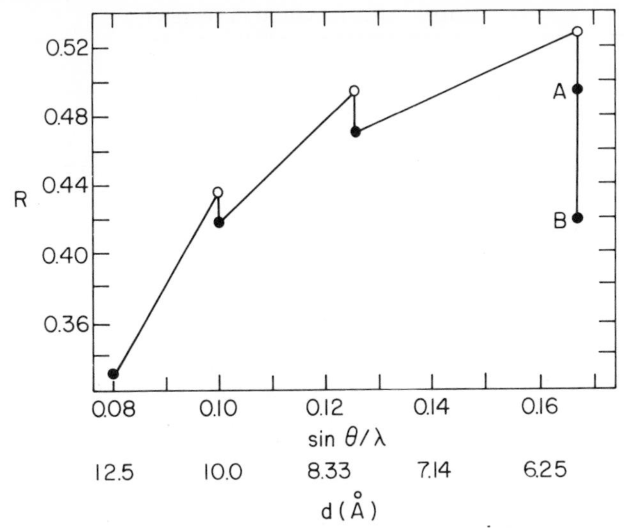

FIG. 7. R factor for the intermediate-resolution domain refinement of $tRNA_f^{Met}$. Each open circle corresponds to the R factor as higher resolution data were included, and the closed circles to the R factor at the same resolution after the structure was refined to convergence as four rigid groups. The vertical drop at 6 Å resolution corresponds to the constrained temperature factor refinement of each of the four rigid groups (i.e., an additional 4 degrees of freedom).

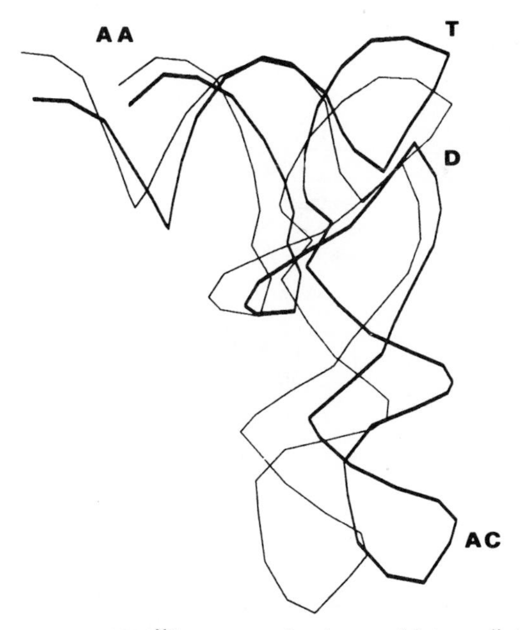

FIG. 8. Change in the $tRNA_f^{Met}$ structure after low- and intermediate-resolution refinement as four rigid groups (see Fig. 5c). The light trace represents the starting coordinates (R = 58%, 12.5–20 Å) and the heavy trace the structure after the initial group refinement (R = 42%, 6–20 Å).

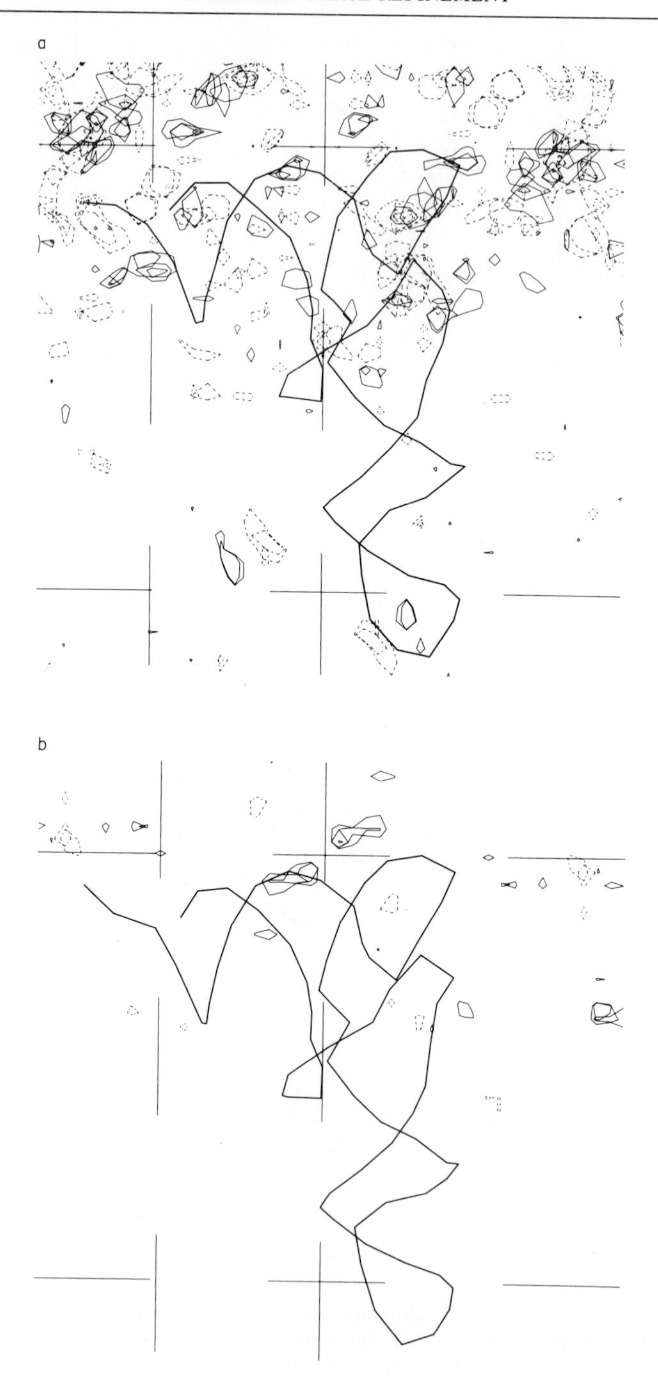
a
b

conformation from $tRNA^{Phe}$ are found. Examples of two such maps at an early stage and near the end of the refinement are shown in Fig. 9.

The *R* factor at the present stage of refinement is 26% based on all 3302 reflections to 4 Å resolution (about 75% of the theoretically possible total number), with a total of 1096 degrees of freedom (rotation, translation, torsion angles, and temperature factors for the 129 groups, and a single overall scale factor) and 2033 chemical restraints between the groups. Thus based solely on the X-ray observations (excluding the chemical restraints), the ratio of the number of observations to degrees of freedom is about 3.0. This is clearly an underestimate, as it is difficult to compare the information contained in X-ray reflections with that introduced by chemical restraints, all of which are observables.

In order to compare the structures of $tRNA_f^{Met}$ and $tRNA^{Phe}$, we have superimposed them and calculated the distances between equivalent points (Fig. 10). To get a feeling for possible crystal packing effects and the error level for a case of two independently refined structures we also compared the $tRNA^{Phe}$ crystal structures in the monoclinic[38] and orthorhombic forms.[25] In the latter case the rms difference between the centers of each corresponding phosphate, ribose, and base moiety is ~0.5 Å, although there is a ~6 Å difference in the position of residue D16. In the orthorhombic form residue D16 points out into the solvent while in the monoclinic form it points into the D loop. This shows that a portion of the D loop is actually quite flexible in $tRNA^{Phe}$, as has been observed in NMR studies.[39] In contrast, the comparison between $tRNA_f^{Met}$ and $tRNA^{Phe}$ shows a much larger difference in the two structures, i.e., an rms difference of 2.0 Å, and several regions of consecutive nucleotides that have a concerted difference in conformation. It is possible that these are primarily due to difference in crystal packing, but if this were the case then it is

[38] B. Hingerty, R. S. Brown, and A. Jack, *J. Mol. Biol.* **124,** 523 (1978).
[39] R. V. Kastrup and P. G. Schmidt, *Biochemistry* **14,** 3612 (1975).

FIG. 9. Difference electron density maps at two different stages in the refinement of the crystal of structure of $tRNA_f^{Met}$, together with a superimposed trace of the backbone. (a) Difference map at 4.5–20 Å resolution, $R = 40.5\%$, where only residues in the four double-helical stems were allowed to vary freely during the refinement (see Fig. 3c), while the loop regions were constrained to the $tRNA^{Phe}$ conformation. There is a large amount of unaccounted-for electron density near the CCA end, in the region of the T and D loops as well as in the middle of the AC loop. (b) The same resolution map, contoured as in (a), at the end of the refinement, treating the entire structure as composed of the groups shown in Fig. 3, $R = 27\%$. Here the map as a whole is much cleaner than in (a) and shows how the conformation of the structure changed during the refinement to fit better the observed X-ray data.

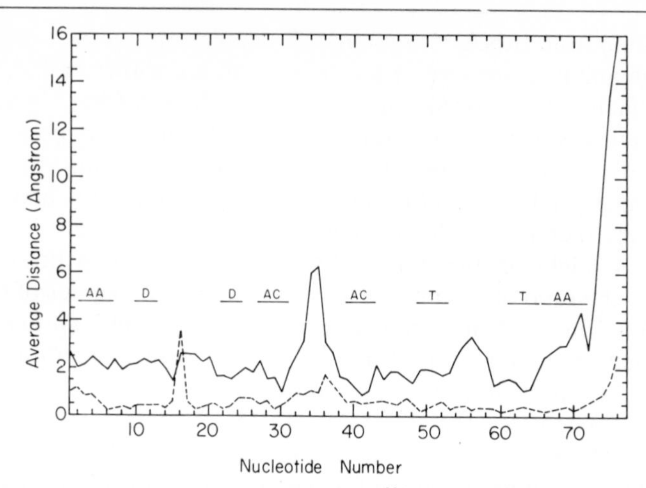

FIG. 10. A quantitative comparison of the $tRNA_f^{Met}$ and $tRNA^{Phe}$ (orthorhombic) crystal structures (solid trace). The two structures were first superimposed by a least-squares procedure that aligned the 75 equivalent phosphorus atoms. The distance between equivalent phosphorus atoms and the centers of each ribose and base were then calculated. For each residue the average of these three distances was plotted as a function of nucleotide number along the sequence. The dashed trace is a control comparing two different $tRNA^{Phe}$ crystal structures (the orthorhombic[25] and the monoclinic[38]) in the same way.

surprising that we do not see as large a difference in the two $tRNA^{Phe}$ crystal structures, especially in the AC loop which has completely different crystal packing in the two crystal forms.

The largest concerted change in structure between the initiator and chain elongator tRNAs is the anticodon loop. It is interesting to speculate on the possible biological significance of this difference especially if it is not just due to the different crystalline packing. In protein synthesis on ribosomes the AC loop is physically close to the AUG initiator codon in the "P" site of the ribosome. Therefore it would not be surprising if this were the region of the molecule which expresses a discriminatory function in its role as an initiator tRNA.

Crystalline DNA-A Octamer: d(GGTATACC)

The structure of this fragment of DNA was determined by a five-dimensional search procedure[40,41] for low-order data only. It incorporated

[40] Z. Shakked, D. Rabinovich, W. B. T. Cruse, E. Egert, O. Kennard, G. Sala, S. A. Salisbury, and M. A. Viswamitra, *Proc. R. Soc. London, Ser. B* **213,** 479 (1981).

[41] Z. Shakked, D. Rabinovich, O. Kennard, W. B. T. Cruse, S. A. Salisbury, and M. A. Viswamitra, *J. Mol. Biol.* **166,** 183 (1983).

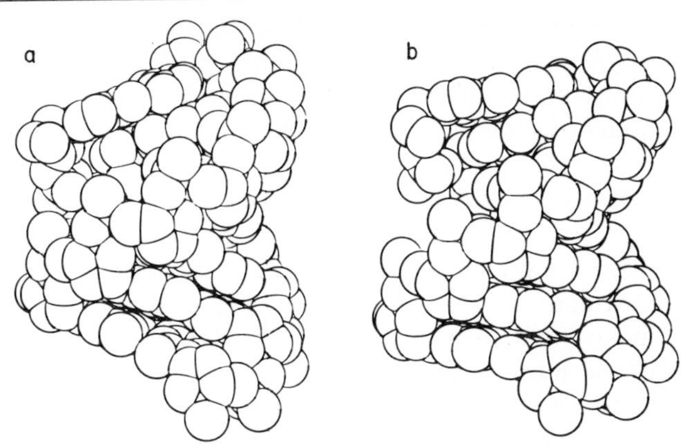

FIG. 11. Space-filling drawings of the crystal structure of the octamer d(GGTATACC): (a) the starting model and (b) the refined structure. The view is perpendicular to the helix axis.[40]

packing criteria and *R* factor calculations assuming models of both B and A types of DNA. The results indicated unambiguously that the octamer in the crystal adapted the right-handed DNA-A double-helical conformation.

The structure was refined by CORELS. In the final refinement cycles, each residue was partitioned into three CORELS units, the bases and phosphates being treated as rigid groups while the conformation of the sugar was allowed to vary.[25,41] For the initial five cycles, only the 8 to 3 Å data were used and the *R* factor dropped from 48 to 28%. Structure factor calculations for 3251 observed reflections (8 to 1.8 Å) at the end of the refinement yielded $R = 19.8\%$.[41] The refinement has resulted in a symmetric bending of the double helix about the internal 2-fold axis of the base sequence of approximately 15°. As a consequence of the bend, the width of the major groove is increased by about 2 Å, whereas the minor groove width is essentially unchanged (see Fig. 11).

Smooth Bending of DNA in Chromatin

There are two extreme possible ways to fold DNA in a nucleosome, the main building block of chromatin. "Kinked" folding, i.e., folding with regularly repeating unstacked base pairs, was initially proposed by Crick and Klug[42] as an attempt to avoid a presumed large cost in energy of the alternate possibility, i.e., smooth deformation.

[42] F. H. C. Crick and A. Klug, *Nature* (*London*) **255,** 530 (1975).

The CORELS model-building procedure was used to explore the possibility that DNA in chromatin can be packed smoothly without breaking of base-stacking interactions.[43] Specifically, we built a stereochemically acceptable model of superhelical DNA which fits the known estimates of the dimensions of the nucleosome structure. The structure of smoothly deformed DNA was constructed in two steps, geometric transformation followed by the refining of stereochemistry (see Fig. 12). As the initial fragment we chose a double-stranded molecule with the sequence of bases along one strand: ATCGGCTAAGATC. . . . in the DNA-B conformation. To make the deformed structure (def-DNA) close to the known parameters of the nucleosome core particle, the radius of the DNA superhelix was chosen to be 45 Å and the pitch 28 Å.[44] For the DNA itself, the number of the base pairs per helical repeat (33.8 Å) was chosen to be 10.375.[45]

This DNA structure was then geometrically transformed, atom by atom, so that the DNA cylinder itself was wrapped around an assumed larger cylinder. The transformation can be described by a series of equations.[43] From it we obtained a geometrical model with the curvature of the DNA axis consistent with the chosen model of nucleosome (Fig. 12b). The detailed stereochemistry of the model, of course, was far from standard. In order to improve the stereochemistry as well as to relieve several unacceptably close contacts, we used CORELS to fit standard nucleic acid building blocks (Fig. 3) to the geometrical model. The refined model is shown in Figs. 12c and 13.

The refined structure has covalent bond lengths and bond angles that differ from standard values by no more than 0.01 Å and 2.6°, respectively. The root mean square deviation of the axis of the stereochemically ideal def-DNA from the helical axis of geometrically transformed DNA is 0.01 Å, which means that the stereochemical refinement did not appreciably change the curvature of the DNA axis. As a result of refinement, the number of extreme close contacts in def-DNA is about the same as in DNA-B.

It should be noted that some parameters of def-DNA are not constant along the molecule, but oscillate; e.g., the torsion angles of the sugar–

[43] J. L. Sussman and E. N. Trifonov, *Proc. Natl. Acad. Sci. U.S.A.* **75,** 103 (1978); E. N. Trifonov and J. L. Sussman, *in* "Molecular Mechanism of Biological Recognition" (M. Balaban, ed.), p. 227. Elsevier/North-Holland Biomedical Press, Amsterdam, 1979 (coordinates deposited in the Protein Data Bank, Brookhaven National Laboratories, Brookhaven, New York).

[44] J. Finch, L. Lutter, D. Rhodes, R. Brown, B. Rushton, M. Levitt, and A. Klug, *Nature (London)* **269,** 29 (1977).

[45] E. N. Trifonov and J. L. Sussman, *Proc. Natl. Acad. Sci. U.S.A.* **77,** 3816 (1980).

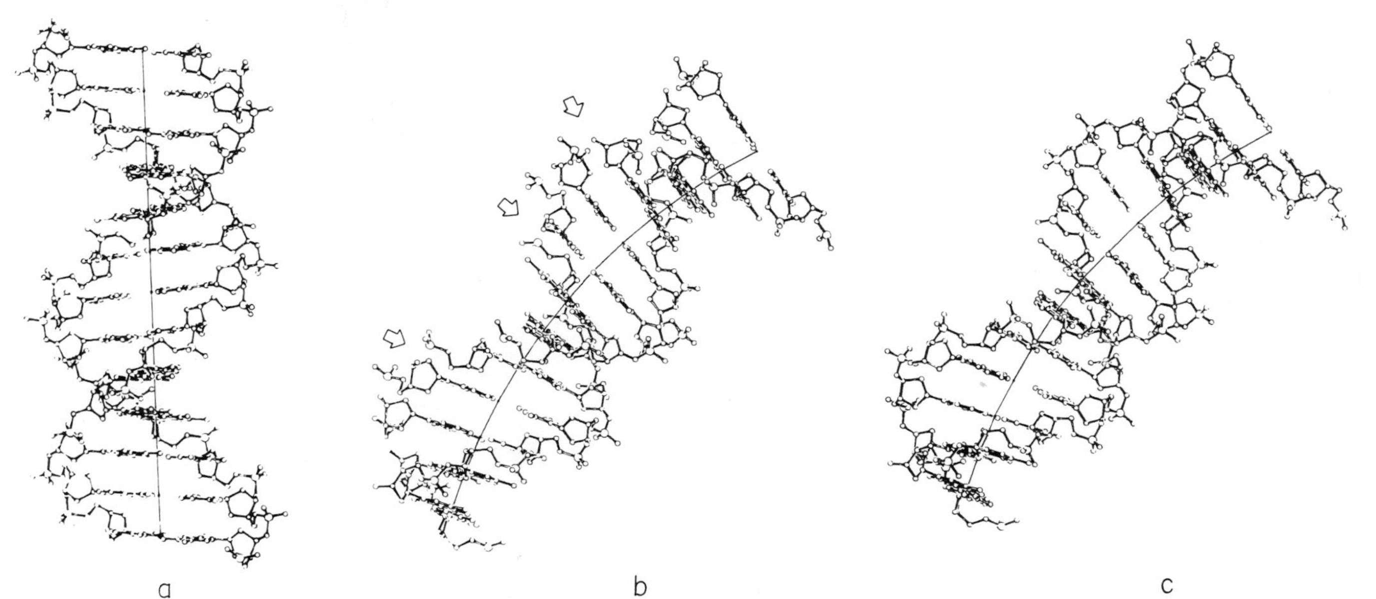

FIG. 12. Smoothly deformed DNA. Steps of model building. (a) Nondeformed DNA-B; (b) result of geometrical transformation as viewed along the axis of the superhelix (arrows show some of the nonstandard bonds "disrupted"); (c) refined model of deformed DNA, in the same orientation (bond previously too long are now "sealed").

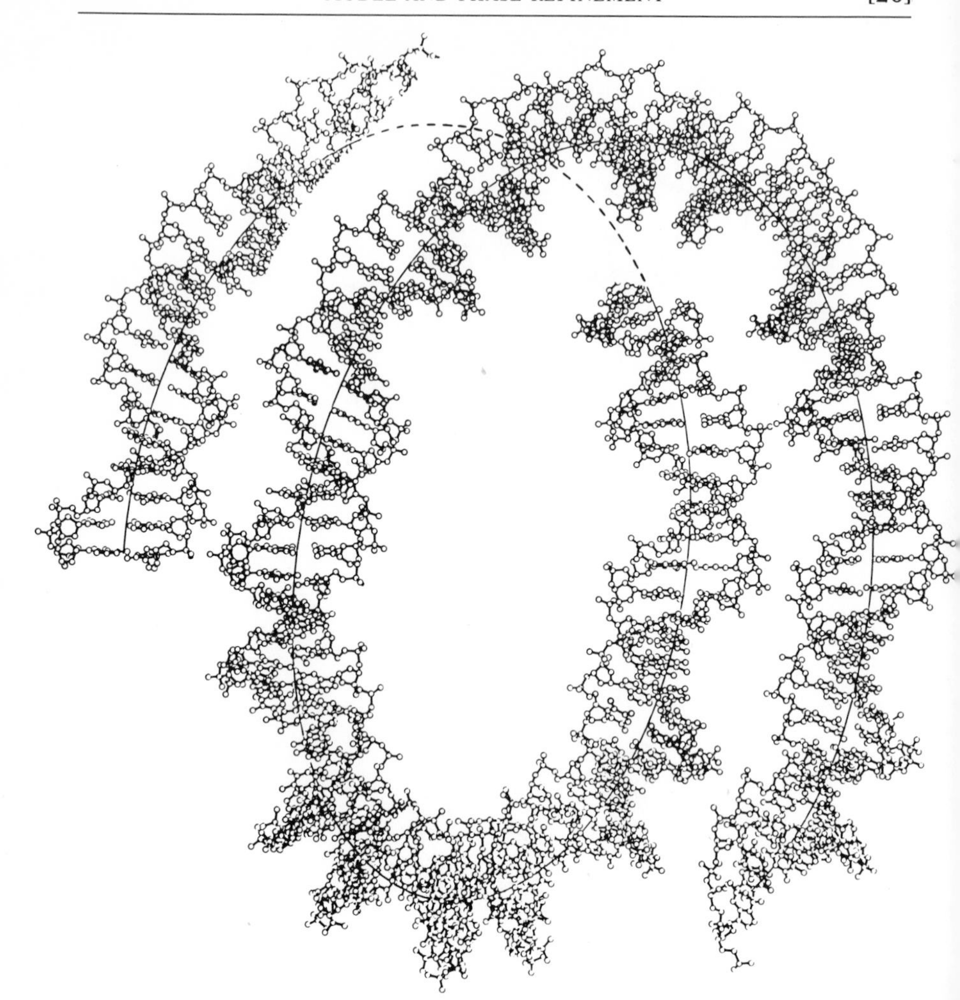

FIG. 13. Smoothly deformed DNA.[43] The fragment illustrated corresponds roughly to the DNA of one chromatin subunit helically wrapped around this protein core. The thin line represents that path of the DNA axis.

phosphate chain oscillate within reasonable limits (see Fig. 14). The angles for adjacent base pairs are more similar to each other than to those several residues away. The sugar pucker in def-DNA has approximately the same value as in DNA-B.[46]

This model-building approach showed that smoothly deformed DNA

[46] S. Arnott and D. W. L. Hukins, *Biochem. Biophys. Res. Commun.* **47,** 1504 (1972).

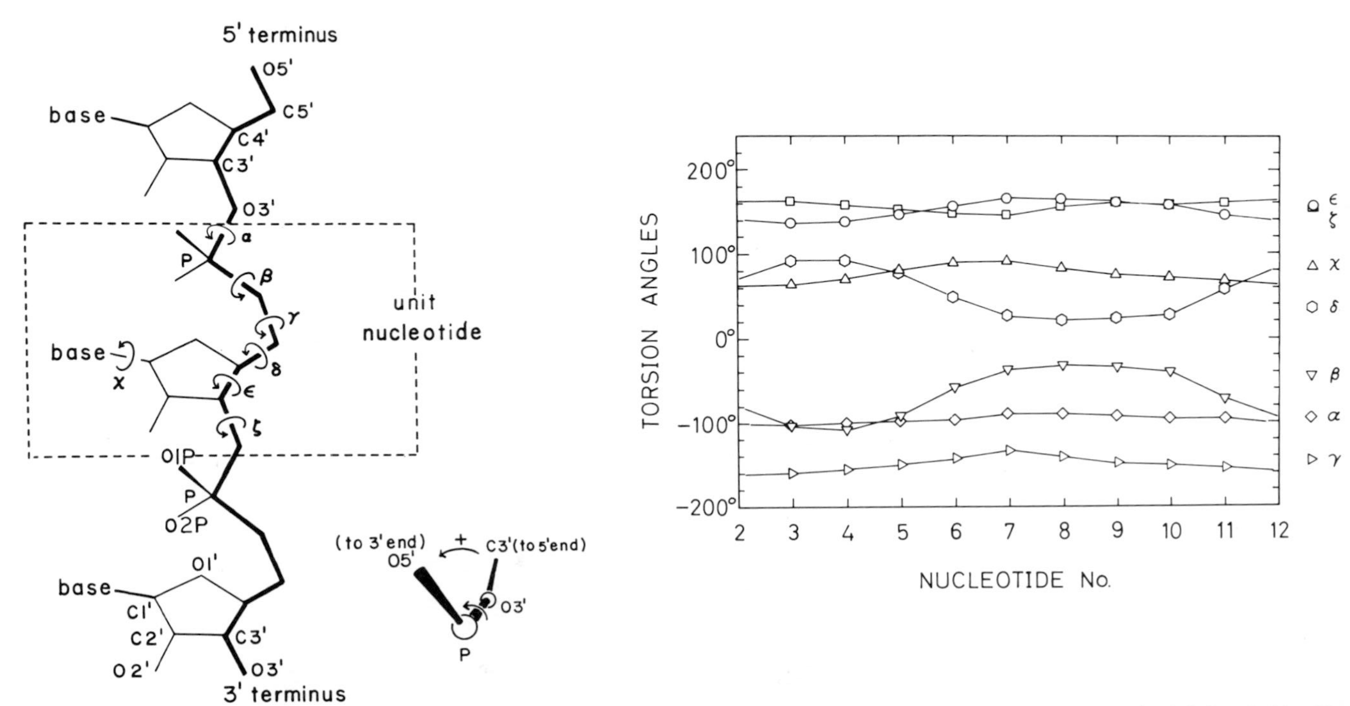

FIG. 14. Torsion angles of the refined def-DNA structure. Definitions of the angles are schematically shown on the left-hand side. The torsion angles corresponding to DNA-B[41] are indicated on the right side.

theoretically can be considered as a viable alternative to models requiring kinks from an energetic standpoint.

Demetallized Concanavalin A

Concanavalin A is a saccharide-binding protein of the Jack bean[47] whose saccharide-binding property in solution has been shown to depend on the presence of two metal ions in the protein. The details of the structural changes in the protein on successive occupation of the two metal-binding sites can be elucidated by comparisons between X-ray structures of metal-free concanavalin A, its transition metal complex, and the native protein containing both metals. Of these, the X-ray structure of the native form has been described in some detail.[48–50]

The crystal structure of demetallized concanavalin A, at a resolution of 3.2 Å, was determined by the molecular replacement method using the known structure of native concanavalin A.[51] After completing the rotation and translation search, the R factor for data to 3.2 Å was 47%. As might be expected from such a starting model, the electron density maps were unclear, and detailed interpretation would have been unreliable (see Fig. 2 of Ref. 51). To overcome this difficulty and to maintain acceptable stereochemistry we refined the structure from the starting model by treating each amino acid as an independent unit with specific distance restraints to neighboring residues in the chain (see Fig. 1c). Within each unit, dihedral and main-chain bond angles as well as the composite temperature factors of groups of atoms were refined. All restraints were implemented by means of flexible "springs" between pairs of atoms. By suitably varying the stiffness of these springs it was possible to make the distance between pairs of atoms as close to standard lengths as desired.

The strategy we used in refining the structure was initially to vary only a minimal number of parameters and gradually to allow more parameters to vary in the following order:

1. Rigid-body movements (rotation and translation) of individual amino acids, with restraints between adjacent residues in the sequence to maintain proper interresidue geometry.

[47] J. B. Sumner, *J. Biol. Chem.* **37,** 137 (1919).

[48] G. M. Edelman, B. A. Cunningham, G. N. Reeke, Jr., J. W. Becker, M. J. Waxdal, and J. L. Wang, *Proc. Natl. Acad. Sci. U.S.A.* **69,** 2580 (1972).

[49] K. D. Hardman and C. F. Ainsworth, *Biochemistry* **11,** 4910 (1972).

[50] J. W. Becker, G. N. Reeke, Jr., B. A. Cunningham, and G. M. Edelman, *Nature (London)* **259,** 406 (1976).

[51] M. Shoham, A. Yonath, J. L. Sussman, J. Moult, W. Traub, and A. J. Kalb (Gilboa), *J. Mol. Biol.* **131,** 137 (1979).

2. Variation of the side-chain dihedral angles.

3. Introduction of "repulsive" springs between nonbonded atoms to minimize repulsive interaction due to close van der Waals contacts.

4. Refinement of two subgroup temperature factors per residue, one for main-chain atoms (N, C_α, C_β, C, and O) and one for the remaining atoms.

This refinement sequence was designed to allow the biggest movements of large groups of atoms to occur first. The refined structure consisted of 3612 atoms of the dimer, and the behavior of equivalent residues in the two monomers provided a check of the accuracy of the results.

Figures 15 and 16 summarize the refinement steps. As we let more parameters vary, the ratio of observations to parameters decreased from 2.4 to 1.9. The refinement remained overdetermined because of the approximately 6000 distance restraints. It should also be noted that when

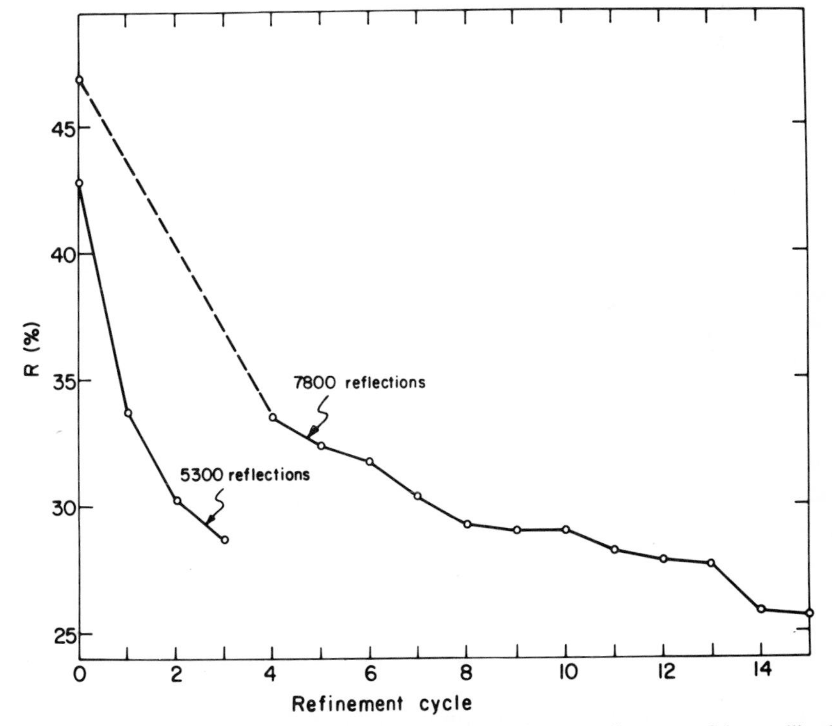

FIG. 15. Decrease of the R factor during the reciprocal space refinement of demetallized concanavalin A. In the first three cycles a partial data set of 5300 reflections was used. From cycle 4 the entire data set of 7800 reflections was used. The initial R factor for the full data set was 47.8% as indicated by the dashed line. For more details see Table 1 of Ref. 51.

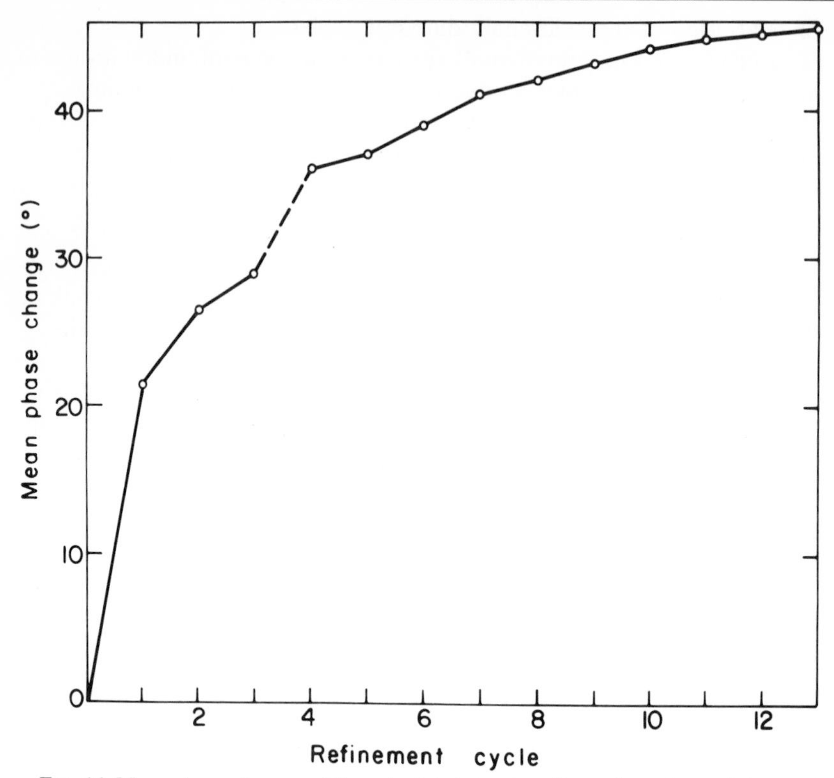

FIG. 16. Mean phase changes of the calculated structure factor from those of the starting model during the reciprocal space refinement of demetallized concanavalin A.[51] The dashed line indicates the incorporation of the entire data set.

nonbonded restraints were introduced in refinement cycle 10, the *R* factor did not increase. The average change in phase angles was 45° (rms 60°), which is about the same as has been found in other protein structures during the course of refinement.[5] The final difference electron density maps for the refined structure were much cleaner than those at the initial stages of refinement (cf. Fig. 7 and Fig. 3 in Ref. 51) and allowed us to compare the structures of native and demetallized concanavalin A in detail.

One very interesting conclusion that emerged from this work was that it was possible to rule out a cis/trans isomerization of the peptide bond between Ala 207 and Ala 208 upon the removal of the metal ions when going from the native to the demetallized structure. Such a model was proposed[52] to explain a kinetic analysis of NMR data of a 22 kcal/mol energy barrier in the activation of the demetallized protein.[53]

Conclusions

The use of a constrained–restrained least-squares procedure has proven to be extremely useful in refining macromolecular structures, especially when the initial model has severe errors. This method inherently has many fewer degrees of freedom than restrained refinement procedures and therefore is applicable at extremely low resolution with a very large radius of convergence. It is superior to a strictly constrained procedure in that the restraints between groups reduce the influence of nearest neighbors on positional shifts. The program is suitable for either structure factor least-squares refinement with distance restraints and/or model building to guide coordinates. It is also very simple to use and converges quickly.

Acknowledgments

This work was supported in part by the United States–Israel Binational Science Foundation and by a Long Term EMBO Fellowship. I wish to thank R. S. Howland and Dr. Alexander Wlodawer for critical reading of this chapter and Mrs. Linda Tripp for the typing.

[52] G. N. Reeke, Jr., J. W. Becker, and G. M. Edelman, *Proc. Natl. Acad. Sci. U.S.A.* **75,** 2286 (1978).
[53] R. D. Brown, C. F. Brewer, and S. H. Koening, *Biochemistry* **16,** 3883 (1977).

[21] Experience with Various Techniques for the Refinement of Protein Structures

By J. DEISENHOFER, S. J. REMINGTON, and W. STEIGEMANN

Introduction

In the field of crystallography of biological macromolecules the primary phases for the observed structure factor amplitudes are generally obtained by multiple isomorphous replacement (MIR) and/or Patterson search methods and molecular replacement. The former method yields experimentally determined phases without prior knowledge of the structure, whereas the latter allows deduction of a phase set of an unknown crystal structure from a known molecular transform (see relevant chapters in this volume). In both cases the phases are inaccurate. Hence the molecular models obtained from the interpretation of electron density maps calculated with such phases can be only approximate. They must be improved in order to obtain sufficiently reliable results for understanding the function of the biological macromolecules studied.

The first step of a refinement procedure is the improvement of the model of the crystal structure while keeping constant the observed and known quantities of the structure factors. For the method of MIR these are the measured structure factors plus the MIR phases, resulting in an observed electron density map. For the method of molecular replacement, these are the measured structure factor amplitudes plus the molecular transforms of partial structures, i.e., the first approximations of the investigated molecule. Initial improvements can therefore be performed by optimally fitting the molecule to the MIR electron density map in case 1 and molecular transform fitting to obtain optimal agreement between calculated and observed structure factors by reciprocal space least-squares refinement in case 2. We are using Diamond's[1] real space refinement technique (RLSP[2]) and Sussman's[3] constrained–restrained least-squares (CORELS[4]) for these purposes.

After these initial steps of improvement of the crystal structures are executed, further refinement has to be performed, and the MIR phases or phase sets from molecular fragments are abandoned. The limited amount of reflection data generally prohibits the application of conventional least-squares (for exceptions and computational techniques, refer to relevant chapters in this volume) because of the poor ratio of observations to parameters. This ratio can be improved either by increasing the number of observations (which need not necessarily be X-ray observations) or by reducing the number of parameters. A method of the latter kind would be Diamond's real space refinement[1] when applied in a cyclical manner in which newly calculated phases are introduced into the electron density map after each cycle.[5,6] Another method we have experimented with is Sussman's constrained–restrained least-squares procedure.[3] The stan-

[1] R. Diamond, *Acta Crystallogr., Sect. A* **A27,** 436 (1971).

[2] R. Diamond, this volume [18].

[3] J. L. Sussman, S. R. Holbrook, G. M. Church, and S. Kim, *Acta Crystallogr., Sect. A* **A33,** 800 (1977).

[4] J. L. Sussman, This volume [20].

[5] R. Huber, D. Kukla, W. Bode, P. Schwager, K. Bartels, J. Deisenhofer, and W. Steigemann, *J. Mol. Biol.* **89,** 73 (1974).

[6] J. Deisenhofer and W. Steigemann, *Acta Crystallogr., Sect. B* **B31,** 238 (1975).

dard refinement technique in this laboratory is now, however, that of Jack and Levitt,[7,8] a procedure which minimizes simultaneously the X-ray residual and potential energy. This procedure may be regarded as belonging in the category of using an increased number of observations which represent chemical and physical properties.

Needless to say, with every method as many X-ray reflection data as possible should be collected. Of crucial importance for the success of any refinement procedure are the manual interventions which eliminate gross errors in the model which lie outside the radius of convergence of the particular procedure and, therefore, cannot be corrected automatically by the selected refinement method. It turns out that these actions are rate limiting in the whole refinement. For this reason it is very important to have powerful computer-assisted tools available to shorten the period required for model building and model correcting. Computer graphics is one of the techniques applied for this purpose.[9]

In the following we will report experiences with the refinement techniques RLSP,[1,2] CORELS,[3,4] and EREF[7,8] obtained until 1982.

Real Space Refinement

Details of the method are described elsewhere.[1,2] The features of the technique essential for this representation are the spacial fitting procedure of electron density in real space and the fixation of all bond lengths and the majority of bond angles to the values imposed on the starting model. The dihedral angles in main and side chains and some selected interbond angles are the only variable parameters. Possible deviations from standard geometry cannot be detected.

This technique, accompanied with the exchange of the electron density map after each round of refinement for a new one calculated with phases from the most recent model, has been used routinely in our laboratory for the refinement of a number of protein structures.[5,6,10–13] From this experience, we have recognized several advantages and disadvantages

[7] A. Jack and M. Levitt, *Acta Crystallogr., Sect. A* **A34,** 931 (1978).

[8] M. Levitt, *J. Mol. Biol.* **82,** 393 (1974).

[9] T. A. Jones, *J. Appl. Crystallogr.* **11,** 268 (1978).

[10] W. Bode and P. Schwager, *J. Mol. Biol.* **98,** 693 (1975).

[11] W. Bode, P. Schwager, and R. Huber, *J. Mol. Biol.* **118,** 99 (1978).

[12] O. Epp, E. E. Lattman, M. Schiffer, R. Huber, and W. Palm, *Biochemistry* **14,** 4943 (1975).

[13] W. Steigemann and E. Weber, *J. Mol. Biol.* **127,** 309 (1979).

which are inherent to the method and which are summarized in the following.

The principal advantage for improvement of the initial model by fitting to an MIR map lies in the fact that an optimum interpretation of such a map can be obtained before the observed phases are replaced by calculated ones. This technique may also be of advantage in cases in which maps are calculated with phases obtained by phase combination of independent sources.[14] The result and success of refinement are more readily visible and traceable than in reciprocal space. Refinement in real space has, in general, a large radius of convergence. These properties, which derive exclusively from the volume fitting feature, make this procedure a powerful technique especially in the initial stages of refinement.

The majority of disadvantages derive from the fact that Diamond's real space procedure fixes standard geometry and does not allow for deviations from it during the refinement process. Therefore it is very important to have good starting models since there is no possibility of controlling and correcting deviations from standard geometry. In this way, it is extremely tedious to perform major revisions of a model by insertion, deletion, or modification of amino acid residues. Very elaborate and careful model building with good standard geometry is necessary before real space refinement can be continued. This fact makes the procedure less attractive for the refinement of proteins with inaccurately known chemical sequences. Similarly, major stereochemical changes in the backbone (e.g., peptide flip) are hard to control with the options supplied by the program and are obtained often only after a number of different trials. The reason for these unfavorable properties lies in the fact that complicated mathematical manipulations are necessary for the projection of Cartesian coordinate space into dihedral angle space with fixed margins at the NH_2- and COOH-terminal ends of the peptide under refinement. This, in turn, is necessary, since the program does not know the stereochemistry of single amino acids and therefore cannot allow deviations from rigidity of bond lengths, bond angles, etc.

In summary, the refinement procedure has its strength in real space fitting, but shows its weak points especially during the breakpoints in manual intervention which are the most complicated and crucial steps in the process of refinement. Therefore, we feel that the procedure has lost much of its attractiveness for routine refinement. However, an extremely useful application is still the primary fitting of an initial model to a map calculated with experimental phases. For this particular application we

[14] W. A. Hendrickson and E. E. Lattman, *Acta Crystallogr., Sect. B* **B26,** 136 (1970).

want to give some recommendations of parameters necessary for refinement:

1. Decision on the grid of electron density map. Since there is no need and reason for refining individual atomic radii (i.e., temperature factors) at this stage, the choice of coarseness of grid is not crucial. Although a spacing of one-third of the resolution has been recommended,[1] we have found that often a value of 0.5 is satisfactory. The procedure is likely to become unstable at a resolution of 3.5 Å or less.

2. Choice of radii for individual atom types. In principle, the apparent atomic radii depend on the resolution of the map and the temperature factors. The actual choice for the initial step of model fitting is not very critical. Perhaps it is advisable to start with radii larger than the true ones at a given resolution in a first cycle, as this even enlarges the radius of convergence. The reduction to values smaller than the true ones in a subsequent cycle allows a more accurate fit. Typical values that have been used for different atom types for a number of structures with resolutions between 3 and 2 Å lie between 1.7 and 1.4 Å.

3. Choice of individual weights for dihedral angle parameters. We have not experimented very much with these. As has been recommended by R. Diamond[1] the ratio between weights assigned to dihedral angle and bond angle parameters should be of the order of 10 : 1. Typical values are 3.7 for main and side chain and 0.3 for the main-chain bond angle τ at the α-carbon atom. In a first round we often have set the τ weight to 0, which introduces absolute rigidity.

4. Eigenvalue filtering. The actual relative flexibility also depends on the filter ratio chosen for the BEND[1] refinement. We have set it empirically so that 80–90% of all possible shifts have been performed in eigenvector space. 100% selection makes the particular choice of stiffness parameters superfluous. For KDAZ[1] refinement (only the overall scale factor K and background density d should be selected) all eigenvalues larger than 0 can be selected from the full spectrum. The same applies to TRANS[1] refinement where the actual volume fitting is performed. In case of danger that fitting will be to the wrong density, exclusion of the last 10% of eigenshifts is advisable. The fit is not significantly degraded at a reduced overall shift of the atomic positions.

5. Zone length. The length of the peptide which glides along the chain residue by residue and which is under refinement should not be too short to allow for reasonable flexibility of the main chain with fixed margins at the end of the peptide. An excessive length, on the other hand, does not result in better convergence, but enlarges computing time considerably. We have found values between 5 and 7 very reasonable.

Rigid-Body Refinement in the Method of Molecular Replacement (CORELS)

The solution of a crystal structure using the methods of Patterson search[15–17] and molecular replacement (for a review, see Rossmann[18] and Lattman,[5] this volume) has three critical stages. The first such stage is the correct orientation of the model in the crystal cell. This is usually solved by rotating the model Patterson map, calculated in an appropriate unit cell, and calculating the product correlation of the model Patterson against the native Patterson map as a function of the rotation angles, either in direct space[16] or reciprocal space.[19] If a reasonable solution to this orientation problem is obtained, one must then solve the translation problem correctly, placing the model in the crystal cell. For this purpose procedures in direct space[16,17] and reciprocal space[20] are available and have been successfully used. The implementation of the modified translation function[20] in a program by E. E. Lattman is particularly fast and efficient, and is probably the most sensitive technique available. If these techniques yield a reasonable solution, an attempt must be made in the third step to interpret an electron density map (usually a "Sim-weighted"[21] $2|F_{obs}| - |F_{calc}|$ map) based on calculated phases. At an R factor of typically 0.35–0.42 at 3.0 Å resolution, this is unlikely to be a trivial problem. Given a solution to the orientation and translation problem, we have found in every instance of the use of these techniques that rigid-body refinement of the rotational and translational parameters by placing the model in the crystal cell resulted in a reduction of the crystallographic R factor, sometimes dramatically, and invariably improved the quality of the electron density map. For the rigid-body reciprocal space refinement step, we use the program CORELS.[3,4]

General Considerations

In order to test the feasibility of the rigid-body refinement procedure, some tests were first carried out using the refined structure of pancreatic trypsin inhibitor (PTI).[6] The PTI molecule was displaced by 0.1 Å and rotated by 5.0°, then subjected to the rigid-body refinement technique.

[15] W. Hoppe, *Acta Crystallogr.* **10,** 750 (1957).

[16] R. Huber, *Acta Crystallogr., Sect. A* **A19,** 353 (1965).

[17] R. Huber, *in* "Crystallographic Computing" (F. R. Ahmed, S. R. Hall, and C. P. Huber, eds.), p. 96. Munksgaard, Copenhagen, 1969.

[18] M. G. Rossmann, ed., "The Molecular Replacement Method," Int. Sci. Rev., No. 13. Gordon & Breach, New York, 1973.

[19] M. G. Rossmann and D. M. Blow, *Acta Crystallogr.* **15,** 24 (1962).

[20] R. A. Crowther and D. M. Blow, *Acta Crystallogr.* **23,** 544 (1967).

[21] G. A. Sim, *Acta Crystallogr.* **12,** 813 (1959).

After three refinement cycles using reflections from 5.0 to 4.0 Å, convergence was essentially complete, with a residual error of about 0.2°, and 0.02 Å from the correct position. Apparently low-resolution data are adequate to determine the few parameters of the rigid-body problem, and the use of low-resolution data would be expected to increase the radius of convergence of the technique. In choosing the resolution range for the rigid-body step, one would like to minimize computing time and maximize the radius of convergence. The largest useful Bragg spacing is set by the density of the solvent in the crystal, as neglect of solvent in the structure factor calculation has a severe effect on reflections with spacings of about 5.0–7.0 Å and greater. Likewise, the smallest spacing will depend on the unit cell parameters, the expected radius of convergence (often estimated to be $D/4$, where D is the spacing of the highest resolution reflections used), and thus, the number of reflections. CORELS is slow, as is any procedure which calculates shifts from analytical derivatives of the structure factors with respect to the parameters. We have found that 1000–2000 reflections is entirely adequate to solve the rigid-body rotation problem. Our version of CORELS requires that each group have at least one temperature factor, and there is an overall temperature factor. These parameters must of course be held fixed during low-resolution refinement.

Since this technique is only likely to be used infrequently, a great deal of time need not be spent on programming new space groups (at the expense of computer time), as a simple expansion of the reflection set to space group P1 is always possible. Also, CORELS is particularly flexible in regard to the composition of the "standard groups" which make up the model being refined. Connections between these standard groups may be made or disregarded as the problem requires, greatly easing the problem of refining independent domains, as will be seen in the following. We have written a small program which converts any part or all of a standard-format coordinate set into a CORELS "standard group" for general applicability; several such "groups" are simply concatenated to make up the standard-group input file. One need only keep track of the number of atoms in a group, its center of mass coordinates, and its relative location in the group input file (i.e., first, second, etc.). In applying shifts calculated by CORELS to generate a new model, care must be taken to follow the same series of steps taken by the program (i.e., rotate about the center of mass, then add positional shifts), as well as to take into account the convention for orthogonalization and deorthogonalization. This and the angular convention used are adequately described by Sussman *et al.*[3,4] In October 1983, a new version of CORELS became available from J. Sussman, in which the definition of rigid groups especially, and the treatment of different space groups, have been simplified significantly.

After domain refinement, consideration might be given to breaking the molecule up into smaller fragments for example helices. In our opinion, this is unlikely to be useful, as a shift of an entire helix (relative to the rest of the molecule) is bound to be accompanied by changes in side-chain orientation. If used with care, the radius of convergence of other refinement techniques discussed in this volume (particularly the EREF technique in the next section, "Crystallographic Refinement") appears to be large enough to allow shifts of up to 1.0 or 2.0 Å in side chains and small groups to occur, given that the molecule as a whole is properly positioned.

Applications

The first real application of this procedure in this laboratory involved the solution of the intact immunoglobulin molecule KOL[22] by using the program CORELS, and the success of this application led to the use of this program as a standard procedure. Five such examples are summarized in Table I, and will be briefly discussed here. All involved breaking the model up into "domains" which are independently rotated and translated as rigid bodies. No constraints or restraints are applied between the domains, as typically the only interest is in determining 6–30 parameters, and often many thousands of reflections can be used in this determination.

The cases discussed are IgG KOL, the intact immunoglobulin molecule KOL[22]; KK, the serine protease kallikrein[23]; TK, the kallikrein–pancreatic trypsin inhibitor complex[24]; TGK, the trypsinogen–PSTI (pancreatic secretory trypsin inhibitor) complex[25]; and CSC4, the third crystal form of citrate synthase.[26] In the first two cases, poor MIR Fouriers were available which led to estimated starting coordinates for the models, which were then refined by domains with CORELS. In the last three cases, the structures were solved entirely by the use of Patterson search and molecular replacement methods.

Starting Models and Domain Structure. In every case there was an obvious choice for the separation of the model into domains. In the case of IgG KOL,[22] a poor MIR Fourier map[27] was available, and had been partially interpreted in terms of the refined structure of KOL Fab.[22] It was

[22] M. Marquart, J. Deisenhofer, R. Huber, and W. Palm, *J. Mol. Biol.* **141,** 369 (1980).

[23] W. Bode, Z. Chen, K. Bartels, C. Kutzbach, G. Schmidt-Kastner, and H. Bartunik, *J. Mol. Biol.* **164,** 237 (1983).

[24] Z. Chen and W. Bode, *J. Mol. Biol.* **164,** 283 (1983).

[25] M. Bolognesi, G. Gatti, E. Menegatti, M. Guarneri, M. Marquart, E. Papamokos, and R. Huber, *J. Mol. Biol.* **162,** 839 (1982).

[26] G. Wiegand, S. J. Remington, J. Deisenhofer, and R. Huber, *J. Mol. Biol.* **174,** 205 (1984).

[27] P. M. Colman, J. Deisenhofer, R. Huber, and W. Palm, *J. Mol. Biol.* **100,** 257 (1976).

natural to divide the molecule into the four domains: heavy chains and light chains of the constant and variable regions (C_H, C_L, V_H, V_L). These regions are connected by extended "hinge" peptides, which were left out of the calculations. The starting model was the refined KOL Fab model. In the case of the serine protease kallikrein, an extremely poor 4.0 Å isomorphous Fourier map indicated two molecules in the asymmetric unit with trypsin-like structures expected on the basis of sequence similarities. The starting model in this case consisted of two totally independent structures which were derived from the refined structure of trypsin[10] by removal of side chains where the sequence differed.

In the next three cases, the orientation and translation of the starting models were determined entirely with the use of Patterson search methods and the modified translation function. As indicated by the final crystallographic *R* values for these structures in Table I, the technique has been entirely successful, leading to interpretable Fourier maps, and successful rounds of restrained crystallographic refinement. In the case of TK, the kallikrein–PTI complex, the starting model was based on the refined structure of the trypsin–PTI complex.[5] The refined kallikrein model was transformed onto the trypsin–PTI complex by fitting the α-carbon positions in a least-squares procedure. A similar starting model served for the solution of the TGK (trypsinogen–PSTI) structure, in which the refined third domain of Japanese quail ovomucoid OMJPQ3[28,29] was fit to PTI in the refined trypsin–PTI model, solely on the basis of the main-chain atoms of the trypsin binding site of PTI. In both of these instances the starting models were refined in terms of two domains, the protease and the inhibitor. In the last example, the C4 crystal structure of citrate synthase (which contains a dimer of M_r 100,000 in the asymmetric unit), it was known that the small domain (SD) of the molecule can undergo large conformational changes, both internal and with respect to the large domain.[30] Since there was some doubt as to the relative orientation of the small and large domains (MD) of the molecule, all calculations proceeded with three starting models in parallel, the "open" form, the "closed" form, and a form consisting solely of the large domain. The search models were in all cases dimers generated on the basis of the crystallographic symmetry axis of the two refined structures.[30] The rigid-body step involved first the entire dimer as a unit, which was then broken down into four domains that were independently refined.

[28] E. Weber, E. Papamokos, W. Bode, R. Huber, I. Kato, and M. Laskowski, Jr., *J. Mol. Biol.* **149,** 109 (1981).

[29] E. Papamokos, E. Weber, W. Bode, R. Huber, M. W. Empie, I. Kato, and M. Laskowski, Jr., *J. Mol. Biol.* **158,** 515 (1982).

[30] S. J. Remington, G. Wiegand, and R. Huber, *J. Mol. Biol.* **158,** 111 (1982).

TABLE I

FIVE EXAMPLES OF THE USE OF CORELS IN THE RIGID-BODY RECIPROCAL SPACE REFINEMENT TECHNIQUE[a]

				Parameter shifts						Crystallographic R factors			
Structure (domains)	Resol.	No. of cycles	No. of refl.	ϕ (°)	θ (°)	ρ (°)	X (Å)	Y (Å)	Z (Å)	R_i	R_f	(Resol.)	Final R
IgG	5.5–4.0	3	—										
IgG (V_L)	6.0–3.5	3	9785	0.79	−1.06	−1.03	−0.02	0.14	0.07				
(V_H)				−1.24	−1.74	0.32	0.06	0.23	0.24				
(C_L)				1.19	0.32	0.70	0.21	0.12	0.20				
(C_H)				0.65	2.57	−0.04	−0.18	0.13	−0.07	0.370	0.350	6.0–3.5	0.24 (3.0 Å)
KK (1)	6.0–5.5	2	490	0.52	0.12	−1.42	0.13	0.08	−0.04				
(2)				−0.01	0.27	0.10	−0.13	−0.02	0.22	0.493	0.461	6.0–5.5	
KK (1)	5.5–4.5	2	1766	−0.26	1.60	−1.73	0.14	−0.01	−0.33				
(2)				−1.27	−0.54	−0.48	0.14	0.04	0.07	0.443	0.434	5.5–4.5	0.21 (2.1 Å)
TK (PTI)	5.5–4.5	2	1635	2.99	−0.13	−0.74	0.38	0.04	−0.12				
(KK)				−0.94	−0.80	0.30	−0.12	0.06	−0.14	0.316	0.298	5.5–4.5	0.23 (2.5 Å)
TGK (Tryp)	5.5–4.0	4	1870	4.20	−0.10	−0.50	−0.06	−0.02	−0.03				
(OMJPQ3)				1.30	−0.10	−2.40	0.52	−0.18	−0.07	0.485	0.392	7.0–3.0	0.19 (1.8 Å)
CSC4 (Dimer)	5.5–5.0	3	1087	0.31	−0.35	−1.20	0.11	−0.09	0.09	0.414	0.361	5.5–5.0	
CSC4 (MD1)	5.5–5.0	3	1087	−0.22	−0.17	−0.12	−0.21	−0.36	0.18				
(SD1)				−0.26	−0.17	−0.10	−0.41	−1.12	−0.45				
(MD2)				0.11	0.04	0.13	−0.04	0.27	−0.32				
(SD2)				0.04	−0.02	0.29	0.26	0.87	−1.17	0.388	0.374	5.5–5.0	0.22 (2.9 Å)

[a] Structure refers to the examples discussed in the text. Resol. is the resolution range for the refinement step; No. of cycles is the number of refinement cycles; and No. of refl. is the number of reflections in the resolution range. ϕ, θ, and ρ are the rotation angles about Z, X', Y'' as defined by Sussman *et al.*[3] R_i and R_f are the initial and final crystallographic R factors after the rigid-body step for the range given by the column labeled Resol., not necessarily the same range as the refinement. Final R gives the final crystallographic R factor after the structural refinement was complete.

Results. The results from various applications of the rigid-body refinement techniques are summarized in Table I. The drop in R factor during the rigid-body refinement is obviously dependent on the quality of the starting model (i.e., how similar the starting model is to the final refined model) and was particularly dramatic in the case of TGK (trypsinogen–PSTI). In this example, the structures of the two molecules were essentially correct, but their relative orientation, as determined by the Patterson search function and the translation function, was inaccurate: the trypsin model was rotated more than 4.0° from the starting orientation, and the OMJPQ3 by about 3.0°. Further, the OMJPQ3 model was shifted by 0.52 Å in X. The R factor was reduced from 48.5 to 39.2% for the resolution range 7.0 to 3.0 Å, which is quite surprising when one considers that only 1870 reflections from 5.5 to 4.0 Å were used in the refinement. This gives some idea about the radius of convergence of the method.

Less dramatic, but equally successful was the application of the method to the CSC4 citrate synthase structure. On the basis of the Patterson search function, we were unable to distinguish between the "open" and "closed" forms of the enzyme, and the translation function was not solvable until the variable small domain was excluded from the calculation. With a dimer consisting of 680 of the 880 total residues in the main domains, a solution to the translation function was found, and this refined with CORELS as a single unit. A $2|F_{obs}| - |F_{calc}|$ Fourier map calculated on the basis of this solution, with the small domain excluded from the structure factor calculation, clearly revealed density at the position expected for the small domain in the "closed" form of the enzyme. A subsequent run of CORELS was then performed with a four-domain model in which the small and large domains were treated separately. "Hinge" regions were rather carefully chosen on the basis of the known conformational changes in the molecule. It is interesting that after reassembly of the four refined domains, no bad contacts, bond lengths, or bond angles were detected, indicating that the overall shifts in the four domains were compatible with local stereochemistry at the hinge and contact regions. Note that the two small domains were each shifted by more than 1.0 Å from their starting (center of mass) positions.

Crystallographic Refinement (Jack–Levitt)

There is no fixed recipe for the crystallographic refinement of an atomic model of a protein. Each case has its individual characteristics. In the following, the system of methods and programs is described that has been found satisfactory for crystallographic refinement of the following protein structures:

1. Human Fc fragment and its complex with fragment B of protein A.[31]
2. Human myeloma protein KOL (IgG) and KOL Fab fragment.[22]
3. Trypsin and trypsinogen.[32]
4. Kallikrein.[23]
5. Kallikrein complex red with pancreatic trypsin inhibitor.[24]
6. Two crystal forms of citrate synthase.[30]
7. Glutathione peroxidase.[33,34]
8. The third domain of Japanese quail ovomucoid.[28,29]
9. The complex of trypsinogen with pancreatic secretory trypsin inhibitor.[25]
10. C3a anaphylatoxin.[35,36]
11. The oxygenated form of the insect hemoglobin CTT III.[37]

Automated Reciprocal Space Refinement with Energy Restraints

We start with a set of atomic coordinates and the corresponding set of calculated structure factors. With the method of Jack and Levitt[7,8] as the central part, a cycle of automated refinement consists of three steps. In step 1, the system of normal equations for diagonal-matrix least-squares refinement of atomic parameters is derived from a difference Fourier map (DF map) (program DERIV). In step 2 the conformational energy of the model is minimized together with a crystallographic residual to refine atomic positions (program EREF). Alternatively, atomic B values can be refined. Step 3 completes a refinement cycle by the calculation of new structure factors from the model produced in step 2.

Setting up the Normal Equations. This method[7] uses a DF map to work out the system of normal equations which minimize

$$X = \sum W(|F_{\text{obs}}| - |F_{\text{calc}}|)^2 \quad (1)$$

[31] J. Deisenhofer, *Biochemistry* **20,** 2361 (1981).

[32] J. Walter, W. Steigemann, T. P. Singh, H. Bartunik, W. Bode, and R. Huber, *Acta Crystallogr., Sect. B* **B38,** 1462 (1982).

[33] R. Ladenstein, O. Epp, K. Bartels, T. A. Jones, R. Huber, and A. Wendel, *J. Mol. Biol.* **134,** 199 (1979).

[34] O. Epp, R. Ladenstein and A. Wendel, *Eur. J. Biochem.* **133,** 51 (1983).

[35] R. Huber, H. Scholze, E. P. Paques, and J. Deisenhofer, *Hoppe-Seyler's Z. Physiol. Chem.* **361,** 1389 (1980).

[36] J. Deisenhofer, to be published.

[37] W. Steigemann and E. Weber, *in* "Interaction Between Iron and Proteins in Oxygen and Electron Transport" (C. Ho, ed.), p. 19. Elsevier/North-Holland, New York, 1982.

These equations can be written

$$A\mathbf{dp} = \mathbf{b} \tag{2}$$

where **dp** is the vector of desired parameter shifts (positions, occupancies, B values); the matrix A is approximated by a diagonal matrix; the diagonal elements are estimated for each atom type. The elements of the vector **b** are obtained by convoluting the Fourier transform of the atomic form factor with the gradient of the DF map around the atomic site (position refinement) or with the DF map itself (occupancy or B refinement). This convolution is calculated in a box of usually $3 \times 3 \times 3$ grid points around the atomic positions; the grid spacing is about one-third of the resolution given in angstroms.

The resolution range in which refinement is to be done, and the weights W applied to the terms of X [Eq. (1)] are determined when the Fourier coefficients for the DF map are selected. In most cases reflections with Bragg spacings greater than 7 Å are omitted because of the poor fit between F_{obs} and F_{calc} due to the omission of the solvent continuum from the model.

A rejection criterion frequently used for weighting is that a reflection is excluded from the Fourier summation if $2||F_{obs}| - |F_{calc}||/(|F_{obs}| + |F_{calc}|)$ is greater than a specified limit. This limit depends on the stage of the refinement and also on the quality of the experimental data. A typical value is 1.2.

Another weighting method which has been applied is Sim weighting.[14,21] Here, reflections with large discrepancies between observed and calculated amplitudes contribute to the DF map, however with reduced weight. In the early stages of refinement this appears to be preferrable to the all-or-nothing rejection scheme described above. Other types of weighting could be easily implemented in the existing programs. An exponential factor comparable to a temperature factor has been used during refinement of C3a anaphylatoxin[36] to increase the weight of high-resolution reflections.

Application of Shifts to Model Parameters. Since the normal matrix obtained in step 1 is a diagonal matrix, the determination of shifts to the model parameters would be trivial if the ratio of observations to parameters were sufficiently high to allow independent refinement of each parameter. For a vast majority of protein structures with experimental data available up to and even beyond 1.5 Å resolution this is not the case. To prevent model parameters from adopting highly improbable values during refinement, constraints or restraints have to be introduced.

Atomic Positions. For refinement of atomic positions we use simultaneous minimization of the crystallographic residual and of the conformational energy as described.[7] The main reasons for the selection of this procedure are, on the one hand, the application of physically reasonable potential parameters, and on the other hand the use of FFT (fast Fourier transform) methods for the calculation of the crystallographic contributions.

The function that is to be minimized is given by

$$R = E + kX \tag{3}$$

where E represents the conformational energy and X [Eq. (1)] the crystallographic term. The factor k controls the contribution of X to the total residual R. Its magnitude is of crucial importance as may be verified when it adopts the extreme values $k = 0$ and $k = \infty$. In the first case pure energy refinement is performed, in the latter pure crystallographic refinement.

The conformational energy term consists of the contributions

$$\begin{aligned} \mathrm{E} = & \sum \tfrac{1}{2} K_b(b_i - b_0)^2 + \text{(bonds)} \\ & + \sum \tfrac{1}{2} K_\tau(\tau_i - \tau_0)^2 + \text{(bond angles)} \\ & + \sum K_\theta[1 + \cos(m\theta_i + \delta)] + \text{(torsion angles)} \\ & + \sum (Ar^{-12} + Br^{-6}) \text{ (nonbonded interactions)} \end{aligned} \tag{4}$$

The potential parameters we have used are those described by Levitt.[38] According to the atom types involved in forming bonds and bond angles, values of the force constants are applied which have been derived from vibration spectra of small molecules. The bond angle and torsion angle force constants are corrected for the omitted H atoms. The coefficients A and B in the Lennard-Jones potential [fourth term in Eq. (4)] for the nonbonded interactions are of a more empirical nature. Attractive terms (negative B) are used for potential hydrogen bond partners (minimum in the energy function at 2.9 Å) and atom pairs with more hydrophobic character (minimum energy between 4 and 5 Å). Apart from this only repulsive forces are applied. (The detailed values of A and B are chosen to give a 20 kcal/mol contribution at an extreme short contact which lies between 2.6 and 3.0 Å.) Atom pairs which are involved in a bond, a bond angle, or a torsion angle do not contribute to the nonbonded interaction energy.

We regard it as a major advantage of this method that the parameters employed depend on the type of atoms involved in a particular interaction. Specific flexibility or rigidity in particular regions of the molecule may be introduced by use of artificial atom types and appropriately chosen potentials. This was particularly useful in the refinement of oxyerythrocruorin.[37]

[38] M. Levitt, *J. Mol. Biol.* **82,** 393 (1974).

The type of residual [sum of independent functions in Eq. (3)] chosen in this refinement procedure indicates that in this method the number of observations (the target values of the stereochemical parameters and their associated energies) is increased rather than the number of parameters decreased. The energy parameters are not used to reparametrize the least-squares problem. Refinement is performed in Cartesian space. We term this procedure, for simplicity, a crystallographic refinement with energy restraints, especially because of the free choice of the parameter k [Eq. (3)].

B VALUES. Target values for model geometry, e.g., bond lengths and bond angles, are easily obtained from X-ray structure determinations of small molecules. An equally well-determined model for B values does not exist. Since it is not known which B values to expect, it is also difficult to decide whether B values resulting from refinement are reasonable. However, comparison of independently refined B values from different crystal forms of trypsin[32] show a high degree of correlation; differences can be explained by the different crystal contacts. If the available resolution is high enough (2 Å or better) incorrect B values are often indicated by peaks or holes in DF maps.

Usually we begin B refinement when the R value has dropped to about 0.30. The individual B values within a residue are then averaged to give one or two (main-chain, side-chain) B values per residue. Depending upon resolution available in the final stages of refinement, after correction of most of the positional errors, an attempt may be made to use individual rather than averaged B values. The resulting values are checked for improbable or unreasonable values, for example, negative B values, or high variation in B values of neighboring main-chain atoms.

Calculation of Structure Factors. This step is a standard procedure in crystallographic refinement. We almost exclusively use the FFT program of Ten Eyck[39,40] (see this volume [22]), which first constructs a model electron density map from which it then calculates the structure factors by Fourier transformation. After a minor modification, this program can also be used to calculate the contribution of the bulk solvent to the low-resolution structure factors. Addition of the solvent contribution to the model structure factors greatly improves[30] the agreement between F_{obs} and F_{calc} at low resolution.

Remarks on Scale Factors. The normal equations [Eq. (2)] are set up in a very fast way by using various approximations, especially for the calculation of diagonal elements of the matrix A. Therefore, the unre-

[39] L. F. Ten Eyck, *Acta Crystallogr., Sect. A* **A29,** 183 (1973).

[40] L. F. Ten Eyck, *Acta Crystallogr., Sect. A* **A33,** 486 (1977).

strained parameter shifts obtained from the normal equations (section on "Setting up the Normal Equations") have to be multiplied by a scale factor in order to give an optimum decrease of the *R* value. This scale factor depends on the resolution and quality of the model, and has to be determined from time to time during refinement of a structure. We do this by calculating structure factors and *R* values for selected reflections (usually the centric zones), using a series of coordinate sets which were obtained with different scale factors. Values of the scale factor of 0.1–1.5 have been obtained for different structures.

The scale factor for optimum shift size was much closer to 1.0 when an exact method for setting up the normal equations was used. For this purpose we used the program MIDAS[41] which calculates a block-diagonal normal matrix (blocks of 3×3 matrix elements for x, y, z coordinates of each atom). This method, applied to the monoclinic crystal form of citrate synthase,[30] converged in fewer cycles than the approximate method described in the section on normal equations; however, it was slower by an order of magnitude.

The scale factor for optimum shift size should not be confused with the factor k in Eq. (3) that determines the relative weights of conformational energy and crystallographic residual. There is no a priori method to determine an optimum value for k. We usually start with a value of 0.0001, which makes the two terms roughly comparable in weight. From the behavior of the refinement a decision can be made whether a change of k is desirable.

We found it helpful to relax restraints temporarily by increasing k, which led to intermediate models with high energy. Usually in such a case the conformational energy does not increase in a uniform way throughout the model; regions where the model is wrong are likely to become highly distorted, which can help in detecting errors. Apparently, relaxation of restraints also helps to overcome some of the shallow local minima of the energy function, leading to faster convergence.

The choice of k influences the *R* value that can be reached during simultaneous minimization of energy and crystallographic residual [Eq. (3)]. The quality of a model resulting from this type of refinement (and other types of restrained refinement) is therefore described by the *R* value and by the deviation of the model from standard geometry. We feel that a deviation of 0.01–0.015 Å rms for bond lengths and 2–3° rms for bond angles is the maximum allowable. However, protein structures may deviate locally from standard geometry. If such deviations are observed in a refined model, they can therefore show a true property of the molecule;

[41] A. Jack, personal communication.

alternatively they can just be a consequence of having used too high a value for k. Of course, the confidence in a deviation from standard geometry increases with the ratio of crystallographic observations to parameters.

Manual Intervention during Refinement

There are classes of errors in protein models which cannot be corrected by automated refinement. In the presence of such errors the refinement procedure converges at a local minimum R value. Such errors might be due to (1) side chains or peptide groups misoriented by more than 90°; (2) incompleteness of the model, e.g., lack of ordered solvent molecules, or presence of uninterpreted electron density; (3) incorrect course of the polypeptide chain; or (4) uncertainty of the chemical sequence: examples are glutathione peroxidase[34] and the early stages of refinement of citrate synthase.[30]

A system of methods which allow detection and correction of such errors is of supreme importance for the progress of refinement. As the most important tool we use a Vector General 3400 interactive graphics display system together with the program FRODO.[9,42]

We try to find errors mainly by looking at DF maps or at electron density maps. The DF map is a classical tool in crystallographic model refinement; however the interpretation becomes very difficult at low resolution. A low-resolution (about 3 Å) DF map often indicates an error without showing what the correct model should be. In such cases it was found helpful to inactivate pieces of the model with doubtful conformation for several cycles of automated refinement, and to look for positive density in a subsequent DF map.

Electron density maps are easier to interpret than DF maps. If phases from isomorphous replacement are not available, we use coefficients of the form

$$(2|F_{\text{obs}}| - |F_{\text{calc}}|) \exp(i\alpha_{\text{calc}}) W_{\text{sim}}$$

W_{sim} is a Sim-type weight for each reflection[14,21]:

$$W_{\text{sim}} = 2F_{\text{obs}}F_{\text{calc}}/I_{\text{u}}$$

with

$$I_{\text{u}} = \langle |F_{\text{obs}}|^2 - |F_{\text{calc}}|^2 | \rangle$$

[42] T. A. Jones, this volume [12].

(I_u = "unknown structure" contribution) calculated in 10 shells of equal thickness in reciprocal space.

Phase information from isomorphous replacement, if available, has been found extremely useful in crystallographic refinement because it is independent of model errors. Three different types of Fourier coefficients are used.

1. At very low resolution (infinity to about 7 Å) F_{calc} values are very incorrect; therefore we use

$$|F_{obs}| \exp(i\alpha_{MIR})m$$

where m is the figure of merit.

2. At medium resolution (7 to 3 Å) both MIR phases and F_{calc} are reasonably accurate, so we use

$$(2|F_{obs}| - |F_{calc}|) \exp(i\alpha_{comb}) W_{comb}$$

where α_{comb} are phases obtained by combination of α_{calc} and α_{MIR}[14]; and W_{comb} are weights resulting from the combination procedure.

3. At high resolution the MIR phases are often inaccurate; in this region we use

$$(2|F_{obs}| - |F_{calc}|) \exp(i\alpha_{calc}) W_{sim}$$

Phase information from isomorphous replacement was used during refinement of Fc fragment,[31] Fab fragment of IgG KOL,[22] glutathione peroxidase,[33,34] kallikrein,[23] and citrate synthase.[30] All these cases are characterized by very bad starting models and/or lack of amino acid sequence information. In retrospect it appears doubtful whether successful refinement of these structures would have been possible without using the technique of phase combination.

A check of the plausibility of the model helps to find errors. Main-chain conformational angles outside the allowed regions of the Ramachandran diagram indicate the possibility of a wrong conformation. Lists of preferred values of side-chain torsion angles[43] can be used to find side chains with improbable conformations. Distorted bond lengths and bond angles after refinement also indicate probable errors in the model.

It has to be emphasized that conformations which violate the rules are improbable, but not impossible. An example is residue His 274 in the active site of citrate synthase with $\varphi/\psi = -123°/-126°$, well outside the allowed region.[30]

[43] J. Janin, S. Wodak, M. Levitt, and B. Maigret, *J. Mol. Biol.* **125,** 357 (1978).

Example

The course of Jack–Levitt refinement of human Fc fragment[31] may serve as an example for low-resolution refinement. Extensive use of constrained real space refinement[1,6] had led to a model with an R value of 0.312. Then Jack–Levitt refinement was begun with an initial value of $k = 0.0001$; this value was later reduced to 0.00003. Table II shows the stages of further refinement at which maps were calculated for checking and manual revisions of the model. Between steps 9 and 11, the energy was lowered at constant R. In the final steps 11 to 15, further corrections of the model reduced both R and the conformational energy. In steps 9 to 15, whenever large manual changes of the model were done, subsequent automated refinement was started at 7 to 4 Å resolution to increase the radius of convergence.[7] In the following cycles, the upper resolution limit was gradually increased to 2.9 Å. The refinement converged at an R value of 0.22 for reflections between 7 and 2.9 Å resolution.

Average residue temperature factors were updated before inspection of each new map. When individual B values of atoms within a residue were averaged, the minimum allowed mean value was set to 6 Å^2. B

TABLE II
COURSE OF FC FRAGMENT REFINEMENT[a]

Stage	Event	R value (7 to 2.9 Å)	Number of active atoms[b]	Energies (kcal/mol) Total	Bond lengths	Bond angles	Torsion angles	Nonbonded interactions
1–8	Real space refinement	0.312	—	—	—	—	—	—
9	$2F_o - F_c$ map; Sim weights	0.265	2805	569	357	1250	985	−2025
10	$2F_o - F_c$ map; combined phases	0.260	2901	−687	159	708	696	−2251
11	As above	0.260	2892	−1102	98	559	693	−2451
12	As above	0.243	2976	−1536	95	520	588	−2740
13	As above	0.231	3048	−1867	59	400	540	−2866
14	As above	0.226	3110	−1967	76	384	501	−2928
15	End	0.220	3182	−2131	56	327	469	−2984

[a] Energy refinement, combined with diagonal-matrix least-squares refinement.
[b] Active atoms are those contributing to calculated structure factors.

refinement was done in spite of the low resolution because a high variation of crystalline order between different parts of the molecule was apparent in early stages of structure determination. The average temperature factor in both C_{H3} domains is 19.5 Å^2; in contrast, C_{H2} of chain 1 has an average B of 25 Å^2; for C_{H2} of chain 2, the average B is 30 Å^2. These values demonstrate that in crystals of Fc fragment, the C_{H2} domains are less well ordered than the C_{H3} domains, and that C_{H2} of chain 2 is more disordered than C_{H2} of chain 1.

$2|F_{obs}| - |F_{calc}|$ maps calculated with combined model and MIR phases provided the clearest guide to necessary changes in the model. In general, corrections to the model made "by hand" were influenced by the inspected map, by stereochemical considerations (plausible conformation, contacts to neighbors), and by comparison of the two crystallographically independent polypeptide chains of the Fc fragment molecule. On several occasions, polypeptide segments of up to five residues were deliberately built with different conformation in both chains to see which one would fit better to the electron density after the next cycles of automated refinement. The MIR map was used throughout the refinement of Fc fragment as an additional guide when the model was changed. After each round of manual corrections to the model, the R value rose by 0.01 to 0.03; the first one or two cycles of automated refinement were needed to bring R back to the previous level.

Observed structure factor amplitudes of 9332 reflections between 7 and 2.9 Å resolution were used for refinement. During Jack–Levitt refinement 10596 coordinates of 3532 atoms (of which only 3182 contributed to calculated structure factors during the final cycles) were restrained by 3646 bond lengths, 4998 bond angles, and 6072 torsion angle potentials. 1598 torsion angles had energy barriers of 20 kcal/mol or more. About 40,000 atom pairs with distances less than the sum of the van der Waals radii plus 2.0 Å contributed to the nonbonded interaction energy; pairs of atoms which occur in a bond, a bond angle, or a torsion angle were excluded from the list of nonbonded interactions.

Bond lengths and bond angles of the refined model are very close to ideal values; the rms deviations are 0.007 Å, and 1.7°, respectively. The rms deviation of ω and similar angles from minimum energy values is 5.9°. For the torsion angles which keep aromatic rings planar, the rms deviation is 1.8°.

The rms difference between C_α coordinates of chain 1 of the starting and refined models is 1.18 Å for 158 of the 206 residues. This value is very large compared to the rms C_α shift of 0.45 Å during refinement of the pancreatic trypsin inhibitor,[6] where the initial model was based on an MIR map at 1.9 Å resolution. In each polypeptide chain 38 residues and the carbohydrate moiety had to be rebuilt during refinement; these are not included in the above differences.

An estimate of the accuracy of the refined Fc fragment coordinates can be obtained from comparison of chain 1 and chain 2. Main-chain atoms (340) from 85 residues of C_{H3} chain 1 with temperature factors below 35 Å^2 were rotated onto the corresponding atoms of chain 2. After least-squares refinement of the transformation parameters, the rms distance of equivalent atom positions was 0.29 Å. This number is a rough estimate of the accuracy of coordinates which could be achieved in regions of high crystalline order. In less well-ordered regions the coordinate errors can be considerably larger. Such regions exist mainly in the C_{H2} domains. At 2.9 Å resolution carbonyl oxygens of the main chain cannot be located unambiguously in the electron density map to determine the peptide orientations. This problem is less serious in regions with regular secondary structure. Outside these regions, especially in zones with high temperature factors, peptide orientations may well be wrong in the refined model.

Summary of Advantages of Jack–Levitt Refinement

The following main advantages of the Jack–Levitt method[7] became apparent during the refinement of the above-mentioned structures: (1) refinement at low resolution is possible, (2) geometric restraints can be relaxed temporarily, (3) treatment of branched chains is easy, (4) the method requires about 30% less computing time than Diamond's[1] real space refinement, (5) distorted geometry can be repaired, and (6) refinement of temperature factors at 2.9 Å resolution is possible.

Acknowledgment

The authors wish to express their thanks to Prof. R. Huber for enabling and supporting this work and for many fruitful discussions and continuous encouragement.

[22] Fast Fourier Transform Calculation of Electron Density Maps

By LYNN F. TEN EYCK

Introduction

This chapter is concerned with the calculation of electron density maps and difference maps from phased structure factors. This problem is intimately bound up with the symmetry of crystals and diffraction patterns. You will find a brief introduction to these subjects in this chapter, but it is no replacement for a textbook on crystallography such as Stout and Jensen.[1]

Diffraction

When a plane wave is scattered by an object, the scattered radiation is described by the formula

$$\mathbf{F}(\mathbf{h}) = \int \rho(\mathbf{x}) \exp(2\pi i \mathbf{h} \cdot \mathbf{x})\, dV \tag{1}$$

where **F** is a complex number which represents the amplitude and phase of the scattered radiation in a direction determined by the vector **h**, $\rho(\mathbf{x})$ is the scattering function at a position **x** in the object, and the integral is taken over the volume of the object. For X-ray diffraction, $\rho(\mathbf{x})$ is the electron density at position **x** (excluding anomalous scattering effects). For neutron diffraction, $\rho(\mathbf{x})$ is the scattering cross section of any atomic nuclei at position **x**. Thus *the scattered radiation is described by the Fourier transform of the object as seen by the incident radiation*. We also have, by the inverse Fourier transform,

$$\rho(\mathbf{x}) = \int \mathbf{F}(\mathbf{h}) \exp(-2\pi i \mathbf{h} \cdot \mathbf{x})\, dV^* \tag{2}$$

where dV^* is a volume element in the space spanned by **h**.

We are interested in the important special case in which the scattering object is a crystal. The fundamental property of a crystal is that $\rho(\mathbf{x})$ is periodic in all three spatial dimensions. It is well known that the Fourier transform of a periodic function is zero except when **h** is an integer multiple of the periodicity. Thus the structure factors are zero except on a three-dimensional lattice.

[1] G. Stout and L. Jensen, "X-ray Structure Determination," Macmillan, New York, 1968.

The natural coordinate system for a crystal is

$$\mathbf{x} = x\mathbf{a} + y\mathbf{b} + z\mathbf{c} \tag{3}$$

where **a**, **b**, and **c** are the basis vectors for the unit cell of the crystal. The basis vectors are *not* necessarily orthogonal, but they do span three-dimensional space. Evaluation of Eqs. (1) and (2) is greatly simplified by choosing the correct coordinate system for **h**. If

$$\mathbf{h} = h\mathbf{a}^* + k\mathbf{b}^* + l\mathbf{c}^* \tag{4}$$

then if $\mathbf{a}^*$, $\mathbf{b}^*$, and $\mathbf{c}^*$ are chosen to obey the Laue relations

$$\begin{array}{lll} \mathbf{a}\cdot\mathbf{a}^* = 1, & \mathbf{a}\cdot\mathbf{b}^* = 0, & \mathbf{a}\cdot\mathbf{c}^* = 0 \\ \mathbf{b}\cdot\mathbf{a}^* = 0, & \mathbf{b}\cdot\mathbf{b}^* = 1, & \mathbf{b}\cdot\mathbf{c}^* = 0 \\ \mathbf{c}\cdot\mathbf{a}^* = 0, & \mathbf{c}\cdot\mathbf{b}^* = 0, & \mathbf{c}\cdot\mathbf{c}^* = 1 \end{array} \tag{5}$$

the vector inner products in the integrals (1) and (2) simplify to

$$\mathbf{h}\cdot\mathbf{x} = hx + ky + lz \tag{6}$$

In this coordinate system $\mathbf{F}(\mathbf{h})$ is nonzero when h, k, and l have integer values. The vectors $\mathbf{a}^*$, $\mathbf{b}^*$, and $\mathbf{c}^*$ are the basis vectors for the structure factor lattice. The space in which the structure factors are defined is often called *reciprocal space,* because the lengths of the basis vectors are inversely proportional to the lengths of the basis vectors of the unit cell.

With these definitions of coordinate systems we can normalize Eqs. (1) and (2) to reflect the contents of one unit cell rather than an arbitrarily sized crystal. The normalized versions of these equations are

$$\mathbf{F}(\mathbf{h}) = \iiint \rho(\mathbf{x}) \exp[2\pi i(hx + ky + lz)]\, dx\, dy\, dz \tag{7}$$

and

$$\rho(\mathbf{x}) = \frac{1}{V}\iiint \mathbf{F}(\mathbf{h}) \exp[-2\pi i(hx + ky + lz)]\, dh\, dk\, dl \tag{8}$$

$$= \frac{1}{V}\sum_h\sum_k\sum_l \mathbf{F}(\mathbf{h}) e^{-2\pi i(hx+ky+lz)}$$

The conversion from an integral to a sum in Eq. (8) follows from the discrete nature of $\mathbf{F}(\mathbf{h})$. This chapter is concerned with the evaluation of Eq. (8).

Symmetry

Crystals commonly contain more than one copy of a molecule within the unit cell. These copies are usually related to one another by symmetry operators, which apply throughout the crystal. A symmetry operator can be represented by a matrix **R** and a vector **t** such that if

$$\mathbf{x}' = \mathbf{R}\mathbf{x} + \mathbf{t} \tag{9}$$

then

$$\rho(\mathbf{x}') = \rho(\mathbf{x}) \tag{10}$$

for *all* **x** in the crystal. The requirement that the symmetry operators apply throughout the crystal, combined with the requirement that the crystal is periodic in three dimensions, leads directly to the restriction that only 232 different combinations of symmetry operators can exist in crystals. These different combinations of symmetry operators are known as *space groups*.

Equation (7) shows that all of the symmetry elements in a crystal have corresponding symmetries in reciprocal space. However, reciprocal space is not periodic. The symmetry of reciprocal space is based on *point groups* which describe symmetries about a single point. There are no translational symmetries in reciprocal space. Since each structure factor has both an amplitude and a phase, any symmetry operators must affect both. I will now show that the symmetry-related structure factors have the same amplitudes, but phases that shift in predictable ways. This kind of symmetry is known as a *point group with color symmetry* and is discussed elegantly by Bienenstock and Ewald.[2]

Specifically, if (**R**, **t**) is a symmetry operator in a particular space group, and we let

$$\mathbf{h}' = \mathbf{h}^{\mathrm{T}}\mathbf{R} \tag{11}$$

where $\mathbf{h}^{\mathrm{T}}$ is the transpose of a reciprocal space vector **h**, we may write

$$\mathbf{F}(\mathbf{h}') = \iiint \rho(\mathbf{x}) \exp(2\pi i\mathbf{h}' \cdot \mathbf{x})\, dx\, dy\, dz \tag{12}$$

But

$$\mathbf{h}' \cdot \mathbf{x} = \mathbf{h} \cdot (\mathbf{R}\mathbf{x}) = \mathbf{h} \cdot \mathbf{x}' - \mathbf{h} \cdot \mathbf{t} \tag{13}$$

and, because (**R**, **t**) is a symmetry operator,

$$\rho(\mathbf{x}') = \rho(\mathbf{x})$$

and

$$dx\, dy\, dz = dx'\, dy'\, dz'$$

which gives

$$\mathbf{F}(\mathbf{h}^{\mathrm{T}}\mathbf{R}) = \exp(-2\pi i\mathbf{h} \cdot \mathbf{t})\mathbf{F}(\mathbf{h}) \tag{14}$$

[2] A. Bienenstock and P. P. Ewald, *Acta Crystallogr.* **15,** 1253 (1962).

by direct substitution. Equation (14) will hold whenever $(\mathbf{R}, \mathbf{t})$ is a symmetry operator throughout the crystal.

In addition to the symmetry of Eq. (14), reciprocal space has *Hermitian symmetry*. Because the electron density function is real rather than complex, direct substitution into Eq. (7) shows that

$$\mathbf{F}(\bar{\mathbf{h}}) = \mathbf{F}^*(\mathbf{h}) \tag{15}$$

where $\bar{\mathbf{h}} = -\mathbf{h}$ and $\mathbf{F}^*$ is the complex conjugate of $\mathbf{F}$.

As an example we will consider two closely related space groups, P2 and $P2_1$. The equivalent positions for P2 are (x, y, z) and $(\bar{x}, y, \bar{z})$. Thus there is one symmetry operator, given by

$$\mathbf{R} = \begin{pmatrix} -1 & 0 & 0 \\ 0 & 1 & 0 \\ 0 & 0 & -1 \end{pmatrix}, \qquad \mathbf{t} = \begin{pmatrix} 0 \\ 0 \\ 0 \end{pmatrix} \tag{16}$$

Then

$$\mathbf{F}(\bar{h}, k, \bar{l}) = \mathbf{F}(h, k, l) \tag{17}$$

and (by Hermitian symmetry)

$$\mathbf{F}(h, \bar{k}, l) = \mathbf{F}^*(h, k, l) \tag{18}$$

For $P2_1$ the equivalent positions are (x, y, z) and $(\bar{x}, y + \frac{1}{2}, \bar{z})$. The symmetry operator is

$$\mathbf{R} = \begin{pmatrix} -1 & 0 & 0 \\ 0 & 1 & 0 \\ 0 & 0 & -1 \end{pmatrix}, \qquad \mathbf{t} = \begin{pmatrix} 0 \\ \frac{1}{2} \\ 0 \end{pmatrix} \tag{19}$$

and we have

$$\mathbf{F}(\bar{h}, k, \bar{l}) = e^{-i\pi k}\mathbf{F}(h, k, l) \tag{20}$$

Since $e^{-i\pi k} = (-1)^k$, we have

$$\begin{aligned} \mathbf{F}(h, \bar{k}, l) &= \mathbf{F}^*(h, k, l), \qquad k \text{ even} \\ \mathbf{F}(h, \bar{k}, l) &= -\mathbf{F}^*(h, k, l), \qquad k \text{ odd} \end{aligned} \tag{21}$$

Cost of Fourier Transform Calculations

This chapter was written because the cost of directly evaluating the electron density function can be very large. Electron density maps are typically calculated on a grid three times finer than the resolution of the data. The number of points in such a grid is $27V/d^3$, where V_0 is the volume of the unit cell and d is the resolution in angstrom units. The number of structure factors at a resolution d is the volume of a hemisphere in reciprocal space of radius $1/d$, divided by the volume of the reciprocal lattice cell (which in turn is the reciprocal of the volume of the unit cell), or $2\pi V/3d^3$ structure factors. The cost of evaluating the sum in Eq. (8) is proportional to the number of terms in the sum multiplied by the number of grid points:

$$\text{cost} \propto 18\pi(V^2/d^6)$$

The volume of a unit cell depends to a certain extent on the amount of solvent in the crystal and the density of the molecule. For a typical protein crystal the volume of the unit cell is roughly 2.4 Å^3 per dalton of mass. The cost of evaluating the electron density function is therefore proportional to the *square* of the molecular weight and inversely proportional to the *sixth* power of the resolution. For a protein of molecular weight 30,000 at 3 Å resolution the cost will be proportional to 4×10^8; at 2 Å resolution the cost increases by a factor of 1.5^6, which is more than an order of magnitude. Efficient calculation of electron density maps for macromolecules obviously requires some thought.

The approach given in most crystallography texts is to note that Eq. (8) can be expanded as a cosine series, and that the dependencies on each index can be factored out, resulting in much less computation per grid point. A much simpler approach is to use the fast Fourier Transform (FFT) algorithm. The FFT is a method for replacing a set of N Fourier coefficients with their Fourier transform at a cost proportional to $N \log_2 N$. The cost of computing an electron density map by these methods depends almost exclusively on the grid size, that is,

$$\text{cost}_{\text{FFT}} \propto (27V/d^3) \log_2(27V/d^3)$$

The FFT cost for a 30,000-dalton protein at 3 Å resolution is proportional to 1.5×10^6 instead of 4×10^8; the FFT cost of 2 Å resolution is proportional to 5.5×10^6 instead of 5×10^9. The cost of the FFT method grows much more slowly than the cost of the direct summation methods both with increasing molecular weight and increasing resolution. The savings possible with the FFT method are so large that I will not discuss direct summation methods any further.

Algebra of the Fast Fourier Transform

The FFT method for replacing a sequence of Fourier coefficients with its Fourier transform depends upon the observation that if the number of points N in the sequence can be factored, the Fourier summation can be factored into a series of shorter Fourier sums. Factoring also reveals some interesting and useful properties of Fourier sums. To describe the algebra of the fast Fourier transform, we introduce the following notation.

1. $R(t)$, $S(t)$, $X(t)$, $Y(t)$, and $Z(t)$, $t = 0, \ldots, N - 1$ are sequences of (possibly) complex numbers. The sequences are assumed to be periodic modulo N, that is $R(N) = R(0)$, etc.
2. $X^*(t)$ is the complex conjugate of $X(t)$, and similarly for any of the other sequences.
3. $\mathrm{Re}(X)$ is the real part of the complex number X.
4. $\mathrm{Im}(X)$ is the imaginary part of the complex number X.
5. Let the function $e(x)$ be defined as

$$e(x) = e^{-2\pi ix} = \cos 2\pi x - i \sin 2\pi x$$

Note that the function $e(x)$ has the following useful properties:

$$\begin{aligned} e(x) &= 1, \qquad \text{if } x \text{ is any integer} \\ e(ab) &= e(a)e(b) \\ e(1 - x) &= e(-x) \end{aligned}$$

6. $\mathbf{R}(\mathbf{t})$, $\mathbf{S}(\mathbf{t})$, $\mathbf{X}(\mathbf{t})$, $\mathbf{Y}(\mathbf{t})$, and $\mathbf{Z}(\mathbf{t})$, $\mathbf{t} = 0, \ldots, N - 1$, are the finite discrete Fourier transforms of R, S, X, Y, and Z, defined by

$$\mathbf{X}(\mathbf{t}) = \sum_{t=0}^{N-1} X(t)e(t\mathbf{t}/N)$$

Note that if $t = N - h$,

$$e(t\mathbf{t}/N) = e(-h\mathbf{t}/N)e(\mathbf{t}) = e(-h\mathbf{t}/N)$$

so $X(N - h)$ is the same as $X(-h)$ as far as evaluating the sum is concerned.

Factoring the Finite Discrete Fourier Transform

Now consider the finite discrete Fourier transform for the case in which N has factors, i.e., $N = AB$. Then the expressions $aB + b$ and $a + bA$ will each generate all of the integers in the range $0 \leq t < N$ once and only once when $a = 0, \ldots, A - 1$ and $b = 0, \ldots, B - 1$. If we let

$$\begin{aligned} t &= aB + b \\ \mathbf{t} &= \mathbf{a} + \mathbf{b}A \end{aligned}$$

the finite discrete Fourier transform of $X(t)$ can be written as

$$\mathbf{X}(\mathbf{a} + \mathbf{b}A) = \sum_{b=0}^{B-1} \left[\sum_{a=0}^{A-1} X(aB + b)e[(aB + b)(\mathbf{a} + \mathbf{b}A)/AB] \right] \tag{22}$$

Expanding and simplifying the complex exponential gives

$$e[(aB + b)(\mathbf{a} + \mathbf{b}A)/AB] = e(a\mathbf{a}/A)e(a\mathbf{b})e(b\mathbf{a}/AB)e(b\mathbf{b}/B) \quad (23)$$
$$= e(a\mathbf{a}/A)e(b\mathbf{a}/AB)e(b\mathbf{b}/B)$$

Substituting back and grouping the terms in the sums yields

$$\mathbf{X}(\mathbf{a} + \mathbf{b}A) = \sum_{b=0}^{B-1} e(b\mathbf{b}/B) \left[e(b\mathbf{a}/N) \sum_{a=0}^{A-1} X(aB + b)e(a\mathbf{a}/A) \right] \quad (24)$$

The inner sum is a Fourier transform of length A, which costs A^2 and must be done B times. The outer sum is a Fourier transform of length B, which costs B^2 and must be done A times. Thus the cost of evaluating the finite discrete Fourier transform is

$$ABB + BAA = N(A + B) \quad (25)$$

or N times the sum of factors of N. (Obviously this reasoning can be applied recursively to the A and B point transforms when A and/or B have factors.) When N is a power of two, the sum of factors of N is $2 \log_2 N$. This is the basis for the statement that the cost of the FFT algorithm is proportional to $N \log N$. *The actual cost will depend strongly on the factoring of N*. For example, an 84-point transform will have a cost proportional to $84(2 + 2 + 3 + 7) = 1176$, while an 85-point transform will have a cost proportional to $85(5 + 17) = 1870$.

Factors of Two

There are two special forms which arise when N is even. If we let $A = 2$, we can expand Eq. (24) to get

$$\mathbf{X}(2\mathbf{b}) = \sum_{b=0}^{B-1} e(b\mathbf{b}/B)[X(b) + X(b + B)] \quad (26)$$

$$\mathbf{X}(2\mathbf{b} + 1) = \sum_{b=0}^{B-1} e(b\mathbf{b}/B)\{e(b/N)[x(b) - x(b + B)]\} \quad (27)$$

These equations show how the even and odd terms of the Fourier transform are derived from the Fourier transforms of the sums and differences of the first and last halves of the Fourier coefficients. It is not a coincidence that there is a corresponding relationship between the first and last halves of the Fourier transform and the even and odd Fourier coefficients. If we let $B = 2$ instead of $A = 2$, we get

$$\mathbf{X}(\mathbf{a}) = \sum_{a=0}^{A-1} X(2a)e(a\mathbf{a}/A) + e(\mathbf{a}/N) \sum_{a=0}^{A-1} X(2a + 1)e(a\mathbf{a}/A) \quad (28)$$

$$\mathbf{X}(\mathbf{a} + N/2) = \sum_{a=0}^{A-1} X(2a)e(a\mathbf{a}/A) - e(\mathbf{a}/N) \sum_{a=0}^{A-1} X(2a + 1)e(a\mathbf{a}/A) \quad (29)$$

The importance of these special cases is that they show how specific portions of the Fourier transform can be calculated economically. The most common example is the calculation of a real Fourier transform from complex data, as is done in the last step of an electron density map calculation. It is easy to show that if the Fourier transform is real, the Fourier coefficients are Hermitian symmetric. Specifically, if $\mathbf{X}(\mathbf{t})$ is real, then $X(t) = X^*(N - t)$. All of the unique information in the Fourier coefficients can be found in the first half of the sequence. Furthermore, since half of a complex sequence of length N occupies exactly as much space as a real sequence of length N, the complete Fourier transform can be written over half of the Fourier coefficients. Thus we never have to deal with the second half of the Fourier coefficients. We proceed as follows:

1. Let

$$R(t) = X(t) + X(t + N/2), \; t = 0, \ldots, N/2 - 1$$

Then

$$R(t) = X(t) + X^*(N/2 - t) \quad (30)$$
$$R(N/2 - t) = X^*(t) + X(N/2 - t) \quad (31)$$

The Fourier transform of R will give the even terms of the Fourier transform of X:

$$\mathbf{X}(2\mathbf{t}) = \mathbf{R}(\mathbf{t}) \quad (32)$$

2. Let

$$S(t) = e(t/N)[X(t) - X(t + N/2)], \; t = 0, \ldots, N/2 - 1$$

Then

$$S(t) = e(t/N)[X(t) - X^*(N/2 - t)] \quad (33)$$
$$S(N/2 - t) = e(-t/N)[X^*(t) - X(N/2 - t)] \quad (34)$$

The Fourier transform of **S** will give the odd terms of the Fourier transform of X:

$$\mathbf{X}(2\mathbf{t} + 1) = \mathbf{S}(\mathbf{t}) \quad (35)$$

3. **X** is real. Therefore we can combine the calculations of **R** and **S**. Let $Z(t) = R(t) + iS(t)$. Then

$$\mathbf{Z}(\mathbf{t}) = \mathbf{R}(\mathbf{t}) + i\mathbf{S}(\mathbf{t}) \quad (36)$$

or

$$\mathbf{X}(2\mathbf{t}) = \mathrm{Re}[\mathbf{Z}(\mathbf{t})] \tag{37}$$
$$\mathbf{X}(2\mathbf{t} + 1) = \mathrm{Im}[\mathbf{Z}(\mathbf{t})] \tag{38}$$

This trick, and a number of others for calculating crystallographic Fourier transforms, is described in Ten Eyck.[3] If a computer system with sufficient memory (real or virtual) to hold an entire unit cell is available, this is the only special symmetry that need be included in the calculation. For large structures, or problems in which maps are to be calculated many times, the cost of writing a program specifically for that space group may be justified. Further details on space-group-specific FFT programs are given in Ten Eyck.[3]

Multidimensional FFTs

So far all of the details on FFT algebra have been presented for one-dimensional problems. Multidimensional Fourier transforms can be calculated as sequences of one-dimensional Fourier transforms. For general crystallographic use we have several alternatives for designing our FFT routines.

1. Use a one-dimensional FFT applied successively along lines in reciprocal space.
2. Use a multidimensional FFT which transforms an entire three-dimensional array.
3. Use a multidimensional FFT which will transform *one* dimension of a three-dimensional array on each call.

Each of these choices has its own particular advantages and disadvantages. The first alternative, successive calls to a one-dimensional FFT, is the most flexible, and allows the maximum economy in not transforming lines which are known to be zero. It has the disadvantage that all of the program overhead of the FFT subroutine—determining factors of N, calculating sines and cosines, determining the permutation cycles to put the transformed data into the correct order, and the general overhead associated with iteration and subroutine calls—must be paid in full for each lattice line.

The second alternative, use of a general multidimensional FFT subroutine, is the simplest to use but the least efficient and the least flexible. It can only be used if the entire unit cell will fit into the computer memory, and cannot take advantage of known zeros in the data.

[3] L. F. Ten Eyck, *Acta Crystallogr., Sect. A* **A29,** 183 (1973).

The last alternative given is quite flexible if properly set up, and can be very efficient. The basic idea is that precisely the same operations must be performed on *all* points lying on a plane normal to the direction being transformed as must be performed on *any* of them. Thus, where the one-dimensional FFT might have a section which does the following,

```
R = X(T+N OVER 2) + X(T)
S = X(T+N OVER 2) − X(T)
X(T) = R
X(T+N OVER 2) = S
```

the multidimensional FFT would have something like

```
DO (FOR EACH LINE)
  R = X(T+N OVER 2) + X(T)
  S = X(T+N OVER 2) − X(T)
  X(T) = R
  X(T+N OVER 2) = S
END DO
```

This greatly reduces the program overhead, provided the multidimensional indexing is done efficiently. As a rough estimate, this method is more efficient than calculating successive one-dimensional Fourier transforms if the overhead in calculating a 100-point FFT is greater than about 5% of the execution time of the subroutine.

Organization of an FFT Electron Density Map Calculation

The most economical calculation of electron density maps is done by using only the unique set of structure factors in the summation, and evaluating only those grid points in the crystallographic asymmetric unit. This is possible, using FFT techniques, for most space groups. The resulting programs are rather complicated mathematically and can be difficult to write. Furthermore the molecules in the unit cell usually do not lie neatly within the boundaries of the asymmetric unit. An asymmetric unit is usually made up of pieces of several molecules. A crystallographic Fourier synthesis program should be able to calculate a volume which contains a whole molecule, not just an asymmetric unit. There are many applications for which the time and effort required to write a space-group-specific Fourier program are well spent—very large problems, such as virus structures, for example, or during structure refinements in which Fourier transforms are part of the refinement process and may be calculated many times. Otherwise the simplest form of the method should be used, which is to expand the structure factors by symmetry and calculate the map as though the crystal had no symmetry. This has the advantage that it is perfectly general, but is somewhat more costly to run. In the simplest case the map for an entire unit cell will fit in the computer memory.

Problems Which Fit Entirely within the Computer Memory

This case presents no particular difficulties. The program has to read through the input data once only. No scratch files are required. The procedure is as follows.

1. Read the control information: the unit cell parameters, the resolution desired, the output grid desired, the bounds of the output volume, and the list of symmetry operators. It is perfectly feasible to derive the symmetry operators from a list of equivalent positions as given in the *International Tables for Crystallography*.[4]
2. Read the structure factors, screen them on resolution (and possibly on other criteria), and expand them by symmetry according to Eq. (14). The expanded structure factors must fill a hemisphere of reciprocal space, for example $h \geq 0$, all k, and all l. The symmetry expansion will often produce positions in the wrong hemisphere, which must be transformed to the correct hemisphere by Hermitian symmetry. In the boundary plane ($0kl$ in this case) both the $0kl$ and $0\overline{kl}$ structure factors must be present, so the Hermitian symmetry must be applied in this case also. As the structure factors are expanded they are placed in the computer memory according to their indices. (Negative indices are converted to positive indices by adding the number of grid points in that direction to the index.)
3. Calculate the Fourier transform in the l direction. Those values of (h, k) for which there are no structure factors do not have to be included in the calculation, which can lead to a substantial saving.
4. Calculate the Fourier transform in the k direction. This transform should be done for those z planes which will appear in the output volume, but need not be done for any others. Some economy is possible if only those values of h for which there are structure factors are transformed.
5. Calculate the Fourier transform in the h direction. Each separate line in the h direction is Hermitian symmetric because the final result must be real. Therefore all of the information in the transform can be extracted from the structure factors for $h \geq 0$. Since each structure factor is a complex number, which requires twice as much memory as a real number, there is sufficient room to replace the $h \geq 0$ structure factors with the

[4] "International Tables for Crystallography," Vol. A. D. Reidel Publishing Company, Boston, Massachusetts, 1983.

final Fourier transform. This transform only needs to be calculated for those values of (y, z) which will appear in the output volume.

6. Write the electron density map from computer memory. If the output volume spans unit cell boundaries, the periodicity of the crystal can be used to locate the correct values in the unit cell.

A program written in this fashion is rather simple. It requires an input routine, symmetry expansion, a basic complex FFT subroutine, a Hermitian symmetric FFT subroutine, and an output subroutine. It is limited by the available computer memory, which may or may not be a problem. Many computers today have virtual memory systems, which permit programs to be very large. (Very large programs can suffer severe performance penalties in such systems.) Like all FFT programs, it is limited to calculations in the crystallographic coordinate system.

Problems Larger than Available Computer Memory

Modification of the method given above for computers in which the entire unit cell will not fit in memory is straightforward. The problem is divided into three phases. The first phase is the same as the input phase described above. The second phase is to sort the expanded structure factors on h. The third phase is a buffered calculation in which the results of the transform on l are stored on a scratch file, then read back piecemeal for calculation of the transforms on k and h.

1. The input phase is the same as that described above, except that the structure factors are written to a file after expansion rather than being placed in the computer memory.

2. The Fourier transform program can only calculate transforms along lines for which it has complete information. By that I mean that the structure factors with both positive and negative indices are required. Since there may well be more structure factors than will fit in the available computer memory, they must be taken in groups. This is done most easily if the structure factors are sorted on h. Most computer systems have a utility sorting program which is perfectly adequate for this purpose. There are also numerous references to good sorting methods in the computer science literature, for example Knuth.[5]

3. The first pass of the Fourier transform program reads the sorted structure factor file and writes intermediate results to a temporary file. The structure factors are read for a block of h values, all k, and all l. The intermediate transform on l is calculated, which gives partial sums for the block of h, all k, and all z. The results are written to the scratch file as

[5] D. Knuth, "The Art of Computer Programming," Vol. 3. Addison-Wesley, Reading, Massachusetts, 1973.

records which contain the intermediate results for a particular z, all k, and the current block of h.

4. The second pass of the program reads intermediate results back from the scratch file and calculates the Fourier transforms in the k and h directions. The records are not read in the same order in which they were written, but instead are collected for all h and k in blocks of z. Writing and rereading the intermediate results is a way of transposing a three-dimensional matrix larger than the memory of the computer. At the end of the second pass, the current block of z sections can be written to the output file.

Compared to the previous program, this program is more complex in only two areas. It requires the additional sorting step, and it requires management of the scratch file. The memory is effectively remapped between passes of the program. If the dimensions of the unit cell are N_x, N_y, and N_z, and the maximum indices for the structure factors are h_{max}, k_{max}, and l_{max}, the effective dimensioning in the first pass is N_z by $k_{max} + 1$ by however many h values can be accommodated. In the second pass the effective dimensions are N_x by N_y by however many z values can be accommodated.

Problems Much Larger than Available Computer Memory

It is sometimes necessary to run large calculations on rather small computers. Such computers often have the attractions of very low price and high availability. If the memory is too small to hold even one plane of the electron density map, the calculation must be even further subdivided. The Fourier transform in the k direction must be treated separately from that in the h direction.

1. The structure factors should be sorted on both h and k. With this modification, the expansion and sorting steps remain the same as given above. The sorting step, however, may be awkward on a small computer. Minicomputer operating systems often do not have decent sorting programs. The unit cell parameters and the resolution of the data are sufficient information to determine the limits on the h, k, and l values. This information greatly simplifies design of an efficient sorting program.

2. The first pass of the calculation should write the scratch file in blocks which contain intermediate results for a single z, all k, and a single h—in other words, one block for each line to be transformed in the k direction.

3. The second pass of the program is divided into two subpasses, both of which are executed for each z value in the output volume. Since our

basic assumption is that the available memory is less than $N_x \times N_y$, the output planes must be calculated piecemeal. If the memory is M words, the first subpass maps it as an array N_y by $M_h = M/N_y$. The second subpass must map the memory as an array N_x by $M_y = M/N_x$.

4. The first subpass of the second pass collects, for a given z value, all k and $M_h h$ values, calculates the Fourier transform on k, and writes the results to a temporary file in blocks containing $M_h \times M_y$ intermediate values. It repeats this process until all h values for this z section have been transformed.

5. The second subpass of the second pass reads the temporary file to collect the intermediate values corresponding to all h and the first set of $M_y y$ values. It calculates the final Fourier transform in the h direction and writes this portion of the electron density map to the output file. This process is repeated until all of the blocks of y have been transformed. At that point the current z section is complete and the first subpass can start on the next z section.

Adding the second level of buffering is not particularly difficult or complex. When the calculation is organized in this fashion the amount of disk storage needed for the second temporary file is less than that required for one section of the density map. Disk space is likely to be the limiting resource in a calculation of this type on a small computer system.

Summary

This chapter has described the mathematical basis of the fast Fourier transform as applied to the calculation of crystallographic Fourier syntheses. The relationship between real space and reciprocal space symmetry operators has been described. Finally, program organizations have been presented for performing general crystallographic Fourier transforms on computer systems ranging from the very largest systems down to minicomputers. Programs are available from the author, written in FORTRAN IV and in Ratfor, which are suitable for building blocks in these program designs.

Acknowledgments

This work was supported in part by NIH Biotechnology Research Resource Grant RR00898 to the Department of Computer Science, University of North Carolina, Chapel Hill, North Carolina 27514.

Section IV

Presentation and Analysis of Structure

[23] Describing Patterns of Protein Tertiary Structure

By JANE S. RICHARDSON

The patterns into which proteins fold are sufficiently asymmetric and complex that it can be difficult to provide brief and comprehensible descriptions of their overall organization. For the first 10 or 15 protein structures solved, each one (except the hemoglobin–myoglobin pair) appeared completely individual and unrelated. In the early 1970s the first relationships among structures were recognized: the "nucleotide-binding domains" of dehydrogenases, flavodoxin, and adenyl kinase,[1] and the "immunoglobulin fold" of immunoglobulin domains and Cu,Zn superoxide dismutase.[2] These have since turned out to represent the two commonest tertiary structure types. Even simpler, although somewhat less common, folding patterns have emerged, such as the four-helix cluster[3] and the eight-stranded parallel β-barrel.[4] Now that the number of solved structures is in the hundreds, repetition is gradually teaching us to recognize those patterns which are natural and obvious for the proteins, even when they do not match human preconceptions of simplicity. Although some structures still elude a unifying description, in most cases one enormously increases comprehensibility of a new protein structure by relating it to the empirically discovered common categories that are illustrated below.

Although protein tertiary structures are clearly hierarchical (see Rose, this volume [29]), so far the most useful regularities have either been in small-scale features such as preferred angles of helix–helix packing,[5,6] or else at the level of what is traditionally called a "domain," assumed to represent a semiindependent unit significant either for folding stability, evolutionary recombination, or functional rigid-body motions. In particular, classification at the level of intact molecules is difficult, because aside from the prevalence of internal duplication they appear to represent

[1] M. G. Rossmann, D. Moras, and K. W. Olsen, *Nature (London)* **250,** 194 (1974).

[2] J. S. Richardson, D. C. Richardson, K. A. Thomas, E. W. Silverton, and D. R. Davies, *J. Mol. Biol.* **102,** 221 (1976).

[3] P. Argos, M. G. Rossmann, and J. E. Johnson, *Biochem. Biophys. Res. Commun.* **75,** 83 (1977).

[4] M. Levine, H. Muirhead, D. K. Stammers, and D. I. Stuart, *Nature (London)* **271,** 626 (1978).

[5] T. J. Richmond and F. M. Richards, *J. Mol. Biol.* **119,** 537 (1978).

[6] C. Chothia, M. Levitt, and D. C. Richardson, *Proc. Natl. Acad. Sci. U.S.A.* **74,** 4130 (1977).

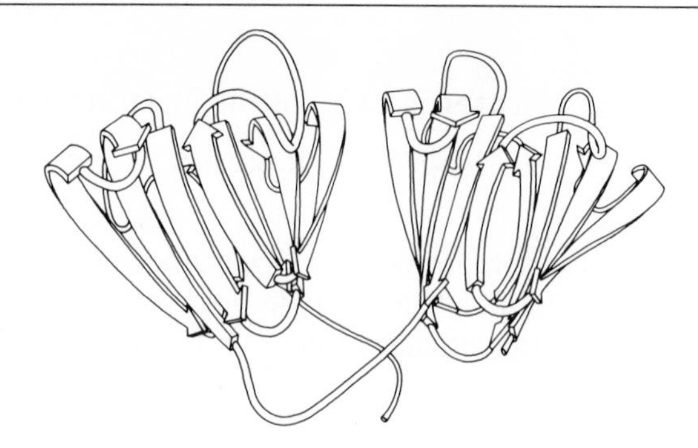

FIG. 1. Schematic backbone drawing of γ crystallin.[10] It has two extremely similar and clearly separated domains, each with eight antiparallel β-strands and related loop conformations.

nearly random combinations of the available domain types.[6a] Most proteins have only a single domain or two similar domains (such as the γ crystallin in Fig. 1),[10] in which case there is no problem. However, fairly often there are two, or even three, quite different types of domain within a single subunit, such as the *p*-hydroxybenzoate hydroxylase in Fig. 2[11] which has a parallel β, an antiparallel β, and an α-helical domain. That structure also illustrates the worst ambiguity in domain definition: occasionally, as here, one is forced to choose between sequence contiguity and spatial contiguity. At the current state of our understanding either choice is reasonable, but one should explicitly say which alternative is being used.

The "mini-atlas" of illustrations in this chapter gives examples of the overall backbone organization found in each of the common types of tertiary structure. Chapter [24] describes how to make this sort of sche-

[6a] The earliest classification scheme[7] was at the level of whole proteins, and included an "α + β" category for cases in which the helices and β-strands were separated in the sequence rather than alternating as in the α/β class. That handled two-domain cases such as papain and thermolysin very well, but the increased variety since then has forced later versions to use the domain level.[8,9]

[7] M. Levitt and C. Chothia, *Nature (London)* **261,** 552 (1976).

[8] G. E. Schulz and R. H. Schirmer, "Principles of Protein Structure." Springer-Verlag, Berlin and New York, 1979.

[9] J. S. Richardson, *Adv. Protein Chem.* **34,** 167 (1981).

[10] T. Blundell, P. Lindley, L. Miller, D. Moss, C. Slingsby, I. Tickle, B. Turnell, and G. Wistow, *Nature (London)* **289,** 771 (1981).

[11] R. K. Weirenga, R. J. de Jong, K. H. Kalk, W. J. G. Hol, and J. Drenth, *J. Mol. Biol.* **131,** 55 (1979).

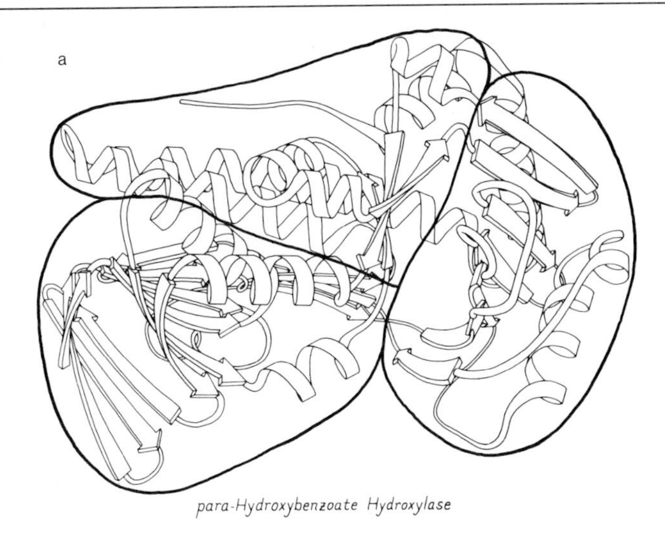

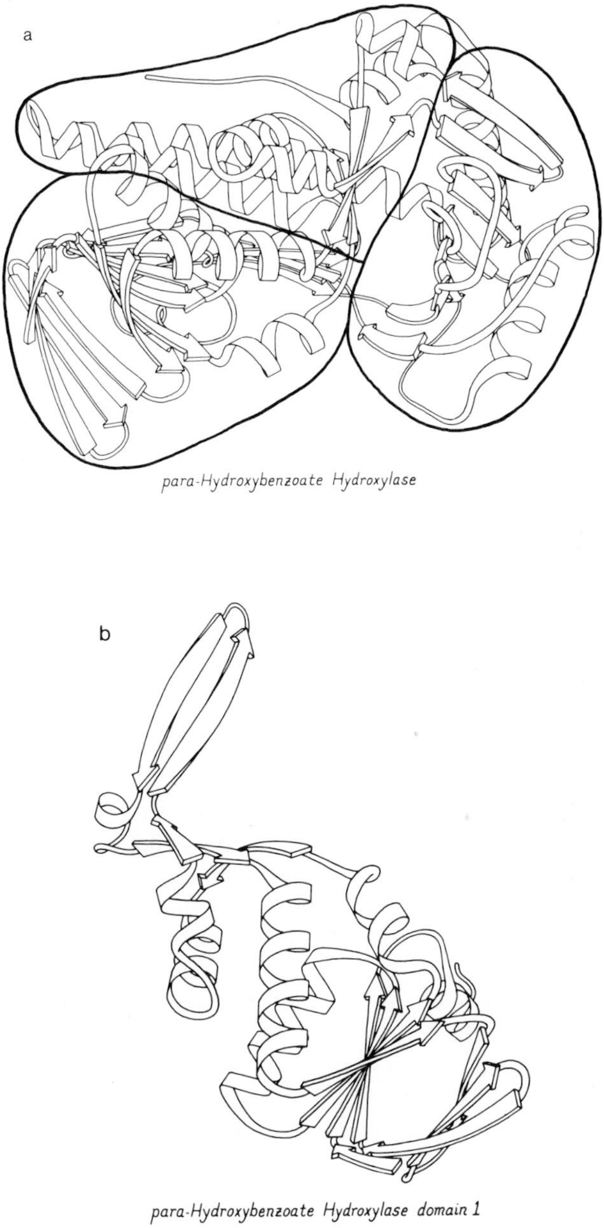

b

para-Hydroxybenzoate Hydroxylase domain 1

FIG. 2. Schematic backbone drawing of *p*-hydroxybenzoate hydroxylase.[11] (a) An entire subunit of the enzyme, with dark lines showing approximate boundaries of the three domains as chosen for spatial contiguity. (b) The parallel α/β domain as chosen for sequence contiguity (rotated view).

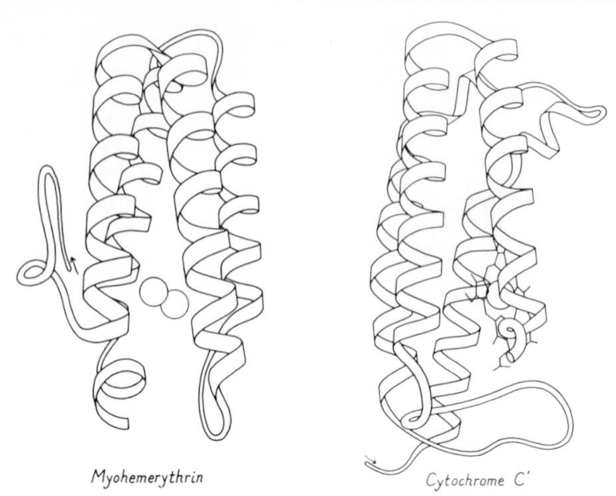

FIG. 3. Up-and-down helix bundle proteins: myohemerythrin[12] and cytochrome c'.[13]

matic drawing, in which β-strands are shown as arrows, helices as spiral ribbons, and nonrepetitive loops as round "ropes." More examples, and a more detailed discussion of each structure type, can be found in Ref. 9.

There are four major categories of tertiary structure, within which we will describe more specific subtypes: (1) antiparallel α; (2) parallel α/β; (3) antiparallel β; and (4) small irregular proteins.

Antiparallel α Structures

Within this first major category, the simplest type of structure is the up-and-down helix bundle: an approximate cylinder of α-helices with each connected to its nearest neighbor. The chain goes up in one helix, moves to the next helix and down it, over by one and up, etc., all around the cylinder. Many of these structures have only four helices, such as the myohemerythrin and the cytochrome c' shown in Fig. 3.[12,13] The prevalence of nearest-neighbor connections between elements of secondary structure, as seen in these up-and-down helix bundles, is one of the most general regularities in protein structure. However, one must always allow for the possibility of exceptions: for instance, the ferritin subunit[14] probably has its four helices connected quite differently.

[12] W. A. Hendrickson and K. B. Ward, *J. Biol. Chem.* **252,** 3012 (1977).

[13] P. C. Weber, R. G. Bartsch, M. A. Cusanovitch, R. C. Hamlin, A. Howard, S. R. Jordan, M. D. Kamen, T. E. Meyer, D. W. Weatherford, N. H. Xuong, and F. R. Salemme, *Nature (London)* **286,** 302 (1980).

[14] D. W. Rice, G. C. Ford, J. L. White, J. M. A. Smith, and P. M. Harrison, *Adv. Inorg. Biochem.* **5,** 39 (1983).

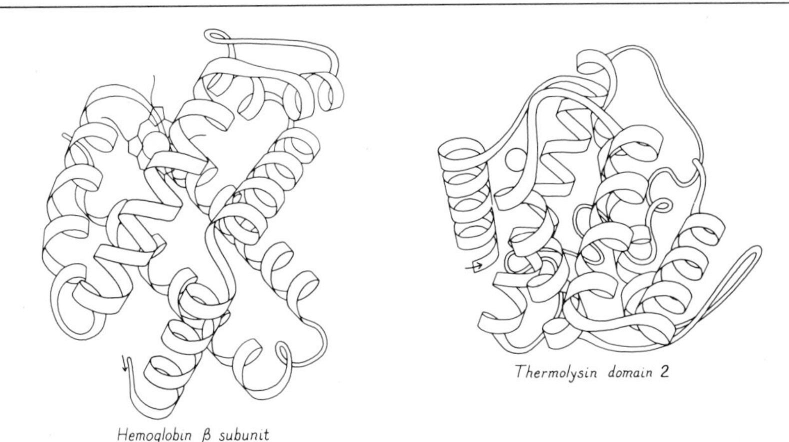

FIG. 4. Greek key helix bundle structures: the β chain of hemoglobin[15] and the second domain of thermolysin.[16]

The second simplest organization of antiparallel helices has at least one non-nearest-neighbor connection in which the chain goes across either the top or bottom of the cylinder of helices. This does not happen in any of the four-helix bundles, but it may with five or more helices, as shown in Fig. 4[15,16] for the hemoglobin β chain and the second domain of thermolysin. In analogy to the β-barrels described below, this second pattern of antiparallel α structure is called a Greek key helix bundle. They are much less common than up-and-down helix bundles, they have no characteristic handedness, and the examples so far each include only one non-nearest-neighbor connection.

In each of the major structural categories there is a small group of miscellaneous proteins that do not fit into a recognized subgroup. Within the antiparallel α category these used to include the carp muscle Ca-binding protein (Fig. 5),[17] whose α-helices are arranged in almost perpendicular pairs (called "E–F hands") with a Ca-binding loop in between them. The structure of the intestinal Ca-binding protein (Fig. 5) confirmed that these are the first representatives from a family of structurally similar calcium-regulated proteins,[18] now also including troponin C[18a,18b] and

[15] R. C. Ladner, E. J. Heidner, and M. F. Perutz, *J. Mol. Biol.* **114,** 385 (1977).

[16] M. A. Holmes and B. W. Matthews, *J. Mol. Biol.* **160,** 623 (1982).

[17] R. H. Kretsinger and C. E. Nockolds, *J. Biol. Chem.* **248,** 3313 (1973).

[18] D. M. E. Szebenyi, S. K. Obendorf, and K. M. Moffat, *Nature (London)* **294,** 327 (1981).

[18a] O. Herzberg and M. N. G. James, *Nature (London)* **313,** 653 (1985).

[18b] M. Sundaralingam, R. Bergstrom, G. Strasburg, S. T. Rao, P. Roychowdhury, M. Greaser, and B. C. Wang, *Science* **227,** 945 (1985).

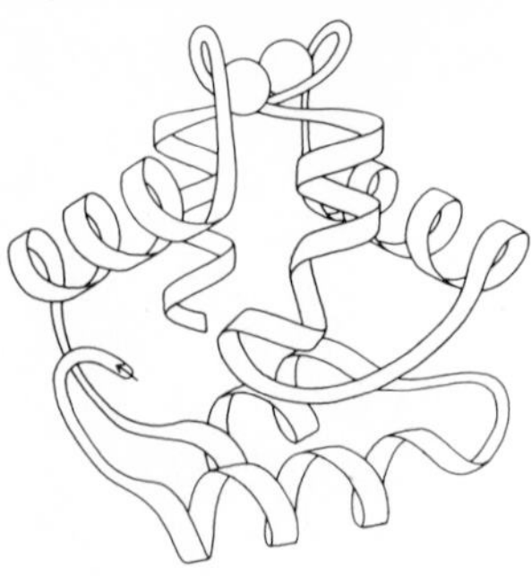

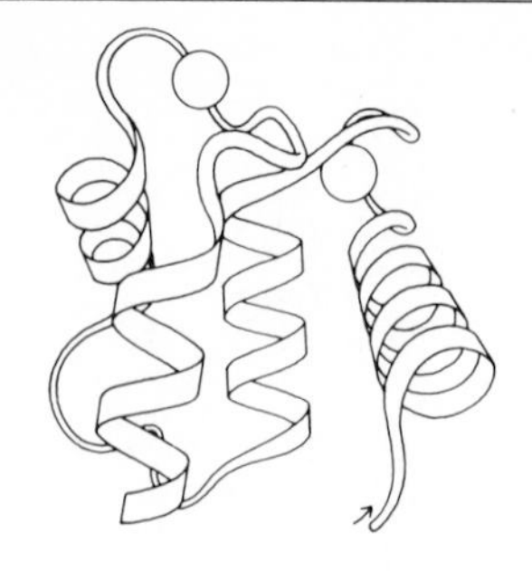

FIG. 5. Calcium-regulated proteins, with pairs of "E–F hands" (perpendicular helices with a Ca-binding loop in between): carp muscle Ca-binding protein[17] and bovine intestinal Ca-binding protein.[18]

calmodulin.[18c] The latter two proteins are both dumbbell shaped, with two quite separate domains each with two Ca-binding E–F hands. The bar of the dumbbell is formed by a long (8- or 9-turn) α-helix which is integrated into the domains at each end but is very thoroughly, and unusually, exposed for several turns in the middle. Citrate synthase,[19] which includes a long pair of buried helices down the middle, may be the first example of larger, three-layer helical domains. In general we can hope that the miscellaneous structures are just those for which our sample is still too small to have included their relatives.

Parallel α/β Structures

In the second major group of tertiary structures, the first subcategory is made up of large but remarkably simple domains with a central, eight-stranded, parallel β-barrel and a concentric outer cylinder of eight α-helices. Each helix connects the top of one β-strand to the bottom of the next strand over, so that the entire structure winds continuously around the barrel in one direction, advancing by one β-strand at a time. This pattern is called the singly wound parallel β-barrel. Figure 6[20] shows end on and side views of triose-phosphate isomerase, the classic example.

[18c] Y. S. Babu, J. S. Sack, T. J. Greenhough, C. E. Bugg, A. R. Means, and W. J. Cook, *Nature (London)* **315,** 37 (1985).

[19] S. Remington, G. Wiegand, and R. Huber, *J. Mol. Biol.* **158,** 111 (1982).

[20] D. W. Banner, A. C. Bloomer, G. A. Petsko, D. C. Phillips, C. I. Pogson, and I. A. Wilson, *Nature (London)* **255,** 609 (1975).

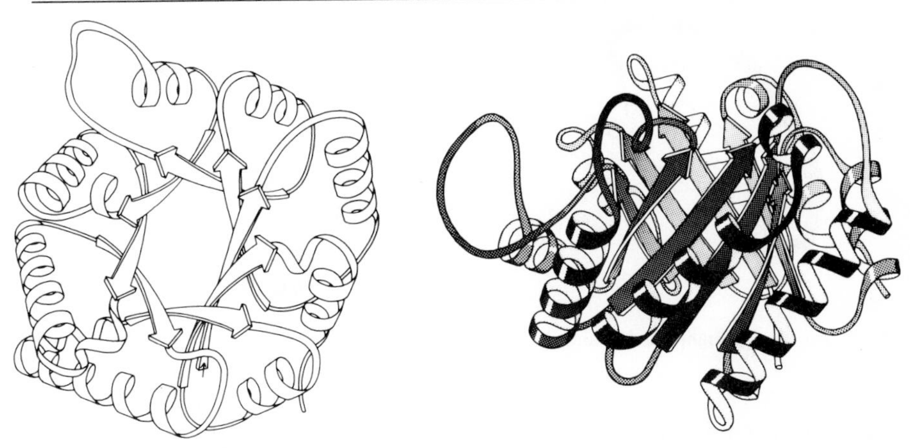

FIG. 6. Triose-phosphate isomerase,[20] a singly wound parallel β-barrel, shown in end view (unshaded) and side view (shaded).

The parallel α/β structures in general, and especially the singly wound barrels, are dominated by the need to satisfy an extremely strong empirical handedness requirement. The chain that connects two parallel strands in a β-sheet must get from one end of the sheet to the other end (as opposed to a hairpin connection between antiparallel strands, which stays at one end of the sheet); such a crossover loop is a handed structure (see Fig. 7). It turns out that 99% of the crossover connections in known protein structures are right-handed, regardless of the number of strands between them in the sheet or of the length and conformation of the connecting loop.[21,22] The connections in a parallel β-barrel must all be on the outside of the barrel, since its interior is filled with hydrophobic side chains and has no room for another piece of backbone. Therefore the need to make right-handed crossover connections forces the chain to move in one continuous direction around the barrel. It would theoretically be possible to skip a strand in this progression and fill it in after going once around the barrel, but that would produce a knot in the chain as well as a less stable structure during folding. Thus a parallel β-barrel needs to have

[21] J. S. Richardson, *Proc. Natl. Acad. Sci. U.S.A.* **73,** 2619 (1976).
[22] M. J. E. Sternberg and J. M. Thornton, *J. Mol. Biol.* **110,** 269 (1977).

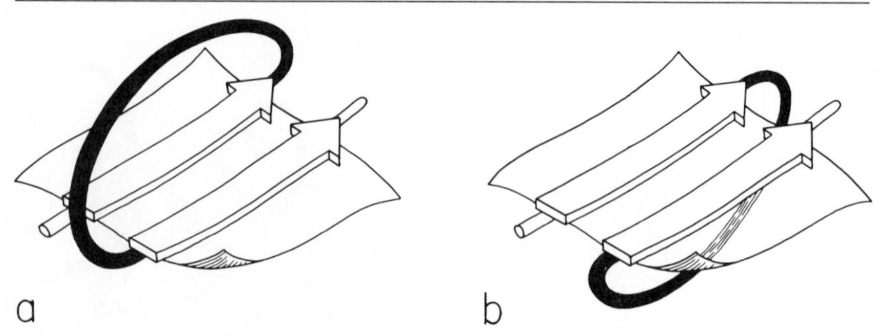

FIG. 7. (a) Right-handed and (b) left-handed crossover connections in a β-sheet.

a singly wound topology, and all of them are very similar to one another.

The second major division within the parallel α/β structures are the doubly wound parallel β-sheets, or "nucleotide-binding domains."[1,9] These domains have a central layer of twisted β-sheet with four to nine strands of all parallel or mostly parallel β structure. The central sheet is flanked on each side by a layer of α-helices. The β-sheet is buried on both sides and therefore presumably hydrophobic on both sides, which is characteristic of parallel β structure in general. The flanking helix layers are formed by the crossover connections between β-strands; in order for them to cover both sides of the sheet using only right-handed connections, it is necessary for the sequential progression of strands along the sheet to change direction at least once. In practice this is accomplished in a remarkably consistent way among most of these proteins: the chain starts near the middle of the β-sheet and winds toward one edge using nearest-neighbor right-handed crossover connections and covering one side of the β-sheet with helices; it then skips back to the center and winds toward the opposite edge of the sheet, burying the other side with helices. Figure 8[23] shows a side view and an edge view of lactate dehydrogenase domain 1, the original and classic example of a doubly wound parallel β sheet. Figure 9[24-27] shows additional doubly wound sheet domains with varying degrees of similarity to lactate dehydrogenase. This is the commonest type of protein structure, with over 30 examples now known.

[23] J. J. Holbrook, A. Liljas, S. J. Steindel, and M. G. Rossmann, *in* "The Enzymes" (P. D. Boyer, ed.), 3rd ed., Vol. 11, p. 191. Academic Press, New York, 1975.

[24] C.-I. Branden, H. Jornvall, H. Eklund, and B. Furugren, *in* "The Enzymes" (P. D. Boyer, ed.), 3rd ed., Vol. 11, p. 104. Academic Press, New York, 1975.

[25] G. E. Schulz, M. Elzinga, F. Marx, and R. H. Schirmer, *Nature* (*London*) **250,** 120 (1974).

[26] D. C. Rees, M. Lewis, and W. N. Lipscomb, *J. Mol. Biol.* **168,** 367 (1983).

[27] R. Thieme, E. F. Pai, R. H. Schirmer, and G. E. Schulz, *J. Mol. Biol.* **152,** 763 (1981).

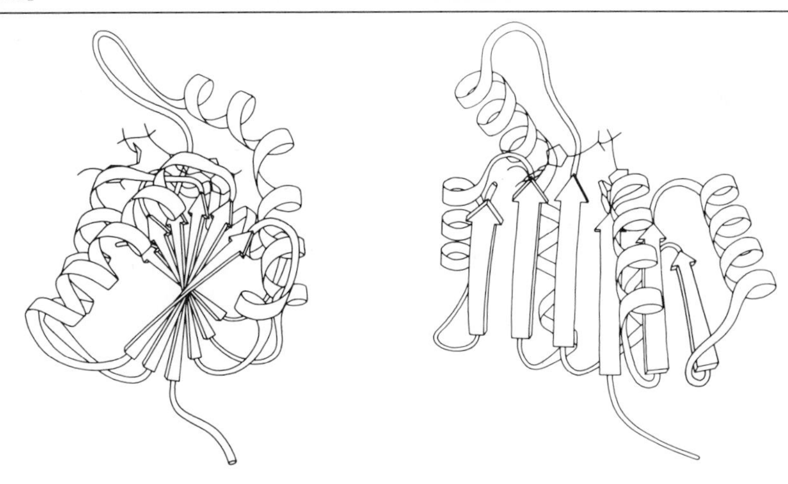

FIG. 8. The first domain of lactate dehydrogenase,[23] a doubly wound parallel β-sheet structure, viewed edge on and from the side.

Many of them have a mono- or dinucleotide binding site at the COOH-terminal end of the β-strands at the center of the sheet where it changes direction, and many of them are enzymes of the glycolytic pathway.

Antiparallel β Structures

The third major category is the antiparallel β structures. As in the antiparallel α proteins, the simplest type of antiparallel β domain forms an approximate cylinder, or barrel, with up-and-down nearest-neighbor connectivity between the strands all around. Figure 10[28] shows bovine catalase domain 1 as an example. In the β-barrels, however, the simple up-and-down topology is actually rather rare while the Greek key β-barrels are the second commonest domain structure, with about 20 examples known so far. Figure 11[29,30] shows several examples of protein domains that are Greek key β-barrels and how their topologies when rolled out flat can resemble the "Greek key" motif common on Greek vases.[31] The

[28] M. R. N. Murthy, T. J. Reid, III, A. Sicignano, N. Tanaka, and M. G. Rossmann, *J. Mol. Biol.* **152,** 465 (1981).

[29] P. M. Colman, H. C. Freeman, J. M. Guss, M. Murata, V. A. Norris, J. A. M. Ramshaw, and M. P. Venkatappa, *Nature (London)* **272,** 319 (1978).

[30] J. A. Tainer, E. D. Getzoff, K. A. Beem, J. S. Richardson, and D. C. Richardson, *J. Mol. Biol.* **160,** 181 (1982).

[31] J. S. Richardson, *Nature (London)* **268,** 495 (1977).

FIG. 9. An assortment of doubly wound parallel β-sheet structures: domain 2 of alcohol dehydrogenase,[24] adenylate kinase,[25] carboxypeptidase A,[26] and domain 2 of glutathione reductase.[27]

simple feature shared by all the Greek key structures is having at least one (or as many as four) connections that cross an end of the barrel.

The Greek key barrel is a handed structure, which is manifested either in the direction of swirl (counterclockwise as in Fig. 11, rather than clockwise) when viewed from the outside, or else in the right-handedness of the crossover connections between opposite sides of the barrel. All but one of the known examples are counterclockwise or right-handed, the sole exception being staphylococcal nuclease,[9] the only β-barrel with as few as

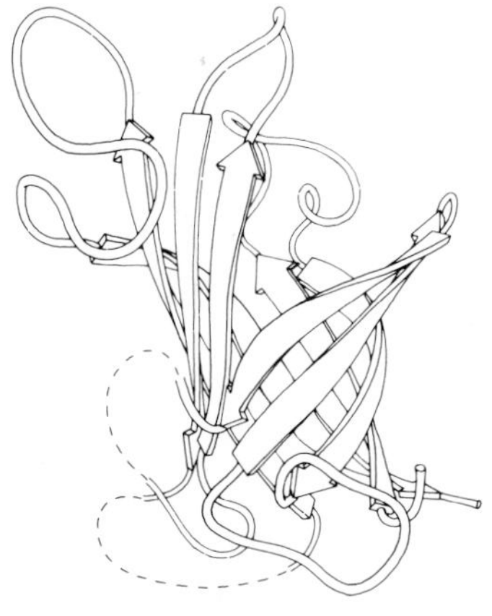

FIG. 10. Bovine catalase[28] domain 2, an up-and-down antiparallel β-barrel.

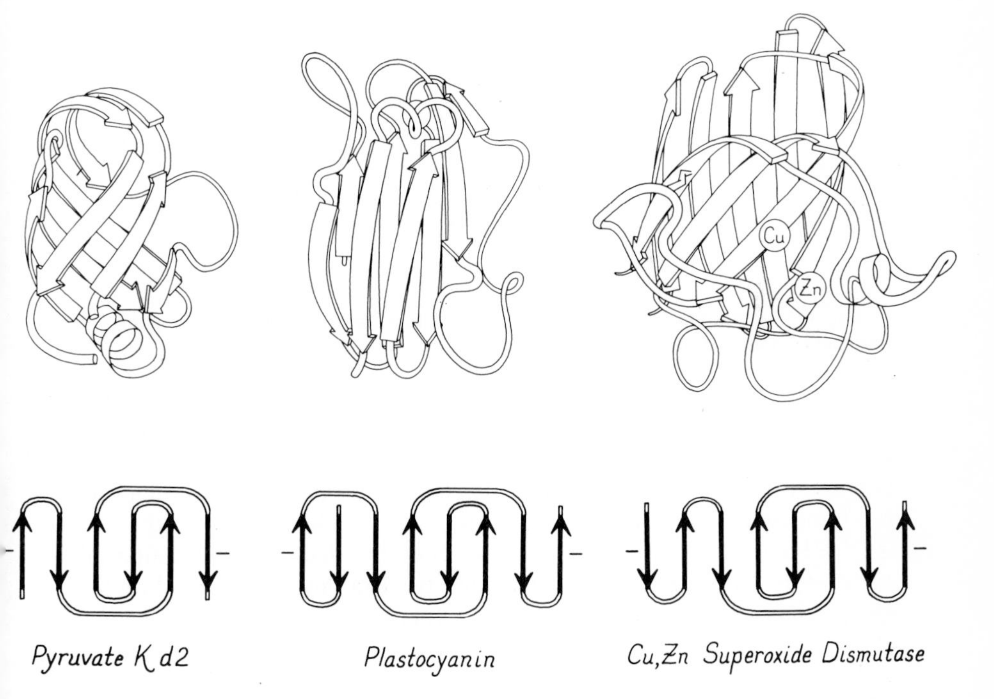

FIG. 11. Greek key antiparallel β-barrels: domain 2 of pyruvate kinase,[4] plastocyanin,[29] and Cu,Zn superoxide dismutase.[30] Below, their topologies are diagrammed as opened up and laid flat.

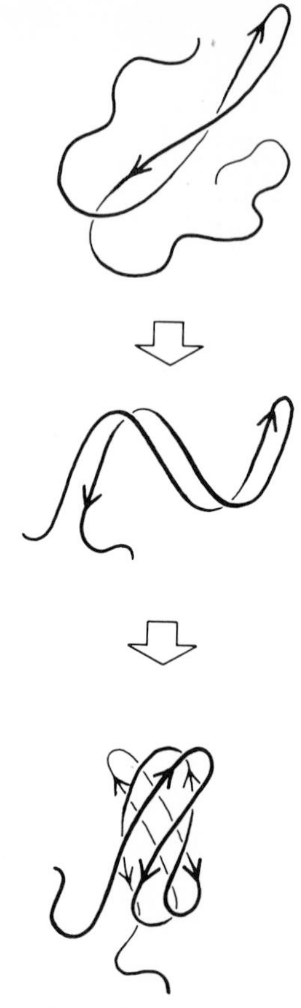

FIG. 12. A proposed folding pathway for Greek key β-barrels, starting from a long, two-stranded ribbon of twisted antiparallel β-sheet.

five strands and very highly twisted. This handedness is quite a large-scale effect, involving a minimum of about 50 or 60 residues in the sequence. It is very likely that both the common occurrence and the preferred handedness of Greek key β-barrels is a result of the way they fold up.[31,32] Figure 12 illustrates how a long two-stranded ribbon of antiparallel β structure with the usual right-handed twist (along the strand direction) can automatically produce a Greek key topology with the correct handedness when it loops over onto itself into a compact form. Figure 13[32a,32b]

[32] O. B. Ptitsyn and A. V. Finkelstein, *in* "Protein Folding" (R. Jaenicke, ed.), p. 101. Elsevier/North-Holland, New York, 1980.

[32a] D. B. McKay, I. T. Weber, and T. A. Steitz, *J. Biol. Chem.* **257,** 9518 (1982).

[32b] T. A. Jones and L. Liljas, *J. Mol. Biol.* **177,** 735 (1984).

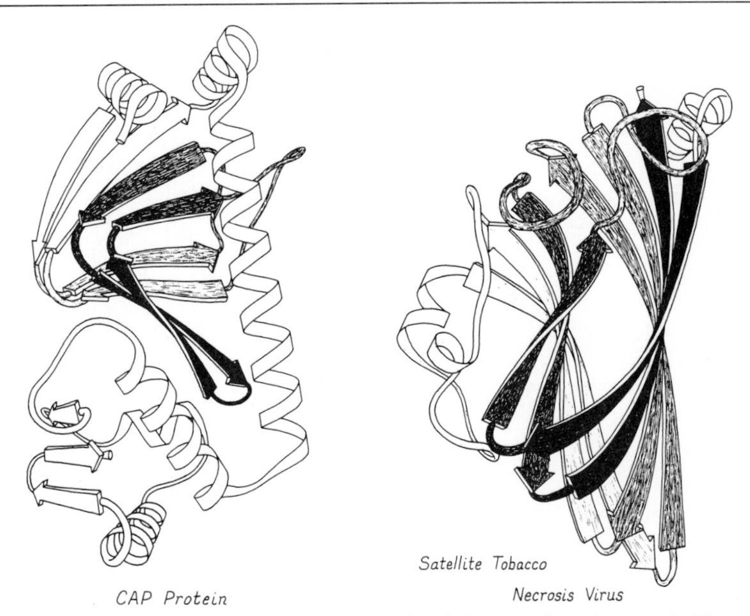

FIG. 13. β-Barrel structures with "perfect" Greek key topology (see text). The strand pairs that wind together around the barrel are progressively shaded, darkest at the central hairpin.

shows two examples of eight-stranded β-barrels with "perfect" Greek key topology: from the middle of the sequence, the β-strands loop around the barrel in pairs all the way to the two adjacent ends (at the top in each drawing). If these structures had been the first Greek key barrels seen, then the proposal of folding as a double ribbon would probably have seemed obvious from the start.

The structures in Figs. 10–13 have been more or less interchangeably described either as β-barrels[31] or as "β sandwiches."[33] The sandwich description is more useful for studying internal packing in a given native protein, since the tightest interactions are between adjacent strands within a sheet; it is also used to distinguish the relatively low-twist "sandwich" packing from the high-twist "foldover" structures. On the other hand, the barrel description produces far simpler strand topologies and is probably preferable for studying folding, since covalent connectivities almost always jump back and forth between the two sides of a sandwich. It cannot yet be shown conclusively which organization is more influential during protein evolution.

The next subgroup within the antiparallel β category is the "open-face sandwich," with a single antiparallel β-sheet and a layer of helices and

[33] C. Chothia and J. Janin, *Proc. Natl. Acad. Sci. U.S.A.* **78,** 4146 (1981).

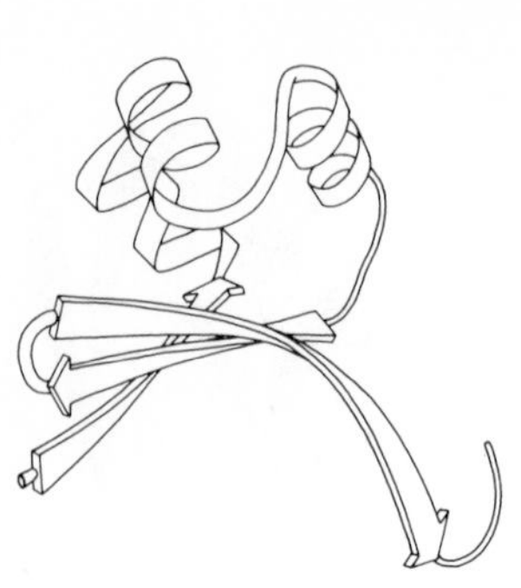

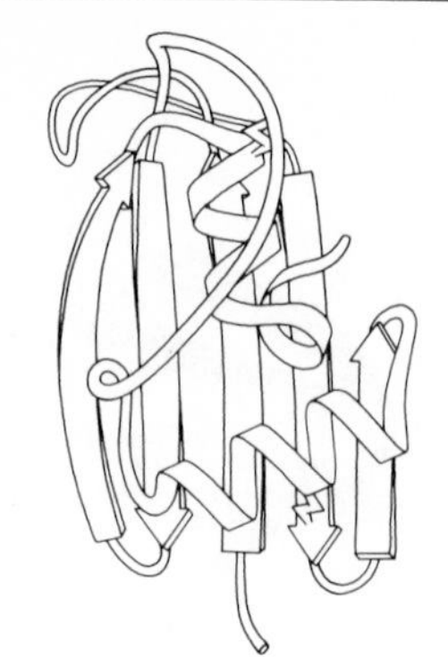

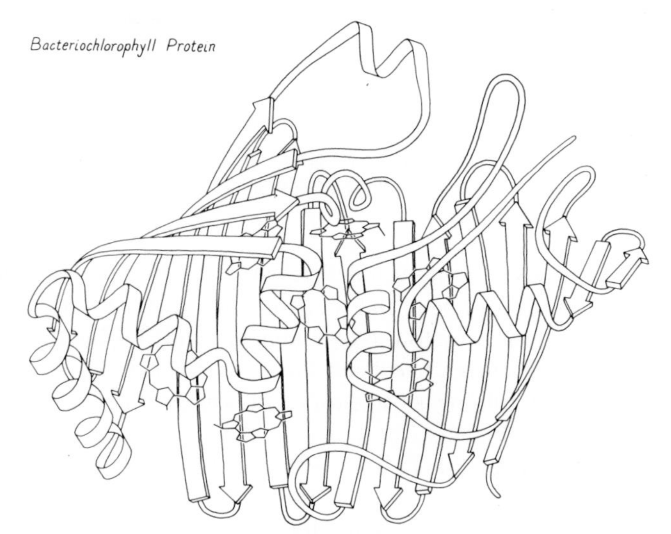

FIG. 14. "Open-face sandwich" antiparallel β-sheet structures: Cro repressor,[34] streptococcal subtilisin inhibitor,[35] and bacteriochlorophyll protein.[36]

loops on only one side of the sheet. These domains are quite common, but vary considerably in size and in topology (see Fig. 14),[34-36] from only three β-strands in Cro repressor protein all the way to 15 strands in bacteriochlorophyll protein, the "super" example of this class. In addi-

[34] W. F. Anderson, D. H. Ohlendorf, Y. Takeda, and B. W. Matthews, *Nature (London)* **290,** 754 (1981).

[35] Y. Mitsui, Y. Satow, Y. Watanabe, and Y. Iitaka, *J. Mol. Biol.* **131,** 697 (1979).

[36] B. W. Matthews, R. E. Fenna, M. C. Bolognesi, M. F. Schmid, and J. M. Olson, *J. Mol. Biol.* **131,** 259 (1979).

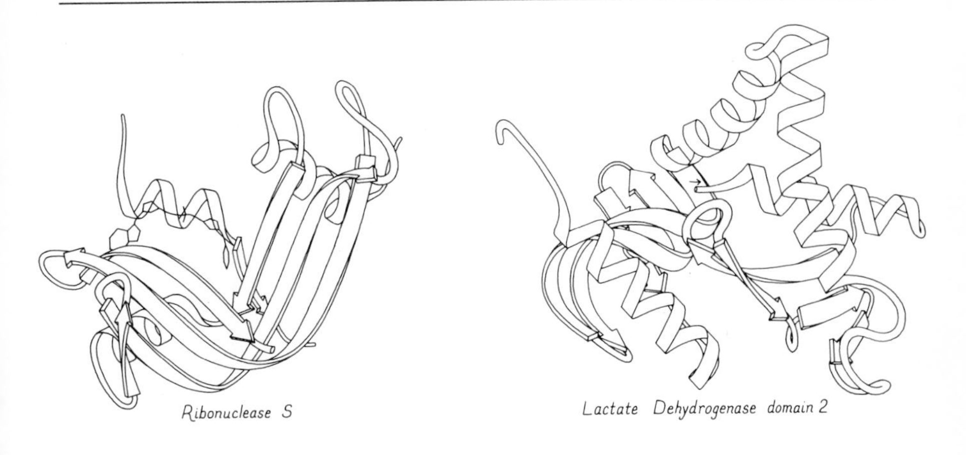

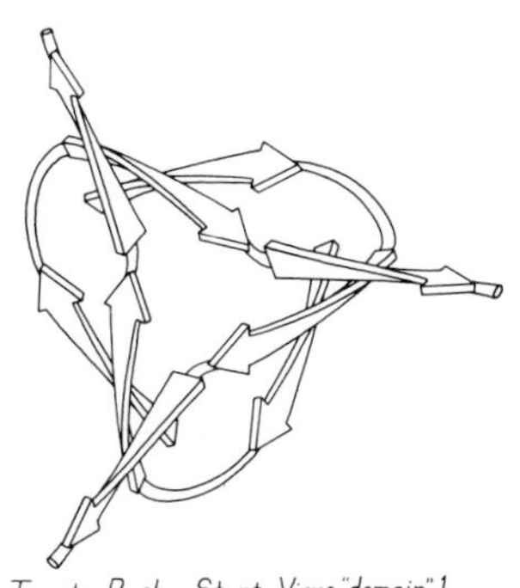

FIG. 15. Miscellaneous antiparallel β structures: ribonuclease S,[37] the second domain of lactate dehydrogenase,[23] and the annulus formed by the NH_2-terminal strands of three subunits of tomato bushy stunt virus protein.[38]

tion to the usual hydrophobic side chains between the β-sheet and the helix layer, bacteriochlorophyll protein has seven bacteriochlorophyll molecules bound asymmetrically but specifically in its large hydrophobic interior.

Then there is an interesting set of miscellaneous structures in the antiparallel β category (see Fig. 15).[37,38] They include sheets that have folded around partway toward making a β-barrel such as pancreatic ribonuclease, structures with several two-stranded β ribbons such as lactate dehydrogenase domain 2, and unusual structures such as the "β annulus"

[37] H. W. Wyckoff, D. Tsernoglou, J. W. Hanson, J. R. Knox, B. K. Lee, and F. M. Richards, *J. Biol. Chem.* **245,** 305 (1970).

[38] A. J. Olson, G. Bricogne, and S. J. Harrison, *J. Mol. Biol.* **171,** 61 (1983).

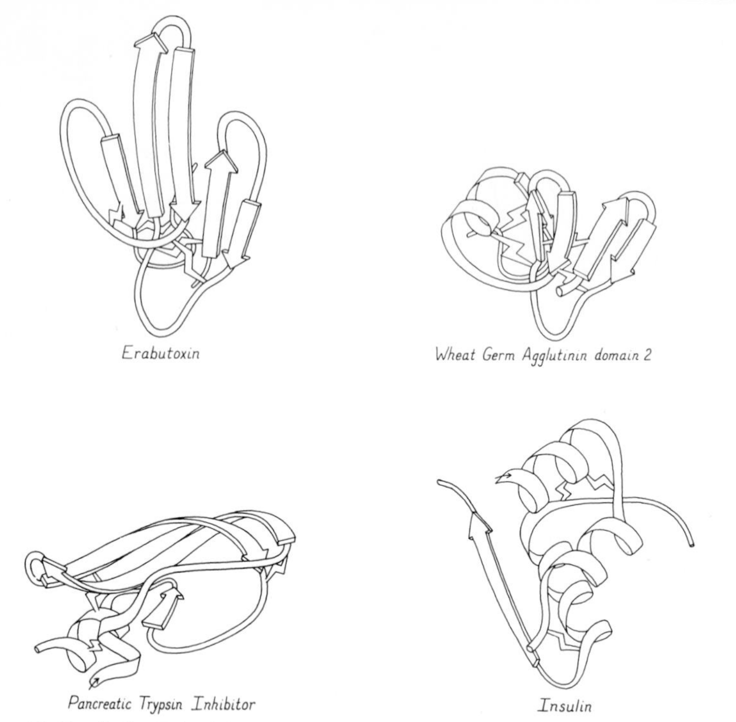

FIG. 16. Small disulfide-rich proteins: erabutoxin,[39] wheat germ agglutinin domain 2,[40] basic pancreatic trypsin inhibitor,[41] and insulin.[42]

formed by association of the NH_2-terminal tails of three subunits of tomato bushy stunt virus protein around a three-fold axis in the spherical virus capsule.

Small Irregular Proteins

The fourth major category of protein structure is different logically from the other three, which are dominated by patterns of secondary structure. The small irregular proteins have relatively little conventional secondary structure, and what they do have is mostly short and irregular. Instead their organization is apparently dominated either by a high content of disulfides (more than 10% half-cystine) or else by bound metals or prosthetic groups. Figure 16[39–42] shows several small, dislulfide-rich pro-

[39] B. W. Low, H. S. Preston, A. Sato, L. S. Rosen, J. E. Searl, A. D. Rudko, and J. S. Richardson, *Proc. Natl. Acad. Sci. U.S.A.* **73,** 2991 (1976).
[40] C. S. Wright, *J. Mol. Biol.* **111,** 439 (1977).

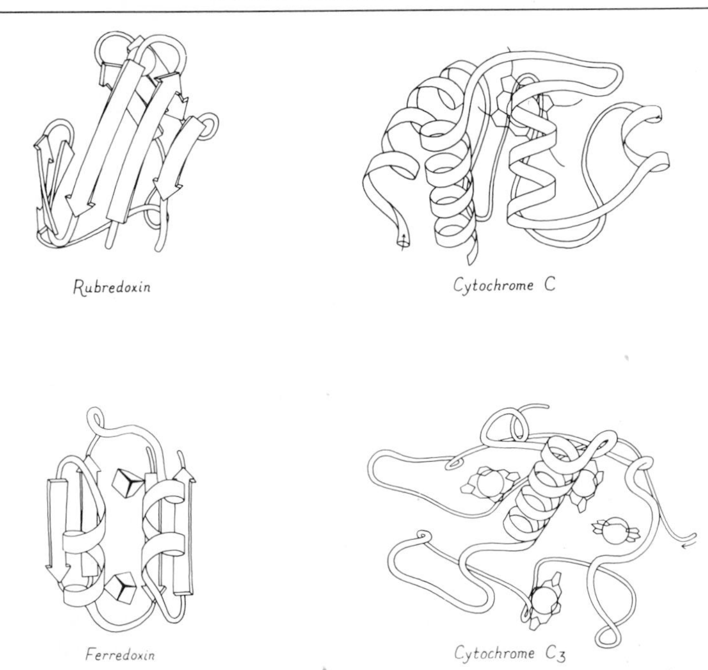

FIG. 17. Small metal-rich proteins: rubredoxin,[43] cytochrome c, [44] ferredoxin,[45] and cytochrome c_3.[46]

teins and Fig. 17[43–46] shows several small, metal-rich proteins. Almost all of the small irregular domains occur by themselves as single-domain, single-subunit proteins, and the few exceptions (insulin, wheat germ agglutinin, and the multidomain protease inhibitors[47]) are self-associating rather than occurring in combination with one of the major categories. Disulfide-rich proteins are typically extracellular, and most function as toxins, lectins, enzyme inhibitors, or hormones. The only two which

---

[41] J. Deisenhofer and W. Steigemann, *Acta Crystallogr., Sect. B* **B31** 238 (1975).

[42] E. J. Dodson, G. G. Dodson, D. C. Hodgkin, and C. D. Reynolds, *Can. J. Biochem.* **57,** 469 (1979).

[43] K. D. Watenpaugh, L. C. Sieker, and L. H. Jensen, *J. Mol. Biol.* **131,** 509 (1979).

[44] R. Swanson, B. L. Trus, N. Mandel, G. Mandel, O. B. Kallai, and R. E. Dickerson, *J. Biol. Chem.* **252,** 759 (1977).

[45] E. T. Adman, L. C. Sieker, and L. H. Jensen, *J. Biol. Chem.* **248,** 3987 (1973).

[46] M. Pierrot, R. Haser, M. Frey, F. Payan, and J.-P. Astier, *J. Biol. Chem.* **257,** 14341 (1982).

[47] E. Weber, E. Papamokos, W. Bode, R. Huber, I. Kato, and M. Laskowski, Jr., *J. Mol. Biol.* **149,** 109 (1981).

share a specific topology are the snake neurotoxins and the domains of wheat germ agglutinin.[48]

Most of the small metal-rich proteins form approximately cylindrical structures with either a simple up-and-down (e.g., rubredoxin, cytochrome *c*) or a Greek key (e.g., ferredoxin) topology, but where the elements forming the cylinder are a mixture of helices and extended strands (which may or may not form any β structure). The ultimate example of a small irregular protein is cytochrome c_3, which has four hemes enclosed by just over 100 amino acids, and which has almost no secondary structure (see Fig. 17).

In general, most of the small irregular proteins give the impression of being small, distorted versions of one of the other structural categories. They may be distorted helix bundles, distorted β-barrels, or distorted open-face sandwiches. However, there are no small, distorted versions of α/β structures, perhaps because those seem to be dependent on large, buried, and quite regular β structure for their stability. Although almost all α/β domains contain either nucleotide prosthetic groups or metals, there are no heme groups in spite of all the α-helices, and no disulfides except the single active-site disulfides in thioredoxin[49] and glutathione reductase.[27] As a general rule, disulfides appear to be incompatible with either metals or prosthetic groups, a dichotomy which allows the separation of the small irregular proteins into the disulfide rich and the metal rich.

Conclusion

Although each protein structure is highly individual in what are probably its most significant aspects, there are three important advantages to be gained from relating its gross morphology to the overall classifications described here. The first advantage is that many of the tertiary structure types are recognizable even at low resolution, while use of the statistical preferences of each category in terms of handedness, twist, topology, etc. can greatly improve the chances of a correct chain tracing at intermediate resolution (see this volume [14]). Second, such classification may often suggest possible evolutionary relationships or rule them out, either within or between proteins. The third advantage is one of communication and memorability: description of the backbone "fold" provides a starting point from which the individual variations of the structure and function can be more meaningfully understood.

[48] J. Drenth, B. Low, J. S. Richardson, and C. S. Wright, *J. Biol. Chem.* **255,** 2652 (1980).
[49] A. Holmgren, B.-O. Soderberg, H. Eklund, and C.-I. Bränden, *Proc. Natl. Acad. Sci. U.S.A.* **72,** 2305 (1975).

[24] Schematic Drawings of Protein Structures

By JANE S. RICHARDSON

An accurate stereo figure is certainly essential to reporting a protein structure. It is aimed at that small but important fraction of readers who have the ability to see the stereo and to interpret what they see, and who are sufficiently interested actually to do so.

However, it is also extremely important to provide accessible, if simplified, three-dimensional information (in mono) for the entire audience. A schematic drawing (such as Fig. 1 or 2) can summarize the overall features of a structure in a quickly graspable and relatively memorable form, and can provide a framework within which to place further details. Such a drawing has inherent dangers, of course: by definition any simplification must omit information, and any representation which aids conceptual understanding must involve interpretation and choice (which is equally true of an automated computer drawing). But even if one is led to miss alternative interpretations something worthwhile is gained, because understanding one conceptualization of a structure is much better than not understanding it at all.

It follows that there are two crucial and nontrivial tasks of a schematic drawing. The first is to portray the overall organization of the structure rather than a collection of details; for example, one should try to draw a β-sheet rather than drawing β-strands. Some degree of interpretation must be done before starting to draw, but feedback from the appearance of the drawing can modify one's interpretation. When the structure is successfully perceived as a unified object, then comparisons and symmetry relationships can also be perceived directly rather than laboriously worked out by matching individual features (e.g., the relation between the two subunits in Fig. 1).

The second major task is to communicate accurate three-dimensional information, by utilizing all available monocular depth cues and, where possible, by mimicking the appearance of a binocular image. For abstract and unfamiliar objects such as these proteins, it is necessary to exaggerate the monocular depth cues in order to achieve a realistic perception of the three-dimensional relationships. Helical ribbons must be made curlier, foreshortening exaggerated, and more than the true amount of the side of an arrow shown as it starts to twist. A direct example of compensating for the lack of the second eye is the case of an arrow viewed edge on as it twists around: the most satisfactory convention shows both faces of the

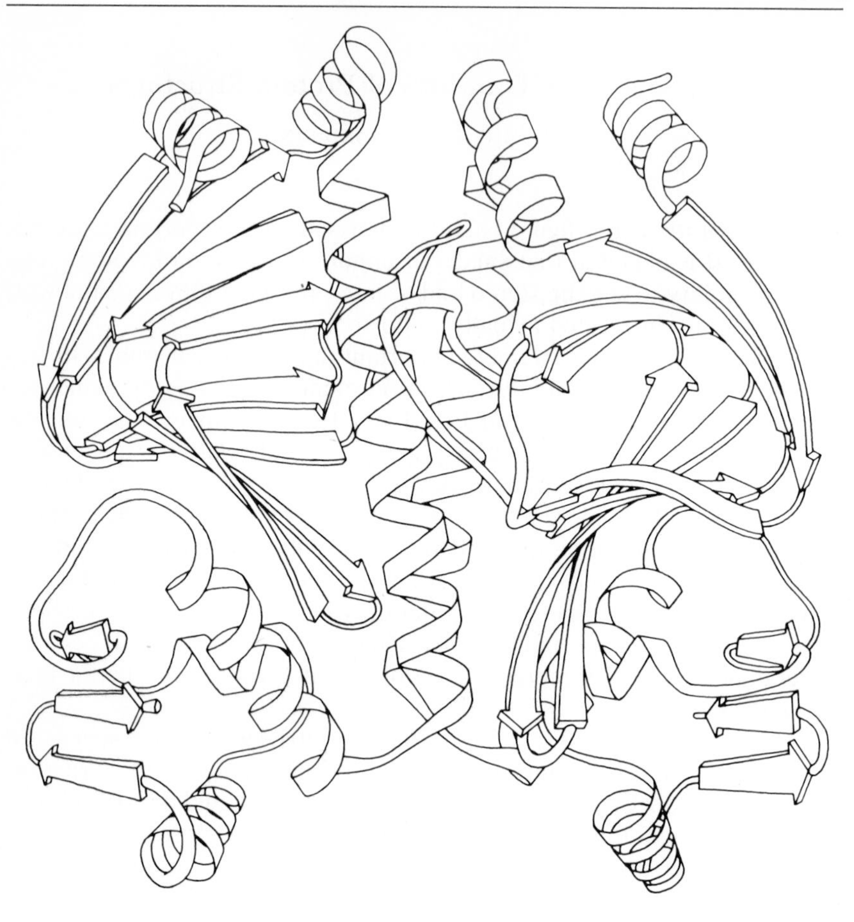

FIG. 1. Schematic drawing of a protein structure, the CAP protein dimer.

arrow as visible around the changeover point (see Fig. 3b), which seems logically peculiar but more nearly matches binocular vision. This can be demonstrated by holding a belt or plastic ruler stretched vertically in front of you with a half twist from top to bottom; look at the crossover point first with just one eye and then with two.

Emphasis on the large-scale features need not compromise accurate positioning. If a fairly favorable view has been chosen, then ambiguities can be resolved and loops smoothed without shifting anything more than 1 or 2 Å. To make a drawing "work" in mono, what needs to be modified is

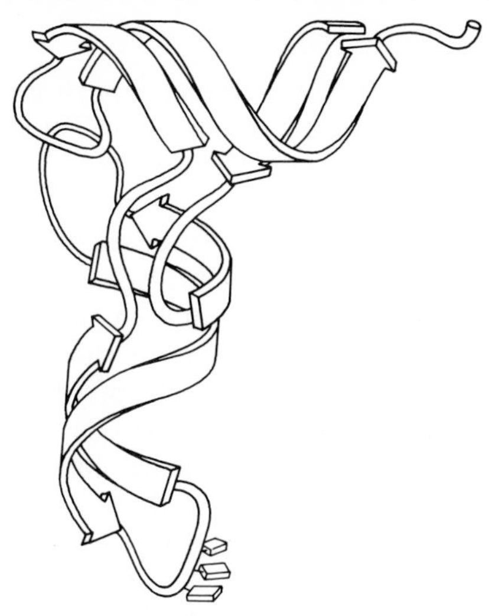

FIG. 2. Yeast $tRNA^{Phe}$: schematic drawing of a nucleic acid structure.

not positions but the local cues to depth and orientation. The overriding criterion for evaluating a schematic drawing is its final overall appearance, judged as a representation of the major patterns and relationships you see in the three-dimensional structure.

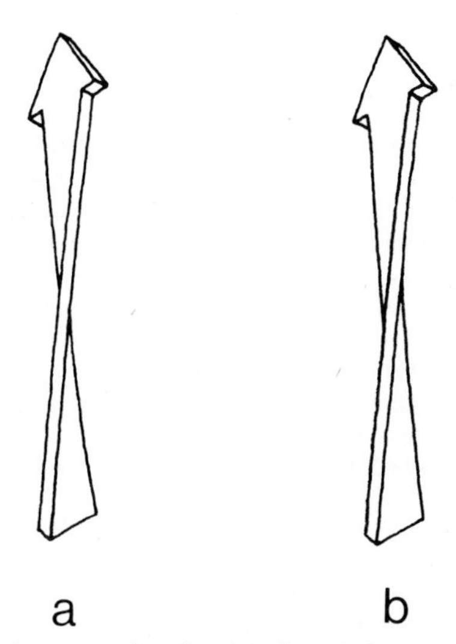

FIG. 3. Comparison of a monocular view (a) of a twisting arrow with a more convincing drawing (b) which imitates a binocular image.

Specific Drawing Methods: Materials and Setup

In this section are provided detailed and explicit instructions for producing schematic drawings of the type shown in Fig. 1 and used for the "mini-atlas" in the previous chapter. There are many different but related types of schematics in the literature, e.g., for myoglobin,[1] carbonic anhydrase,[2] thioredoxin,[3] immunoglobulin,[4] PGM,[5] and tRNA.[6] Along with the specific methods explained here, one could also add or substitute conventions from some of those other representations if they were especially suitable for showing the features of a given structure. For example, in showing large multisubunit structures it is helpful to simplify further to cylindrical helices and entire β-sheets or β-barrels.[7–9]

It is generally desirable to make an original ink drawing two to three times larger than it will be reproduced, so that irregularities in the lines will be less visible. One must then remember to use a wide enough pen for the lines to show well when reduced. The originals for the drawings shown here were made at a scale of approximately 7 Å/inch, and some specific measurements quoted below refer to this scale. The drawings are reproduced at a 3 : 1 reduction in "Advances in Protein Chemistry,"[10] at 2 : 1 in the figures of the previous chapter, and at 1 : 1 in the "Protein Structure Coloring Book."[11] At this scale, a No. 1 size Leroy pen produces a reasonable line width.

Producing the C_α plot to the right scale and from a good viewpoint is a critical step in this process. The best method is to view a backbone model

[1] R. E. Dickerson, *in* "The Proteins" (H. Neurath, ed.), 2nd ed., Vol. 2, p. 634. Academic Press, New York, 1964.

[2] A. Liljas, K. K. Kannan, P.-C. Bergsten, I. Waara, K. Fridborg, B. Strandberg, U. Carlbom, L. Jarup, S. Lovgren, and M. Petef, *Nature* (*London*), *New Biol.* **235,** 133 (1972).

[3] A. Holmgren, B.-O. Soderberg, H. Eklund, and C.-I. Branden, *Proc. Natl. Acad. Sci., U.S.A.* **72,** 2307 (1975).

[4] M. Schiffer, R. L. Girling, K. R. Ely, and A. B. Edmundson, *Biochemistry* **12,** 4628 (1973).

[5] J. W. Campbell, H. C. Watson, and G. I. Hodgson, *Nature* (*London*) **250,** 302 (1974).

[6] S.-H. Kim, F. L. Suddath, G. J. Quigley, A. McPherson, J. L. Sussman, A. Wang, N. C. Seeman, and A. Rich, *Science* **185,** 436 (1974).

[7] G. A. Clegg, R. F. D. Stansfield, P. E. Bourne, and P. M. Harrison, *Nature* (*London*) **288,** 299 (1980).

[8] R. E. Dickerson and I. Geis, "Structure and Action of Proteins," p. 88. Harper, New York, 1969.

[9] J. S. Richardson, *Nature* (*London*) **268,** 497 (1977).

[10] J. S. Richardson, *Adv. Protein Chem.* **34,** 181 (1981).

[11] J. S. Richardson, "The Protein Structure Coloring Book." Little River Institute, Bahama, North Carolina, 1979.

on computer graphics: study it well from a variety of directions to decide what features to emphasize, choose a precise viewing direction, and then either obtain plotter output (in stereo) or photograph the screen directly. It is also quite possible to plot C_α coordinates by hand or with a batch computer program, or to photographically enlarge an existing stereo diagram, but this allows less choice, or no choice, of viewpoint. A computer program is now available[12] which produces arrow-and-cylinder drawings in stereo from complete atomic coordinates; it should provide an excellent starting point for making a mono drawing, as well as being very useful in its own right for making stereos.

The following criteria should be considered when choosing an optimal viewing direction:

1. Look through the minimum, or nearly minimum, depth of the structure.
2. Put features of special interest near the front.
3. Do not view either helices or β-strands end on.
4. Minimize the places where features lie behind one another so extensively that their continuity is lost, especially when more than two layers are involved.
5. Favor slight overlaps, which provide valuable hidden-line depth cues.
6. A β-sheet looks best if at least one corner turns over, which helps perception of the shape.
7. If similar structures are to be compared, the viewpoints should be identical, not just similar.
8. For real clarity, it may be worth drawing more than one view (see triose-phosphate isomerase and lactate dehydrogenase in the previous chapter).

If the scale is being adjusted photographically, then either an α-helix or a β-strand lying in the plane of the paper can serve as a standard, using Fig. 4 to adjust the repeat distances. As well as a single C_α plot at standard scale, you need a smaller stereo pair that must be referred to constantly while doing the sketch drawing. This is absolutely essential to producing a mono drawing which accurately represents the three-dimensional relationships.

The first pencil sketch is made on heavy tracing paper taped onto a C_α plot at the correct scale, and the final ink drawing is then traced from the sketch onto translucent Mylar drafting film. Another sheet of tracing paper can be used, but the Mylar does not curl up with heavy ink applica-

[12] A. M. Lesk and K. D. Hardman, *Science* **216,** 539 (1982).

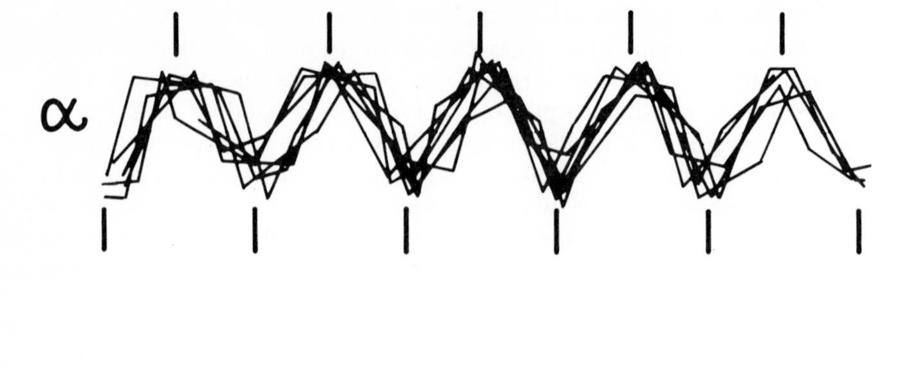

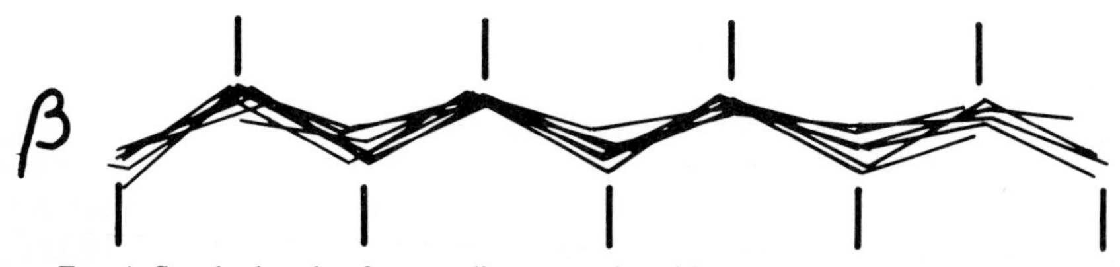

FIG. 4. Standard scale of repeat distances using either (a) an α-helix or (b) a β-strand which lies in the plane of the drawing.

tion, does not become yellow or brittle with age, and takes corrections more easily. Corrections are done by inking in the corrected version and then scraping away the unwanted ink lines with a razor blade or Exacto knife (it is worth changing the blade frequently). Mylar film (usually in large sheets, which can be cut to size), tracing paper, ink, and Leroy or other drafting pens are available at drafting supply stores. It is best to use the heaviest, most expensive tracing paper available, since the lightweight types crumple up. Inks need to be reasonably free flowing in pens, and thus unfortunately none of them are sufficiently opaque black to make high-quality reproduction simple. A fairly acceptable one that is generally available is Higgins "Black Magic."

α-Helices

Helices, in this convention, are shown as spiral ribbons with a cylindrical diameter just a little larger than the C_α positions. Figure 5 lays out the steps in sketching a straight helix lying in the plane of the page. From the stereo, decide where the helix begins and ends. Using a transparent ruler, judge the line of the helix axis and put a tick mark at each end of the axis. Mark a 3/4-in. width at each end perpendicular to the axis, and draw

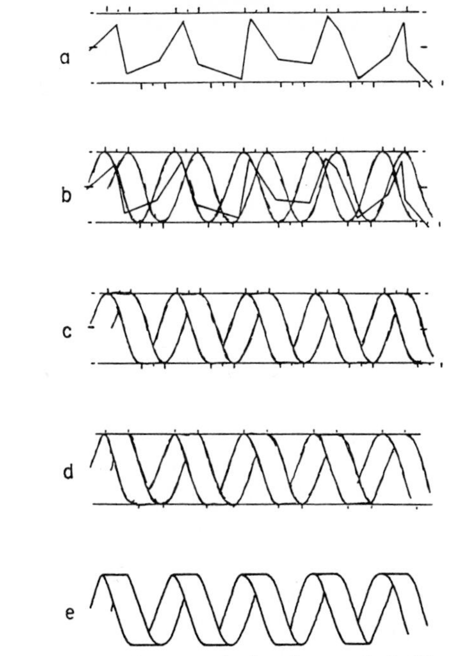

FIG. 5. Successive steps in drawing a spiral ribbon α-helix.

a pair of lines 3/4-in. apart to show the sides of the helix spiral. Along the top line, make a series of equally spaced marks which give the best average locations of the turns in the C_α plot (these marks will be fairly close to 3/4-in. apart). Along the bottom line, put marks centered between the top ones; adjust all of them if necessary. Lay out 1/4 in. surrounding each mark to show where the edges of each turn of the ribbon should fall. You now should have a sketch resembling Fig. 5a. As in Fig. 5b, draw a smooth sine curve joining the left edges of the 1/4-in. marks. Draw a second sine curve through the right edges, keeping the spacing even. Remove hidden lines to produce a right-handed spiral as in Fig. 5c, either by erasing the unwanted lines or by darkening the desired ones. For a helix longer than about three turns, it helps to show that one can see a little farther inside those spiral turns near the ends; this is done by progressively rounding the tops and pointing the bottoms of the sine curves toward the right end and vice versa toward the left, as shown in Fig. 5d. If used, this effect is easiest incorporated between steps b and c. To make the final ink drawing (5e), first use a straightedge to draw the short segments along the sides. Then draw the curves smoothly in freehand, starting at the top and drawing each line in one smooth, slow motion. Tilt the whole drawing if needed to get a favorable angle for the hand motion (one can draw most smoothly going downward, and from the concave side of a curve). Do the front section of each turn first, and then the back pieces.

In general, I have chosen to show all helices as regular except for bending (see below) and a few cases in which the irregularity is particularly significant (such as the open turn at the active disulfide in glutathione reductase). Varying the size or pitch tends to be confused with depth changes, and apparently irregular helices in initial tracings often become more regular with refinement. However, if you have, for instance, a clear 3_{10} helix, by all means try showing it as such.

Some helices bend significantly and should be shown bent (otherwise the position at an end can easily be off by 5 Å). If several H bonds are missing, a good representation is two separate, straight helices, in which their spiral ribbons meet smoothly (e.g., see the A and B helices of hemoglobin in the "mini-atlas"). If only one H bond is missing (almost always at a proline), the bend can be made by curving the outline for two successive turns, as in the lactate dehydrogenase domain 1 drawing in the previous chapter. Sometimes long helices have a continuous gentle curve, either circular (as in *p*-hydroxybenzoate hydroxylase, see this volume [23]) or spiral (see Figs. 1 and 6). To draw these, simply make a smooth curve along the helix axis with parallel curves 3/8 in. on either side, and proceed as above. If the helix curves in or out of the plane of the paper, then adjust the amount of view inside each turn according to the appropriate local angle (as in the long helices of Fig. 6). The lines forming the top and bottom of each turn should be straight, but tangent to the curve. Curved helices would probably be difficult to show convincingly at highly oblique angles.

To draw straight helices at oblique angles to the line of view, the procedure is very similar to Fig. 5, except that the outer guide lines should taper to show foreshortening and the ribbon edges are smoothly looping curves rather than sine curves. The "curliness" of the curves needs to be shown a bit greater than it would really be in a helix at the desired apparent angle, and the curve is more open toward the back because the visual angle is steeper at the back than the front if the object is not infinitely distant. Figure 7 provides a gallery of helix examples at varying apparent angles. Fairly good results can be obtained by simply tracing the example nearest to the appropriate angle. Directly end-on helices should be avoided if at all possible; the nearest acceptable version I have managed attains an apparent angle of 75–80° by using a "binocular" distortion. In general, the lengths and relationships of helices are perceived best when they are at low angles to the plane of the paper.

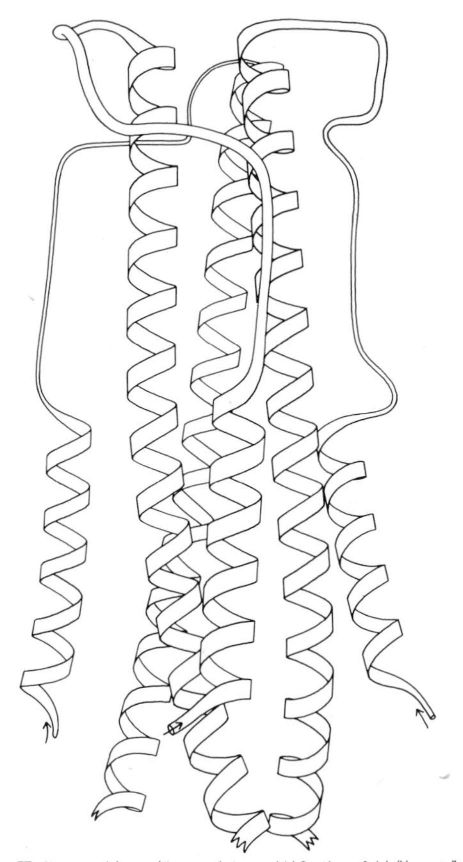

FIG. 6. Long, curved α-helices which spiral around a 3-fold axis (influenza virus hemagglutin HA2).

β-Sheet

In this style of representation, β-strands are shown as arrows with thickness (about one-quarter as thick as they are wide), which gives valuable cues to orientation and twist of the strand. For helices the ribbons are drawn without thickness, since there are already sufficient cues to orientation and the picture would be confusingly complex with those extra lines. The loops, in contrast, are drawn as round "ropes," because none of the simple conventions for showing peptide plane orientation in nonrepetitive structure correspond to meaningful structural features (for helices and β-strands, the ribbon plane really represents the hydrogen bonds

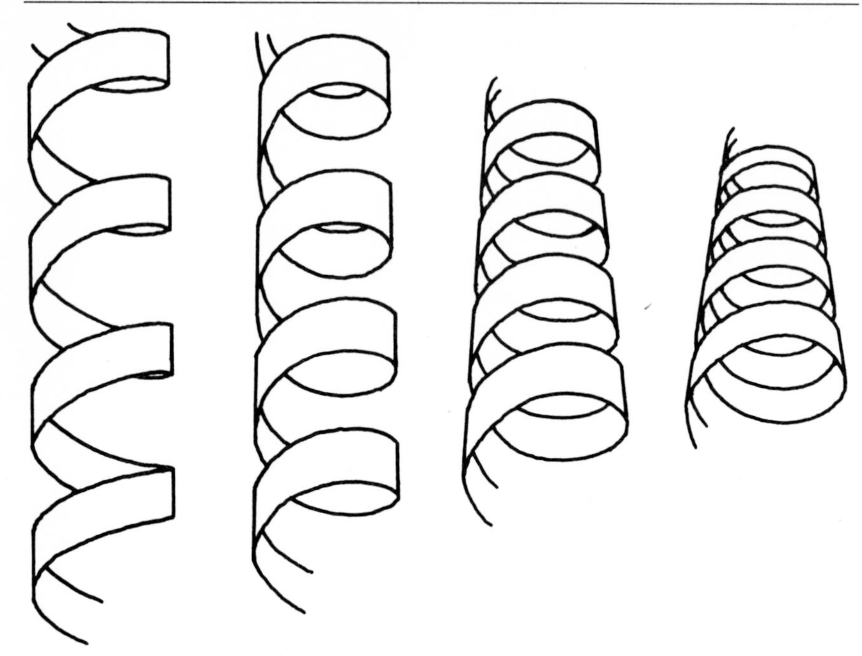

FIG. 7. A collection of α-helices at varying angles to the page.

rather than the peptide direction, since successive peptides flip by 180° in a β-strand but not in a helix). This inconsistent set of conventions surprisingly produces a visually acceptable, unified appearance, and seems to work better than forcing a single convention which is bad for some of the structural features.

The β-strand arrows are about half the width of the strand-to-strand separation. This is a compromise: if the arrows are much wider it is difficult to see features behind them, and if they are narrower there is not enough visual continuity between H-bonded strands. That continuity, which is mainly a matter of making the adjacent strands locally coplanar, is the single most crucial factor in drawing comprehensible β-sheets. The local plane of an arrow (perpendicular to its length) is determined by the average direction to its two H-bonded neighbors or by the direction between it and a single H-bonded neighbor. If it makes no H bonds for a gap of more than two residues, or more than one at an end, then it is no longer shown by an arrow. β-Strands are very seldom completely flat for their entire length, and the correlation of twist and bend from one strand to the next is a powerful signal to perceive them as part of a unified structure.

Since this representation emphasizes the hydrogen-bonding relationships, the extent of a strand (or arrow) can be defined from explicit

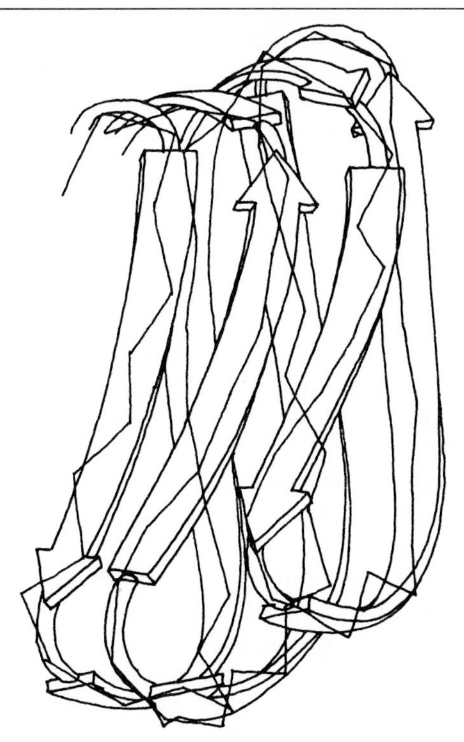

FIG. 8. Sketch of β-strand arrows superimposed on the α-carbon backbone.

H-bonding information if it is available. Otherwise it is done by careful evaluation (in stereo) of which regions maintain appropriate strand-to-strand separation and direction, including consideration of whether the backbone "pleats" of neighboring strands are correlated. I prefer to think of β-strands as continuing through β bulges[13] and other local irregularities, and to be rather generous at the ends or edges of large sheets even if the strand may be separated by a bit more than a good H-bond distance (this must especially be considered when working from a list of H bonds, which are sometimes specified by quite strict criteria). When you have decided where the β-strands are, mark the beginning and end of each on the C_α plot. One should also decide in which cases H bonding is lost on one side before the end of the strand.

Figure 8 shows a sketch of β arrows superimposed on the C_α plot. First identify the NH_2 and COOH termini of the protein and follow

[13] J. S. Richardson, E. D. Getzoff, and D. C. Richardson, *Proc. Natl. Acad. Sci. U.S.A.* **75,** 2574 (1978).

through the entire chain in the stereo, drawing an arrowhead on the C_α plot at the end of each β-strand. Where the plane of the entire β-sheet is closer than about 45° to the plane of the paper, lightly sketch the two edges of each strand about one-quarter of the way from its midline to its neighbors (so the strand and the gaps are about the same width). Where the plane is more tilted, the controlling factor is determining which strands turn over from this viewpoint, and where. This must be done by following the orientation of the strand and its neighbors in stereo. If a strand begins or ends at right angles to the viewer, then turn the end of it slightly one way or the other (it is usually better to let it turn over if it almost does). This is another case in which avoidance of confusing special positions actually mimics the real binocular image, which is dominated by whichever eye has the more informative view and therefore never appears exactly edge on.

Draw the double line for the front edge of the arrow as it crosses over, and then add the line for the lower edge coming in on either side. Be sure to make both sides visible in a region around the crossover (as in Fig. 3b). Taper the arrow width smoothly in between crossover points and full-width flat regions. In general, arrow width is mainly an orientation cue, and is used for foreshortening only at extreme angles (such as the end-on view of the triose-phosphate isomerase barrel in the previous chapter). Smoothly taper the width of arrow edge visible, always making at least a narrow band show on one side or the other.

Sketch in the angle forming the tip of the arrowhead; it should be almost 90° if the arrow is lying flat, obtuse if you are seeing that arrow more nearly from an end, and acute if from one side. The very tip should lie exactly on the smoothly curving continuation of the arrow midline. Then sketch the line of the arrowhead base and the line across the tail end of the arrow; visualize them as lying in the local H-bond plane of the sheet and perpendicular to the local axis of the arrow. Alter those angles by trial and error until they all look correct and give a consistent idea of the orientation of the sheet. Add the double lines and corner shapes for the edges of the arrowheads and tails. Again, either the top or the bottom of each arrowhead should be made visible. It is good to use a straightedge for drawing the arrowhead and tail (at least on the final ink drawing), but it is better to draw the rest of the arrow freehand, because it almost always has at least a slight curve.

The prealbumin dimer of Fig. 9 includes examples of β-strand arrows with a variety of orientations and twists, and shows how they visually form sheets. Examples of more cylindrical barrels can be seen for triose-phosphate isomerase, catalase, or pyruvate kinase in the previous chap-

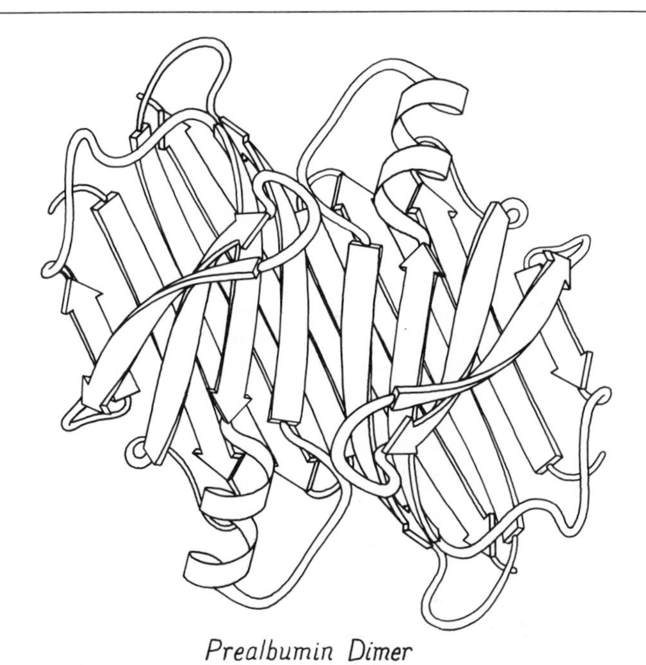

FIG. 9. Prealbumin dimer. β-Strands cross diagonally under other strands; the edges must be offset in order to appear continuous and straight. Front sheet illustrates the feeling of depth produced when a sheet turns over at the corners.

ter, plus examples of twisted sheets viewed edge on in the doubly wound α/β category.

When drawing double sheets that cross at shallow angles as in prealbumin, one should allow for another side effect of binocular vision that produces a perceptual illusion of offset in the back lines. In ordinary perception of such a structure in three dimensions, the two eyes would see slightly around behind the edges of the front arrows, so that in a binocular image the edges of the back arrows would appear offset outward where they emerge from under the top ones. In a mono drawing those back lines must be slightly offset in order to appear straight and continuous; this was done in the prealbumin figure, as can be verified by viewing the page at a glancing angle. Figure 10 shows the appearance of straight and optimally offset lines, and a superposition of the two to show how to achieve the correct offset. Unfortunately, a greater offset in mono merely looks wrong rather than increasing the apparent depth; probably that is because one sees a fused binocular image only for relatively mild

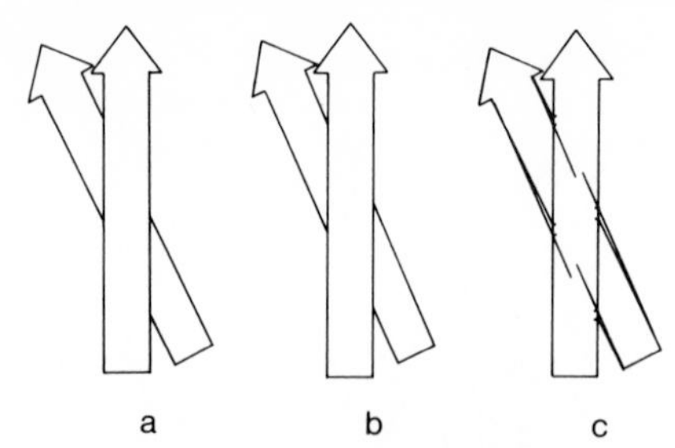

FIG. 10. Offset of rear β-strand necessary to appear straight. (a) Rear arrow edges exactly straight, but appear offset because of optical illusion; (b) edges offset enough to appear straight; (c) superposition of (a) and (b) with offset marked.

offsets and a double image for more extreme ones. Figure 10a is a classic example of the "Poggendorf illusion," which has been known to perceptual psychology since before 1860[14] but has not traditionally been explained in the manner treated here.

Loops and Miscellaneous

The nonrepetitive loops are shown as round "ropes," made fatter in the foreground and thinner toward the back. Sometimes the taper is locally exaggerated to clarify the orientation of an individual loop. In general these loops are simple to draw, with only two factors needing care: smoothing, and treatment of places where they disappear behind other pieces of structure.

A great deal of simplification is necessary to produce a comprehensible schematic in mono: we have already omitted the side chains and the "pleat" of the β-strands, and smoothing of the loops is another important part of that process. If the detailed zigzags are left in the loops, one loses two essential features of a schematic: the ability to unambiguously follow the backbone from one end to the other, and the perception of the structure as a unified whole. Figure 11 shows a C_α plot and the smoothed loops as sketched on top of it. The position drawn for a point on the loop depends mainly on the nearest C_α, but also to some extent on its neighbors up to two away in each direction along the chain. (This is also true of the midline and the orientation of the β-strand arrows.)

The smoothing is critical for unambiguous perception of the continuity of a loop where it passes behind another piece of chain. The loop should be straight or have a gentle, continuous curve that is readily unified by the

[14] R. L. Gregory, "Eye and Brain," 2nd ed. McGraw Hill, New York, 1973.

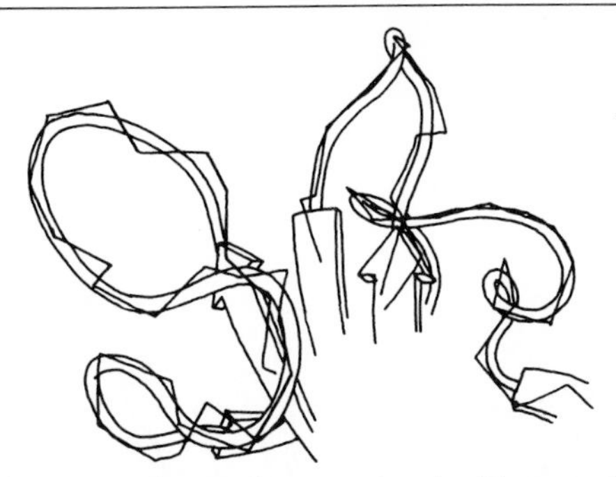

FIG. 11. Sketch of loops superimposed on α-carbon backbone, to illustrate smoothing.

eye. If the angle is oblique, that may require a slight offset (as shown in Fig. 10) in order to *appear* straight. If three chains cross in one spot, a very slight offset of two of them in opposite directions can make the figure unambiguous. Where a loop curls over on itself, the set of representations shown in Fig. 12 all provide a clear visualization of orientation and handedness.

The junctions between loops and helices are simply smooth tapers from one to the other (see Fig. 13 and others); they can be thought of as though the round rope gradually flattened out into the thin helical ribbon. Junctions between loops and arrows are shown best in Figs. 13 and 15. Be sure not to cover very much of the tip of an arrow or the shape becomes confusing.

Other Features

Building on the basic drawing described above, there are a number of possible additions and embellishments. It can be very useful to add selected side chains (e.g., Fig. 14), as long as the total drawing is still kept fairly simple.

The NH_2 and COOH termini must be identifiable. In proteins with a β-strand near one of the termini the directionality of the arrow is sufficient.

FIG. 12. Ways of showing loop ends that curl over by varying amounts.

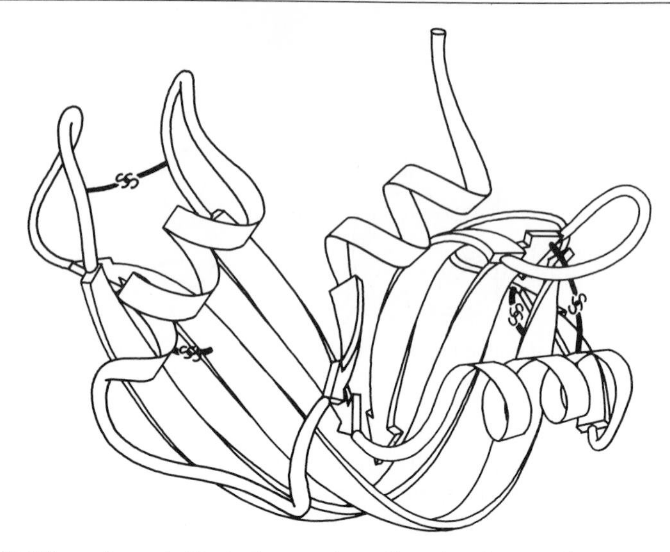

FIG. 13. Ribonuclease A, illustrating a variety of junctions between loops and helices and between loops and β-strand arrows.

In other cases, one or both of the termini can be identified either by small arrows (see Fig. 15) or by letters. For disulfides, I have used either an interlocked SS symbol (Fig. 13) or else a zigzag rather like a conventionalized lightning stroke (see Fig. 14a here and Fig. 16 of previous chapter).

Prosthetic groups or inhibitors can be shown as stick figures (e.g., lactate dehydrogenase domain 1 in the previous chapter) and metals as spheres, either drawn as a single circle or with a couple of partial circles on the lower right to add depth (see Fig. 15). An ordinary drafting compass works well for circles down to about 3/8 in. diameter, but for smaller circles (down to about 1/16 in.) a type of compass called a drop bow pen is extremely useful.

After making a complete sketch of a structure, one should carefully compare again with the stereo to check for errors, especially handedness and whether the correct one is in front where two parts of the chain cross.

If the drawing is to be photographed, it is important to examine it (preferably under a low-power microscope) for line width and blackness. Remember that the eye is very good at compensating for local variations in contrast but the camera is not. Retouch where necessary, using a narrower pen for control, but do not rework the lines too extensively because if the ink is built up thickly or is deeply scratched, that can produce reflections that degrade its blackness when illuminated for photography.

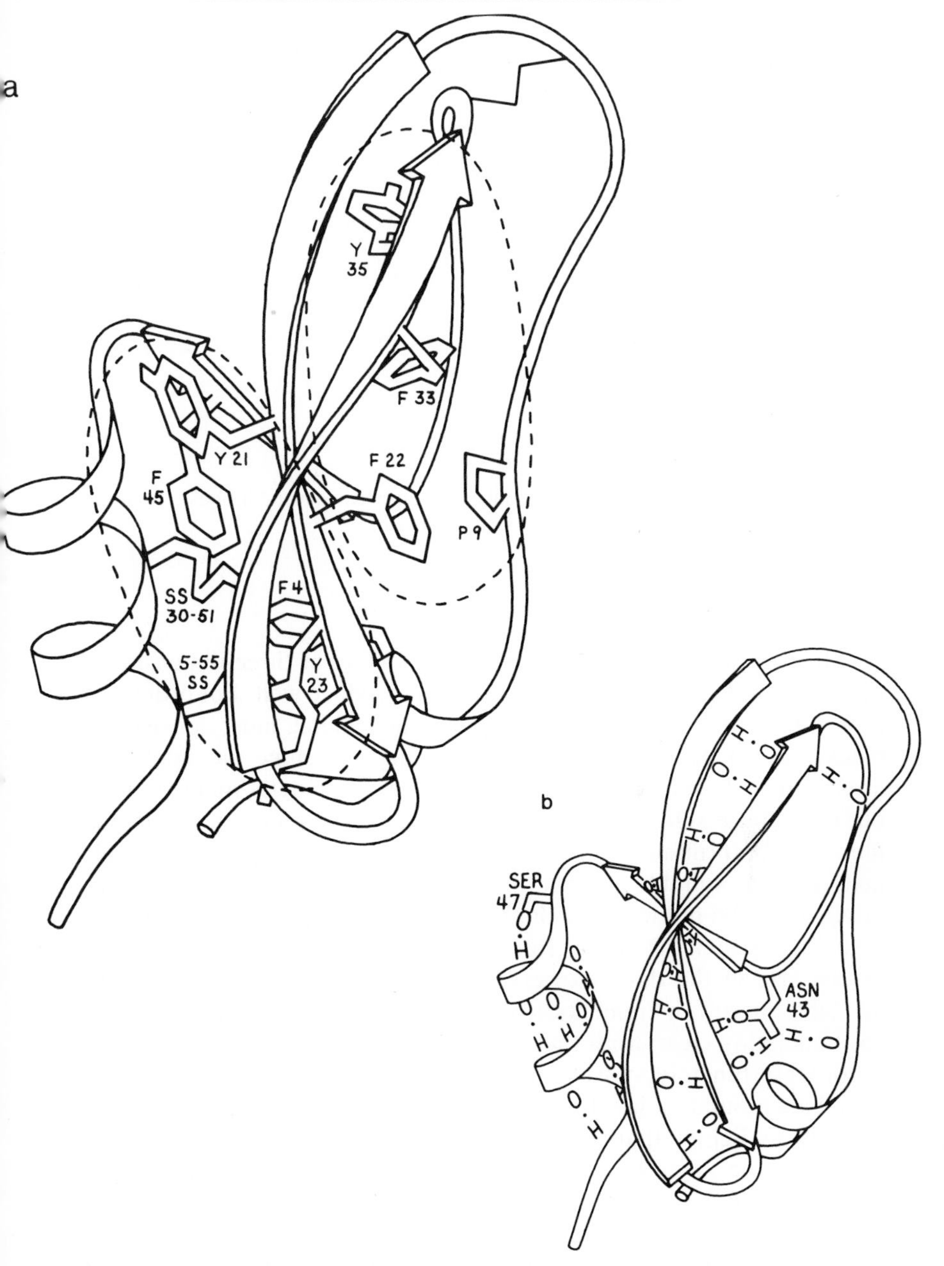

FIG. 14. (a) One way of adding selected side chains to a schematic drawing: the hydrophobic clusters in pancreatic trypsin inhibitor. (b) Pancreatic trypsin inhibitor with its major hydrogen bonds illustrated. The view point had to be slightly different from (a).

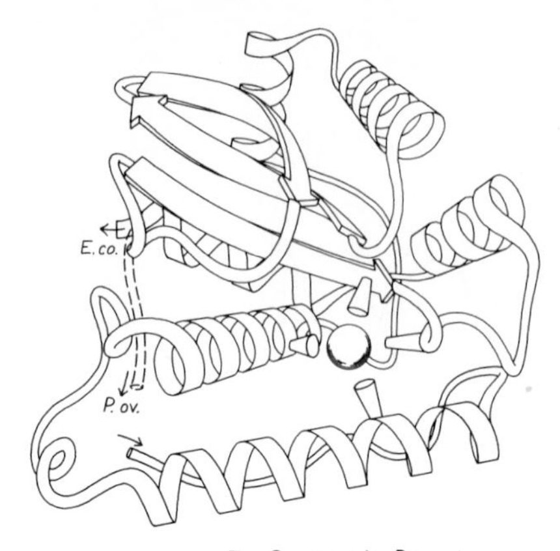

FIG. 15. Ways of shading a sphere, indicating unspecified metal ligands, chain termini, and alternate conformations (Fe superoxide dismutase).

Shading and Color

Shading helps greatly in conveying a feeling of three-dimensionality; the only problem is that it often does not reproduce well. For ordinary line reproduction (as opposed to half-tone), the shading must consist of fine black lines or dots whose size and spacing determine the shade. Small areas can be done by hand, such as the edge of the metal sphere in Fig. 15, but if an entire schematic is to be shaded one should use stick-on plastic film with preprinted dot patterns, which is a standard commercial art material. The suitable number of dots per inch depends on how much the original is to be reduced; for the 7 Å/inch scale it is reasonable to use film with 30 dots/inch (as in Fig. 16). The darkness of the shade is determined by the size of the dots, expressed in percentage area of black, usually available in 10% increments from 10 to about 70% (darker ones would have such tiny clear areas that they would inevitably be lost in reproduction). Unfortunately, most available brands have dark gray dots on a light gray background rather than black on white, and often a "50%" may actually be lighter than a "40%." The only acceptable brand we have discovered so far is Formatt (see notes at end for mail-order sources if you cannot find a good brand locally).

These films are used by cutting out an oversized area with an Exacto knife or Olfa cutter, peeling it off its backing, sticking it over the part of the drawing to be shaded, carefully cutting along the outline desired, and

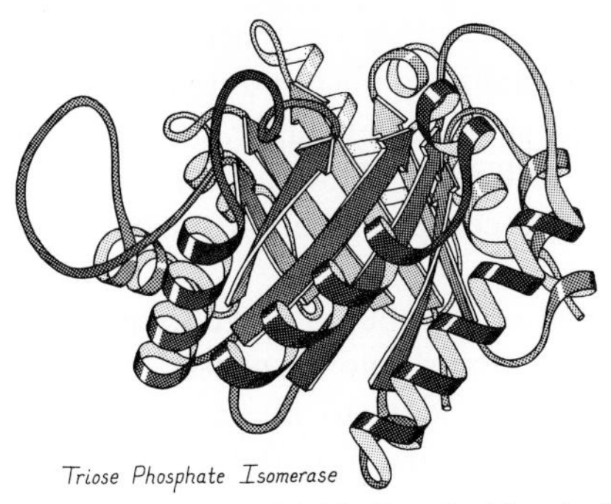

FIG. 16. Triose-phosphate isomerase, with black-and-white shading to show depth, and highlights on the α-helices.

peeling away the excess from around the edges. Light shadings should be cut along the inside of the ink line (since if left across it they may make it a lighter gray), while dark shades should be cut near the outside edge, both to avoid leaving a thin line of white and also because some of the ink may peel off with the film.

When using any sort of shading or color, one has the opportunity to show highlights, which add considerably to the three-dimensionality and appeal of the schematic. The simplest highlights to make, and probably the most effective, are those on the outside of the helices (see Fig. 16). The highlight is a narrow strip of brightness slightly high and/or to the left of the helix midline. For dot or color overlay films, use a straightedge to guide a pair of parallel cuts, and peel off the highlight strips. Exact distance in from the edge is not very critical, but it is important that it be accurately parallel to the edge of the helix (or at the correct intermediate angle if the helix tapers) and that it line up correctly from one turn to the next. If coloring with markers, make the highlight edges accurate in average position and direction, but irregular.

Figure 16 shows a dot-shaded schematic carefully produced with high-quality dot film, retouched, and reproduced from a 1:1 Kodalith 4 × 5 negative so that only a single reproduction step is involved. Of course, I do not know in advance whether it is an example of successful or unsuccessful reproduction; however, in the best cases the shading can be extremely effective, especially for the more complicated multilayer structures. For journal reproduction it is probably safer to stay with unshaded, simple-line schematics, but for making slides, which is an inherently simpler process over which you can have direct control (see below), shading is definitely worthwhile. The most effective simple system of shading involves making features in front the darkest, gradually becoming lighter (and lower contrast) toward the back. This presumably works because it emphasizes what you would see best and it suggests the fade-out into the misty distance that one sees, for example, in mountain landscapes. Since there are not enough different shading levels available to differentiate everything, a reasonable compromise is to use 10% for the back (inner) side of all helices.

Even more information can be conveyed with the use of color, while the corresponding difficulties and expense of reproduction are even greater. The simplest method is to photocopy your original, then use markers or colored pencils to color the copy, and take a slide of it. Colored pencils are very easy to control and shade but tend to give pale results. Markers provide striking colors, but they fade after several months. They also should be tested to find types and colors that blend fairly smoothly between new and dry ink.

Stick-on plastic film is also made in colors (called color overlay film), and produces very uniform, brilliant colors. It is best used directly on the ink original, since it lifts off the black lines on copies. Some companies (e.g., Pantone) make varying percentage shades of a single color (produced with an extremely fine dot screen, usually in 20% increments) with which one can make shaded color drawings.

If a drawing is to be done in color for publication, the best system is to use what are called "color separations." Using a special dark orange plastic film on a clear plastic background (or black ink on paper for pure line work), make a separate positive for each shade of each color. A set of three registration marks (available as stick-on cross hair symbols) is exactly lined up on the original and each separation. Tape the orange film (on its backing) over the original, cut around the relevant edges, and peel off the orange film everywhere except where that particular color shade is to be. The printer will then make one master negative for each color, by combining the separations for all shades of that color. This process is time-consuming and expensive, but it does produce excellent results as long as the registration of the different colors is done carefully in the final printing step.

Reproduction

Many of the steps described here involve photography, and in a very real sense such a drawing has not been finished until it is translated into a reproducible medium. Some of this can be done by handing it to a professional photographer, but it is faster and can produce much more reliably high-quality results if you do it yourself.

The main criteria for both color and black-and-white photography are sharp line reproduction and very high contrast (so that the background is completely clear and the lines are really black or the colors really saturated). This cannot be achieved with normal snapshot films and ordinary development. If you are in a hurry (as one often is!) you can photograph a color original in *bright* sunlight with Kodachrome or illuminate it on a copy stand and use tungsten-light Ektachrome film. Best exposures will be about two stops more open than the meter reading (because of the large white spaces), but be sure to bracket them by at least a stop on each side, in half-stop increments. The results will be worth showing, but the colors will be a bit pale and the background usually slightly tinted or, at best, light grey. Slide films are usually balanced to come out very slightly yellowish, so one can use color-compensating filters to obtain a neutral background. The exact balance varies with the batch of film, but for Ektachrome 50 we have used something between 5C 5M and at most 20C 20M.

The only system I know of for achieving really high-contrast color slides was worked out by David Richardson. It uses Kodak photomicrography film (type 2483), which requires a strong green filter (about 90C 50Y) to produce true colors but is inherently highly saturated and has a clear base. It is developed in E-4 chemistry with doubled times for both the first and color developers, and fresh chemicals for each run. Unfortunately, he has not found an E-6 process film that will give a really clear background even with push-processing.

For any copy photography, but especially for color, evenness of illumination is very critical. It cannot be judged by eye, but should be checked out with a light meter to be within one-half stop over the entire area the original will occupy. For large originals, this usually requires four lights. They must be set so that no reflective highlights are visible through the camera, since they will locally wash out the image. The backing behind the original should be a bright white, and if the drawing has any tendency to curl it should be taped down.

For good black-and-white slides or prints, probably the simplest system to handle is Kodak high-contrast copy film (type 5069) developed in D-19. It seems to be identically replaceable with the 35-mm movie film type 5369. Both are very high contrast, fine grain, and have a completely clear base. Correct exposure for the initial negative will be two to three stops more open than the meter reading. When choosing the optimum exposure, examine the negatives under magnification to look for narrowing or unevenness of the (white) ink lines or a brownness of narrow black areas if exposure is too long, and for spreading of the white areas or a gray background if exposure is too short. A low-power microscope is also useful for judging color slides or black-and-white positives, but they should also always be projected.

To make a positive slide from a 35-mm negative it is useful to have a slide copier with bellows, to adjust magnification; a light box makes a reasonable light source. Exposure latitude is wide enough with high-contrast copy film so that once you have worked out the correct *f*-stop and exposure for your setup, you should not need to bracket exposures each time.

Another useful black-and-white film is Kodalith, either for 35-mm negatives or positives, or for 4 × 5 high-quality negatives for publication. The exposure latitude for producing a good negative is much narrower than for copy film, but once achieved it will produce good prints over a very broad range since the blacks are totally opaque. They may, however, show tiny clear pinhole defects which need to be blacked out on the negative.

Sources of Materials

For graphics supplies, including Formatt dot screen overlays, Pantone color overlay films both solids and % shades, Amberlith masking film (for color separations), stick-on register marks, Olfa or HT cutters with snap-off blades, etc.:

Graphic Supplies and Services
2166 Faulkner Road, N.E.
Atlanta, Georgia 30324

Charrette Corporation
31 Olympia Avenue
Woburn, Massachusetts 01801

A. I. Friedman
25 West 45th Street
New York, New York 10036

For Leroy lettering pens, high-quality tracing paper, drop bow pens, etc.:

Keuffel and Esser Co.
Morristown, New Jersey 07960

[25] Computer-Generated Pictures of Proteins

By ARTHUR M. LESK and KARL D. HARDMAN

Goethe wrote, "Architecture is frozen music." Molecular biologists find this an apt remark as they investigate themes and their variations and interactions in the architecture of proteins and nucleic acids. In this chapter we describe the applications of a computer program that draws stereoscopic pictures of biological macromolecules.[1] The program makes it easy to explore various ways of depicting a structure, including the free mixing in a single drawing of different styles of representation of different portions of a molecule, together with labels and comments.

Computer Graphics in Molecular Biology

The visual display of structures is a principal tool of computational molecular biology. The initial motives for applying computer graphics were the unwieldliness and instability of physical models and their readiness to denature, together with the fact that the space and material requirements increase as more structures appear. Recently, it has become clear that the real struggle is for intelligibility against the pressure of growing complexity. The structures being solved are increasing in size and accuracy as well as in number, and the questions we want to ask are increasing in subtlety.[2] Furthermore, although not all calculations are intrinsically pictorial, it is important to be able to *integrate* graphics into the operation of many computer programs. For example, it is frequently very useful to superpose pictures of two or more molecules in order to compare their structures, or to look at a "movie" of the course of a molecular dynamics simulation.

To design software it must be recognized that no single type of representation of a protein or nucleic acid structure is adequate for all purposes.[3,4] Various modes of representing structures have been tried, both by draughtsman and computer. Initially there were attempts to reproduce the appearance of the physical models which the computer programs were

[1] A. M. Lesk and K. D. Hardman, *Science* **216,** 539 (1982).

[2] A. M. Lesk and K. D. Hardman, *in* "Supercomputers in Chemistry" (P. Lykos and I. Shavitt, eds.), p. 143. Am. Chem. Soc., Washington, D.C., 1981.

[3] A. M. Lesk, *Trends in Biochem. Sci.* **6,** XIV (1981).

[4] A. M. Lesk, "Introduction to Physical Chemistry," pp. 466 and 542. Prentice-Hall, Englewood Cliffs, New Jersey, 1982.

superseding. Thus some representations correspond to a fairly direct translation—atom by atom or bond by bond—of the structure into a picture.[5,6] These are most useful for detailed analysis of a small portion of a large molecule, or the environment of a substrate or other small molecule bound to a large one.

The complexities of large structures have demanded an evolution of other representations, including abstractions. These include the schematic diagrams used first by A. Rossmann, A. Liljas, and B. Furugren, in which cylinders represent α-helices and arrows represent strands of β-sheet, and related representations which have been applied extensively by J. Richardson. Questions of the gross folding pattern, or topology, of a macromolecule are better served by these more abstract representations.

In order to gain the ability to produce stereoscopic diagrams, and to provide the flexibility essential to mix different representations in one picture, we have compiled into one coherent computer program the many ways of depicting biological macromolecules that have proved their utility. The options range from wire or skeletal models to schematic diagrams, or "cartoons." Any structural segment may be assigned a color, or may be drawn with broken rather than solid lines. The program can produce stereo pairs, and remove hidden lines whenever appropriate and desired.

Basic Operation of the Program

Line Drawings

Execution of the line-drawing version of the program may be divided into four phases: (1) Input of coordinates and picture composition information; (2) Creation of the geometry of the picture, in three-dimensional space; (3) Hidden-line removal, if desired; and (d) Projection into two dimensions, character generation, and output. For stereo pairs, steps 2, 3, and 4 are repeated for two orientations of the molecule (rotated by approximately 6° around an axis perpendicular to the line of sight).

The Input Phase. The program begins by reading the two input files: the coordinates, and the description of the contents and appearance of the picture. The picture composition file selects regions of a structure to draw, and specifies how they are to appear. This file is read in free-format

[5] C. K. Johnson, "ORTEP: A Fortran Thermal-Ellipsoid Plot Program for Crystal Structure Illustrations," ORNL-3794 revised. Oak Ridge Natl. Lab., Oak Ridge, Tennessee, 1965.

[6] C. K. Johnson, *in* "Crystallographic Computing" (F. R. Ahmed, S. R. Hall, and C. P. Huber, eds.), p. 227. Munksgaard, Copenhagen, 1970.

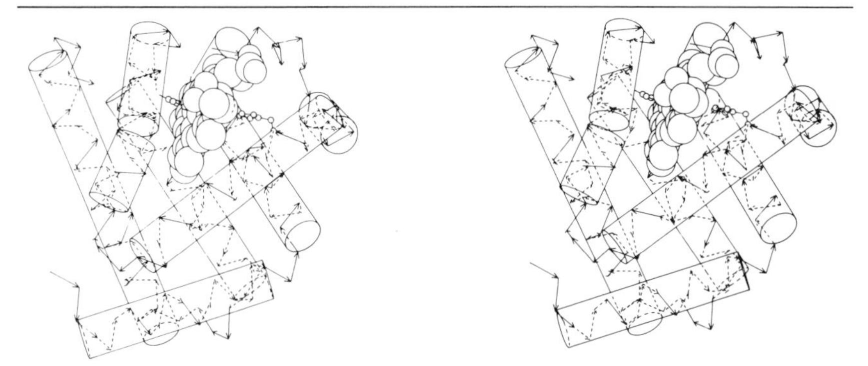

FIG. 1. Sperm whale myoglobin, showing the representation of helices as translucent cylinders, the tracing of the backbone by arrows linking successive C_α atoms, the heme group as an opaque space-filling model, and proximal and distal histidines as ball and stick models.

style, with keywords introducing lists of parameters. Sensible defaults are provided whenever possible. A user's manual describes the conventions.

Picture Generation. Next, the program prepares a picture by appropriate geometric transformations of coordinates into picture elements. For example, for each α-helix that is to be represented by a cylinder, the program generates a cylinder of appropriate size and orientation, by superposing the coordinates onto a standard polyalanine helix. Two sets of information about each cylinder are retained. First, the endpoints of the line segments that define its outline are placed in a file of draw commands.

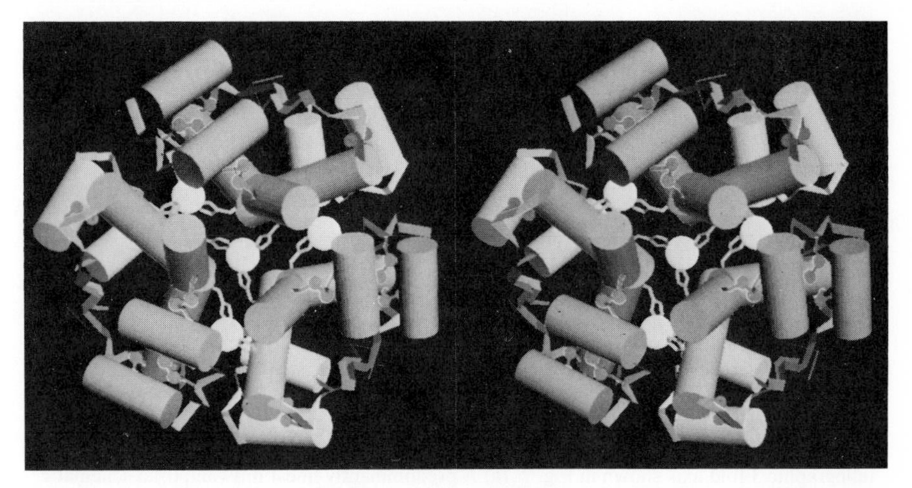

FIG. 2. A halftone picture of the four-zinc insulin hexamer, viewed down the 3-fold axis.

Second, the surfaces of the cylinder, which are potentially opaque, are saved in a file of "windows" for a subsequent hidden-line removal step. (Windows may be open = transparent = inactive in hidden-line removal, or screened = active in hidden-line removal). Both drawing commands and window data are stored by pointers to a list of coordinates extracted from, or computed from, the initial atomic coordinates.

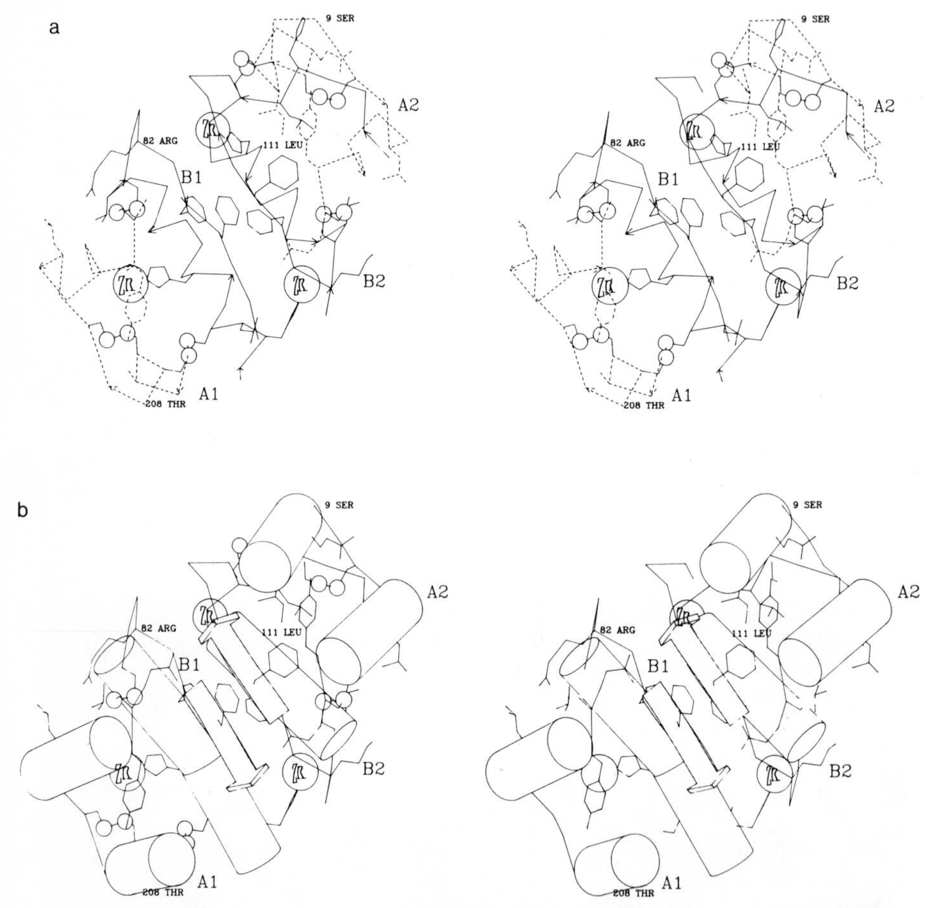

FIG. 3. Comparison of two representations of the dimer of four-zinc insulin, showing the backbone, selected side chains, disulfide bridges, and zinc atoms. Each monomer contains two polypeptide chains: A chains drawn with broken lines and B chains and disulfide bridges drawn with solid lines. The A1-B1 and A2-B2 monomers are related by an approximate (noncrystallographic) 2-fold axis perpendicular to the paper, and perpendicular to the crystallographic 3-fold axis shown in Fig. 2. (a) A predominantly linear drawing; (b) a schematic drawing.

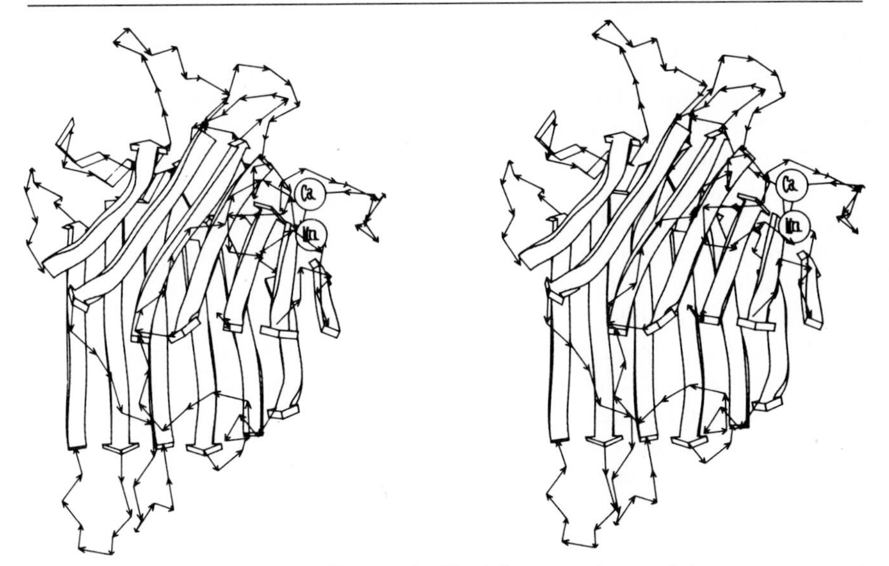

FIG. 4. Jack bean concanavalin A, a double-β-sheet protein containing manganese and calcium.[7]

All of the data at this stage are still in three-dimensional form. Indeed, we can "tap off" these data at this point in the computation, and send them directly to a three-dimensional display device.

Hidden-Line Removal. Some of the possible representations of sections of protein structures may have the appearance of solid objects, such as cylinders representing α-helices, or arrows representing strands of β-sheet. Only these representations generate windows for the hidden-line removal step; skeletal models do not. Associated with every potentially

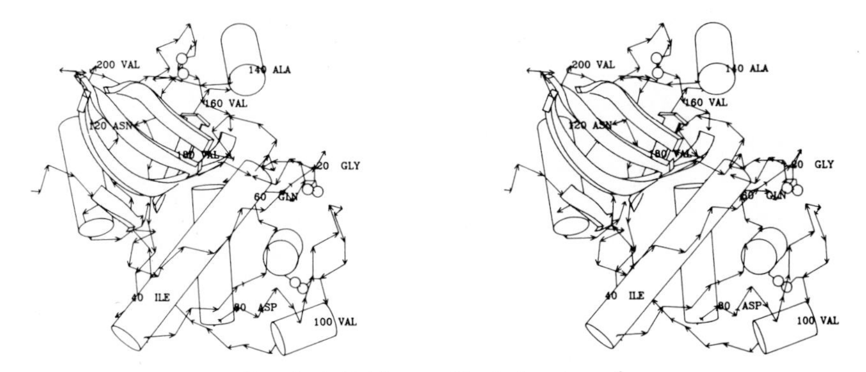

FIG. 5. Actinidin, a sulfhydryl protease.[8]

opaque element of the picture is an effective "optical density": the object may be regarded as *transparent, translucent,* or *opaque*. If it is transparent, lines passing behind it are left unchanged. If it is translucent, lines passing behind it are converted to dashed lines. If it is opaque, lines passing behind it are deleted entirely.

It is useful to be able to assign different optical densities to different portions of a protein. For example, to show the geometrical relationship between two occluding helices while deleting everything lying beyond the further helix, the front helix might be made translucent and the back helix opaque. It is possible to override hidden-line removal selectively: any object may be assigned the BURN attribute, in which case it will appear in the drawing even if it lies behind one or more opaque objects (these lines will be "burned through" the opaque material that lies in front of them).

In saving the data for hidden-line removal, it is important to ensure that objects do not obliterate themselves. To avoid this, and also to facilitate such options as BURN, there is a system of assignments of "immunity," by which certain lines are rendered immune from clipping by certain windows. This is a useful general technique for manipulating pictures.

The output from the hidden-line removal step is a modified set of draw commands. Certain "drawto's" have been converted to "moveto's" to effect the deletion of a line segment; conversion of solid lines to dashed lines (when a line segment passes behind a translucent object) is accomplished by changing an attribute of the line segment.

Output. After the hidden-line removal step, the line segments are projected to two-dimensional space. At this stage character strings are expanded to sets of line segments, via a set of stroke tables based on the Hershey fonts. The final output file contains only commands of the following three types: "MOVETO X Y," "DRAWTO X Y," and

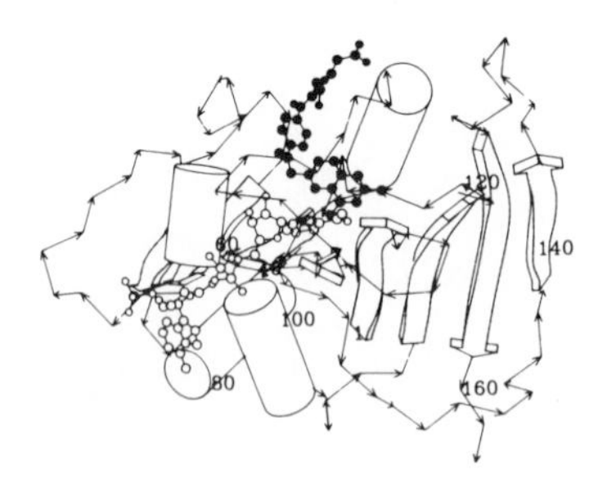

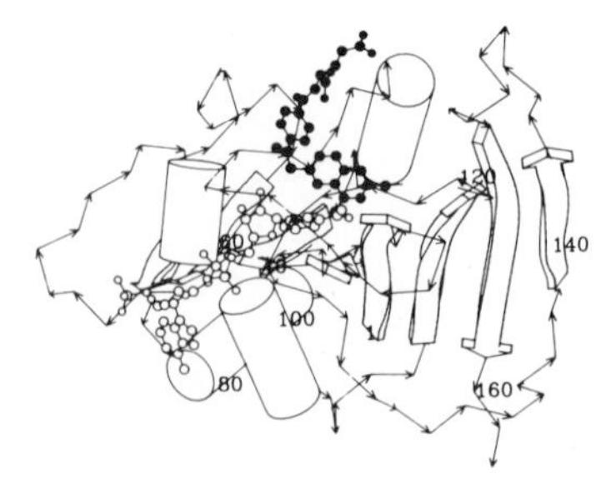

FIG. 6. The ternary complex of dihydrofolate reductase, cofactor NAD (light atoms), and inhibitor methotrexate (dark atoms).[9]

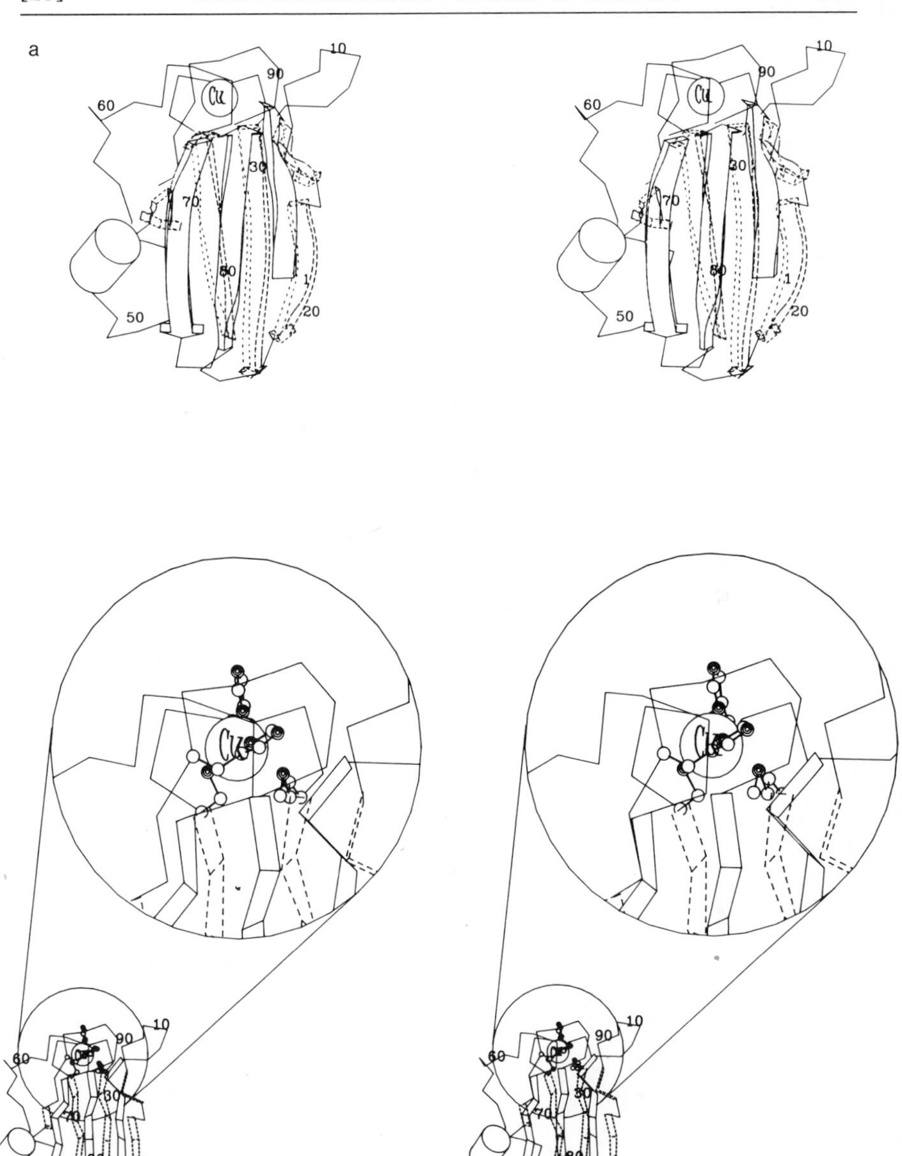

FIG. 7. Poplar leaf plastocyanin, a double-β-sheet protein containing a copper atom.[10] The front sheet is drawn with solid lines, the back sheet with dashed lines. (a) Representation of sheets by curved arrows. (b) Representation of sheets by ribbons, with a close-up view of the copper binding site. The copper ligands are His 37, Cys 84, His 87, and Met 92. Nitrogen and sulfur atoms of these side chains are darkened.

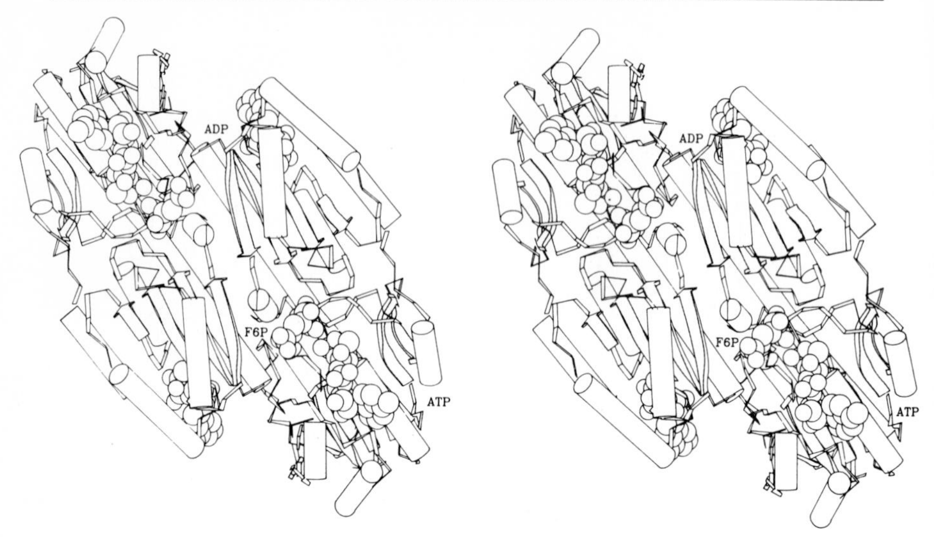

FIG. 8. Phosphofructokinase (R state), binding substrates ATP, and fructose 6-phosphate; and activator, ADP.[11] This view shows one dimer of a tetramer. There is a 2-fold axis perpendicular to the page.

"CHANGE PEN COLOR to n." The coordinates X and Y are scaled to a frame $0.0 \leq X,\ Y \leq 100.0$.

Color-Raster Output

To create color-raster output, an alternative step parallel to the hidden-line removal step is executed, in which the drawn lines are ignored

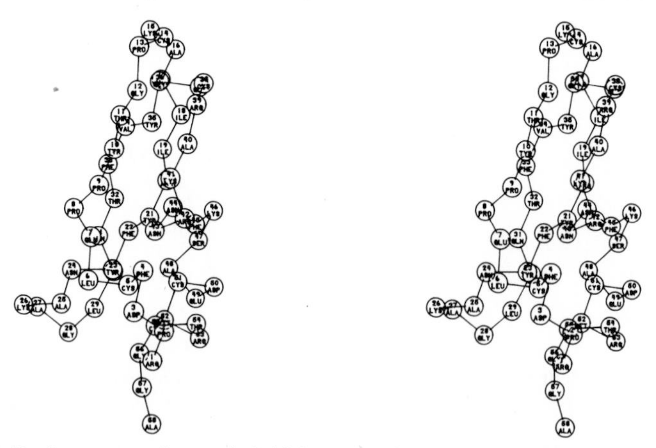

FIG. 9. Bovine pancreatic trypsin inhibitor, showing the amino acid sequence and backbone conformation.[12]

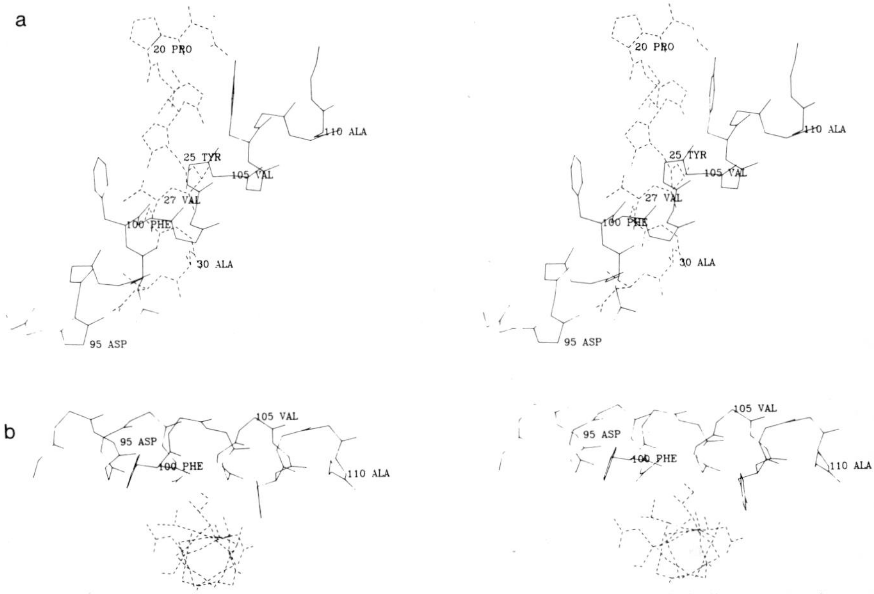

FIG. 10. The interdigitation and packing of side chains in the B–G helix contact in *Chironomus* erythrocruorin.[13] B helix residues are drawn with solid lines, and G helix residues with broken lines. (a) and (b) Skeletal drawings of the two helices viewed in directions perpendicular and parallel to the contact between the helices. (c) Space-filling representation of the packing at the contact. Slices are drawn through spheres of the van der Waals radius around each atom. Several parallel planes are assembled in this picture. This figure may be compared to the hand-drawn version in Ref. 14, p. 248. (d) Overlay of the skeletal model onto the space-filling representation.

and the *windows* are painted in, on the basis of the user-assigned color and the reflected light intensity computed from the orientation of the window polygon relative to a simulated light source.

Examples

Figures 1–10 illustrate the use of various combinations of representations to portray a number of interesting structural features of proteins.[6a–14]

[6a] All pictures were photographed directly from computer output.

[7] K. D. Hardman and C. F. Ainsworth, *Biochemistry* **11,** 4910 (1972).

[8] E. N. Baker, *J. Mol. Biol.* **141,** 441 (1980).

[9] D. A. Matthews, R. A. Alden, J. T. Bolin, D. J. Filman, S. T. Freer, R. Hamlin, W. G. J. Hol, R. L. Kisliuk, E. J. Pastore, L. T. Plante, N.-H. Xuong, and J. Kraut, *J. Biol. Chem.* **253,** 6946 (1978).

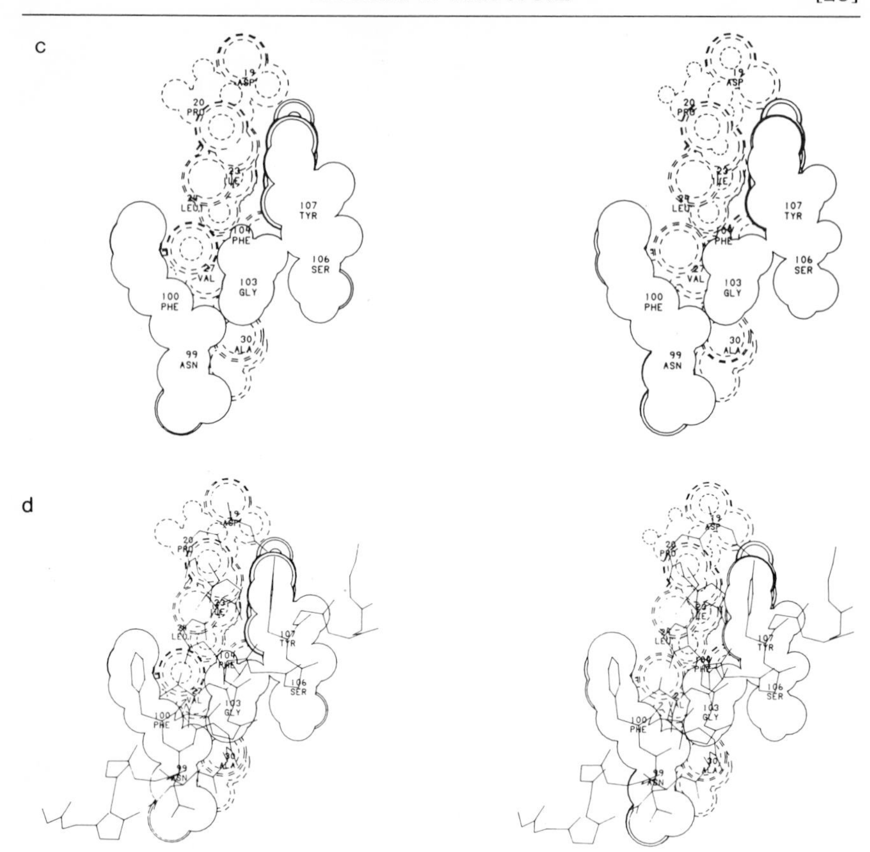

FIG. 10. (*continued*)

Acknowledgments

Work supported in part by the National Science Foundation Research Grant (PCM83-20171), the North Atlantic Treaty Organization Division of Scientific Affairs (RG275.80), the European Molecular Biology Organization (SF4-1980), and the Thomas J. Watson Research Center, IBM Corporation. A. M. L. is grateful to the Medical Research Council Laboratory of Molecular Biology for their hospitality during part of the time this work has been carried out. We thank Drs. P. R. Evans and M. C. Lawrence for their picture of phosphofructokinase.

[10] P. M. Colman, H. C. Freeman, J. M. Guss, M. Murata, V. A. Norris, J. A. M. Ramshaw, and M. P. Venkatappa, *Nature* (*London*) **272,** 319 (1978).

[11] P. R. Evans, G. W. Farrants, and P. J. Hudson, *Philos. Trans. R. Soc. London, Ser. B* **293,** 253 (1981).

[12] J. Deisenhofer and W. Steigemann, *Acta Crystallogr., Sect. B* **B31,** 238 (1975).

[13] E. Weber, W. Steigemann, T. A. Jones, and R. Huber, *J. Mol. Biol.* **120,** 327 (1978).

[14] A. M. Lesk and C. Chothia, *J. Mol. Biol.* **136,** 225 (1981).

[26] Macromolecule Backbone Models

By BYRON RUBIN

There has long been a need for a simple method for easily visualizing the conformations of biological macromolecules. X-Ray crystallographic reports generally provide lists of atomic coordinates and several diagrams of each protein. Without some means for using the atomic coordinates to construct other representations, however, the perception of a protein's structure is limited to those diagrams. Graphical representations, ribbon diagrams, cylinder and arrow figures, stereo drawings, and even xograph[1] illustrations (now Nimslo photographs)[2] are limited to a small number of views. Computer graphics systems[3–6] overcome this problem, but the difficulty of stereo viewing and limited accessibility of these systems to the general scientific community suggest the need for a simpler alternative. The AMSON[7] approach provides a large number of mono and stereo views of each protein in microfiche format. Its modest cost makes it attractive for a number of applications but it and the other graphical approaches to macromolecule structure display presently available fail to provide the ease of visualization which comes from a solid, three-dimensional model.

Construction of comprehensive models in which all of the nonhydrogen atoms in a macromolecule are shown,[8–12] however, is a large and

[1] R. A. Harte and J. A. Rupley, *J. Biol. Chem.* **243,** 1663 (1968).

[2] Nimslo Inc., Atlanta, Georgia 30340.

[3] R. Diamond, "Bilder: A Computer Graphics Program for Biopolymers." International Symposium of Biomolecular Structure, Conformation, Function, and Evolution, Madras, Vol. 1. (Srinivasan Ramachandran, ed.), pp. 567–588. Pergamon, Oxford.

[4] Evans and Sutherland Computer Corporation, 580 Arpeen Drive, Salt Lake City, Utah 84108.

[5] R. Langridge, T. E. Ferren, I. D. Kuntz, and M. L. Connolly, *Science* **211,** 661 (1981).

[6] E. F. Meyer, Jr., C. N. Morimoto, J. Villareal, H. M. Berman, H. M. Carrell, R. K. Stodola, T. F. Koetzle, L. C. Andrews, F. C. Bernstein, and H. J. Bernstein, *Fed. Proc., Fed. Am. Soc. Exp. Biol.* **33,** 2402 (1974).

[7] R. J. Feldmann, "AMSON: Atlas of Macromolecular Structure on Microfiche." Tracor Jitco, Inc., Rockville, Maryland, 1976.

[8] J. C. Kendrew, H. C. Watson, B. E. Strandberg, and R. E. Dickerson, *Nature (London)* **190,** 666 (1960).

[9] Cambridge Repetition Engineers Ltd., Green's Road, Cambridge, England.

[10] F. M. Richards, *J. Mol. Biol.* **37,** 225 (1968).

[11] Nicholson Models Labquip Inc., 18 Rosehill Park Estate, Caversham, Reading R64 8XE, England.

[12] The Ealing Corporation, South Natick, Massachusetts 01760.

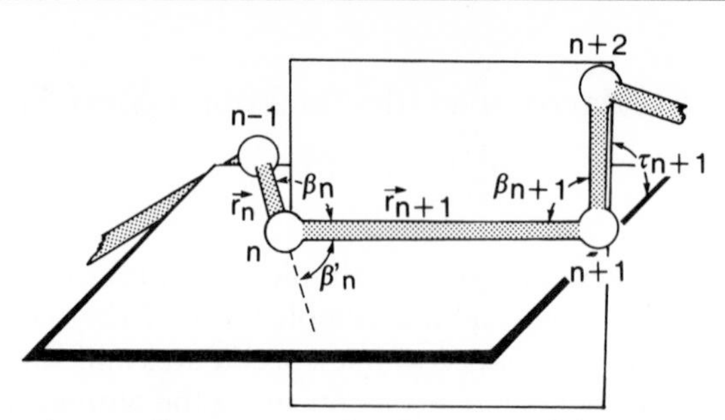

FIG. 1. Progressive coordinate system. The distance r and the angles β and τ are defined for atoms $n - 1$ through $N + 2$.

time-consuming undertaking. Moreover, for many purposes these complete molecular models provide a level of complicating and unneeded detail. When the needs of an investigation can be met with a simpler model, in which only the polypeptide α-carbon or ribose-phosphate backbone are shown (with perhaps a few side chains), several alternatives are now available.

Coordinate System

Each of the two model-building systems to be described share the feature that they use a coordinate system in which the need for a rigid lattice work on which to construct the model is eliminated. Model construction is thus simplified and the models produced can be held in one's hands and viewed from any angle. The model-building coordinate system transformation takes coordinates from an X-ray structure determination, either Cartesian or fractional, and converts them to the (r, β, τ) coordinates of the progressive coordinate system first introduced for this purpose by Haas.[13] In the bent-wire model coordinate system shown in Fig. 1, $|\mathbf{r}_{n+1}|$ corresponds to the distance between consecutive atoms in the model; τ_{n+1} is the torsion angle about the vector $\mathbf{r}_{n+1}$ joining atoms n and $n + 1$ in the model; and β_{n+1} is the angle between two consecutive vectors $\mathbf{r}_{n+1}$ and $\mathbf{r}_{n+2}$. In this coordinate system there is an ambiguity in specifying the first and last atoms of a linear macromolecule. Determination of the τ for these terminal atoms requires additional information about the molecule which is usually available or which can be assumed, with little consequence in the resulting model. The advantage of this coordinate system for model building is that the reference frame for the next atom to be

[13] D. Haas, *Biopolymers* **9**, 1547 (1970).

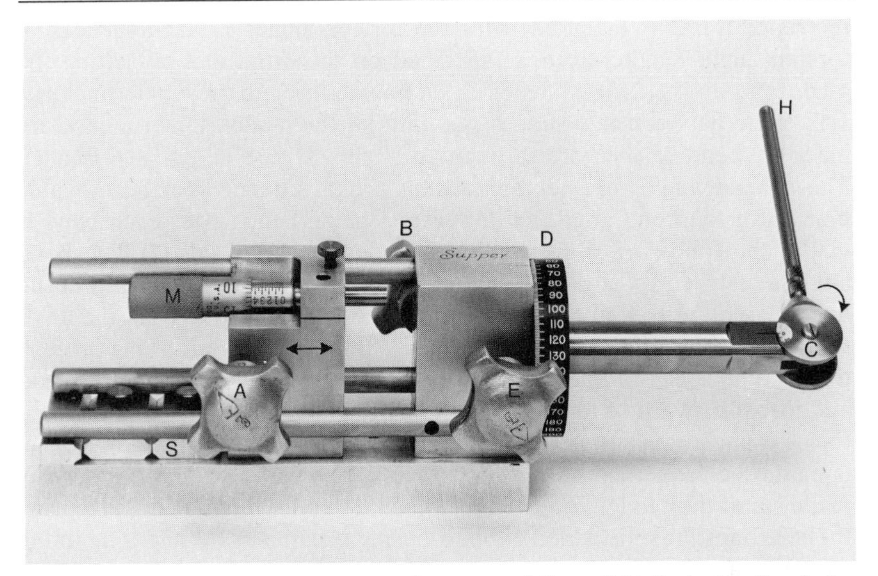

FIG. 2. "Byron's Bender." The parts of a commercially available[15] wire bending device are identified. (A) Locking control on the sliding rod clamp with the direction of motion (↔) shown; (B) locking control on the fixed rod clamp; (C) and (H) bending anvil and handle showing the dial on which the bending angle β' is read and the direction of the handle motion; (D) and (E) drum dial and locking control for the angle τ'; (M) micrometer with which the length of each segment, proportional to $|\mathbf{r}|$, is set; and (S) adjustable backstop for sliding clamp A.

added to the chain is determined by the previous two atoms. Although this eliminates the need for a rigid lattice on which to build the model, it also causes errors introduced at any stage of model construction to be propagated throughout the rest of the model.

Constructing Bent-Wire Models

Bent-wire models[14] are formed from a continuous length of 3-mm-diameter steel rod. A bending device, shown in Fig. 2, on which a multiple of each distance $|\mathbf{r}_n|$ and two angles related to τ_n and β_n can be set, is commercially available.[15] With this device a continuous length of wire is passed first through a *sliding clamp,* then through a *fixed clamp,* and finally through a *bending anvil.* The drum dial attached to the bend anvil rotates around the wire while the wire is held in at least one of the two clamps. By always keeping the steel rod secured in one of the clamps, a

[14] B. Rubin and J. S. Richardson, *Biopolymers* **11,** 2381 (1972).
[15] Charles Supper Co., Natick, Massachusetts 01760.

reference frame is maintained for the torsion angles τ. Each successive torsion angle τ_n, therefore, is specified on the drum dial by setting the angle $\tau'_n = \text{mod}_{360} \sum_i^n \tau_i$. The wire as it passes through the bender makes a 180° angle between two successive units of the model. Other angles are made by bending the wire through an angle $\beta'_n = 180 - \beta_n$ (see Fig. 1). These bend angles are set by reading a dial, attached to the movable portion of the bend anvil, as the wire is being bent. After each bend is made, the following sequence of operations is followed. (1) The bend anvil is restored to its zero position; (2) with the steel rod secured in the fixed clamp the sliding clamp is released and moved away from the anvil to its backstop and the clamp tightened; (3) the appropriate value proportional to $|\mathbf{r}_n|$ is set on the micrometer dial; (4) the fixed clamp is released and thrust forward to the bend anvil; (5) the fixed clamp is tightened and the next angle τ'_{n+1} is set on the drum dial. After the bend in the rod, β'_{n+1}, is made the sequence of operations 1–5 is repeated and another bend is made, until the model is complete. Disulfide bonds, side chains, or prosthetic groups must be added either by using a clamp or by soldering to the steel backbone. This hardware, however, is not commercially available.

Models built in this way (one is shown in Fig. 3) generally need to be tuned by comparing and then bending the model to agree with a computer-generated graphic. Tuning compensates for bending errors introduced during the course of the model building. Wetlaufer and Rose[16] have made a detailed study of these bending errors, and have identified sources and corrections for local errors involving single bends and global errors in which disparate parts of the model lie in positions which deviate significantly from their intended position. A major source of local error was wear of the bend anvil, which caused significant discrepancies between the bends produced and the readings on the bend anvil dial. These errors are not linear and increase in magnitude with continued use. They are easily compensated for, however, by regularly preparing and using a calibration curve for the bend angles. Any error in the r, β', or τ settings are cumulative and lead to global errors. These errors can be corrected by inserting a few reinforcing elements of correct length between disparate parts of the model. A more complex tuning procedure involves using a lattice work, setting the positions of few atoms with their correct Cartesian coordinates in the lattice, and then adjusting the model so that these atoms are in the correct relative positions. Either of these procedures or their combination leads to models of modest fidelity and durability.

[16] D. Goldberg, S. Saliterman, D. B. Wetlaufer, G. Rose, and T. E. Hopkins, *Biopolymers* **14,** 633 (1975).

FIG. 3. Wire model of the enzyme ribonuclease S.

Shapely Models

Light-weight plastic models, based on a coordinate system similar to the (r, β, τ) system above, can be constructed using the molded parts shown inset in Fig. 4.[17] These cleverly designed model-building parts, which are available commercially,[18] have the advantages (1) of needing no special tools for their construction, and (2) of being compatible with another, commercially available model-building system.[11] As one can see in Fig. 4, each ball and stick is inscribed with a vernier dial with which the torsion angle τ and the bend angle β can be set with an accuracy of approximately 2°. The scale of the models and the distance between consecutive bonds are fixed. The errors which this fixed length introduces are small and can be compensated for by recalculating the β and τ angles to minimize the deviation. In constructing models using the Shapely system, the torsion angles for each residue in a short segment (~10 residues) of

[17] R. J. Fletterick and R. Matela, *Biopolymers* **21,** 999 (1982).

[18] Shapely Molecules, 352 S. Morningside Ave, Mill Valley, California 94941.

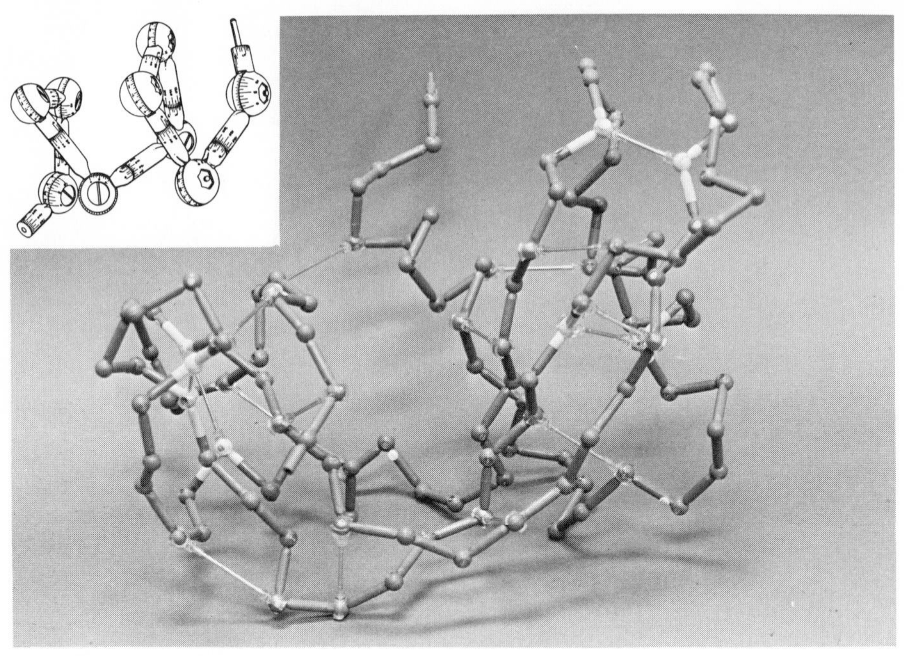

FIG. 4. Shapely model of ribonuclease A. The inset diagram shows the vernier dials on which the bend angle β and torsion angle τ are shown.

the molecule are set one at a time. Quick-setting cyanoacrylate cement is applied to the pair of mating "stick" pieces and then twisted to the correct torsion angle. Bend angles are set in similarly by adjusting the mating "ball" pieces to the correct angle and, after one has checked that the angles were correctly set, using the same glue to fix their position. When several short segments have been constructed, these segments are then assembled. As the molecule grows in size, distances between nonconsecutive residues can be checked. Inserting structural cross-links between residue for which distances have been calculated brings the model closer to the desired conformation and makes the model stronger. A transparent plastic fixture which can be snapped over a "ball" and subsequently fixed in place with glue serves as a means for forming these structural cross-links or for attaching Lab-Quip models. A model of ribonuclease A built using this system is shown in Fig. 4. Molecules of substantial size (800 residues) can be built without the need for an external supporting structure, thereby making them easy to view in any orientation. The models are somewhat more fragile than those bent from wire, and one is limited to a single scale. The individual amino acids can be easily labeled. A selection of colors is available for highlighting or identifying particular residues or residue types.

Acknowledgments

The work reported here has been supported in part by Research Grants GM 26905 and GM 27907 from the National Institutes of Health.

[27] Comparison of Protein Structures

By BRIAN W. MATTHEWS and MICHAEL G. ROSSMANN

Introduction

As more and more protein structures are determined there has arisen considerable interest in studying the relation between them. Such comparative studies reveal a great deal about the evolution of protein structure and suggest factors which are important in the polypeptide folding of these complicated macromolecules.

At the present time perhaps 200 distinct protein structures are known. This provides a reasonably diverse sample from which to draw, but, as it has turned out, the correlation and interpretation of this mass of data have proved to be difficult and, in some instances, controversial. Most protein structures are complicated and can undergo substantial change during evolution. To recognize structural similarity in different proteins may not be trivial. To assess the probability that such structural similarity is due to divergence from a common precursor (rather than convergence toward a pattern of folding dictated by functional requirement or by a facile mode of folding) may be even more difficult.

In some cases, such as in the original structure determinations of the α and β chains of horse hemoglobin[1] and sperm whale myoglobin,[2] two structures are seen to be obviously similar. Later it was shown that the structures of globins from lamprey, marine worms, larval insects, and

[1] M. F. Perutz, M. G. Rossmann, A. Cullis, H. Muirhead, G. Will, and A. C. T. North, *Nature* (*London*) **185,** 416 (1960).

[2] J. C. Kendrew, R. E. Dickerson, B. E. Strandberg, R. G. Hart, and D. R. Davies, *Nature* (*London*) **185,** 422 (1960).

even lupine root nodules are all very similar.[3,4,5] In addition, the amino acid sequences are all homologous. Similarly, the structures of the serine proteases chymotrypsin, elastase, and trypsin were all seen to be practically identical except for minor alterations on their respective surfaces. In such cases there is no difficulty in recognizing structural similarity; nor is there any doubt that each family of homologous proteins diverged from a common precursor. However, as the amino sequences of two proteins become less similar, more substantial changes in structure can occur. For example, James and co-workers[6,7] have shown from studies of the microbial serine proteases, whose amino acid sequences have about 20% homology with the pancreatic serine proteases, that large surface loops may be added to or deleted from a protein during its evolution. Similarly, cytochromes from sources ranging from horse to photosynthetic bacteria show substantial variations in structure, even though the basic underlying protein fold is still conserved.[8,9] The fact that these families of proteins retain a common underlying structure, even though their amino acid sequences have become very different, shows clearly that tertiary structure changes much more slowly than amino acid sequence during evolution. In other words, proteins which are distantly related may have similar three-dimensional structures but have amino acid sequences which have changed to the point that they are no longer recognizably similar. Over a long enough time, the protein structure may also change substantially.

In some cases, proteins have been found to have striking structural similarities, but no apparent functional or amino acid sequence similarity. One such example is copper, zinc superoxide dismutase, whose structure resembles that of an immunoglobulin domain.[10]

The purpose of this chapter is to give an account of various techniques which have been used to establish similarity in protein folds. We discuss ways in which structural correspondence can be measured, and how such measures can be used to attempt to estimate the probability that a given correspondence might have arisen through convergence or divergence.

[3] B. W. Matthews, *in* "The Proteins" (H. Neurath and R. C. Hill, eds.), 3rd ed., Vol. 3, p. 403. Academic Press, New York, 1977.

[4] G. E. Schulz and R. H. Schirmer, "Principles of Protein Structure." Springer-Verlag, Berlin and New York, 1979.

[5] M. G. Rossmann and P. Argos, *Annu. Rev. Biochem.* **50,** 497 (1981).

[6] L. Delbaere, W. L. B. Hutcheon, M. N. G. James, and W. E. Thiessen, *Nature* (*London*) **257,** 758 (1975).

[7] M. N. G. James, L. T. J. Delbaere, and G. D. Brayer, *Can. J. Biochem.* **56,** 396 (1978).

[8] F. R. Salemme, *Annu. Rev. Biochem.* **46,** 299 (1977).

[9] R. J. Almassy and R. E. Dickerson, *Proc. Natl. Acad. Sci. U.S.A.* **75,** 2674 (1978).

[10] J. S. Richardson, D. C. Richardson, K. A. Thomas, E. W. Silverton, and D. R. Davies, *J. Mol. Biol.* **102,** 221 (1976).

Structural Classification

Rao and Rossmann[11] observed that there were structural components comprising a few α-helices or β-strands which were frequently repeated within structures. These they called "supersecondary structures" (being intermediate to secondary and tertiary structure) and suggested that these might be due to convergence. A variety of recurring structures were subsequently recognized such as the "Greek key."[12] These structural motifs are often represented schematically by "topological" diagrams in which the overall patterns of α-helices and β-sheets are laid out in a manner which makes it relatively easy to see correspondences between different proteins.

The first comprehensive attempt to classify proteins on the basis of structural comparisons was due to Levitt and Chothia[13] who used four major classifications: (1) proteins containing mostly α-helix; (2) proteins containing mostly β-sheet; (3) proteins that contain α-helices and β-strands in an irregular sequence; and (4) α/β proteins with alternate segments of α-helices and β-strands. Another classification scheme has been attempted by Richardson[14] by counting the number of structural layers within a protein.

It soon became apparent that a polypeptide could frequently be separated into two or more domains with quite different structural characteristics. Furthermore, evolution apparently proceeded by suitable recombination of domains[15] and adaption of domains to the requirements of function. Combination of functional elements at the boundaries between domains gives an enzyme its particular properties. That this concept is one principle in the evolution of protein function has been emphasized by the correlation of domains with exons in DNA expression.[16,17]

No quantitative measures have been used in any of the above classification schemes. Similarities of protein folds were based on gross simplifications. Nevertheless, very helpful insights have been obtained with such procedures (cf. Ptitsyn *et al.*[18]). A number of authors have used "diagonal" or "distance" plots for the rapid visual recognition of structural

[11] S. T. Rao and M. G. Rossmann, *J. Mol. Biol.* **76,** 241 (1973).
[12] J. S. Richardson, *Nature (London)* **268,** 495 (1977).
[13] M. Levitt and C. Chothia, *Nature (London)* **261,** 552 (1976).
[14] J. S. Richardson, *Adv. Protein Chem.* **34,** 167 (1981).
[15] M. G. Rossmann and A. Liljas, *J. Mol. Biol.* **85,** 177 (1974).
[16] P. J. Artymuik, C. C. F. Blake, and A. E. Sippel, *Nature (London)* **290,** 287 (1981).
[17] B. W. Matthews, M. G. Grutter, W. F. Anderson, and S. J. Remington, *Nature (London)* **290,** 334 (1981).
[18] O. B. Ptitsyn and A. V. Finkelstein, *Q. Rev. Biophys.* **13,** 339 (1980).

domains.[15,19] The distances between pairs of atoms, usually only the C_α atoms, are plotted as a symmetric square matrix. Domains and other features may show up as recognizable features on such two-dimensional plots.

Quantitative methods for structure comparison have been developed by Rossmann and co-workers[11,20–22] and by Remington and Matthews.[23,24] The method of Rossmann and co-workers is quite general and can systematically determine the position of insertions and deletions. However, it is dependent on a somewhat complicated algorithm which performs slightly differently according to the choice of a small set of starting parameters. In contrast, the method of Remington and Matthews is "clean" and straightforward, independent of any starting conditions, but is less able to cope with major differences in structure, particularly in cases in which there are present insertions in one compound relative to the other.

The Technique of Rossmann and Co-Workers

Determination of Spatial and Topological Equivalence

To facilitate the discussion, the following terms are useful.

Structural equivalence indicates that, on superposition of two molecules, a residue or structural element such as an α-helix or one strand of a β-pleated sheet of one molecule coincides within defined limits with that of the other molecule. If the residues or secondary structural elements run in a similar direction in space, then the term *positive structural equivalence* will be applied; if the chain segments are oppositely directed, the structural equivalence will be termed *negative*.

A *run* is a sequential series of structurally equivalent residues.

Topological equivalence requires a run of positive structural equivalences. Thus, topologically equivalent residues necessarily possess a similar fold in three dimensions. Topologically equivalent residues are therefore a subset of those that have positive structural equivalence, which in turn are a subset of the complete set of structural equivalences. If the size and number of insertions or deletions between topologically equivalent runs are small, and if the number of topological equivalences is large,

[19] K. Nishikawa and T. Ooi, *J. Theor. Biol.* **43,** 351 (1974).
[20] M. G. Rossmann and P. Argos, *J. Biol. Chem.* **250,** 7525 (1975).
[21] M. G. Rossmann and P. Argos, *J. Mol. Biol.* **105,** 75 (1976).
[22] M. G. Rossmann and P. Argos, *J. Mol. Biol.* **109,** 99 (1977).
[23] S. J. Remington and B. W. Matthews, *Proc. Natl. Acad. Sci. U.S.A.* **75,** 2180 (1978).
[24] S. J. Remington and B. W. Matthews, *J. Mol. Biol.* **140,** 77 (1980).

divergence from a common ancestor seems probable, particularly if reinforced by equivalence of substrate binding.

Let us now consider how well two structures compare when the second structure is superimposed on the first by means of the rotation **C** and translation **d**. The rotation **C** can be defined with respect to the chosen axial systems of the two molecules by three Eulerian angles θ_1, θ_2, θ_3. The determination of structural equivalence proceeds in two steps: first, assignment of probabilities P_{ij} relating the ith C_α atom in the first molecule with the jth C_α atom in the second molecule and, second, the imposition of the requirements for topological equivalence.

The probability P_{ij} is defined as the product of two probabilities

$$P_{ij} = P_1 P_2 \tag{1}$$

where

$$P_1 \propto \exp(-d_{ij}^2/2E_1^2) \tag{2}$$
$$P_2 \propto \exp(-S_{ij}^2/2E_2^2) \tag{3}$$

P_1 depends upon the distance d_{ij} between the α-carbon atoms $C_{\alpha i}$ and $C_{\alpha j}$ in the superimposed molecules and P_2 expresses similarity of orientation of the peptide chain in the vicinity of $C_{\alpha i}$ and $C_{\alpha j}$. S_{ij} is the rms scatter of the distances $d_{i-1,j-1}$, d_{ij}, $d_{i+1,j+1}$ from their mean value. Values of E_1 and E_2 are determined as the corresponding rms values for those residues previously considered to be spatially and topologically equivalent. Usual values are $E_1 \simeq 2.5$ Å and $E_2 \simeq 1.0$ Å.

The probabilities P_{ij} thus relate a residue $R_{1,i}$ in the first molecule, and any residue $R_{2,j}$ in the second. They constitute a symmetric matrix from which a sequence must be selected which satisfies the topological requirement of a "similar fold" and the genetic requirement of gene linearity. Thus, it is required not only that

$$R_{1,i} \equiv R_{2,j} \tag{4}$$

but also that "the progression rule" applies, i.e.,

$$R_{1,i+n} \equiv R_{2,j+m} \tag{5}$$

where $n \geq 1$ and $m \geq 1$. In cases in which there are no insertions or deletions, $n = m$.

In the second step of assigning structural equivalences, the probabilities P_{ij} are sorted into descending order for each value of i. If the molecules are exceedingly similar, then the maximum probabilities for each value of i should also satisfy the progression rule. If this is not the case, then there will be, nevertheless, local "runs" in which the progression rule is satisfied. These runs (a run being two or more consecutive residues where the maximum probabilities obey the progression rule) are identified and associated with their total probability.

Adjacent runs can then be compared in pairs. Both runs are accepted if they obey the progression rule, but if this is not the case, the run with the smaller total probability is rejected. Emphasis is then given to the longer and more similar folds of the protein. This process of rejection is continued until every run, and hence every residue, obeys the progression rule. Finally, the runs are extended at both ends by applying the progression rule. Extension is stopped when either the individual probabilities fall below a given value (for example, 0.05) or the extension meets another run. When two extensions meet, the extension with the larger individual probability is accepted. The purpose of run extension is to include those parts of the protein fold for which there may be amino acid homology in the absence of precise structural equivalence, particularly at bends. The overall objective is to align the respective proteins so that N, the total number of equivalences, is made as large as possible.

Refinement of a Set of Equivalenced Residues

With the N equivalences thus established it is now possible to obtain better Eulerian angles and translational vector. A nonlinear least-squares iterative refinement is used to minimize $\Sigma_N (\mathbf{r}_{1,i} - \mathbf{r}_{2,j})^2$ for the current set of equivalenced residues. Here $\mathbf{r}_{1,i}$ is the position vector of $C_{\alpha i}$ in the first molecule equivalenced to $C_{\alpha j}$ in the second molecule at $\mathbf{r}_{2,j}$. The sum is taken over the N equivalenced C_α atoms. The necessary algebra has been given by Rossmann and Argos[20] although faster methods have been discussed by Kabsch,[25] Hendrickson,[26] and McLachlan.[27] As this least-squares operation occurs frequently, it is useful to minimize the time consumed.

When convergence has been reached it is possible to redetermine the set of equivalences in the manner previously described, using the refined Eulerian angles and translation vector to superimpose the molecules. New equivalences might be found and some old ones rejected. A further least-squares cycle can then commence and so forth. Convergence is nearly always achieved and occurs when there is no change in assignment of equivalences between successive cycles. In some cases oscillation occurs between slightly different assignments.

[25] W. Kabsch, *Acta Crystallogr., Sect. A* **A32,** 922 (1976).

[26] W. A. Hendrickson, *Acta Crystallogr., Sect. A* **A35,** 158 (1979).

[27] A. D. McLachlan, *J. Mol. Biol.* **128,** 49 (1979).

Search for Structural Similarity

The above procedure can be used when some prior knowledge of structural similarity is already available and can be used to provide a set of initial equivalences which act as a starting point. Such equivalences may be obtained by superposition of hydrogen bonding networks[28] or simply by visual inspection of models or diagrams. However it is frequently unknown whether there is any similarity of structure, or there might be a dispute whether an apparent similarity exists or is unique. In these cases it is necessary to perform a complete search of all orientations and positions.

The superposition of two three-dimensional rigid bodies is a six-parameter problem, three rotational and three translational. However, it may not be necessary, in the first instance, to use all six parameters to determine whether two proteins have similar folding topologies. When the molecules are similarly oriented, vectors between structurally equivalent residues will be parallel and will result in a small value of S_{ij}. In order to emphasize this similarity of orientation without regard to superposition, it is necessary to make $E_1 >> E_2$. For convenience, the mass centers of the two molecules might be superimposed initially, but once a set of equivalences has been found under these conditions, better translational components can be determined by a simple linear least-squares procedure. Subsequently E_1 and E_2 can be set roughly equal, giving emphasis to both spatial superposition and similarity of orientation.

The search procedure requires stepping through all three Eulerian angles on a three-dimensional grid. At each grid point, the number of topologically equivalent residues, N, is found. The three-dimensional plot of N is best represented as a series of contoured sections. A large peak should occur at those angles which best relate the whole, or part, of the two molecules. The background will be due to occasional equivalence of helices or extended β-pleated sheets which do not continue into other regions of the molecule. Hence the background can be examined, random noise determined, and the significance of a peak (or structural equivalence) established and quantitatively expressed. The procedure is described in more detail by Rossmann and Argos.[21]

An example of a search function is given in Figs. 1 and 2 showing the comparison of T4 phage and hen egg-white lysozyme.[21,23] These two lysozymes have similar enzymatic activity, but phage lysozyme with 164 amino acids is a bigger molecule than hen egg-white lysozyme, which has

[28] M. G. Rossmann, D. Moras, and K. W. Olsen, *Nature (London)* **250,** 194 (1974).

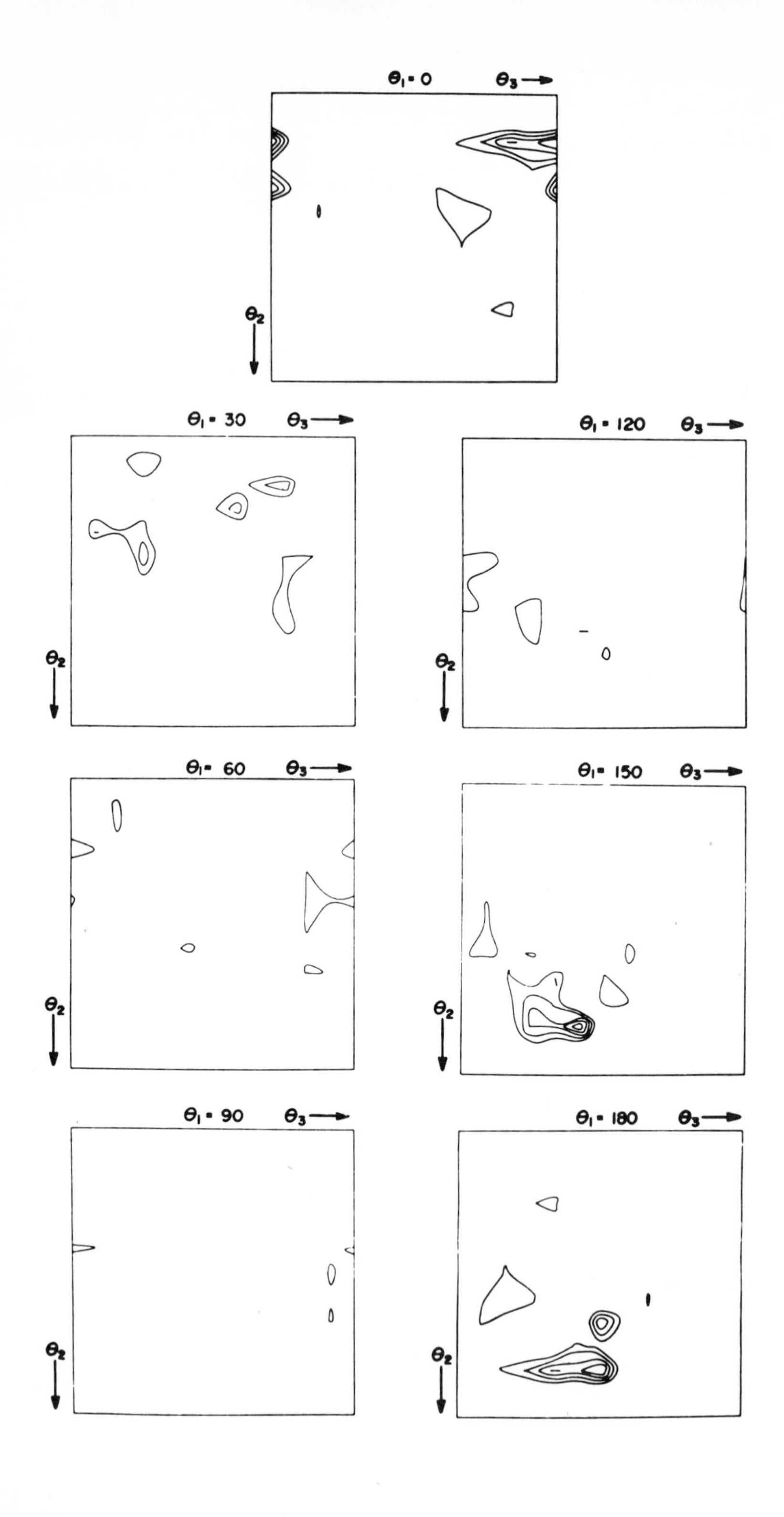

θ1 = 0
θ3
θ2
θ1 = 30
θ3
θ2
θ1 = 120
θ3
θ2
θ1 = 60
θ3
θ2
θ1 = 150
θ3
θ2
θ1 = 90
θ3
θ2
θ1 = 180
θ3
θ2

129. The amino acid sequences of the two lysozymes are not homologous. Table I gives an analysis of the lysozyme search procedure and three quite different search comparisons as well. For the phage–hen egg-white lysozyme comparison the peak height in the search function is 8 standard deviations above background and the next highest peak is only 2.7 standard deviations. The significance of the similarity of these two proteins is thus clearly established in a case in which the correspondence was not obvious on first inspection of the structures. However, search functions such as shown in Fig. 1 only establish that a statistically significant correspondence exists between different structures but leave unanswered whether a given relationship is due to convergence or divergence.

Method of Remington and Matthews

Remington and Matthews[23,24] have developed a method of structure comparison in which all possible segments of one protein of a chosen length are compared, in turn, with all possible segments of the same length from the other protein. In this way an agreement between any part of the first protein with any part of the second protein is automatically noted. At the same time, the large number of individual comparisons provides a "data pool" which can be used to assess the significance of the best comparisons, similar in concept to the background observations discussed in the previous section. The method is based on Fitch's[29,30] method of comparing amino acid sequences.

Method of Comparison

To compare two proteins, a probe length L is chosen, e.g., 40 residues. The first backbone segment, 1–40, of the first protein is then compared in turn with all possible segments of the second protein, i.e., 1–40, 2–41, 3–42, ... The procedure is then repeated, in turn, for each 40-residue segment of the first protein. If protein 1 has N residues and proprotein 2 has M, the total number of individual comparisons is $(M - L + 1) \times (N - L + 1)$.

[29] W. M. Fitch, *J. Mol. Biol.* **16,** 9 (1966).
[30] W. M. Fitch, *J. Mol. Biol.* **49,** 1 (1970).

FIG. 1. Rotational search for structural similarity between hen egg-white lysozyme and phage lysozyme. Contours indicate the number of equivalent amino acids and are drawn at levels of 29, 34, 39, ... equivalences. (From Rossmann and Argos.[21])

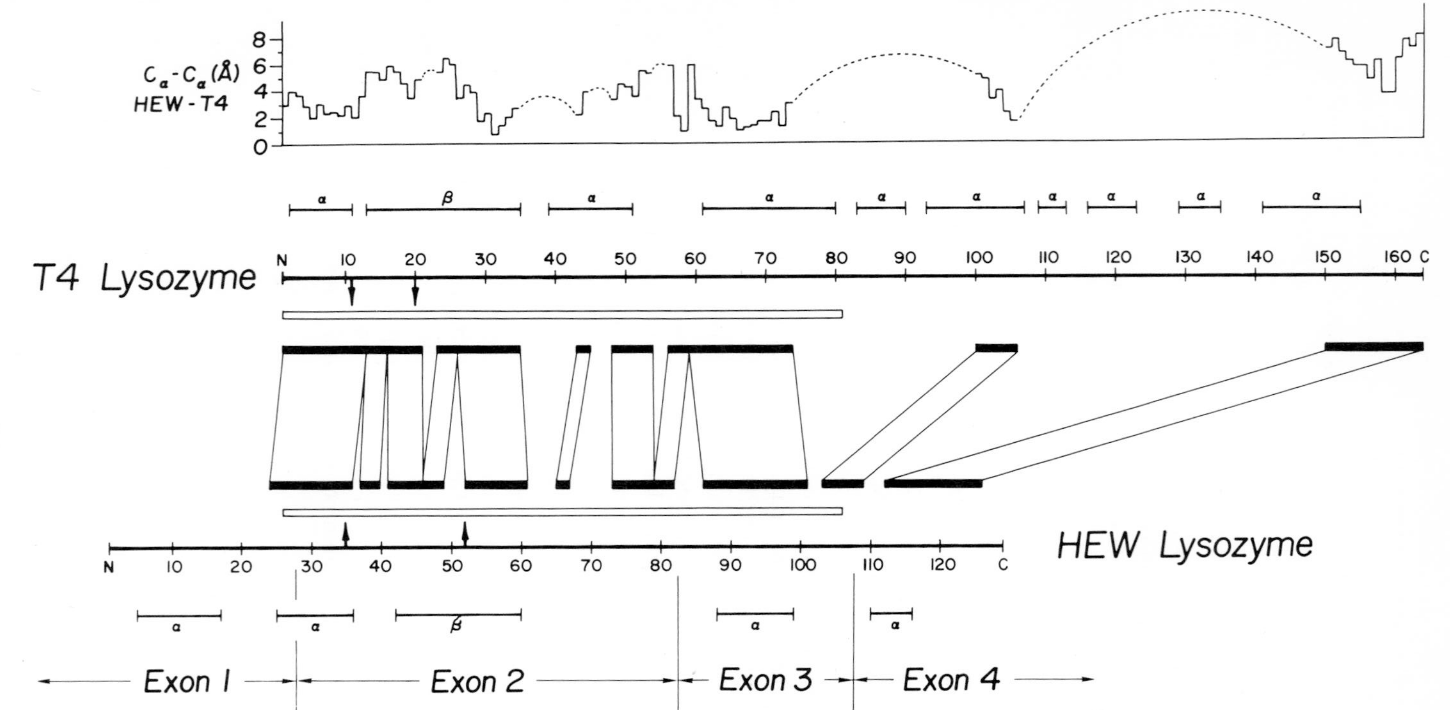

FIG. 2. Schematic diagram illustrating the structural correspondence between T4 phage lysozyme and hen egg-white (HEW) lysozyme. The solid connected bars show the α-carbon atoms of the respective molecules which approximately superimpose and which were determined to be "equivalent."[21] The open bars show the respective 80-residue segments of the two lysozymes shown to correspond by the procedure of Remington and Matthews.[24] Also shown are the locations of the catalytically important acid groups and the boundaries of the exons in the hen egg-white lysozyme gene.[16,17,31]

TABLE I

SUMMARY OF FOUR SEARCHES FOR STRUCTURAL EQUIVALENCE[a]

Comparison	Number of amino acids found structurally and topologically equivalent						
	Mean level	rms	Peak height	Peak height after refinement	Largest noninterpreted feature	E_1 (Å)	E_2 (Å)
LDH–GPD							
(NAD binding domain only)							
1. Major peak	15.8	13.4	91	96	33	2.9	4.0
2. Minor peak			45	52			
LDH–GPD							
(whole molecule)	29.4	16.4	101	—	51	3.5	4.2
Horse hemoglobin β–cytochrome b_5	17.0	8.9	60	58	33	3.9	4.4
Phage–hen egg-white lysozyme	11.1	8.3	55	78	33	4.1	1.1

[a] LDH, Lactate dehydrogenase; GPD, glyceraldehyde-3-phosphate dehydrogenase; NAD, nicotinamide adenine dinucleotide. (Data taken from Rossmann and Argos.[21])

The structural agreement between individual segments is defined as R_{C_α} the minimum value of the rms distance between the respective α-carbon atoms of the two segments:

$$R_{C_\alpha}^{ij} = \left(\frac{1}{L}\sum_{k=0}^{L-1} |\mathbf{r}_{1,i+k} - \mathbf{C}\mathbf{r}_{2,j+k}|^2\right)^{1/2} \quad (6)$$

where $\mathbf{r}_{1,i}$ and $\mathbf{r}_{2,j}$ are, respectively, the coordinate vectors of the ith α-carbon atom of protein 1 and the jth α-carbon atom of protein 2 and $\mathbf{C}$ is the rotation and translation matrix which minimizes R_{C_α}.

Because the number of individual comparisons is usually large (typically 10^4–10^5), it is very desirable that a rapid method be used to minimize R_{C_α}, and the matrix method described by McLachlan[27] has been found very suitable.

After the values of R_{C_α} have been obtained for all segments of the proteins being compared, the mean and standard deviation are calculated and the values of R_{C_α} plotted as a "structure comparison map."

Hen Egg-White Lysozyme vs T4 Phage Lysozyme

An example of the comparison of two structures which are not obviously similar is provided by hen egg-white lysozyme and bacteriophage T4 lysozyme[21,23] (also see the previous section). The correspondence of these two proteins as determined by the Rossmann procedure is illustrated in Fig. 2. The connected solid bars show the residues of the respective lysozymes which are "equivalent," as described in the preceding section.

The structure comparison maps for the two lysozymes, calculated by the Remington–Matthews procedure with probe lengths of 40 and 80 residues, are shown in Fig. 3. The resemblance between the respective lysozymes is indicated by the extended diagonal features. In the structure comparison map with a probe length of 80 residues (Fig. 3b) the relation between the two lysozymes becomes obvious, with the maximum peak (R_{C_α} = 6.2 Å) being 3.8σ better than the average value.

Following Fitch[29,30] one can plot the observed R_{C_α} values as a cumulative frequency distribution. When plotted in this manner (Fig. 4) a Gaussian distribution will appear as a straight line with slope σ and zero intercept equal to the mean. In cases in which the observed distribution of R_{C_α} departs significantly from Gaussian, as is the case for the lysozyme comparison with a probe length of 80 residues (Fig. 4b), this is characterized by a nonlinear distribution.

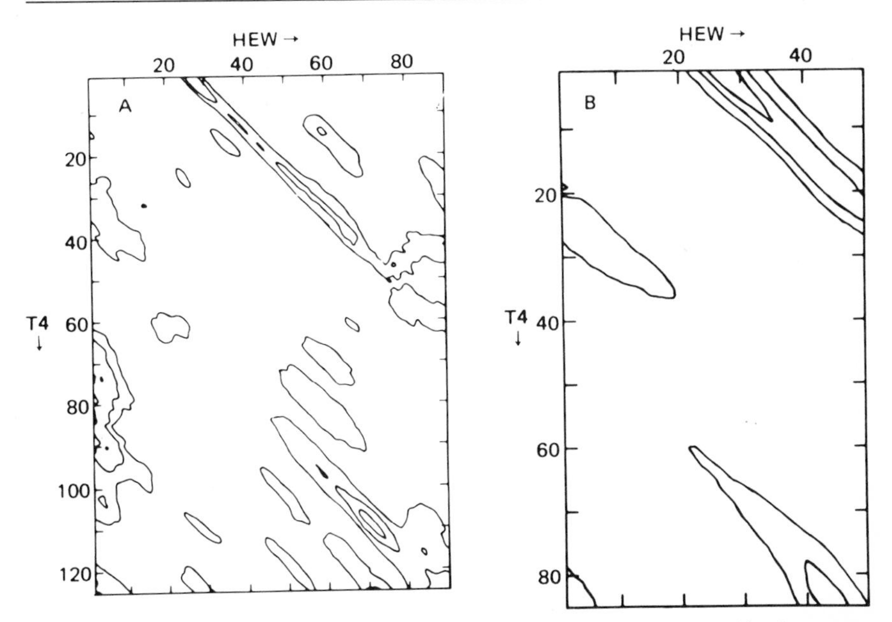

FIG. 3. Structure comparison maps for phage lysozyme and hen egg-white lysozyme compared with a probe length of (A) 40 residues and (B) 80 residues. In each case the contour levels are drawn at intervals of 1 standard deviation below the mean value for all comparisons. (From Remington and Matthews.[23])

As shown elsewhere,[17,31] in addition to the structural similarity of the two lysozymes, there are correspondences in the modes of binding substrates and the presumed mechanisms of catalysis which increase the probability that hen egg-white and phage lysozymes derived from the same evolutionary precursor, notwithstanding their unrelated amino acid sequences.

Choice of Probe Length

In comparing two proteins it is necessary to decide how long the amino acid probe length should be. This length is the one adjustable parameter in the Remington–Matthews procedure (except for special "tricks" which can be used to deal with insertions and deletions). A short probe length such as $L = 20$ residues might be used to search for a special structural element (see the example below of the α-helical DNA-binding

[31] B. W. Matthews, S. J. Remington, M. G. Grütter, and W. F. Anderson, *J. Mol. Biol.* **147,** 545 (1981).

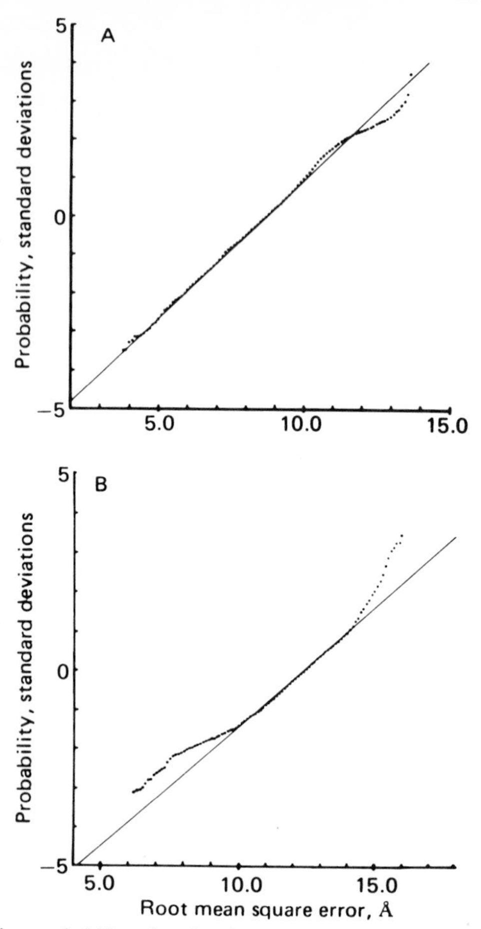

FIG. 4. Cumulative probability plots for the comparisons of T4 phage lysozyme and hen egg-white lysozyme shown in Fig. 3. (A) Probe length of 40 residues. (B) Probe length of 80 residues. (From Remington and Matthews.[23])

fold) but is, in general, of little use for the overall comparison of two different proteins. Longer probe lengths are necessary to locate regions of extended structural similarity, but, as the probe length gets longer, it is more likely that structural correspondences may be "missed" because of insertions or deletions in one protein relative to the other. (Small insertions or deletions up to about five amino acids may not be serious, but larger ones will be.)

In practice, a pair of proteins can usually be compared by using both a medium length and a longer probe, e.g. $L = 40$ and $L = 80$ residues.[24] Then, if circumstances warrant, other probe lengths can be used as well.

Searching for Structural Motifs

A special case of the Remington–Matthews procedure is a systematic search through a given protein, or series of proteins, for a particular structural element.

One trivial example might be a search through a protein to locate all the α-helices. In this case the desired structural motif would be an idealized α-helix of 10 α-carbons in length, for example, which would be systematically compared with all 10-residue segments in the protein.

A more interesting example was provided recently by the comparison of two DNA-binding proteins, *cro* repressor of bacteriophage λ and catabolite gene activator protein of *Escherichia coli*.[32] Each of these proteins was found to contain two contiguous α-helices within which 22 α-carbon atoms agreed with an R_{C_α} value of 1.1 Å. Although it was possible to use the methods described in the next section to estimate the a priori significance of this correspondence, it was desirable to show whether or not this particular two-helical conformation might have occurred in these or other proteins because it happened to be a stable structural element which was favored during protein folding. To answer this question, we took the 22 α-carbon segment of *cro* repressor and compared it in turn with every 22-residue segment of every protein structure listed in the Brookhaven Data Bank. The clear-cut result (Fig. 5) is that the two-helical fold seen in *cro* repressor and catabolite gene activator proteins does not occur in any other protein.

Structural Correspondence Generalized: Analysis of R_{C_α}

It would be useful to develop general criteria to assess the degree of structural similarity between polypeptide segments. Whether structural correspondence is measured by the number of structurally equivalent residues or by the value of R_{C_α}, it is necessary to establish a standard against which the respective values can be compared. In this section we deal with the assessment of the number of equivalences.

Comparison of segments from many different proteins shows that the distribution of the values of R_{C_α} varies with the probe length of the segment, but does not depend very much on the type of protein being compared (i.e., α-helical, β-sheet, or some combination of these.) $\overline{R}_{C_\alpha}$, the average value of R_{C_α} for many comparisons of a given probe length L,

[32] T. A. Steitz, D. H. Ohlendorf, D. B. McKay, W. F. Anderson, and B. W. Matthews, *Proc. Natl. Acad. Sci. U.S.A.* **79**, 3097 (1982).

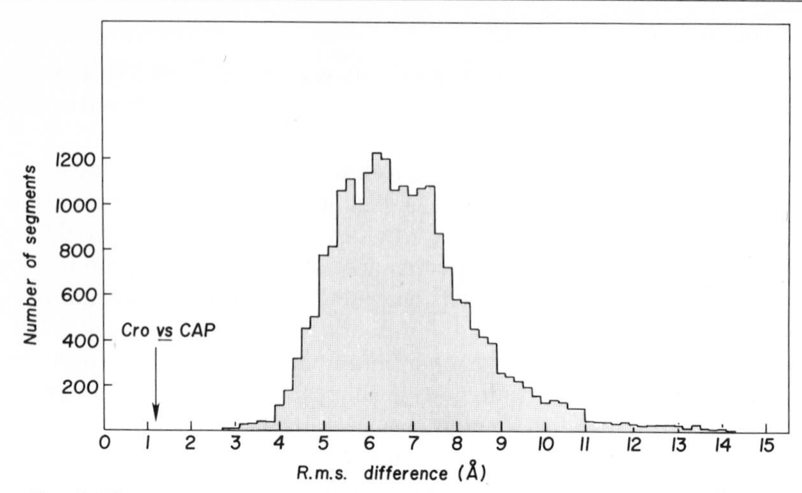

FIG. 5. Histogram summarizing the comparison of a 22-amino acid two-helical segment of *cro* repressor protein with every 22-amino acid segment in the Brookhaven Data Bank. (Data based on the comparison reported by Steitz *et al.*[32])

increases in proportion to the square root of the probe length, and can be fitted reasonably well by the equation

$$\bar{R}_{C_\alpha} = 1.55 \sqrt{L} \tag{7}$$

In contrast, the standard deviation of the distribution of R_{C_α} is essentially independent of probe length, at least for probe lengths greater than 20 residues. The dependence of the standard deviation of R_{C_α} on probe length can be approximated by the empirical relation

$$\sigma(R_{C_\alpha}) = 2.2 \tanh(L/19) \tag{8}$$

For probe lengths of 30 or longer

$$\sigma(R_{C_\alpha}) \simeq 2.2 \quad \text{Å} \tag{9}$$

By combining Eqs. (5) and (6) it is possible to calculate the a priori significance of any structural correspondence which happens to be observed. For example, a 22-residue correspondence has an expected average value of 7.3 Å and a standard deviation of 1.8 Å so that the correspondence of 1.1 Å between *cro* repressor and catabolite gene activator protein is 3.4σ below the average value and is expected to occur by chance with a frequency of 0.03%. This is the frequency with which the value of R_{C_α} calculated for 22-residue segments chosen at random would be expected to equal 1.1 Å or better. Values of R_{C_α} and $\sigma(R_{C_\alpha})$ deter-

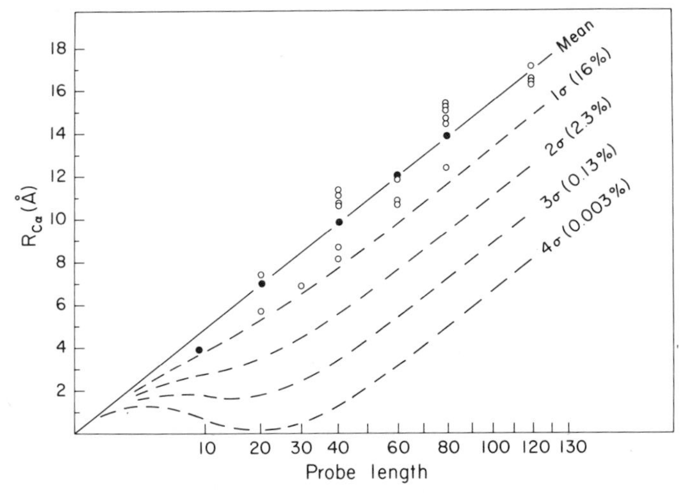

FIG. 6. Generalized structure agreement probability diagram. The solid line gives the average value of R_{C_α} as a function of probe length and the successive broken lines correspond to values of R_{C_α} which are better than average by 1, 2, 3, and 4 standard deviations. The frequencies with which these values are expected to occur by chance in a random population are also shown. The scale of the abscissa is proportional to the square root of the probe length. (From Remington and Matthews.[24])

mined directly from Fig. 5, 6.9 and 1.7 Å, agree well with those quoted above (7.3 and 1.8 Å).

The generalized structure agreement probability distribution derived from Eqs. (7) and (8) is shown in Fig. 6.[24] By using a scale proportional to the square root of the probe length along the abscissa, Eq. (7) becomes a straight line giving the average value expected when comparing two protein segments of length L residues. The broken lines drawn below the average value line show values of R_{C_α} which are 1σ, 2σ, 3σ, and 4σ better than average, for a given probe length. The expected frequency of occurrence for each line is also included. Figure 6 illustrates very clearly that structure comparisons based on short segments require a very low value of R_{C_α} to be statistically significant. For longer probe lengths, on the other hand, structural correspondence at high levels of significance can readily be demonstrated, even when using relatively crude coordinates.

Convergence or Divergence? Analysis of the Number of Equivalent Residues

The determination of relationships among species by the study of structural similarities has been a subject of philosophical and scientific discussion by Aristotle, Lamarck, Darwin, and many others. The validity of such comparisons has received strong support through the analysis of extant amino acid and nucleotide sequences, and, in cases such as the globins and the *c*-type cytochromes, through three-dimensional structure determinations as well. Such homology is fundamental to the previously observed external characteristics.

However, some three-dimensional protein structures are remarkably similar, yet the amino acid sequences have no apparent relation to each other. Hence it is not necessarily obvious whether these structural relationships are due to divergence from a common ancestral protein or convergence caused by functional necessity. In other cases the degree of similarity is sometimes far from obvious, may be missed entirely, or, if observed, may be questioned for its significance. A limited region of correspondence might include only one or two α-helices and such a similarity could be a pure coincidence (but see the previous section for a suitable test). It is apparent, then, that a comparison of protein structures in order to determine their evolutionary relationships (and hence that of the organisms in which they are found) has precisely the same problems as have been encountered in the comparison of gross features of species.

Numerical taxonomy (cf. Sokal and Sneath[33]) has established useful procedures. Each object to be compared is analyzed for a set of characteristics (frequently binary attributes). The degree of similarity between any two objects can be expressed in terms of the number of characteristics that are the same. If most of the characteristics are the same, and only a few are different, it is most probable that the two objects have diverged from a common precursor. Similarly, in the comparison of protein structure, the more attributes in common, and the fewer different, the more likely it is that the two proteins have a common heritage. It is also reasonable to assume that the closer is the correspondence of individual attributes, the more probable is divergence.

In Table II, six measures have been used to assess similarity of structure. The first four, namely the number of equivalent residues, percentage of equivalent residues, rms distance between C_α atoms, and minimum base change per codon will be discussed in some detail. The comparisons are ranked according to the percentage of residues in the smaller protein (or domain) which are equivalent. However, the total number of equivalent residues must also be taken into account. Similarity based on percentage is not adequate for a small structure which might, for example, be purely an α-helix, and cannot be used to differentiate between conver-

[33] R. R. Sokal and P. H. A. Sneath, "Principles of Numerical Taxonomy." Freeman, San Francisco, California, 1963.

TABLE II STRUCTURAL COMPARISONS

Proteins[a]		Number of residues		Number of equivalences	Percentage of equivalences		rms (Å)	MBC/C	F	σ_t	Reference
Mol. 1	Mol. 2	Mol. 1	Mol. 2		Mol. 1	Mol. 2					
Hb (α)	Hb (α)	146	141	139	95	99	1.9	—			42
SBMV (C)	TBSV (C)	221	198	179	81	90	2.2	—			43, 44, M. G. Rossmann *et al.*, unpublished results
Rhod. (1)	Rhod. (2)	142	135	117	82	87	2.0	1.27			45
V_H	V_L	110	110	77	70	70	1.7	—			10
GPD (NAD)	LDH (NAD)	148	144	96	65	67	2.9	1.24	15	6.7	21, 28
Hb (β)	Cyt. b_5	146	86	58	40	67	3.9	1.29	33	3.1	20, 21, 42
TBSV (S)	TBSV (P)	167	110	69	41	63	3.8	—			46
Con A	TBSV (P)	237	110	68	29	62	3.4	—			46
SBMV (A)	STNV	187	173	104	56	60	3.7	1.36			47, M. G. Rossmann *et al.*, unpublished results
T4L	HEWL	164	129	78	48	60	4.1	1.53	23	4.6	21
Cyt. b_5	Cyt. b_{551}	86	82	41	48	50	4.9	—			42
CAP (DNA)	Cro	73	66	31	42	47	3.1	—			32
SOD	V_L	150	110	51	34	46	2.7	—			10
LDH (NAD)	Flavod.	144	138	39	27	28	2.4	1.23			11

[a] Hb(α) and Hb(β), α and β chains of horse hemoglobin; SBMV (A) and (C), A and C subunits of southern bean mosaic virus; TBSV (S) and (P), S and P domains of tomato bushy stunt virus; TBSV (C), C subunit of S domain of TBSV; STNV, satellite tobacco necrosis virus; Rhod. (1) and (2), first and second domain of rhodanese; V_H and V_L, variable domains of heavy and light immunoglobulin chains; Con A, concanavalin A; GPD (NAD), nucleotide adenine dinucleotide-binding domain of glyceraldehyde 3-phosphate dehydrogenase; LDH (NAD), NAD-binding domain of lactate dehydrogenase; Cyt. b_5, cytochrome b_5; Cyt. b_{555}, cytochrome b_{555}; T4L, bacteriophage T4 lysozyme; HEWL, hen egg-white lysozyme; CAP (DNA), DNA-binding domain of catabolite gene activator protein; Cro, *cro* repressor protein; SOD, superoxide dismutase; Flavod., flavodoxin.

[b] Minimum base change per codon.

42. P. Argos and M. G. Rossmann, *Biochemistry* **18,** 4951 (1979). 43. C. Abad-Zapatero, S. S. Abdel-Meguid, J. E. Johnson, A. G. W. Leslie, I. Rayment, M. G. Rossmann, D. Suck, and T. Tsukihara, *Nature* (*London*) **286,** 33 (1980). 44. S. C. Harrison, A. J. Olson, C. E. Schutt, F. K. Winkler, and G. Bricogne, *Nature* (*London*) **276,** 368 (1978). 45. J. H. Ploegman, G. Drent, K. H. Kalk, and W. G. J. Hol, *J. Mol. Biol.* **123,** 557 (1978). 46. P. Argos, T. Tsukihara, and M. G. Rossmann, *J. Mol. Evol.* **15,** 169 (1980). 47. L. Liljas, T. Unge, K. Fridborg, T. A. Jones, S. Lovgren, U. Skoglund, and B. Strandberg, *J. Mol. Biol.* **159**, 93 (1982).

gence and divergence in such cases. At the other extreme, the total number of equivalent residues will tend to increase for larger structures because of random coincidences, and in such cases the percentage of equivalences is a better measure of structural correspondence. The rms distance between C_α atoms, as in Table II, is not a reliable measure because it includes only "equivalent" atoms and therefore, by definition, excludes any pairs of C_α atoms which happen to be far apart. Only for very similar structures is the rms distance a useful measure of structural correspondence. The minimum base change per codon is also unreliable except for cases in which the structures being compared have clearly homologous amino acid sequences.

Table II does not make any explicit mention of function, although it is obvious that the comparisons mainly refer to proteins whose functions might be similar. The various viral coat proteins all envelop RNA, although it is not entirely clear how the apparently conserved folds conserve function, as the interaction of the subunits (and hence their enveloping function) is rather different among these viruses.[34] The NAD cofactor binds in a very similar fashion to LDH and GPD as well as to other proteins with a rms difference between equivalent NAD atoms of less than 3.0 Å when the polypeptide chains are superimposed.[35,36] Similar comments are true for the binding of polysaccharide to the two lysozymes,[17,21,31] heme to globin and cytochrome b_5, or nucleotides to LDH and flavodoxin.

There is little doubt about the divergence of the hemoglobin α and β genes from a common ancestor on account of similarity of amino acid sequence, function, and gene arrangement. It is usually accepted that nucleotide (as opposed to nicotinamide adenine dinucleotide) binding proteins with $(\beta\alpha)_3$ structure are the consequence of convergence to attain parallel oriented α-helices with appropriate dipoles.[37] The similarity of phage and hen egg-white lysozyme has been attributed to divergence both because of function and structural similarity[17,31] and because of the organization of exons in relation to the domain structure.[16,17] Thus Table II represents a spectrum of possibilities, with the probability of divergence close to 100% at the top and rather low for the last entries in the table.

Attempts have been made to measure the difference between divergence and convergence more precisely by the use of structural compari-

[34] M. G. Rossmann, C. Abad-Zapatero, M. R. N. Murthy, L. Liljas, T. A. Jones, and B. Strandberg, *J. Mol. Biol.* **165,** 711, 1983.

[35] M. G. Rossmann, A. Liljas, C.-I. Bränden, and L. J. Banaszak, *in* "The Enzymes" (P. D. Boyer, ed.) 3rd ed., Vol. 11, p 61. Academic Press, New York, 1975.

[36] W. Eventoff and M. G. Rossmann, *CRC Crit. Rev. Biochem.* **3,** 111 (1975).

[37] W. G. J. Hol, L. M. Halie, and C. Sander, *Nature (London)* **294,** 532 (1981).

sons.[22] Let l^+ and l^- be the lengths of structurally equivalent runs with the + and − sign indicating whether they occur in the same or opposite direction of the polypeptides. In cases of convergent evolution there are perhaps as many positive (l^+) as negative (l^-) structurally equivalent runs. Thus if

$$F = \frac{l^-}{l^+ + l^-} \times 100 \tag{10}$$

a convergent case is represented by F approaching 50%.

Structures which have diverged from a common precursor are more likely to have long runs. On the other hand, convergent structures are more likely to have insertions or deletions between functionally important secondary structural elements. Thus the second moment of topological equivalence can be defined as

$$\sigma_t = \left(\frac{\sum l^2}{t}\right)^{1/2} \tag{11}$$

where there are t runs of length l which are topologically equivalent. The application of these criteria is shown for some examples in Table II and is discussed in more detail by Rossmann and Argos.[22]

By using measures such as these, future comparisons can be placed on scales, as shown in Table II, which provide a crude method to assess the probability of divergence or convergence.

Phylogenetic Trees

A further question is whether measures such as given in Table II can provide the basis for constructing a network which expresses the process of evolution, whether it be divergent or convergent. This has been done frequently from differences in amino acid sequences (cf. Fitch and Margoliash[38]). An attempt in this direction has been made by Eventoff and Rossmann.[36] A matrix can be constructed representing the degree of dissimilarity between each pair of structures. Although Eventoff and Rossmann used only the number of residues that did not have topological equivalence (Table IIIa), a weighted sum of all the various measures discussed above could, in principle, be used. A tree can now be constructed which is the most parsimonious with the data in Table IIIa. It contains the implied assumption that evolutionary processes proceed by the least number of steps. Visual inspection suggests a tree as shown in

[38] W. M. Fitch and E. Margoliash, *Science* **155,** 279 (1967).

TABLE III
DISSIMILARITY MATRIX[a]

	LDH	LADH	GPD	Flavodoxin	Subtilisin	AK
			a. Observed			
LDH		44	50	58	70	76
LADH			64	72	79	82
GPD				78	81	88
Flavodoxin					84	82
Subtilisin						82
Adenylate kinase (AK)						
			b. Calculated			
LDH		44.0	51.6	60.5	69.2	72.5
LADH			62.4	71.6	79.8	83.3
GPD				75.6	83.8	87.3
Flavodoxin					81.2	84.7
Adenylate kinase						81.9

[a] Each entry equals the number of amino acids in the LDH nucleotide binding protein (143) minus the number of equivalent residues between any pair of proteins. LADH, liver alcohol dehydrogenese.

Fig. 7a (Tree 1) where A, B, C, D, E, F, G, H, I are the lengths of the arms. It is now possible to set up 15 observational equations of the form $A + B = 44$ (relating LDH and LADH), or $B + C + D = 64$ (relating LADH and GPD), etc. A linear least-squares solution can then be found for the nine parameters A, B, C, ... , I. When this procedure is applied to the data in Table IIIa, it produces Tree 2 (Fig. 7b). Tree 2 can then be translated into a "calculated" dissimilarity matrix (Table IIIb). When the observed matrix is subtracted from the calculated matrix the differences, Δ, can be used to compute a criterion of agreement such as $\Sigma|\Delta|/\Sigma$ (observed elements). It was found to be 2.2%. All other trees gave larger values, a result which shows the tree in Fig. 7 to be the most parsimonious.

An alternative approach to constructing a tree is the ancestral structure technique (corresponding to the ancestral sequences technique). This

[39] M. O. Dayhoff and R. V. Eck, eds., "Atlas of Protein Sequence and Structure," Vol. 5. Natl. Biomed. Res. Found., Washington, D.C., 1972.

[40] D. Boulter, J. A. M. Ramshaw, E. W. Thompson, M. Richardson, and R. H. Brown, *Proc. R. Soc. London Ser. B* **181,** 441 (1972).

[41] J. S. Farris, *Syst. Zool.* **19,** 83 (1970).

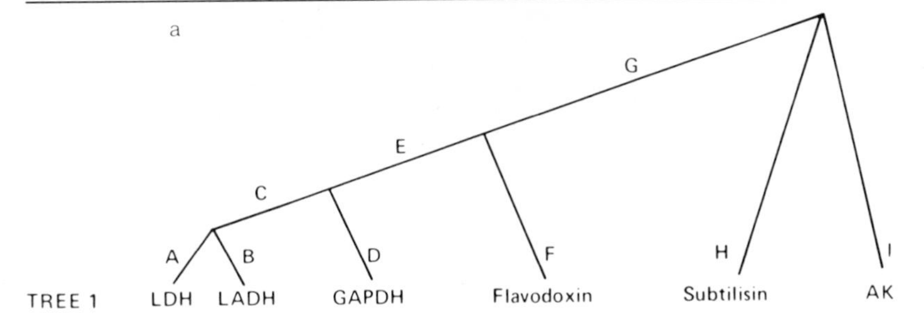

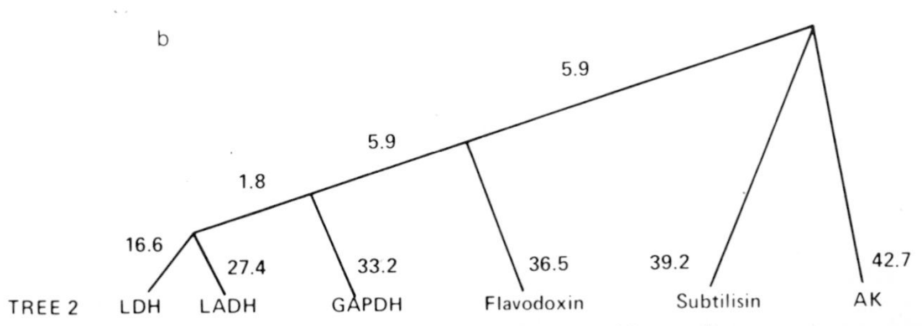

FIG. 7. (a) Tree 1 showing a proposed relationship, due either to divergent or convergent evolution, between a variety of nucleotide binding proteins. (b) In Tree 2, the lengths of the arms in the tree were computed from the data in Table IIIa.

has been used in the analysis of amino acids sequences by Dayhoff and Eck[39] and by Boulter *et al.*[40] Farris[41] has also used a similar approach to studying classical taxonomic data.

Conclusions

Quantitative methods are available to assist in the comparison of protein structures. Two methods are described here, one based on the comparison of structural segments, the other based on overall structural and topological equivalence. In both cases, previous structural comparisons provide a scale of values with which comparisons of new proteins can be compared. By using different measures of structural correspondence, and taking previous examples as a guide, it is becoming possible to assess the probability that a given pair of proteins are related by divergence or by convergence.

Acknowledgments

We are grateful to our colleagues, in particular Drs. S. J. Remington and P. Argos for their respective contributions to the methods described here. This work was supported in part by grants from the National Science Foundation (PCM8312151; PCM78-16584), the National Institutes of Health (GM20066; GM21967; GM10704), and the M. J. Murdock Charitable Trust.

[28] Domains in Proteins: Definitions, Location, and Structural Principles

By JOËL JANIN and CYRUS CHOTHIA

Introduction

Evidence for the presence of globular substructures within proteins goes back to the early descriptions of the polypeptide chain fold in hen lysozyme by Phillips[1] and in papain by Drenth *et al.*[2] The authors observed that these single-chain enzymes form pairs of globular units separated by a cleft and that the cleft constitutes the binding site for the substrate: a polysaccharide for lysozyme, a peptide for papain. Thus, X-ray analysis revealed a remarkable structural property of proteins that is evidently linked to their function. This could be related to even earlier results of protein chemistry. Limited proteolysis and sequence analysis of immunoglobulins by Porter and collaborators[3,4] and by Edelman's group[5] proved that immunoglobulin chains can be cleaved into stable fragments and that the fragments correspond to repeats in the amino acid sequence. The X-ray structure of immunoglobulins eventually confirmed that the repeat is also found in the three-dimensional structures, each being a

[1] D. C. Phillips, *Sci. Am.* **215,** 78 (1966).

[2] J. Drenth, N. Jansonius, R. Koekoek, H. Swen, and B. Wolthers, *Nature (London)* **218,** 929 (1968).

[3] R. R. Porter, *Biochem. J.* **73,** 119–126 (1959).

[4] R. R. Porter, *Science* **180,** 713 (1973).

[5] G. M. Edelman, *Science* **180,** 830 (1973).

globular unit (see reviews by Davies *et al.*,[6] Amzel and Poljak,[7] and Marquart *et al.*[8]).

Putting together the results of X-ray crystallography and protein solution studies, Wetlaufer[9] proposed that globular substructures, called domains, play a central role as intermediates in protein folding. Experimental studies of protein renaturation by Goldberg[10] had already demonstrated that large proteins refold in fragments. Wetlaufer defined domains as stable units of protein structure, which could fold autonomously.

The examples of lysozyme and papain, soon to be followed by many other enzymes, indicated that domains also play a role in function. In many enzymes, the active site is placed at the interface between domains and, in enzymes with more than one substrate, individual domains are often associated with the binding of a different substrate. This was first observed in NAD-dependent dehydrogenases by Rossmann *et al.*[11] Domains thus appear as *units of protein function as well as protein structure, and possibly units of protein evolution.* Immunoglobulin domains contain similar amino acid sequences and must derive from the duplication of a common ancestral gene. Divergence has given different functions to variable domains, which bind antigen, and to constant domains. In this case, evolution uses similar structures to different ends. Alternatively, the basic function may be maintained while evolution shuffles domains to different environments to provide new possibilities. Thus the similar NAD-binding domains found in alcohol dehydrogenase, lactate dehydrogenase, and glyceraldehyde-3-phosphate dehydrogenase are associated with a quite different second domain that binds the different substrates.[12] A common domain also binds the three-carbon substrates of pyruvate kinase and of triose-phosphate isomerase,[13] though the other domains of the former protein are very different. Association of domains performing individual function may thus be a general way of generating complex enzymatic activities. The activity actually carried out by a multidomain protein need not be related to the hypothetical function of its ancestral components.

[6] D. R. Davies, E. A. Padlan, and D. M. Segal, *Annu. Rev. Biochem.* **44,** 639 (1975).

[7] L. M. Amzel and R. J. Poljak, *Annu. Rev. Biochem.* **48,** 961 (1979).

[8] M. Marquart, J. Deisenhofer, and R. Huber, *Immunol. Today* **6,** 160 (1982).

[9] D. B. Wetlaufer, *Proc. Natl. Acad. Sci. U.S.A.* **70,** 697 (1973).

[10] M. E. Goldberg, *J. Mol. Biol.* **46,** 441 (1969).

[11] M. G. Rossmann, D. Moras, and K. W. Olsen, *Nature (London)* **250,** 194 (1974).

[12] M. G. Rossmann, A. Liljas, C.-I. Bränden, and L. J. Banaszak, *in* "The Enzymes" (P. D. Boyer, ed.), 3rd ed., Vol. 11, p. 61. Academic Press, New York, 1975.

[13] M. Levine, H. Muirhead, D. K. Stammers, and D. I. Stuart, *Nature (London)* **271,** 626 (1978).

Thus, the acid proteases are made up of two well-defined lobes or domains, and each lobe contributes an aspartic acid residue to the active site as proton donor or acceptor during the catalytic cycle. An isolated domain with a single acid group cannot have a hydrolytic function, though a dimer conceivably could. One may therefore propose that the present single-chain acid proteases have evolved from a precursor[14] whose gene was duplicated and fused.

An additional, yet important, role for domains in the dynamics of proteins has been suggested in enzymes such as yeast hexokinase[15,16] and horse alcohol dehydrogenase.[17] Both these proteins fold into two well-defined domains and substrate binding is associated with different relative positions of the domains. More precisely, the proteins seem to close over their substrates by one domain rotating relative to the other. This movement makes it possible for the substrates to diffuse in and out of the binding sites, while keeping the active center completely shielded from the surrounding water during catalysis. Other kinases show this type of movement.[18–20] Though shielding of the active center can be achieved by lesser movements in some enzymes, the presence of domains as autonomous globular structures gives particular flexibility to the protein.

In this chapter we shall describe how the presence of domains can be identified from the three-dimensional structure of proteins, and go on to review the major principles that govern the structure of the domains themselves.

Definition and Location of Domains in Proteins

Visual Inspection

An early survey of protein structures was carried out by Wetlaufer[9] relying on visual inspection of molecular models derived from X-ray analysis. He identified domains as compact globular units formed by regions

[14] J. Tang, M. N. G. James, I. N. Hsu, J. A. Jenkins, and T. L. Blundell, *Nature* (*London*) **271,** 618 (1978).

[15] W. S. Bennett, Jr. and T. A. Steitz, *J. Mol. Biol.* **140,** 183 (1980).

[16] W. S. Bennett, Jr. and T. A. Steitz, *J. Mol. Biol.* **140,** 211 (1980).

[17] H. Eklund, J. P. Samama, L. Wallen, C.-I. Bränden, A. Åkeson, and T. A. Jones, *J. Mol. Biol.* **146,** 561 (1981).

[18] R. D. Banks, C. C. F. Blake, P. R. Evans, R. Haser, D. W. Rice, G. W. Hardy, M. Merrett, and A. W. Phillips, *Nature* (*London*) **279,** 773 (1979).

[19] C. A. Pickover, D. B. McKay, D. M. Engelman, and T. A. Steitz, *J. Biol. Chem.* **254,** 11323 (1979).

[20] C. M. Anderson, F. N. Zucker, and T. A. Steitz, *Science* **204,** 375 (1979).

of the polypeptide chain, either a single stretch (continuous domains), or several stretches coming together (discontinuous domains).

Many proteins can be divided into such domains on simple inspection of a physical model of their structure. However, this definition is subjective and the estimates of the globular and autonomous nature of the substructures are only qualitative. Thus, though crystallographers do still characterize domain structures from visual inspection, computer algorithms using atomic coordinates produced from X-ray data have been developed in recent years to give more objective assessments.

Distance Maps

Distance maps were first introduced by Phillips[21] and by Ooi and Nishikawa.[22] To draw such maps, the distances r_{ij} between the α-carbon atoms of residues numbered i and j in the amino acid sequence are calculated and plotted against the two indices as a matrix which, for convenience, is usually represented as a two-dimensional contour map with a line joining points where r_{ij} has a given value, say, 10 Å. Regions of the chain that stay together in space are within this contour line. Elements of secondary structure and other specific features of the protein can be identified by inspection of the map. Domains are recognized as clusters of short distances. Continuous domains form triangles of short distances near the diagonal of the plot and are separated from regions away from the diagonal where few short distances occur. Visual inspection of a two-dimensional representation of the protein structures thus replaces visual inspection of physical models. Rao and Rossmann[23] were the first proponents of distance maps as a systematical way of locating domains in lactate dehydrogenase and flavodoxin, while Ooi and Nishikawa[22] applied it to α-chymotrypsin and myoglobin, and Rossmann and Liljas[24] to a large sample of other proteins.

A variant of the distance map has been proposed by Gō,[25] who identifies regions of the polypeptide chain that are remote in space (large values of r_{ij}) rather than close together.

[21] D. C. Phillips, *in* "British Biochemistry Past and Present" (T. W. Goodwin, ed.), p. 11. Academic Press, London, 1970.

[22] T. Ooi and K. Nishikawa, *in* "Conformation of Biological Molecules and Polymers" (E. Bergmann and B. Pullmann, eds.), p. 173. Academic Press, New York, 1973.

[23] S. T. Rao and M. G. Rossmann, *J. Mol. Biol.* **76,** 241 (1973).

[24] M. G. Rossmann and A. Liljas, *J. Mol. Biol.* **85,** 177 (1974).

[25] M. Gō, *Nature* (*London*) **291,** 90 (1981).

Automatic Clustering and Domain Hierarchies

In distance maps, long-range interactions are described as short r_{ij} values occurring between residues far apart in the sequence. Crippen[26] also uses a distance criterion to estimate the extent to which part of the polypeptide chain interacts. He first divides the polypeptide chain into short segments that do not fold over themselves in the three-dimensional structure. Such segments of about 10 residues may or may not be elements of secondary structure. Pairwise interactions between segments are then estimated by defining a "packing density": C_α to C_α distances shorter than a given cutoff are counted and normalized to the size of the fragments. The set of fragments and interfragment "packing densities" is then subjected to a clustering algorithm which builds an ascending hierarchy of segment clusters culminating in the whole protein. In this hierarchy, clusters placed at the higher levels may be considered as domains, those at lower levels as subdomains. Fragments grouped together should have more interactions with one another than with fragments in other clusters. Domains formed by these clusters are almost always discontinuous and can include many pieces of the polypeptide chain. As a consequence, there is a rather poor agreement between the clusters defined by Crippen[26] and the domains of Wetlaufer[9] or Rossmann and Liljas,[24] which are mostly continuous.

In contrast to the algorithm of Crippen,[26] which creates an ascending hierarchy, the procedure proposed by Rose[27] generates a descending hierarchy of continuous protein fragments through a succession of cleavages introduced in the polypeptide chain on a topological basis. The polypeptide chain, reduced to its α-carbon atoms, is projected onto a "disclosing plane," containing two of its principal axes of inertia. A line is then drawn through the two-dimensional projection (equivalent to a plane in three dimensions) in order to divide it into continuous chain segments placed on either side of the line. In most cases, separation of the chain into two continuous segments by a single cleavage plane is not possible, or is possible only near its NH_2- or COOH-terminal end. Crossing the dividing plane at some places other than the cleavage point must therefore be tolerated and a bias introduced to favor cleavage near the middle of the chain. The algorithm actually cleaves the protein into fragments of similar sizes at each step of the process, from the complete protein down to smaller and smaller fragments.[27]

Results presented by Rose[27] show that the higher levels of his descending hierarchies correspond in many cases to domains identified by

[26] G. M. Crippen, *J. Mol. Biol.* **126,** 315 (1978).
[27] G. D. Rose, *J. Mol. Biol.* **134,** 447 (1979).

visual inspection. At lower levels, the significance of the cleavage points is often questionable as the procedure is designed to cleave any structure and gives no indication of the extent to which the segments interact. Thus the bovine pancreatic trypsin inhibitor, usually considered as a single-domain protein, is cleaved just as well as typical two-domain proteins such as papain.

One of the limitations of the original procedure is the requirement that the three-dimensional structure can be cleaved by a plane (actually a plane parallel to an axis of inertia). This requirement is relaxed in a later algorithm proposed by Lesk and Rose[28] who use ellipsoids instead. For each possible segment of polypeptide chain the volume of its ellipsoid of inertia is calculated and compared to its molecular weight, thus defining a "partial specific volume" for the ellipsoid of inertia. Compact units are selected on the basis of their low partial specific volumes. The smaller compact units are then grouped together into larger ones through a scheme of hierarchic condensation. The process is applied to myoglobin and to ribonuclease S and the hierarchies of compact units compared to folding intermediates proposed by other authors for these two proteins.

Surface Area Measurements

Surface area measurements are another way of estimating the extent to which parts of the protein interact in its three-dimensional structure. The solvent accessible surface of a protein atom is defined by Lee and Richards[29] as the surface over which the center of a water molecule can be placed so that it is in van der Waals contact with that atom without penetrating other protein atoms. One may also calculate the area of the interface between two groups of protein atoms:

$$B = A_1 + A_2 - A_{12}$$

where A_1 and A_2 are the accessible surface areas of the two groups when they are separated and A_{12} the accessible surface area of the two together. B is the surface area that becomes buried when the two groups interact. Buried surface areas (or contact areas that are closely related to buried surface areas[30]) measure interatomic contacts without relying on an arbitrary cutoff distance such as from distance maps.

Surface area measurements have been used by Wodak and Janin[31,32] to

[28] A. M. Lesk and G. D. Rose, *Proc. Natl. Acad. Sci. U.S.A.* **78,** 4304 (1981).
[29] B. K. Lee and F. M. Richards, *J. Mol. Biol.* **55,** 379 (1971).
[30] F. M. Richards, *Annu. Rev. Biophys. Bioeng.* **6,** 151 (1977).
[31] S. Wodak and J. Janin, *Proc. Natl. Acad. Sci. U.S.A.* **77,** 1736 (1980).
[32] S. Wodak and J. Janin, *Biochemistry* **20,** 6544 (1981).

identify domains in proteins. The polypeptide chain is assumed to be cleaved at a position i. The area B of the interface between the NH_2- and COOH-terminal fragments is measured and plotted against the position i. The plot, called interface area scan, may show the presence of minima delimiting regions having relatively fewer contacts with the rest of the chain. Such regions are taken to be domains. Each interface area scan necessitates a large number of surface area calculations. These can be performed with the geometrical algorithm of Lee and Richards,[29] or with the much faster analytical approximation of Wodak and Janin[31] which relies only on interresidue distances to estimate buried surface areas.

A comparison of the results presented by Wodak and Janin[32] with those of visual inspection reveals that single-domain proteins such as the bovine pancreatic trypsin inhibitor or isolated immunoglobulin domains yield bell-shaped interface area scans with no significant minimum of B except at the NH_2 and COOH termini of the chain where B drops sharply to zero. Scans of typical two-domain proteins have two such bell-shaped features separated by a minimum where B represents the area of the domain-to-domain interface. More complex domain structures can be analyzed by a repeated cleavage algorithm which, like Rose's algorithm,[27] creates a descending hierarchy of continuous protein fragments: the chain is cleaved first at the deepest minimum of B and the scan is repeated on each of the two fragments until scans with no significant minima are obtained.

An extension of the procedure is applied to proteins that have discontinuous domain structures.[32] In these cases, continuous domains embedded in discontinuous ones are identified as follows: a protein segment of L residues is moved along the polypeptide chain and the area B of the interface between the segment and the rest of the protein is measured. B is plotted in two dimensions against the length L of the segment and the position i of its NH_2 terminus. Minima of B for given values of L represent fragments with few outside interactions for their size, as is to be expected for domains.

Globularity Indices

Interface area scans identify domains as autonomous protein regions which have a minimum of interactions to the rest of the structure. They need not be globular structures. Yet globularity may easily be estimated on the basis of surface area measurements. Small globular proteins follow the law:

$$A_G = 11.1\ M^{2/3}$$

which relates their molecular mass M (in daltons) to their solvent-accessible surface area A_G in square angstroms. This law applies to nearly all monomeric proteins in the range 6000–25,000 daltons.[33,34] Perfect smooth spheres would have about half the surface predicted from this law, ellipsoids with the same dimensions as the proteins somewhat more.[30] In contrast, large proteins and subunits of many oligomeric proteins have larger accessible surface areas than predicted by the formula giving A_G.[40,59] Thus, for a protein or protein fragment, the ratio A/A_G of its actual accessible surface area A to that predicted from its molecular weight is an index of how globular its three-dimensional structure is.

Wodak and Janin[32] observe that domains identified from the minima in area scans have, in general, values of A/A_G close to unity. Thus they constitute globular substructures as well as autonomous ones, in accordance with the original definition of Wetlaufer.[9] It is possible to search just for globular fragments that have the minimal values of A/A_G. Rashin[35,36] has done such a search, though his definition of a globularity index is based on buried surface areas rather than on the accessible. In his analysis, compact globular protein fragments are identified as burying a maximum of surface area for their size when they fold.

Principles Determining the Structure of Domains

In small proteins and in the domains of larger proteins, the polypeptide chain tends to run back and forth across the structure. Usually each run contains an α-helix or a strand of β-sheet, with the links between these pieces of secondary structure on the surface.[37,38] Analysis of observed protein structures has revealed a set of general principles that govern how secondary structures pack together, the nature of their buried and accessible surfaces, and the topology of the polypeptide chain.[38a,38b]

Packing of Helices and β-Sheets

Two factors have a dominating influence on the packing of secondary structures in proteins. First, residues in the interior are usually close

[33] J. Janin, *J. Mol. Biol.* **105,** 13 (1976).
[34] D. C. Teller, *Nature (London)* **260,** 729 (1976).
[35] A. A. Rashin, *Stud. Biophys.* **77,** 177 (1979).
[36] A. A. Rashin, *Nature (London)* **291,** 85 (1981).
[37] I. D. Kuntz, *J. Am. Chem. Soc.* **94,** 8568 (1972).
[38] M. Levitt and C. Chothia, *Nature (London)* **261,** 552 (1976).
[38a] J. Janin, *Bull. Inst. Pasteur (Paris)* **77,** 339 (1979).
[38b] C. Chothia, *Ann. Rev. Biochem.* **53,** 537 (1984).

packed,[39–41] and second, the packed secondary structures have a conformation close to the minimum free energy conformation of the isolated secondary structures.[42–45] Thus the manner in which secondary structures pack will depend upon the shape of their surfaces.

Models have been used to describe and explain secondary structure packings. They assume that, to a first approximation, the surface shape arises from the main-chain conformation of α-helices and β-sheets. Size, shape, and conformation of side chains determine which particular mode or class of packing will occur and modulate its exact geometry. Comparison of the geometrical predictions of the models with the observed geometry of α-helix and β-sheet packing shows that this assumption is generally valid.

Helix–Helix Packing. Crick[46] proposed a model for helix packing, as part of his work on α-keratin, in 1953. Not until 1977 did other workers begin to investigate helix packing. These later workers[47–51] used the observed atomic structure of globular proteins to derive and test models, and a comparative account of the different models is given in Chothia *et al.*[51]

Crick's model[46] had residues on helix surfaces intercalating as "knobs into holes." The more recent models have residues forming "ridges" separated by "grooves" and the ridges of one helix packing into the grooves of the other and vice versa. Different sets of residues can form the ridges to give different packing orientations.

Helix–Sheet Packing. A model for α-helices packing onto β-sheets is discussed in Refs. 47, 52, and 53. Their authors agree on the central point: α-helices normally pack on β-sheets with their axes parallel to the strands because in this orientation they bring together surfaces of complementary twist. The more elaborate models[52,53] have some differences of detail.

[39] F. M. Richards, *J. Mol. Biol.* **82,** 1 (1974).
[40] C. Chothia, *Nature (London)* **254,** 304 (1975).
[41] J. L. Finney, *J. Mol. Biol.* **96,** 721 (1975).
[42] G. N. Ramachandran and V. Sasisiekharan, *Adv. Protein Chem.* **23,** 284 (1968).
[43] R. Chandrasekaran and G. N. Ramachandran, *Int. J. Protein Res.* **2,** 223 (1970).
[44] J. Janin, S. Wodak, M. Levitt, and B. Maigret, *J. Mol. Biol.* **125,** 357 (1978).
[45] B. R. Gelin and M. Karplus, *Biochemistry* **18,** 1256 (1979).
[46] F. H. C. Crick, *Acta Crystallogr.* **6,** 689 (1953).
[47] C. Chothia, M. Levitt, and D. Richardson, *Proc. Natl. Acad. Sci. U.S.A.* **74,** 4130 (1977).
[48] A. V. Efimov, *Dokl. Akad. Nauk SSSR* **235,** 699 (1977).
[49] A. V. Efimov, *J. Mol. Biol.* **134,** 23 (1979).
[50] T. J. Richmond and F. M. Richards, *J. Mol. Biol.* **119,** 537 (1978).
[51] C. Chothia, M. Levitt, and D. Richardson, *J. Mol. Biol.* **145,** 215 (1981).
[52] J. Janin and C. Chothia, *J. Mol. Biol.* **143,** 95 (1980).
[53] F. E. Cohen, M. J. E. Sternberg, and W. R. Taylor, *J. Mol. Biol.* **156,** 821 (1982).

Sheet–Sheet Packing. Inspection of protein structures[47] shows that β-sheets pack together face to face with the main-chain direction inclined at either ∼−30 or ∼90°. Models for the first type of packing are discussed by Cohen *et al.*[54] and by Chothia and Janin.[55] These authors argue that the angle of ∼−30° is due to the right-handed twist of the β-sheets. The characteristics of the second type of packing arise from it being formed by β-sheets that are bent and twisted.[56]

Buried and Accessible Surfaces

An analysis of the first known protein structures by Perutz *et al.*[57] showed the accuracy of Kauzmann's[58] prediction that the surfaces buried in their interior would be hydrophobic. Quantitative descriptions of the accessible and buried surfaces in proteins have been given by Lee and Richards,[29] Shrake and Rupley,[60] Chothia,[61] Janin,[33,62] Teller[34] and Sprang *et al.*[59]

The surfaces of α-helices and β-sheets have been analyzed in detail.[51–56,61,63–65] The surfaces which β-sheets use to pack against other secondary structures have an unusual residue composition with just three, Val, Leu, and Ile, forming half the total.[52–56]

Chain Topology

In observed protein structures the connections between α-helices and β-sheets are significantly nonrandom:

1. Pieces of secondary structure adjacent in sequence are also often in contact in three dimensions.[66]

2. For β–X–β units (where the β is a parallel strand in the same β-sheet, though not necessarily adjacent, and X is an α-helix, a strand in a different β-sheet, or an extended polypeptide), the connections are right-handed in nearly all cases.[67–70]

[54] F. E. Cohen, M. J. E. Sternberg, and W. R. Taylor, *J. Mol. Biol.* **148,** 253 (1981).
[55] C. Chothia and J. Janin, *Proc. Natl. Acad. Sci. U.S.A.* **78,** 4146 (1981).
[56] C. Chothia and J. Janin, *Biochemistry* **21,** 3955 (1982).
[57] M. F. Perutz, J. C. Kendrew, and H. C. Watson, *J. Mol. Biol.* **13,** 669 (1965).
[58] W. Kauzmann, *Adv. Protein Chem.* **14,** 1 (1959).
[59] S. Sprang, D. Yang, and R. J. Fletterick, *Nature (London)* **280,** 333 (1979).
[60] A. Shrake and J. A. Rupley, *J. Mol. Biol.* **79,** 351 (1973).
[61] C. Chothia, *J. Mol. Biol.* **105,** 1 (1976).
[62] J. Janin, *Nature (London)* **277,** 491 (1979).
[63] F. E. Cohen, M. J. E. Sternberg, and W. R. Taylor, *Nature (London)* **285,** 378 (1980).
[64] S. Lifson and C. Sander, *Nature (London)* **282,** 109 (1979).
[65] D. Eisenberg, R. W. Weiss, and T. C. Terwilliger, *Nature (London)* **299,** 371 (1982).
[66] M. Levitt and C. Chothia, *Nature (London)* **261,** 552 (1976).
[67] J. S. Richardson, *Proc. Natl. Acad. Sci. U.S.A.* **73,** 2619 (1976).

3. Crossing of secondary structure connections and knotting of the polypeptide rarely occur.[71-74]

Other regularities in topology, for example the "Greek key" in β-sheets,[71] can be shown to arise from these three.

The Classification of Protein Structures

The restrictions on the ways in which helices and β-sheets can pack and on chain topology mean that proteins, quite unrelated by evolution or function, may have similar structures.[66,71,75,76] A structural classification system for proteins and their domains, based on the observed arrangements of helices and β-sheets, is described by Levitt and Chothia[66] and extended versions are given by Janin[38a] and Richardson.[77]

[68] M. J. E. Sternberg and J. M. Thornton, *J. Mol. Biol.* **105,** 367 (1976).
[69] M. J. E. Sternberg and J. M. Thornton, *J. Mol. Biol.* **110,** 269 (1977).
[70] K. Nagano, *J. Mol. Biol.* **109,** 235 (1977).
[71] J. S. Richardson, *Nature (London)* **268,** 495 (1977).
[72] O. B. Ptitsyn and A. V. Finkelstein, *Proc. FEBS Meet.* **52,** 105 (1979).
[73] M. L. Connolly, I. D. Kuntz, and G. M. Crippen, *Biopolymers* **19,** 1167 (1980).
[74] M. H. Klapper and I. S. Klapper, *Biochim. Biophys. Acta* **626,** 97 (1980).
[75] G. E. Schulz, *Angew. Chem., Int. Ed. Engl.* **16,** 23 (1977).
[76] O. B. Ptitsyn and A. V. Finkelstein, *Q. Rev. Biophys.* **13,** 339 (1980).
[77] J. S. Richardson, *Adv. Protein Chem.* **34,** 168 (1981).

[29] Automatic Recognition of Domains in Globular Proteins*

By GEORGE D. ROSE

Introduction

Numerous protein crystallographers have observed that their molecules can be neatly dissected into two or more spatially distinct, compact, contiguous-chain segments called *domains* or *structural domains* (see Fig. 1). These structural units have come to be regarded as familiar components in proteins and have been variously proposed as folding interme-

* Dedicated to Professor Harry Goheen and to Molly Goheen on the occasion of their retirement.

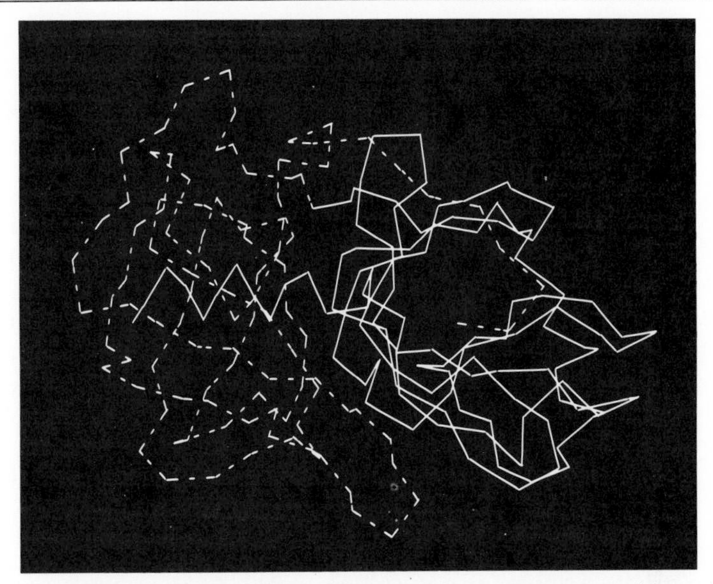

FIG. 1. The domains of bovine trypsin. The broken line connects sequential α-carbons in the NH_2-terminal domain, and continues as a solid line in the COOH-terminal domain. Viewing orientation is along the β-barrel in the COOH-terminal domain.

diates, functional modules, and perhaps exon products.[1-5] Several strategies to identify domains by inspection of X-ray-elucidated protein structures have been presented.[1,2,6] More recently, computer algorithms have been formulated to recognize domains automatically from X-ray coordinates by measuring geometric correlates of the conformational energy.[7-13] This geometric approach to structural issues owes much to the

[1] D. B. Wetlaufer, *Proc. Natl. Acad. Sci. U.S.A.* **70,** 697 (1973).
[2] A. Liljas and M. G. Rossmann, *Annu. Rev. Biochem.* **43,** 475 (1974).
[3] O. B. Ptitsyn, *FEBS Lett.* **93,** 1 (1978).
[4] W. Gilbert, *Nature (London)* **271,** 501 (1978).
[5] C. C. F. Blake, *Nature (London)* **277,** 598 (1979).
[6] G. E. Schulz and R. H. Schirmer, "Principles of Protein Structure." Springer-Verlag, Berlin and New York, 1979.
[7] G. M. Crippen, *J. Mol. Biol.* **126,** 315 (1978).
[8] G. D. Rose, *J. Mol. Biol.* **134,** 447 (1979).
[9] A. M. Lesk and G. D. Rose, *Proc. Natl. Acad. Sci. U.S.A.* **78,** 4304 (1981).
[10] S. J. Wodak and J. Janin, *Proc. Natl. Acad. Sci. U.S.A.* **77,** 1736 (1980).
[11] S. J. Wodak and J. Janin, *Biochemistry* **20,** 6544 (1981).
[12] A. A. Rashin, *Nature (London)* **291,** 85 (1981).
[13] C. Sander, *in* "Structural Aspects of Recognition and Assembly in Biological Macromolecules" (M. Balaban, ed.), p. 183, BALABAN Int. Sci. Serv., Rehovot, Israel, and Philadelphia, Pennsylvania, (1981).

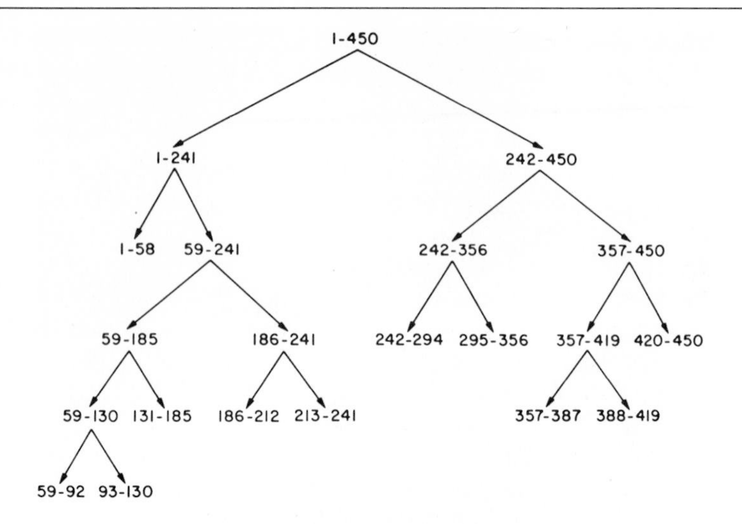

FIG. 2. The hierarchic organization of hexokinase taken from a computer analysis of 22 proteins.[8] Each branch represents the algorithmically discovered best cut of a parent segment into spatially distinct, contiguous-chain strands. The subdivision process is arrested when a segment no longer closes back upon itself.

fundamental work of Richards and co-workers,[14,15] Shrake and Rupley,[16] Finney,[17] and Chothia.[18]

It is now apparent that the large domains reported in the literature are special cases of a more general phenomenon: the definition of a domain as a contiguous-chain segment folded into a spatially discrete region is satisfied not only by large segments but also by smaller ones. Upon repeated application, the definition results in a structural hierarchy wherein the protein monomer can be iteratively subdivided into subdomains of decreasing size until individual segments of secondary structure are realized.[8] Conversely, the monomer can be built up from units of increasing size.[7] This hierarchic architecture is illustrated in Fig. 2.

Thus, the formal definition of a domain is satisfied equally well by a range of segment sizes that spans each protein molecule. This observation suggests a revised view of the large familiar domains: these need no longer be regarded as strictly autonomous units, but instead may arise as

[14] F. M. Richards, *Annu. Rev. Biophys. Bioeng.* **6,** 151 (1977).

[15] B. Lee and F. M. Richards, *J. Mol. Biol.* **55,** 379 (1971).

[16] A. Shrake and J. A. Rupley, *J. Mol. Biol.* **79,** 351 (1973).

[17] J. L. Finney, *J. Mol. Biol.* **96,** 721 (1975).

[18] C. Chothia, *J. Mol. Biol.* **105,** 1 (1976).

the inevitable structural by-products of a concluding step in chain folding.[19] Such an interpretation derives added support from a recent analysis showing that simulated protein chain folds resemble authentic molecules in their hierarchic organization.[20] Apparently, the observed spatial compartmentation of continuous segments arises quite naturally in the course of folding a self-avoiding polypeptide chain toward its internal close-packing limit.

In light of the preceding comments, we distinguish between continuous-chain structural *units* and the folding *pattern* that deploys these units during successive stages in the self-assembly process. This review describes a general iterative method to reveal the folding pattern, but the analysis stops short of introducing additional physical criteria that would be needed to identify particular units within the pattern.[9,11,12] Nevertheless, initial steps in the current analysis often result in chain divisions that correspond to the typical large domains as seen by crystallographers.[8] The reason for this correspondence will be discussed.

Method

The algorithm described here was devised to optimize sequential subdivisions of the folded polypeptide chain into spatially distinct regions without regard for chemical particulars such as residue type, hydrogen bonding, or disulfide loops. The method is based on the following formal definition:

> Let S be a chain segment, cut into two strands A and B, and call V_A the molecular volume occupied by atoms of strand A and V_B the volume occupied by atoms of B. Then A and B yield discrete subdivisions with respect to a plane ρ if V_A resides entirely on one side of ρ and V_B resides entirely on the other side of ρ.

Conceptually, this definition is equivalent to taking a single planar cut through the chain so as to divide it into two long, continuous, spaghetti-like strands, and not into rice. In principle, a planar cut is too stringent because protein chains cannot pack together like books on a shelf. Undoubtedly, a curvilinear cutting surface would provide a more realistic model, but a plane has the virtue of geometric simplicity and, in practice, only small deviations from planar packing are observed to occur.[8]

Exploring the protein structure for a likely cutting plane ρ is an arduous process in three dimensions. The search strategy outlined here simplifies the problem by reducing it to two dimensions. First, a particular viewing plane, called the *disclosing plane,* is chosen. Next, the protein

[19] G. D. Rose, *Biophys. J.* **32,** 419 (1980).
[20] T. J. Yuschok and G. D. Rose, *Int. J. Pep. Protein Res.* **21,** 479 (1983).

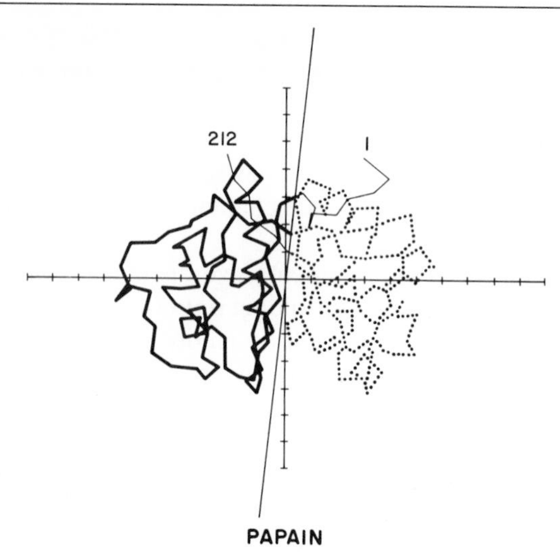
212
1
PAPAIN

FIG. 3. The subdivision algorithm consists of three steps: (1) the disclosing plane is calculated for the segment under consideration; (2) the α-carbon coordinates are projected onto the disclosing plane and connected sequentially; and (3) the cutting line that optimizes division of the projection into distinct, contiguous-chain strands is identified from a grid search. Here, the α-carbons of papain are projected onto their disclosing plane, and the discovered cutting line is shown. Disclosing plane axes are graduated in units of 4 Å. The NH_2-terminal segment is represented as a solid line, and the COOH-terminal segment as a broken line. Short overlapping chain ends, a common feature in protein domains, are represented by faint lines.

structure is projected onto the disclosing plane. Finally, the disclosing plane is searched to discover a line, called the *cutting line,* that optimizes division of the protein's two-dimensional projection into distinct, continuous-chain regions. These operations determine ρ, which is taken to be perpendicular to the disclosing plane through the cutting line. The strategy is illustrated in Fig. 3.

Hierarchy arises upon repeated application of this search strategy. In schematic format:

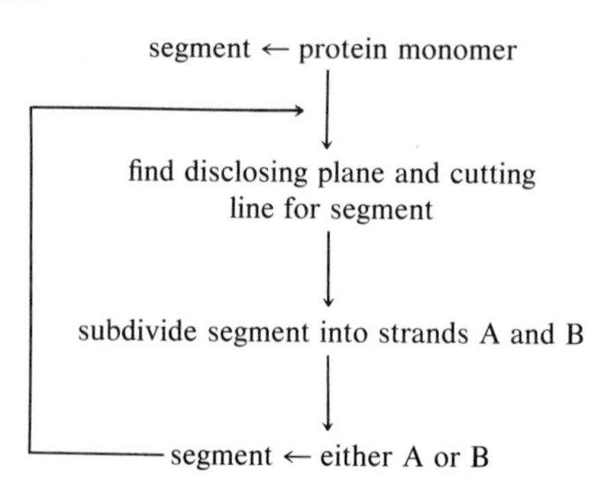
segment ← protein monomer
find disclosing plane and cutting line for segment
subdivide segment into strands A and B
segment ← either A or B

The process can be terminated after a few major subdivisions, or it can be prolonged until supersecondary structure or even secondary structure segments are realized.

The Disclosing Plane

The disclosing plane of any segment is given by its two "best" principal moments of inertia. All three principal moments are obtained as the eigenvectors of the inertia tensor for the segment, and the two with smaller eigenvalues are taken to define the disclosing plane. Following classic mechanics,[21] the components of the inertia tensor are represented by a symmetric matrix **T**,

$$\mathbf{T} = \begin{pmatrix} \Sigma W(2,n_i)^2 + \Sigma W(3,n_i)^2 & & \\ -\Sigma W(1,n_i)W(2,n_i) & \Sigma W(1,n_i)^2 + \Sigma W(3,n_i)^2 & \\ -\Sigma W(1,n_i)W(3,n_i) & -\Sigma W(2,n_i)W(3,n_i) & \Sigma W(1,n_i)^2 + \Sigma W(2,n_i)^2 \end{pmatrix} \quad (1)$$

in which $W(j,k)$ is a center of mass coordinate for the jth component of the kth α-carbon. Summation over each n_i ranges from 1 to N for a segment of N residues. Differences in mass from residue to residue have been neglected.

The ensuing illustrative block of Fortran code calculates **T** from α-carbon coordinates and uses its eigenvectors to transform the segment into disclosing plane coordinates:

```
      Dimension V(3,N), W(3,N), SM(3), CV(6), T(6)
      Dimension RTR(3), U(3,3), WK(14)
C
C     N is the number of residues.
C     V is the original set of alpha-carbon coordinates.
C     W is the set of coordinates translated to have their origin at the center of mass.
C     SM holds the coordinates for the center of mass.
C     CV are cross products of the center of mass coordinates, or covariances.
C     T holds the components of the inertia tensor, in symmetric storage mode, i.e.,
C            T(1)
C            T(2)  T(3)
C            T(4)  T(5)  T(6)
C     The code is set up for use with the IMSL subroutine EIGRS in order to find eigen-
C       vectors and eigenvalues of T.
C     Arrays RTR,U, and WK are set by EIGRS as follows:
C     RTR holds the eigenvalues of T,
C     U holds the eigenvectors of T,
C     WK is a work area required by EIGRS.
C
C
C     Initialize CV and SM to 0.
      Do 1 I=1,6
```

[21] H. Goldstein, "Classical Mechanics." Addison-Wesley, Reading, Massachusetts, 1950.

```
1     CV(I)=0
      Do 2 I=1,3
2     SM(I)=0
C
C     Calculate center of mass.
      Do 3 I=1,N
      Do 3 J=1,3
3     SM(J)=SM(J)+V(J,I)
      Do 4 J=1,3
4     SM(J)=SM(J)/N
C     Coordinates for center of mass now stored in SM.
C
C     Form components of the inertia tensor.
      Do 8 I=1,N
C       Convert to center of mass coordinates.
        Do 7 J=1,3
7       W(J,I)=V(J,I)-SM(J)
C
C       Form cross products.
        L=0
              Do 6 J=1,3
              Do 6 K=J,3
              L=L+1
6             CV(L)=CV(L)+W(J,I)*W(K,I)
8     Continue
C
C     Form components of inertia tensor from cross products.
C     T is arranged in symmetric storage mode, as shown above.
      T(1)=CV(4)+CV(6)
      T(2)=CV(2)
      T(3)=CV(1)+CV(6)
      T(4)=CV(3)
      T(5)=CV(5)
      T(6)=CV(1)+CV(4)
C
C     Call the IMSL routine EIGRS to calculate the eigenvectors and eigenvalues of T.
C       Eigenvalues are returned in RTR and eigenvectors are returned in U in order of
C       ascending eigenvalue.
      Call EIGRS(T,3,1,RTR,U,3,WK,IER)
C
C     Transform the coordinates in array V into principal axis coordinates. V(1,N) and
C       V(2,N) will be the coordinates in the disclosing plane.
      Do 200 I=1,N
        Do 175 J=1,3
        Z=0.
          Do 150 K=1,3
150       Z=Z + W(K,I)*U(K,J)
        V(J,I)=Z
175     Continue
200   Continue
```

The code as shown uses the IMSL routine[22] EIGRS to find the eigenvectors of **T**, but any equivalent subroutine will serve. EIGRS returns the eigenvectors, U, in ascending order by eigenvalue. Upon completion, $V(1,N)$ and $V(2,N)$ will hold disclosing plane coordinates for the segment.

The disclosing plane, defined in this way, is the chain cross section of maximal surface area, a vantage plane that capitalizes upon the overall organization of the protein. In protein molecules, structural segments such as helices and strands tend to run back and forth,[23,24] burying surface between them. A maximal area cross section will cut through these secondary structures, causing their residues to cluster upon planar projection. Secondary structure termini are interconnected by turns and loops,[25–28] which tend to extend across the molecule, and hence to parallel the disclosing plane. With structural segments and their interconnections disposed in this fashion, the disclosing plane is suitably chosen to follow the chain connectivity from cluster to cluster.

The disclosing plane projection for successively smaller segments will preserve this desirable viewing orientation, until the segment size approaches the length of individual helices and strands. Around this limit, the best principal axis tends to parallel the secondary structure segments instead of slicing through them, and the disclosing plane strategy deteriorates. To circumvent this problem, we terminate the process of subdivision when the disclosing plane projection of a segment no longer closes back upon itself (around 25–35 residues). Arresting the algorithm at this threshold is not a serious limitation because the continuing subdivision of such small segments can be achieved by visual inspection.

The Cutting Line

Once the disclosing plane is determined, a grid search is performed to locate the cutting line. In this process, the plane is divided into 2 Å squares, with 36 trial lines (one every 10°) assigned to each vertex. Every trial line is tested to measure how effectively it divides the two-dimensional chain projection into separable, contiguous-chain regions. A refined grid search is then conducted in the neighborhood of the best trial line, resulting in the chosen cutting line.

[22] International Mathematical and Statistical Libraries, Inc., 7500 Bellaire Blvd., Houston, Texas 77036.

[23] M. Levitt and C. Chothia, *Nature (London)* **261,** 552 (1976).

[24] D. B. Wetlaufer, G. D. Rose, and L. Taaffe, *Biochemistry* **15,** 5154 (1976).

[25] G. D. Rose and J. Seltzer, *J. Mol. Biol.* **113,** 153 (1977).

[26] G. D. Rose and S. Roy, *Proc. Natl. Acad. Sci. U.S.A.* **77,** 4643 (1980).

[27] G. D. Rose, *Nature (London)* **272,** 5654 (1978).

[28] J. S. Richardson, *Adv. Protein Chem.* **34,** 168 (1981).

Each trial line l is evaluated by the quality of its intersection with the peptide chain segment under consideration. The segment is comprised of N sequential residues, labeled 1 through N. When l intersects the segment projection, it will divide it into two or more piecewise connected smaller segments. Let

$$R = 1, r_i, r_j, r_k, \ldots, N \quad (2)$$

be the series of residues cut by l, together with the first residue (1) and the last residue (N), and ordered such that $r_i < r_j$ when $i < j$. Then for every two consecutive residues in R, evaluate

$$d_j = r_k - r_j \quad (3)$$

the length of the linear interval from r_j to r_k, and form the set

$$D = d_1, d_2, \ldots, d_m \quad (4)$$

of differences for each consecutive pair in R. When l intersects the projection at one residue only, r_x, then D will consist of the two members $r_x - 1$ and $N - r_x$. In any case, choose $\bar{d}$, the largest member of D; $\bar{d}$ is associated with some linear interval (x, y). Then the chain scission corresponding to l is defined to be $(1, y)$ and $(y + 1, N)$. In the event that D does not have a unique largest member, the $\bar{d}$ with residues nearest the NH_2 terminus is selected.

The procedure described in the previous paragraph defines a unique chain cut for an arbitrary trial line. The actual cutting line is chosen to be the "best" trial line, based upon two criteria:

1. The extent to which either interval crosses over to the wrong side of l. In the ideal case, l effects a perfect splitting of $(1, y)$ and $(y + 1, N)$ into disjoint parts P_a and P_b. Any deviation from ideality is measured by summing the length of the excursions that $(1, y)$ takes in P_b, and conversely, the length of the excursions of $(y + 1, N)$ in P_a.
2. The similarity in length between intervals $(1, y)$ and $(y + 1, N)$. In the ideal case, the intervals are of identical length. Deviation from ideality is measured as a nonlinear function of the difference in lengths such that small differences are of minor significance. This criterion precludes trivial divisions such as $(1, 1)$ and $(2, N)$.

In practice, the cutting line is found after a grid search of the disclosing plane as the minimal value of the function

$$E = \Phi + C \quad (5)$$

where

$$\Phi = \sum_{i=1}^{y} f(i) + \sum_{i=y+1}^{N} g(i) \tag{6}$$

$$C = k(N + 1 - 2y)^2/2N \tag{7}$$

E is a measure of nonideality that includes both of the criteria mentioned above. Φ is a linear function of the excursions that each interval makes across l and into the other interval, with

$$f(i) = \begin{cases} 0 & \text{if residue } i \text{ in } P_a \\ \Delta & \text{if residue } i \text{ in } P_b \end{cases} \tag{8}$$

$$g(i) = \begin{cases} 0 & \text{if residue } i \text{ in } P_b \\ \Delta & \text{if residue } i \text{ in } P_a \end{cases} \tag{9}$$

Here Δ is the Euclidean distance from residue i to l. C is a quadratic penalty that equals 0 when the intervals are of equal length, i.e., when $y = (N + 1)/2$, and grows as the square of the difference in lengths. The constant k is usually set to 25.

The formal definition of a chain subdivision has now been satisfied with intervals $(1, y)$ and $(y + 1, N)$ corresponding to segment strands A and B, and with ρ perpendicular to the disclosing plane and through the discovered cutting line. The device of searching an analytically chosen disclosing plane for the optimal ρ serves to reduce the problem to two dimensions.

Discussion

The disclosing plane of a protein monomer is the chain cross section of maximal surface area, as discussed in the section "The Disclosing Plane." This molecular slice tends to cut through the secondary structure segments, which then cluster upon planar projection, simplifying the road map of chain connectivity. Crystallographers have typically picked a similar slice of the electron density to reveal contiguous-chain structural domains in their molecules. Viewed from this vantage plane, the protein backbone is readily subdivided into spatially distinct segments. Repeated observation of this partitioning phenomenon in unrelated molecules has engendered the notion of a domain as an autonomous structural entity.

This chapter has related the crystallographer's vantage plane (disclosing plane) to an intramolecular partitioning plane (ρ), and described an algorithm to find both. When this is done, it becomes apparent that domains as such are not limited to the topmost echelons of molecular structure; instead, they are reiterated in hierarchic fashion ranging from the whole protein monomer through supersecondary structures down to individual helices and strands.

The concept of a domain is subject to reinterpretation when considered in the broader context of its molecular hierarchy. The conventional large domains cited in the literature (near the top of the hierarchy) are seen to be structural composites, with subparts that are domains in their own right. Thus, the large, spatially distinct chain segments that can be resolved within a protein may not represent strictly autonomous units, but may arise instead in consequence of a concluding step in the sequential folding process.

Acknowledgments

X-Ray coordinates were provided by the Brookhaven Data Bank.[29] The work was supported by PHD Grant 29458 and by an NIH Research Career Development Award.

[29] F. C. Bernstein, T. G. Koetzle, G. J. B. Williams, E. F. Meyer, Jr., M. D. Brice, J. R. Rogers, O. Kennard, T. Shimanouchi, and M. Tasumi, *J. Mol. Biol.* **112,** 535 (1977).

[30] Calculation of Molecular Volumes and Areas for Structures of Known Geometry

By FREDERIC M. RICHARDS

Introduction

At the macroscopic level the concepts of area and volume are quite clear, and the methods of measurement straightforward in principle. At the level of individual molecules the definitions become less obvious. When different methods of measurement are used, it is not clear that area and volume are single valued characteristics. For a one-component system of known molecular weight, the mass density is sufficient to unambiguously define the mean volume per molecule. For a system of two or more components, thermodynamic parameters, the partial molar volumes, can be defined uniquely, but these values bear no necessary simple relation to the actual physical volumes of the individual components which, in turn, may not be uniform throughout the mixture. The concepts of total molecular area are even more elusive (see below).

For the purpose of this discussion we shall concentrate on the definition and calculation of certain geometrical volumes and areas that can be derived from high-resolution structural data. No attempt will be made to give a critical discussion of the use (or potential misuse) of these values.

The following basic input data are required:

1. The list of Cartesian coordinates of the atom centers.
2. One or more lists of attributes for each atom:
   a. Assigned van der Waals radii for each atom (required in most procedures).
   b. Assigned covalent radius for each atom (use depends on the volume algorithm selected).
   c. Covalent connectivity (needed if b is used, and also to assemble atoms into reasonable packing groups)
3. Choice of radius for a spherical probe.

In the presently used procedures no allowance is made for errors or fluctuations in the coordinates. The structure is simply a collection of points in three-dimensional space. Sensitivity to error, if required, is estimated by repeating the calculation with appropriately altered coordinate lists and comparing the results. None of the algorithms place any restrictions on the position of the atom centers.

The van der Waals Envelope

Because of the diffuse radial distribution of the electron density surrounding any atomic center, the apparent position of the surface of a molecule will depend on the technique used to examine it. For chemically bonded atoms the distribution is not spherically symmetric nor are the properties of such atoms isotropic. In spite of all this the use of the hard sphere model has a venerable history and an enviable record in explaining a variety of different observable properties. As applied specifically to proteins, the work of G. N. Ramachandran and colleagues has provided much of our present thinking about permissible peptide chain conformation.[1] Different approaches using more realistic models, complex mathematics, and even quantum mechanical approximations have improved the details but have not altered the basic outline provided by the hard sphere approximation. The steepness of the repulsive term in the potential function for nonbonded interactions is responsible for the success of "hard" in the hard sphere.

In spite of the general success of the hard sphere approximation, the van der Waals envelope of a molecule is not unique and is defined differ-

[1] G. N. Ramachandran and B. Sasisekharan, *Adv. Protein Chem.* **23,** 284 (1968).

ently for different purposes. The bases on which the radii of the individual atoms are derived and the uses to which they are put differ. There may be no simple set of "correct" values. The radii are closely connected to the nonbonded potential energy function. Given this function for a pair of atoms the sum of the radii may be equated either to the value of the interatomic separation at the minimum or to the smaller value where the potential is zero. In the biochemical literature the former has been more commonly used. For the Lennard–Jones 6–12 potential the two values differ by about 12%.

The parameters of the nonbonded potential functions can be derived by fitting the functions to the observed packing in molecular crystals and making use of the structural and thermodynamic data available for a large number of such crystals.[2] The derived functions, however, will depend on how the lattice energy is partitioned. A full expression for the lattice energy may include the electrostatic interaction of all partial charges, bond lengths and angles, special treatment of hydrogen bonds, etc. The nonbonded terms will then account only for the residual energy. On the other hand the partial charges can be omitted and the total energy portioned among the various pair-additive, nonbonded terms. The expressions, and thus the derived radii, will, of course, be different even for the same input data. The proper use of these numbers requires that the basis for their derivation be known and appropriately accounted for.

When working with macromolecules it is frequently convenient not to deal with individual atoms but with small groups of atoms which are considered to be adequately represented by a sphere and characterized by a single radius. These groups usually consist of a single heavy atom and one or more hydrogen atoms. (The relation between the group radii and the individual atom radii are not always obvious.) Such groups have been referred to as "unified atoms" by Dunfield *et al.*[3] and as "extended atoms" by Karplus and colleagues.[4] Some group radii that have been used in various studies are listed in the table.

The complex surface that results from the intersection of a number of spheres is referred to as the van der Waals surface. This surface has a defined area and it encloses a defined volume. Although the construction is easy to visualize and is logically consistent, it should be recognized that no chemical procedure ever directly measures this particular area or volume. However, the various areas and volumes computed by algorithms

[2] F. A. Momany, L. M. Carruthers, R. F. McGuire, and H. A. Scheraga, *J. Phys. Chem.* **78,** 1595 (1974).

[3] L. G. Dunfield, A. W. Burgess, and H. A. Scheraga, *J. Phys. Chem.* **82,** 2609 (1978).

[4] B. R. Gelin and M. Karplus, *Biochemistry* **18,** 1256 (1979).

SOME LISTS OF VAN DER WAALS RADII FOR SELECTED GROUPS OF ATOMS

Symbol	Designation	Bondi[a]	Lee and Richards[b]	Shrake and Rupley[c]	Richards[d]	Chothia[e]	Richmond and Richards[f]	Gelin and Karplus[g]	Dunfield *et al.*[h] and Nemethy *et al.*[i]
$—CH_3$	Aliphatic, methyl	2.0	1.80	2.0	2.0	1.87	1.9	1.95	2.13
$—CH_2—$	Aliphatic, methyl	2.0	1.80	2.0	2.0	1.87	1.9	1.90	2.23
>CH—	Aliphatic, CH	—	1.70	2.0	2.0	1.87	1.9	1.85	2.38
≥CH	Aromatic, CH	—	1.80	1.85	*j*	1.76	1.7	1.90	2.10
>C=	Trigonal or aromatic	1.74	1.80	1.5/1.85	1.7	1.76	1.7	1.80	1.85
$—NH_3^+$	Amino, protonated	—	1.80	1.5	2.0	1.50	.7	1.75	
$—NH_2$	Amino or amide	1.75	1.80	1.5	—	1.65	1.7	1.70	
X (O/N)	Amide (N or O unknown)	1.75	1.55	1.5	1.6	1.65	1.7	—	
>NH	Peptide, NH or N	1.65	1.52	1.4	1.7	1.65	1.7	1.65	1.75 (N)
=O	Carbonyl oxygen	1.5	1.80	1.4	1.4	1.40	1.4	1.60	1.56
—OH	Alcoholic hydroxyl	—	1.80	1.4	1.6	1.40	1.4	1.70	
—OM	Carboxyl oxygen	—	1.80	1.89	1.5	1.40	1.4	1.60	1.62
—SH	Sulfhydryl	—	1.80	1.85	—	1.85	1.8	1.90	
—S—	Thioether or —S—S—	1.80			1.8	1.85	1.8	1.90	2.08

[a] A. Bondi, "Molecular Crystals, Liquids and Glasses." Wiley, New York, 1968. Radii assigned on the basis of observed packing in condensed phases.

[b] Lee and Richards.[25] Values adapted from A. Bondi, *J. Phys. Chem.* **68,** 441 (1964).

[c] Shrake and Rupley.[26] Values taken from L. C. Pauling, "The Nature of the Chemical Bond," 3rd ed. Cornell Univ. Press, Ithaca, New York, 1960.

[d] Richards.[6] Minor modification and extension of Bondi (1968) set (see footnote a, above). Rationale not given.

[e] Chothia.[15] From packing in amino acid crystal structures. Personal communication from T. Koetzle quoted.

[f] Richmond and Richards.[14] No rationale given for values used.

[g] Gelin and Karplus.[4] Origin of values not specified.

[h] Dunfield *et al.*[3] Detailed description of deconvolution of molecular crystal energies. Values represent one-half of the heavy-atom separation at the minimum of the Lennard–Jones 6–12 potential functions for symmetrical interactions.

[i] G. Nemethy, M. S. Pottle, and H. A. Sheraga, *J. Phys. Chem.* **87,** 1883 (1983).

[j] See original paper.

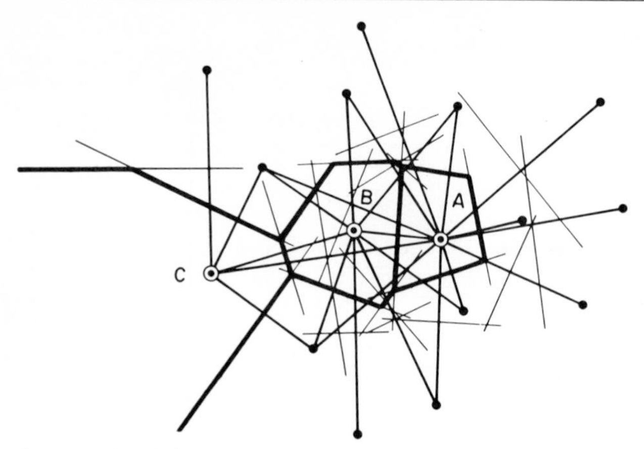

FIG. 1. An example of the Voronoi construction in two dimensions. The heavy lines outline the limiting polygons. The middle weight lines are interatomic vectors between the atom centers represented by dots. The fine lines are the perpendicular bisectors of the vectors. The circled points labeled A and B are interior points surrounded by unique closed polygons. The point labeled C is on the surface of the set of points and a closed polygon cannot be defined. (Adapted from Richards.[16])

discussed in this chapter are very sensitive to the values of the van der Waals radii chosen, and especially to their ratios. When reporting area or volume calculations, the assumed radii should be carefully listed.

Volume

For an infinite set of arbitrary points in space, a geometrical procedure was introduced by Voroni in 1908[5] which divides up space with a unique volume assigned to each point. An example of the construction in two dimensions is shown in Fig. 1. Perpendicular bisectors are drawn for each interatomic vector. The intersections of these bisectors provide the vertices of a polygon surrounding each point (see A and B in Fig. 1). The polygons share common edges and collectively account for the total area. In three dimensions the edges are replaced by planes and the resulting polyhedra account for the total volume. There is an unambiguously defined limiting polyhedron surrounding each point. (See below for the problems of points at the surface of a finite set.) The main computational problem for all volume calculations of this type is to establish this limiting polyhedron.[6–9]

[5] G. F. Voronoi, *J. Reine Angew. Math.* **134,** 198 (1908).
[6] F. M. Richards, *J. Mol. Biol.* **82,** 1 (1974).
[7] J. L. Finney, *J. Mol. Biol.* **96,** 721 (1975).
[8] J. L. Finney, *J. Comput. Phys.* **32,** 137 (1979).
[9] N. Brostow, J. P. Dussault, and B. Fox, *J. Comput. Phys.* **29,** 81 (1978).

The Limiting Polyhedron

Consider a collection of point atoms where the specific atom, *i*, is surrounded by a group of neighbors *j*, *k*, *l*, ... The limiting polyhedron will consist of a set of intersecting planes perpendicular to the interatomic vectors drawn from *i* to *j*, *k*, *l*, ... Each plane containing a face of the polyhedron may be identified by a single index, *j*, *k*, etc. The edges of the polyhedron are identified by pairs of indices from the planes whose intersection produces the edge *jk*, *kl*, etc. The vertices of the polyhedron require three indices, *j*, *k*, *l*, which can be taken either from the three planes whose intersection gives the vertex or alternately from the unique set of indices embedded in the index pairs which represent the three lines which also intersect to define the same vertex (see Fig. 2).

There are at least two somewhat different approaches to selecting the limiting polyhedron. In the first, possible vertices are both generated and selected iteratively on the basis of their position with respect to a developing set of planes which ultimately represents the faces of the final polyhedron. The intersections of the planes representing the faces yields the final list of vertices. In the second, an array of possible vertices, one for each set of four atom centers, which include *i*, is generated on the basis of defined distance criteria. This list is then checked to eliminate all vertices

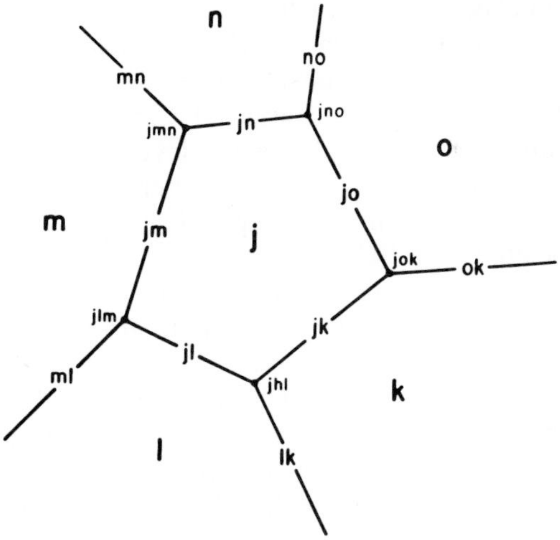

FIG. 2. Top view of a portion of the limiting polyhedron around the central atom *i*. The neighboring atoms are labeled *j*, *k*, *l*, *m*, *n*, *o*, as are the faces drawn perpendicular to the *i*, *j*, ... vectors. The lines representing the intersection of two of these planes are labeled with the corresponding index pairs. The vertices, as points representing the intersection of three planes, are identified by index triples. The lines and vertices surrounding a given face all share a common index. (Adapted from Finney.[8])

that are closer to any other atom center than they are to i. The reduced list provides the vertices of the limiting polyhedron from which, in turn, the faces and edges can be derived.

Approach Based on the Convex Nature of the Polyhedron. Since the limiting polyhedron surrounding atom center i is convex, all vertices are either in a given face plane or on the same side of that plane as i.[6] Any potential vertex on the opposite side of any of the faces is not a member of the final set of vertices. In an iterative procedure vertices are then retained or rejected on the basis of position with respect to each plane in the current set of faces.

The algorithm for implementing this procedure starts by setting up the equations for all planes in the set $\{A\}$ of atoms around i. (See below for the selection of $\{A\}$ and for the various equations that may be used to define these planes.) An arbitrary but very large tetrahedron (i.e., four vertices and four face planes) is set up around i. The position of each of the four vertices with respect to the planes in $\{A\}$ is then examined. A vertex (vertices) not on the same side as i of a given plane is (are) eliminated and replaced by new vertices produced by the selecting plane and the planes contributing to the eliminated vertex or vertices. The index designation for planes, lines, and vertices discussed above make it easy to do the bookkeeping at this stage. The original file of planes for $\{A\}$ is arranged and searched in order of increasing distance from i to the plane. After each plane is checked, the vertex list is changed as required. One pass through the list of planes is then sufficient to yield the limiting polyhedron. The file now contains the position of each vertex and the equation for each face. This procedure is general and is independent of the formula by which the equations for the planes are developed. While the limiting polyhedrons are uniquely defined, the full set of polyhedrons may or may not accurately account for all space depending on the definition used for the planes associated with $\{A\}$.

Approach Based on Distance Selection. For the Voronoi and Radical Plane procedures (see below) each potential vertex can be located at a defined distance from each of four atom centers, one of which is i.[7,8,10] All potential vertices from the atom set $\{A\}$ are calculated. For a vertex to be part of the limiting polyhedron, it must be no closer to any other atom center that it is to i. The distance of each vertex to all atoms in $\{A\}$, other than the four defining atoms, is tested against the distance to i, and is accepted or rejected on this basis. From the indices identifying each vertex in the final list the faces of the polyhedron can be established by searching the list for all sets of triplets having a common index. Thus a

[10] B. J. Gellatly and J. L. Finney, *J. Mol. Biol.* **161,** 305 (1982).

final list is obtained giving each face, the number of vertices in each face, and the total number of faces and vertices as in the first procedure. This approach requires that the distance relations be specified as equalities, but the final set of Voronoi or radical plane polyhedrons do account accurately for all space.

The Atom Set $\{A\}$. For either procedure the efficiency of the calculation depends on making the set of atoms, $\{A\}$, as small as possible. For a completely arbitrary set of points, the total list would have to be surveyed, in principle, since there would be no way in advance of knowing how asymmetric any limiting polyhedron might be. In practice with macromolecules this is not a problem since the points are reasonably uniformly distributed and the resulting polyhedra are quite compact. In the program used by Richards[6] it was found by trial that all necessary positions were included with a comfortable margin of safety if only atoms less that 6.5 Å from i were selected for $\{A\}$. The time for atom selection from the coordinate list is minimized if the list is loaded into a coarse cubic lattice permitting a grid search.

A more systematic and less arbitrary procedure for the set selection was given by Brostow *et al.*[9] The algorithm used for the construction of the limiting polyhedron is comparable to the convex polyhedron approach, but somewhat different criteria are used in selecting and limiting the atom list. The authors suggest that their algorithm is more efficient than that used by Richards or Finney, although no one seems to have made benchmark runs with all three programs on the same data set.

Volume of the Polyhedron

Once the vertex list is complete, the volume of the resulting limiting polyhedron is readily computed. The area of each face can be calculated from the component triangles. The length of the face normal to the center i is already known. The cone volumes associated with each face are summed. The individual atom volumes can be examined or more commonly combined into packing units such as side chains or whole residues.

Although the principal use of this procedure to date has been to assign individual atom volumes, Finney[11] has pointed out that the polyhedrons represent a wealth of information about the surroundings of each atom. The extent of interactions with each neighbor is reflected in a clearly defined way by the area of the shared face. The direction of the interaction is defined by the normal to the face. The neighbors may be other protein atoms or potential solvent molecules.

[11] J. L. Finney, *J. Mol. Biol.* **119,** 415 (1978).

Selection of Position of the Plane along the Interatomic Vector

The Voronoi Construction. In the original Voronoi procedure the planes are drawn as bisectors of the lines between the points, as in Fig. 3 (see Finney[7] and Richards,[6] method A). The points are considered intrinsically equal and no special characteristics are assigned. If d_{ij} is the interatomic distance and p_{ij} the distance from atom i to the intersection of the plane and the vector, then

$$p_{ij}/d_{ij} = 1/2 \tag{1}$$

The procedure gives a single unique face between each atom pair. This face is part of the limiting polyhedrons about both atoms. The method is exact in that all of the space is precisely accounted for without error. The appropriate distance relations are

$$(x_i - x)^2 + (y_i - y)^2 + (z_i - z)^2 = L_c^2, \qquad i = i, j, k, l \tag{2}$$

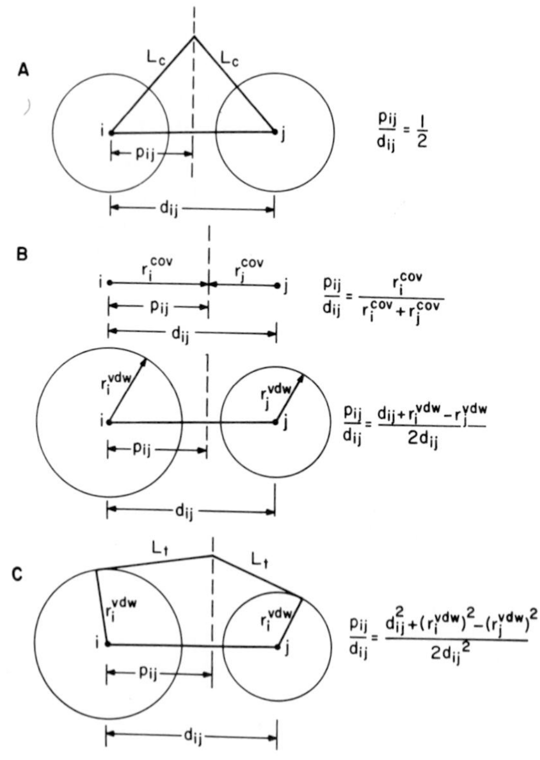

FIG. 3. Definitions of the planes dividing the interatomic vectors in the different partitioning procedures: (A) Voronoi; (B) Richards' method B; (C) radical plane.

where $(x_iy_iz_i)$ and (xyz) are the coordinates of the four atoms and corresponding vertex, and L_c is the common vertex to atom center distance.

Richards' Construction B: While mathematically rigorous, the Voronoi procedure does not make much physical sense since different atoms do have different intrinsic sizes and clearly are not equal. In an attempt to overcome this difficulty, Richards[25] (method B) suggested a modified procedure in which the planes defining the polyhedron do not bisect the interatomic vector but cut it in a ratio which depends on the van der Waals or covalent radii of the two atoms involved (Fig. 2):

Covalent interaction:

$$p_{ij}/d_{ij} = r_i^{\text{cov}}/(r_i^{\text{cov}} + r_j^{\text{cov}}) \tag{3}$$

Noncovalent interaction:

$$p_{ij}/d_{ij} = (d_{ij} + r_i^{\text{vdW}} - r_j^{\text{vdW}})/2d_{ij} \tag{4}$$

This procedure assigns more space to those atoms which are intrinsically larger and less to the smaller atoms. While physically reasonable, the method has lost the mathematical rigor of the strict Voronoi procedure. All of the space is not accounted for. A shared face between two adjacent atoms will not necessarily contain the same vertices for the two polyhedrons surrounding the atoms, thereby leading to the vertex error problem. The little error polyhedrons are variable in size and position, but may, in aggregate, represent considerable volume. A careful comparison has recently been made by Gellatly and Finney[10] who conclude that for ribonuclease S the modified procedure underestimates the total volume of the molecule by about 4%. The use of the procedure will thus depend very much on the purpose for which the numbers are to be used. While it is capable of including both covalent and noncovalent characteristics, the absolute volume errors are variable and may be substantial. Caution is indicated.

The Radical Plane Construction. A different space partitioning method has been developed by Fischer and Koch[12] and applied by Gellatly and Finney[10] to the protein volume problem. The procedure is mathematically accurate and provides a rational basis for handling unequal spheres (Fig. 3). The radical plane is the locus of points from which the tangent lengths L_t to the two spheres are equal. The distance equations are now

$$(x - x_i)^2 + (y - y_i)^2 + (z - z_i)^2 - (r_i^{\text{vdW}})^2 = L_t^2, \qquad i = i, j, k, l \tag{5}$$

[12] W. Fischer and E. Koch, *Z. Kristallogr.* **150,** 245 (1979).

The definition is equivalent to a division of the interatomic vector by the plane as follows:

$$\frac{p_{ij}}{d_{ij}} = \frac{d_{ij}^2 + (r_i^{\mathrm{vdW}})^2 - (r_j^{\mathrm{vdW}})^2}{2d_{ij}^2} \tag{6}$$

The procedure employs only a single atom characteristic, which normally will be taken as the van der Waals radius, r^{vdW}. Thus a distinction between different types of interactions is not possible. This is rarely important as the division of space between covalently bonded atoms is often not required and probably not meaningful. The method is mathematically exact. All space is accounted for.

The Problem of Surface Atoms

The discussion given so far refers to atoms in an infinite set or to interior atoms in a finite set. A surface atom can be seen in position C in Fig. 1. The limiting polyhedron is undefined since there is no outside plane(s) to close the figure. The only proper procedure to get around this problem is to make use of the correct solvent structure surrounding the macromolecule (however, see discussion by Finney[7]). Unfortunately this is not known with certainty in any instance. Thus all procedures are more or less ad hoc and arbitrary. Both Richards[6] and Finney[7] have set up hypothetical solvent molecules around the protein whose sole function is to define positions to which vectors can be drawn to provide planes for closing off the polyhedrons around surface atoms. The details of the procedures are somewhat different.

Richards set up a coarse cubic lattice with an edge of 2.8 Å and inserted the protein molecule. All lattice positions inside the van der Waals envelope of the protein were flagged. Lattice positions adjacent to flagged sites were labeled as solvent unless any of the 13 pairs of opposing sites of the 26 surrounding positions were both occupied by protein. In this case the site was labeled interior and either used as a cavity position (see below) or considered part of the protein depending on the purpose of the calculation. Upon completion of the lattice survey the protein was surrounded with a complete "shell" of solvent positions which were added to the coordinate list and used to define planes for protein atoms. For Richards' method B, the solvent positions were merely used to define a plane normal. The plane was drawn tangent to the van der Waals envelope of the atom regardless of the actual coordinates of the lattice position.

Finney[7] set up spheres in all positions where they would contact the van der Waals envelopes of three surface atoms. The radii were either 1.7

or 1.4 Å for these probes. Initially overlap of the probe envelopes was disregarded. This led to the maximum number of polyhedral faces and thus the smallest assigned volume, V_{min}. In a second procedure probe overlap was eliminated by using a selected probe list. The number of defining planes was thus smaller and the derived volumes slightly larger. For both volume sets the standard deviations were significantly less than those reported by Richards using the cubic lattice procedure where even fewer defining planes were used. Finney *et al.*[13] in 1980 used yet another procedure in which the hypothetical water molecules were set up at a fixed density around the surface atoms. This procedure has been used in the latest study by Gellatly and Finney[10] (see Fig. 4).

The structure of water itself renders an appropriate surface definition of the protein uncertain. As pointed out by Finney[7] in his detailed discussion, water is dominated by the directional H bonding and has a very low formal packing density. It is not a van der Waals liquid and is not dominated by packing considerations. A "loose" solvent layer will assign more volume to the protein than a "dense," close-packed layer. The right compromise to yield reliable total protein volumes is unknown at this time.

Summary on Volumes and Choice of Method

The volumes of various groups in ribonuclease S, as calculated by the three procedures discussed above, are shown in Fig. 4 taken from Gellatly and Finney.[10] There may be no "best" answer on how to go about the volume calculations. The following discussion is taken directly from that paper.

> In the light of the properties of the partitioning schemes examined, we can make some comments on the application of the various methods to calculations of volume in proteins. Such calculations are made at several levels: protein molecule or subunit total volumes; mean volumes and distributions for atom groups (e.g. main chain, side-chains); mean volumes and distributions of single atom groups.
>
> For total volume calculations, both Richards' method B and the radical planes method give results that are almost independent of the surface probe radius used; this seems to us sufficient argument to reject the use of Voronoi's method. (This discussion is wholly within the framework of van der Waals'-type surface probes to handle the surface problem. Other more chemically consistent methods that are under development may cause the conclusions of this section to be modified (B. J. Gellatly, J. P. Bouquiere, and J. L. Finney, unpublished data).) In comparing the use of Richards' B and radical methods for total volumes, all differences in covalent and non-bonded partitioning are cancelled in the summation; the only remaining difference is the vertex error inherent in Richards' method. This error amounts to about 4% for RNAse-S, an error that is outside the errors of experimental volume measurement. Therefore, unless

[13] J. L. Finney, B. J. Gellatly, I. C. Golton, and J. Goodfellow, *Biophys. J.* **32,** 17 (1980).

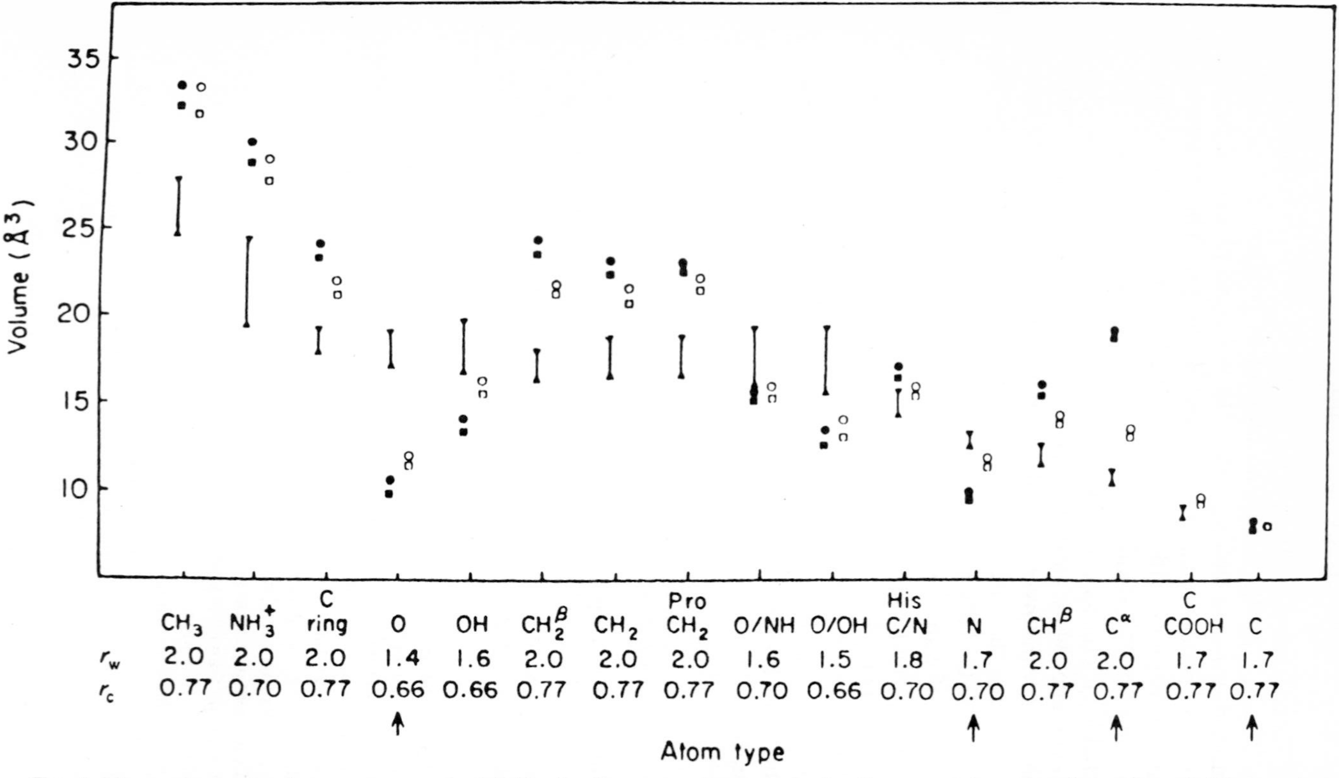

FIG. 4. Mean volumes for atom groups as calculated by the three procedures, and using the same data set for ribonuclease S. The van der Waals radii and the covalent radii used are given below each atom type. Arrows indicate main-chain atoms. The surface was defined by the uniform density probe procedure employing probes with either a 1.7 or 1.4 Å radius. (▼) Voronoi, 1.7 Å; (▲) Voronoi, 1.4 Å; (●) Radical, 1.7 Å; (■) Radical, 1.4 Å; (○) Richards B, 1.7 Å; (□) Richards B, 1.4 Å. Radical and Voronoi values for C carbonyl groups are superimposed. (Reproduced with permission from Gellatly and Finney.[10])

the vertex errors are individually calculated and corrected for (including sub-allocating the surface-involved error tetrahedra between protein and solvent) we argue in favour of the use of the radical method.

When considering volume calculations over atom groups, however, the choice is less clear: absolute volumes are of less interest than deviations of occupied volumes from the mean, and therefore as long as the vertex error is reasonably uniformly distributed this problem is less significant. Again we would reject the use of Voronoi's method because of the non-physical partitioning of non-bonded interactions, even though in some cases the consequent volume spread is masked by the local packing variations. Provided groups are chosen, with constant covalent environments (e.g. main chain atoms except glycine, whole side-chains) the differences in covalent treatment between radical and Richards' (and also Voronoi's methods) are completely cancelled, and as both Richards' and radical methods partition non-bonded interactions reasonably, there is little to choose between them.

For calculations of occupied volumes and volume distributions for single atoms (or atom groups such as CH_2), we would argue that no procedure is satisfactory unless the atoms are grouped together with a constant covalent environment, in which case the same considerations apply as for the larger groups such as main chain and side-chains. For a variable covalent environment, the spread of the resulting volume distributions will be significantly influenced by the placing of the covalent partitioning planes. The effect will be present for both Richards' and radical methods, though the form of the equation is such that the effect will be greatest for radical.

Clearly, it is impossible to devise a volume partitioning procedure that is both rigorous and consistent with the different chemical constraints in proteins. We can handle a system of interacting van der Waals' atoms rigorously, using radical planes, but as soon as we have to deal with covalent interactions, we must *either* use van der Waals' criteria to partition a covalent bond *or* abandon geometrical rigour.

We argue that a discussion of the preferability of using radical or Richards' method for examining packing efficiency of an atom or group of atoms with variable covalent environment would be largely academic and of little value. If we ask questions about *packing* efficiency, then covalent-bond partitioning is physically irrelevant, the identity of the repulsive electron shell between the two atoms having been lost in the covalent interaction. Therefore, any discussion of packing efficiency and variations for atoms or groups with a variable covalent environment must necessarily consider data that are perturbed by volume variations that are *not* due to the packing constraints being investigated. The perturbations will be smaller for Richards' than for the radical method, so if such comparisons are required, then Richards' method B is to be preferred over the radical planes method.

Other discussions and applications of the volume calculations are given by Richards[14] and Chothia.[15]

Addendum on Cavities

Large Grid Approximation. In the above discussion all of the space inside the hypothetical solvent shell is assigned to the protein atoms. If

[14] T. J. Richmond and F. M. Richards, *J. Mol. Biol.* **119,** 537 (1978).
[15] C. Chothia, *Nature (London)* **254,** 304 (1975).

there is a hole in the structure, the volume that it represents is assigned to the surrounding atoms as specified by their limiting polyhedrons. Such a hole only appears as a lowering in the packing density for this group of atoms. If the hole is modest in size and the number of protein atoms large, the variation from the mean of the packing density may not be obvious and location of the hole not easy to derive. There is no unique and mathematically satisfactory solution to this problem so far, but several approximate procedures have been suggested.

In an attempt to focus on the cavity structure Richards[16] made use of the large grid that he had set up in defining the solvent shell (see above). The grid positions outside of the van der Waals envelope of the protein but inside the solvent shell were used to define the cavities. The centers of these empty cubes were used as pseudoatoms in a modified Voronoi calculation (Fig. 5). The limiting polyhedrons were defined by neighboring grid positions and the van der Waals envelope of neighboring protein atoms (see Fig. 1b). Each empty grid position thus had a volume associated with it. The connectivity of these positions could be evaluated to get an idea of the volume and shape of the cavities. The procedure is crude and unlikely to give more than a very rough idea of the cavity distribution. Nonetheless, the volumes assigned to the protein atoms are lowered a little and the standard deviation of their volume distributions markedly reduced. The general cavity distribution can be visualized easily. The purpose of defining and examining the cavities was to consider volume fluctuations in the dynamics of proteins. Such fluctuations must reflect changes in cavity volume, since the van der Waals envelope is essentially incompressible under normal conditions.

The large grid (2.8 Å) used in the above calculation missed many of the smaller cavities which did not happen to include a grid position. If completely empty, a single grid cube is almost big enough to hold an entire water molecule.

Small Grid Approximation. A. Perlo and F. M. Richards (unpublished) tried to improve on the estimate of total cavity volume by using a much finer grid (0.5 Å). The algorithm does not use the Voronoi procedure at all. When the van der Waals envelope of the protein is inserted into the lattice of this small grid, a single atom may cover 100 or more positions. The lattice is checked sequentially in each of the three principal lattice directions. Each empty position is given a number representing its minimum distance in grid units to the van der Waals envelope (Fig. 6a). The problem now is to decide which of these positions are truly external, and thus bulk solvent, and which are cavities either internal or non-solvent-

[16] F. M. Richards, *Carlsberg Res. Commun.* **44,** 47 (1979).

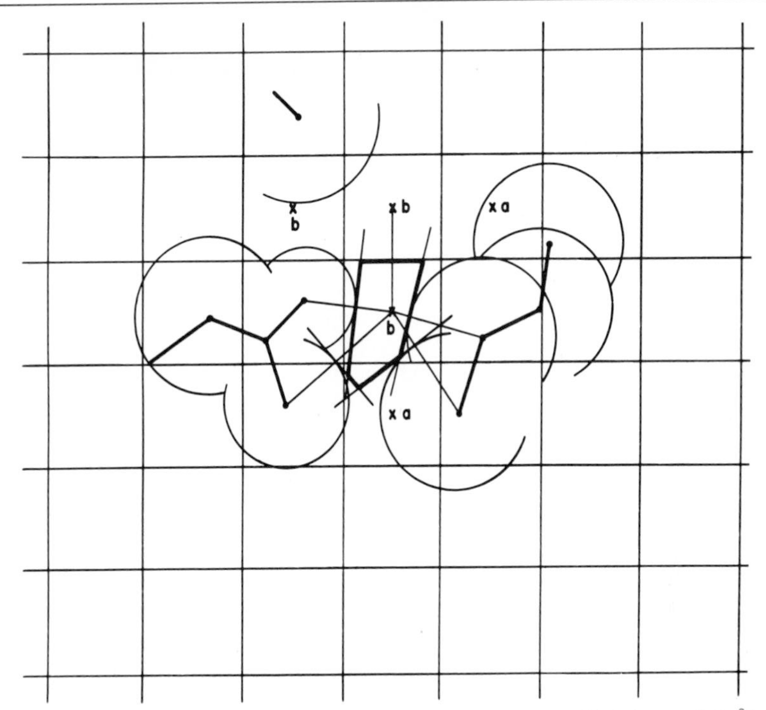

FIG. 5. Volume construction for cavities using the coarse grid (edge = 2.8 Å) and Richards Method B. The van der Waals envelopes of a few atoms are shown. Grid positions whose centers are inside the van der Waals envelope of the protein are labeled a, those outside b. The b locations are used as pseudoatom positions in the volume calculation (see text for discussion). (Reproduced with permission from Richards.[16])

accessible on the surface. The probe is a string of digits against which the lattice positions are checked. With a 0.5 Å grid a water molecule is a sphere with a diameter of about seven grid positions. The closest grid approximation of a sphere would be represented as 1, 3, 3, 4, 3, 3, 1. For ease in subscript manipulation the actual approximation used is 1, 2, 3, 4, 3, 2, 1.

Starting from the edge of the lattice box, known to be "outside" the protein, each group of seven consecutive lattice positions is tested against the probe. If each lattice position is not a protein position and is characterized by a number equal to or greater than that of the probe, then all of these positions could be part of a solvent molecule and are so designated. Otherwise they are classed as potential cavity positions (Fig. 6b). As the probe moves along an axial direction the number of protein surfaces passed is counted in order to assign cavities as external or internal. The

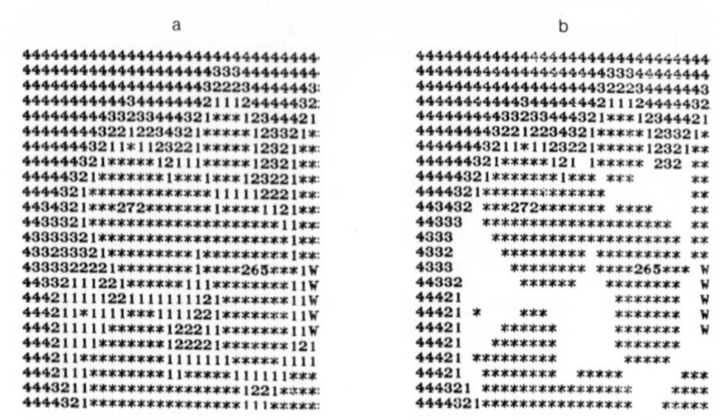

FIG. 6. Cavity definition by the small grid procedure of Perlo and Richards. Grid spacing is 0.5 Å. A small section of a single plane through pancreatic trypsin inhibitor is shown. The asterisks identify those grid positions which are inside the van der Waals envelope of the protein. The numbers in the asterisk area are the serial numbers from the coordinate list of atoms whose centers happen to lie in or close to this plane. In (a) each grid position outside of the asterisk area contains a number which represents the distance in grid units to the nearest part of the van der Waals envelope in any of the three axial directions. Distances equal to or greater than 4 are also listed as 4. The test of this filled lattice for potential cavity positions is described in the text. Such cavity positions located by the algorithm are shown as clear areas in (b). The "W"s are part of the van der Waals envelope of one of the four interior water molecules identified in this structure.

initial cavity list is large. This is reduced as the probe check is carried out in the other two axial directions. From the final list the sum of the number of cavity positions gives directly the total cavity volume. The total protein volume is given by the sum of this volume and the volume (number of grid positions) inside the van der Waals envelope. A cavity as small as a single grid position (0.125 $Å^3$) will be recognized as will internal cavities large enough to hold a solvent molecule. A very similar procedure has been used by Kossiakoff.[17,18]

A complete test of the approximations in this algorithm has not been made, but it has been used for estimating the volume fluctuations during a molecular dynamics simulation of pancreatic trypsin inhibitor (A. Perlo, F. M. Richards, N. Swaminathan, and M. Karplus, unpublished). It has also been used by Pickover and Engelman[19] in their study of the extended low angle X-ray scattering curves of solutions of several proteins.

The procedures of Connolly[20] for depicting protein surfaces (referred to below) can also be used to identify cavities and to estimate their vol-

[17] A. Kossiakoff, *Nature* (*London*) **296,** 713 (1982).

[18] A. Kossiakoff, *Brookhaven Symp. Biol.* **32,** 281 (1983).

[19] C. A. Pickover and D. M. Engelman, *Biopolymers* **21,** 817 (1982).

[20] M. Connolly, Ph.D. Dissertation, University of California, Berkeley (1981).

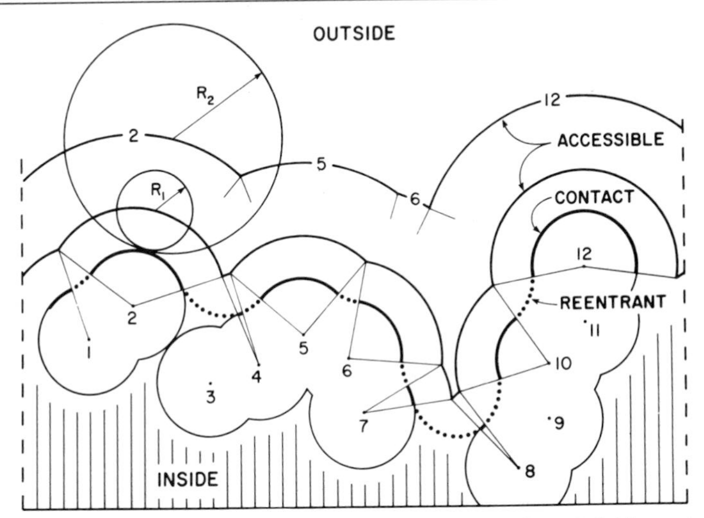

FIG. 7. Schematic representation of possible molecular surface definitions. A section through part of the van der Waals envelope of a hypothetical protein is shown with the atom centers numbered. The accessible surfaces generated by two probes of difference size, R_1 and R_2, and the geometrical definition of contact and reentrant surfaces are shown. (Reproduced with permission from Richards.[22])

ume. In this procedure cavities appear only if they are large enough to contain at least one water molecule. In his survey of a number of proteins with this algorithm, Connolly found that the total cavity volume was of the order of 3% or less of the total volume of the protein.

A different approach developed for glass structures but not yet applied to proteins has been described by Finney and Wallace.[21]

Area

Definitions

On the molecular scale any conceivable probe has dimensions comparable to the features of the surface being examined. Consider the cross section of part of the surface of the hypothetical macromolecule shown in Fig. 7. The trace of the van der Waals envelope of some of the atoms of the structure is shown. A spherical probe of radius R_1 is allowed to roll on the outside while maintaining contact with the van der Waals surface. It will never contact atoms 3, 9, or 11. Such atoms are considered not to be part of the surface of the molecule and are referred to as interior atoms.

[21] J. L. Finney and J. Wallace, *J. Non-Cryst. Solids* **43**, 167 (1981).

The question of how to define and quantitate the surface is a matter of convenience. One straightforward procedure is simply to use the continuous sheet defined by the locus of the center of the probe, the "accessible surface." Another alternative would be to consider the "contact surface," those parts of the molecular van der Waals surface that can actually be in contact with the surface of the probe. This would provide a series of disconnected patches. The "reentrant surface" is also a series of patches defined by the interior-facing parts of the probe when it is simultaneously in contact with more than one atom. Considered together the contact and reentrant surfaces represent a continuous sheet, which might be called the "molecular surface."[22]

By the nature of the geometrical construction there are no reentrant sections of the accessibility surface, i.e., viewed from the molecule each spherical segment is convex. This does entail a possible loss of information as the ratio of contact-to-reentrant surface may be a useful measure of molecular surface roughness. This can be seen qualitatively by inspecting Fig. 7. The molecular surface also has the advantage that the area approaches a finite limiting value as the size of the probe increases. To date most reports have calculated and discussed the accessible area.

With any of the surface definitions the actual numbers derived will depend on the radius chosen for the probe. An example of the change that is produced by probe size is shown in Fig. 7. In going from R_1 to R_2 the number of noncontact or interior atoms increases from three to eight. The accessible surface becomes much smoother (as does the molecular surface, not shown); there is only a slight dimple replacing the deep crevice revealed by the R_1 probe. The appearance of deeply convoluted features or actual holes in the interior of the protein becomes very sensitive to the choice of probe radius. The smaller the probe the larger the number of feature that will be revealed. About the smallest physically reasonable probe is a water molecule considered as a sphere of radius of 1.4 or 1.5 Å. The ratio of this number to the van der Waals radii assumed for the individual atom or atom groups will markedly affect the calculated areas for individual atoms.

The accurate calculation of the surface area is a complex geometrical problem. This problem has in fact been solved rigorously and a closed form analytical expression has been derived.[23,24] Various approximate methods, whose accuracy varies but is adequate for many purposes, have been more widely used.

[22] F. M. Richards, *Annu. Rev. Biophys. Bioeng.* **6,** 151 (1977).
[23] T. J. Richmond, *J. Mol. Biol.* **178,** 63 (1984).
[24] M. Connolly, *J. Appl. Crystallogr.* **16,** 548 (1983).

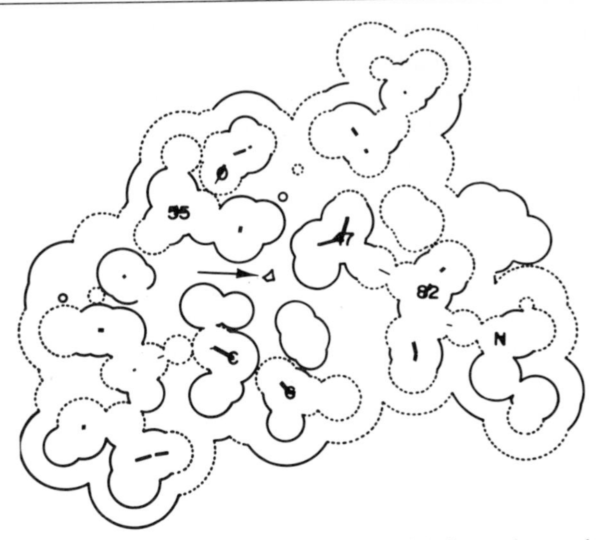

FIG. 8. Superposition of sections through the van der Waals envelope and the accessible surface of ribonuclease S. The arrow indicates a cavity inside the molecule large enough to accommodate a solvent molecule with a radius of 1.4 Å, although it appears to be unfilled in the electron density map. In places the accessible surface is controlled by atoms above or below the section shown. The dashed outline is the surface of N or O atoms, the solid outline C or S atoms. (Reprinted with permission from Lee and Richards.[25])

Procedure of Lee and Richards

The procedure reported by Lee and Richards[25] developed from a program used to graphically portray the van der Waals surface of a protein. For the area calculation the radius chosen for each atom in the structure was the van der Waals radius for that particular atom plus the radius of the hypothetical probe, most often set at 1.4 Å. The structure was then sectioned by a series of planes perpendicular to one of the principal axes. The intersections of the enlarged atom spheres with this plane gave a set of circles of varying size. The outer arcs defined by the intersections of these circles represented the trace of the accessible surface of the protein on that plane (Fig. 8). Some internal surface appeared on occasion, and represented cavities in the structure that were large enough to hold one or more probe spheres. Such cavities were recognized by hand inspection of the lists of surface arcs. The total length of the trace of the accessible surface multiplied by the spacing of the planes gave an approximation to the area of the surface associated with that plane. The sum of such surface

[25] B. Lee and F. M. Richards, *J. Mol. Biol.* **55,** 379 (1971).

increments over the whole set of planes provided the total accessible area of the molecule. This number approached a limiting value as the spacing of the planes was decreased. A practical balance of computing time against numerical accuracy suggested a spacing of between 0.25 and 0.5 Å as being appropriate.

Richmond[14] modified this program so that accessible areas are calculated for each atom separately. Any atom list could thus be processed without calculation over the whole protein each time. An example of a section through one atom and its overlapping neighbors is shown in Fig. 9. Normally, only the accessible area of the atom is recorded as a single scalar quantity. However, the segment lists have much more information potentially available. The shape of each accessible patch on the atom can be visualized along with its vector directions with respect to the coordi-

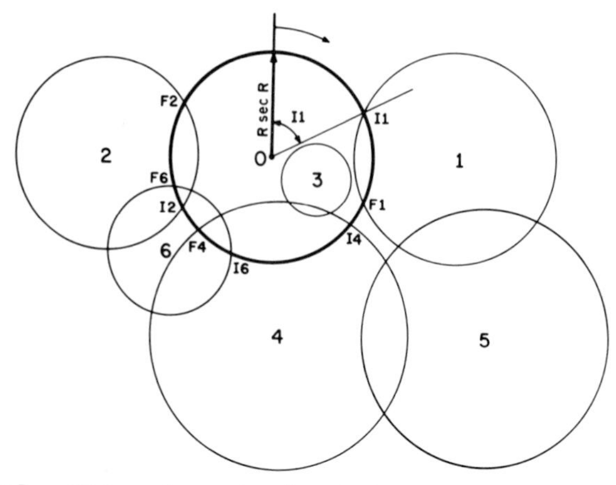

FIG. 9. The Lee–Richmond procedure for calculating the accessible area of an individual atom. The heavy circle is the intersection of the sphere of the target atom (0) with a particular plane. The intersection of the spheres of the neighboring atom, 1 through 6, are shown as lighter circles of differing sizes which are determined by their expanded radii and vertical position with respect to the plane. The intersection check is started at the top as shown by the arrow. For each circle the initial (I) and final (F) intersections (if any) of each circle are accumulated in two lists as their angular position on the 0 circle measured from the starting position. Any atom from the full coordinate list that is possibly close enough to intersect the 0 circle is checked, but the order of checking is arbitrary. The full list of intersections is then ordered on the basis of increasing values of the I intersections. The I and F lists are then tested against each other to get all continuous occluded segments of 0 (in this example I1 to F1 and I4 to F2). Any remaining segments of 0 are part of the accessible surface by definition (i.e., F1 to I4 and F2 to I1). Multiplication of the summed segment lengths by the radius RSECR and by the interplanar spacing gives the approximate accessible area for atom 0 in this plane. Such numbers are summed for the full set of planes to get the total accessible area of 0.

nate axes. An example of such a presentation is shown in Fig. 10. While this format may be useful for visualization, it is likely that the solvent-adjacent polyhedral faces and face normals of the limiting polyhedrons are easier to deal with computationally as suggested by Finney.[7]

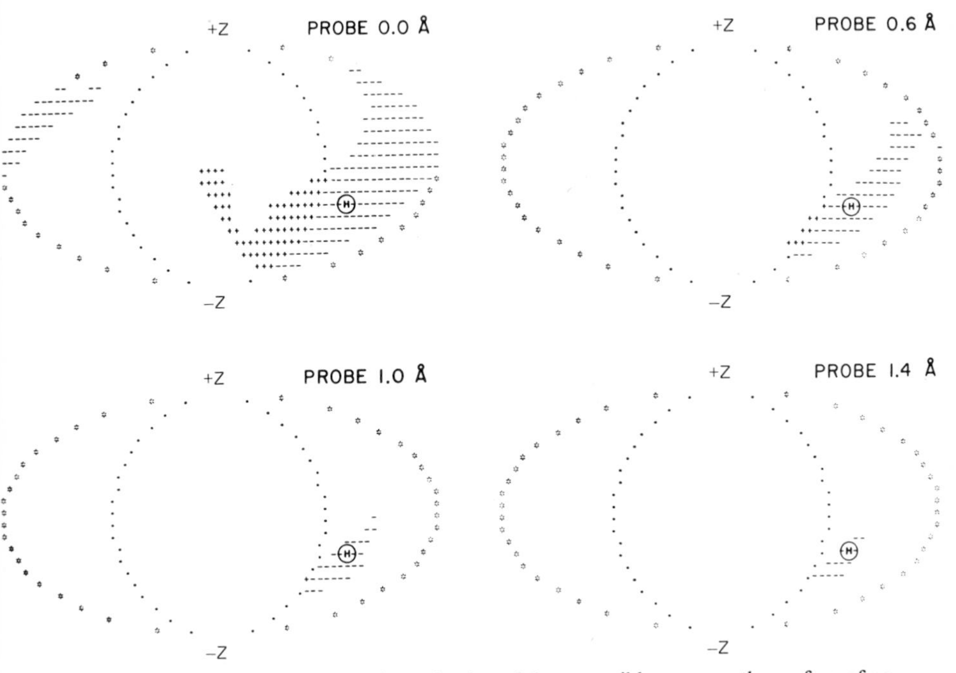

FIG. 10. A modified stereographic projection of the accessible areas on the surface of an atom drawn from segment lists of the type described in Fig. 9. The inner circle of points represents the equator of the sphere, and surrounds the hemisphere above the plane of the paper. The hidden hemisphere is imagined cut along the furthest meridian and then opened up on each side to give the two areas inside the ellipse of asterisks. The right and left lines of asterisks represent the cut meridian. Accessible areas are shown as + on the front hemisphere and − on the back hemisphere. In this example the Z axis of the coordinates is vertical. The $+X$ axis comes toward the viewer. The $+Y$ axis is to the right in the plane of the paper. For illustration the amide nitrogen atom of Phe 120 in ribonuclease S is shown. The H in the small circle gives the N–H vector direction for the amide group (the same direction for each panel). The area patterns are shown for different values of the probe radius. For $r = 0.0$ Å the indicated area is the actual van der Waals surface of the NH group. The clear space representing nonaccessible area is that part of the NH sphere occluded by covalent attachment to the C and C_α atoms of the main chain and by any van der Waals overlap caused by misplacement of nonbonded atoms in the X-ray structure. The decrease in accessible area with increase in probe size is seen in the other panels. At $r = 1.40$ Å there is still a small accessible region around the N–H vector which suggests that this particular proton could exchange with solvent with little or no alteration in the structure of the protein.

Procedure of Shrake and Rupley

A different algorithm for calculating the solvent-exposed areas of atoms was developed independently by Shrake and Rupley.[26] Again, a sphere of expanded radius equal to the van der Waals radius plus the probe radius (taken as 1.4 Å) is set up around each atom. The central atom, whose area is to be calculated, is represented by a set of 92 points distributed nearly uniformly over the surface of the sphere. Each point is then checked against surrounding atoms to find out if it is within any of the spheres. Points outside the spheres of all surrounding atoms lie on the accessible surface and their number is a direct measure of the accessible area of the central atom. For the occluded points of the central atom, the test atom closest to any particular point is credited with occluding the point. Thus the neighboring atoms collectively provide the environment of the central atom and can be scored quantitatively for their influence on the central atom.

These programs provided actual areas in $Å^2$, and these are the values frequently reported. In the original paper Lee and Richards[25] also defined the term *accessibility* which is a dimensionless quantity varying between 0 and 1. It represents the ratio of the accessible area in a particular structure to the accessible area of the same group in a reference compound. The latter is normally taken as gly-X-gly, where the group of interest is in the residue X. Accessibility, so defined, is being used at this time, particularly in electrostatic calculations where interactions are modified by these dimensional factors.[27]

Procedure of Wodak and Janin

Wodak and Janin[28] have proposed an approximate analytical expression for the accessible area rather than the numerical calculation described above. The equations are differentiable and can be used directly as a factor to incorporate solvent influences in energy minimization procedures. The derivation assumes a random distribution of spheres surrounding the target atom and includes a correction for excluded volumes. Although the expression is not accurate for a specific atom, averages taken over all, or large parts, of a structure become very good approximations of total surface area. The original paper should be consulted for the derivation of the following equations, where r_o = van der Waals radius of the

[26] A. Shrake and J. A. Rupley, *J. Mol. Biol.* **79,** 351 (1973).

[27] J. B. Matthew, G. I. H. Hanania, and F. R. N. Gurd, *Biochemistry* **18,** 1919 (1979).

[28] S. J. Wodak and J. Janin, *Proc. Natl. Acad. Sci. U.S.A.* **77,** 1736 (1980).

target atom; r_i = van der Waals radius of a neighboring atom; r_w = radius of the spherical probe; d_i = interatomic distance between atoms o and i; b'_i = maximum area of target atom covered by atom i; b_i = minimum area of target atom covered by atom i; n = number of atoms which occlude any area surrounding the target atom; S = area of expanded target atom = $4\pi(r_o + r_w)^2$; A = approximate value of accessible area of target atom.

$$b = \pi(r_o + r_w)(r_0 + r_i + 2r_w - d_i)[1 + (r_i - r_o)/d_i] \quad (7)$$

$$b' = \pi(r_o + r_w)(r_0 + r_i - d_i)[1 + (r_i - r_o - 2r_w)/d_i] \quad (8)$$

Define

$$A' = S \sum_{i=1}^{n} [1 - (b_i - b'_i)/S] \quad (9)$$

$$B' = \sum_{i=1}^{n} b'_i \quad (10)$$

Then

$$A_c = A' - B' \qquad (A_c = 0 \quad \text{if} \quad A' < B') \quad (11)$$

$$A_c/d_i = A'/d_i - B'/d_i \quad (12)$$

where

$$A'/d_i = \frac{-A'(b_i/d_i - b'_i/d_i)}{(S - b_i + b'_i)} \quad (13)$$

$$B'/d_i = b'_i/d_i \quad (14)$$

The original paper and references therein should be consulted for possible further approximations using single spheres for entire residues and for use of these functions in defining domain structures.

(Note that a mathematically accurate description of accessible areas for any collection of spheres has been developed by T. J. Richmond[23] and M. Connolly.[24])

Surface Representations

The computer presentation of van der Waals surfaces in the form of packing models has been highly developed by R. Feldman at the National Institutes of Health. However, the algorithms have not been reported to generate numbers related to the area or volume of these figures.

A particularly effective presentation of the continuous molecular surface of molecules, including both the contact and reentrant sections, has been developed by M. L. Connolly. Interior cavities can be recognized and enumerated. A brief overview of the computer presentations has been given by Langridge *et al.*[29] and Bash *et al.*[30] Both areas and volumes are provided in newer programs.[24,31] Unfortunately, no details of any of these algorithms are available in published form. The latter are particularly important in providing analytical expressions for the area which can be differentiated and built into energy minimization procedures, as described in detail by Richmond.[23]

[29] R. Langridge, T. E. Ferrin, I. D. Kuntz, and M. L. Connolly, *Science* **211,** 661 (1981).

[30] P. A. Bash, N. Pattabiraman, C. Huang, T. E. Ferrin, and R. Langridge, *Science* **222,** 1325 (1983).

[31] M. L. Connolly, *Science* **221,** 709 (1983).

Author Index

Numbers in parentheses are footnote reference numbers and indicate that an author's work is referred to although the name is not cited in the text.

A

B

H

I

J

K

L

M

N

O

P

Q

R

S

T

U

V

W

Subject Index

A

B

C

D

E

F

G

H

I

K

L

M

S

T

U

V

W

X

Y